Make Geography

COME ALIVE

with GLENCOE'S

TECHNOLOGY TOOLS

Study tools are available online 24 hours a day.

Review lesson content, take notes, and build vocabulary at Study Central™!

Download tools to help you study anywhere!

Bring the sights and sounds of geography alive with Section Spotlight Videos!

Keep up-to-date with current events on Glencoe's btw . . . site.

Find In-Motion Animations, interactive graphic organizers, and more on your StudentWorks™ Plus Online at mybooks.glencoe.com.

QuickPass™

More Than a Textbook

Find it faster.

Visit Geography ONLINE at glencoe.com and enter a QuickPass™ chapter code to go directly to the chapter resources you need.

WGC9952c1

Enter this code with the appropriate chapter number.

Find what you need.

- StudentWorks™ Plus Online
- Section Spotlight Video
- Chapter Overview
- Study Central™
- Section Audio
- Self-Check Quiz

Find extras to help you succeed.

- Download Study-to-Go™ applications
- Interact with In-Motion Animations
- Access current events articles at btw events.glencoe.com
- View video at the Media Library
- Review with ePuzzles and Games
- Explore Student Web Activities
- Multilingual Glossary and more . . .

You can easily launch a wide range of digital products from your computer's desktop with the McGraw-Hill widget.

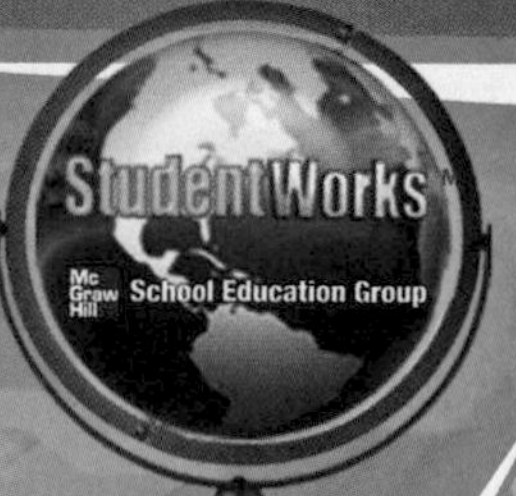

GLENCOE

World Geography and Cultures

Richard G. Boehm, Ph.D.

NATIONAL GEOGRAPHIC

McGraw Hill Glencoe

About the Authors

Senior Author

RICHARD G. BOEHM

Richard G. Boehm, Ph.D., was one of seven authors of *Geography for Life,* national standards in geography, prepared under the Goals 2000: Educate America Act. He was also one of the authors of the *Guidelines for Geographic Education,* in which the five themes of geography were first articulated. Dr. Boehm has received many honors, including "Distinguished Geography Educator" by the National Geographic Society (1990); the "George J. Miller Award" from the National Council for Geographic Education (NCGE) for distinguished service to geographic education (1991); and "Gilbert Grosvenor Honors" in geographic education from the Association of American Geographers (2002). He was President of the NCGE and has twice won the *Journal of Geography* award for best article. He has received the NCGE's "Distinguished Teaching Achievement" award and presently holds the Jesse H. Jones Distinguished Chair in Geographic Education at Texas State University in San Marcos, Texas.

NATIONAL GEOGRAPHIC SOCIETY

The National Geographic Society, founded in 1888 for the increase and diffusion of geographic knowledge, is the world's largest nonprofit scientific and educational organization. Since its earliest days, the Society has used sophisticated communication technologies, from color photography to holography, to convey geographic knowledge to a worldwide membership. The School Publishing Division supports the Society's mission by developing innovative educational programs—ranging from traditional print materials to multimedia programs including CD-ROMs, videos, and software.

Contributing Author

Dinah Zike, M.Ed., is an award-winning author, educator, and inventor known for designing three-dimensional hands-on manipulatives and graphic organizers known as Foldables™. Foldables are used nationally and internationally by teachers, parents, and educational publishing companies. Dinah has developed over 150 supplemental educational books and materials. She is the author of *The Big Book of Books and Activities,* which was awarded Learning Magazine's Teachers' Choice Award. In 2004 Dinah was honored with the CESI Science Advocacy Award. Dinah received her M.Ed. from Texas A&M, College Station, Texas.

The McGraw·Hill Companies

Send all inquiries to:
Glencoe/McGraw-Hill
8787 Orion Place
Columbus, OH 43240-4027

ISBN: 978-0-07-879995-2
MHID: 0-07-879995-3

Printed in the United States of America.

6 7 8 9 10 RJE 15 14 13 12

Make Geography Come Alive with Glencoe's Technology Tools. Study tools are available online 24 hours a day. Review lesson content, take notes, and build vocabulary at Study Central™! Download tools to help you study anywhere! Bring the sights and sounds of geography alive with Section Spotlight Videos! Keep up-to-date with current events on Glencoe's btw… site. Find In-Motion Animations, interactive graphic organizers, and more on your StudentWorks™ Plus Online at mybooks.glencoe.com.

Consultants

Geographic Education Consultant
SARI J. BENNETT, Ph.D.
Director, Center for Geographic Education
University of Maryland, Baltimore County
Baltimore, Maryland

Cultural Geography Consultant
JOSEPH P. STOLTMAN, Ph.D.
Professor of Geography
Western Michigan University
Kalamazoo, Michigan

Foundations of Geography
RUTH I. SHIREY, Ph.D.
Professor of Geography
Indiana University of Pennsylvania
Indiana, Pennsylvania

United States and Canada
CATHERINE M. LOCKWOOD, Ph.D.
Associate Professor of Geography
Chadron State College
Chadron, Nebraska

HAROLD M. ELLIOTT, Ph.D.
Professor of Geography, Chair
Weber State University
Ogden, Utah

Latin America
PAUL NAGEL, Ph.D.
Assistant Professor of Geography/
Coordinator, Louisiana Geography
Education Alliance
Northwestern State University
Natchitoches, Louisiana

DENNIS CONWAY, Ph.D.
Professor of Geography
Indiana University
Bloomington, Indiana

Europe
GEORGE W. WHITE, JR. Ph.D.
Associate Professor of Geography
Frostburg State University
Frostburg, Maryland

Russia
GRIGORY IOFFE, Ph.D.
Professor of Geography
Radford University
Radford, Virginia

SHANNON O'LEAR, Ph.D.
Assistant Professor of Geography
University of Kansas
Lawrence, Kansas

North Africa, Southwest Asia, and Central Asia
AMY MILLS, Ph.D.
Assistant Professor of Geography
University of South Carolina
Columbia, South Carolina

Africa South of the Sahara
TADESSE KIDANE-MARIAM, Ph.D.
Assistant Professor of Geography
Edinboro University of Pennsylvania
Edinboro, Pennsylvania

SETH APPIAH-OPOKU, Ph.D.
Assistant Professor of Geography
University of Alabama
Tuscaloosa, Alabama

South Asia
BIMAL K. PAUL, Ph.D.
Professor of Geography
Kansas State University
Manhattan, Kansas

PRATYUSHA BASU, Ph.D.
Assistant Professor of Geography
University of South Florida
Tampa, Florida

East Asia
LIN WU, Ph.D.
Professor of Geography
California State Polytechnic University,
Pomona
Pomona, California

Southeast Asia
RALPH LENZ, Ph.D.
Professor of Geography
Wittenberg University
Springfield, Ohio

Australia, Oceania, and Antarctica
ELIZABETH J. LEPPMAN, Ph.D.
Adjunct Professor of Geography
Eastern Kentucky University
Richmond, Kentucky

Teacher Reviewers

DANIEL BERRY
University High School
Morgantown, West Virginia

BRAD BOWERMAN
Lakeland Junior-Senior High School
Jermyn, Pennsylvania

JUDY CAMDEN
Izard County Consolidated School District
Brockwell, Arkansas

LOU CAMILOTTO
McCutcheon High School
Lafayette, Indiana

RICK FARNEY
Rhea County High School
Evensville, Tennessee

MARY SUE FRALEY
Easley High School
Easley, South Carolina

PAUL T. GRAY, JR.
Russellville High School
Russellville, Arkansas

CONNIE ROSENLIEB HUDGEONS
Albuquerque High School
Albuquerque, New Mexico

DENISE GERARDINE MCCUTCHEON LAM
Spotswood High School
Penn Laird, Virginia

MATTHEW NEIGHBORS
Galveston Ball High School
Galveston, Texas

KATHERINE OTTERBOURG
East Forsyth High School
Kernersville, North Carolina

MELANIE D. RIEGER
Grace King High School
Metairie, Louisiana

PHILIP RODRIGUEZ
Earl Warren High School
San Antonio, Texas

JAMES A. SCHMIDT
Penn High School
Mishawaka, Indiana

NADINE WRIGHT
Elbert County Comprehensive High School
Elberton, Georgia

QuickPass™—More Than a Textbook
Find it faster. Visit Geography Online at glencoe.com and enter a ***QuickPass™*** chapter code to go directly to the chapter resources you need. Enter this code with the appropriate chapter number. WGC9952c1. Find what you need including StudentWorks™ Plus Online, Section Spotlight Video, Chapter Overview, Study Central™ Section Audio, and Self-Check Quiz. Find extras to help you succeed such as downloading Study-to-Go™ applications, interacting with In-Motion Animations, accessing current events articles at btw events.glencoe.com, viewing video at the Media Library, reviewing with ePuzzles and Games, exploring Student Web Activities, the Multilingual Glossary, and more. You can easily launch a wide range of digital products from your computer's desktop with the McGraw-Hill widget.

Reference Atlas

Celebrating Hungary's membership in the EU (p. 324)

Cherokee woman making beadwork (p. 119)

Adélie penguin (p. 54)

A Tuareg man (p. 75)

UNIT 2 The United States and Canada

Welcome sign in Quebec (p. 159)

Trombonist in a parade in New Orleans (p. 119)

Factory worker in Mexico (p. 255)

Quechua man (p. 202)

UNIT 4 Europe

Eiffel Tower, Paris, France (p. 335)

Farmer in Cornwall, England (p. 294)

St. Basil's Cathedral and the Kremlin (pp. 388–389)

A businesswoman in St. Petersburg (p. 390)

Turkish girl in traditional dress (p. 452)

Burj al Arab Hotel in Dubai, United Arab Emirates (p. 411)

Shop owner in Ghana (p. 564)

Blue wildebeest on the Serengeti Plain, Tanzania (p. 515)

UNIT 8 South Asia

Bhutanese girl (p. 581)

A Pakistani merchant (p. 581)

Victoria Harbor in Hong Kong (pp. 696–697)

Chinese farmer on a terraced field (pp. 660–661)

Thailand celebrates the Water Festival (p. 753)

Petronas Towers, Kuala Lumpur, Malaysia (p. 754)

A Samoan man and his granddaughter (p. 824)

Athletes play Australian rules football (p. 791)

Reference Section

Special Features

NATIONAL GEOGRAPHIC

Why Geography Matters

NATIONAL GEOGRAPHIC

Connecting to the United States

NATIONAL GEOGRAPHIC

Case Study

NATIONAL GEOGRAPHIC

Maps

Reference Atlas

UNIT 1 The World

UNIT 2 United States and Canada

UNIT 3 Latin America

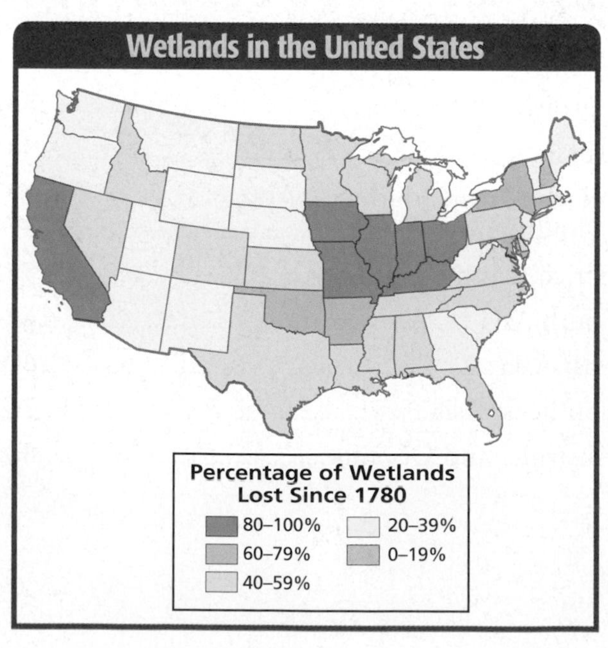

UNIT 4 Europe

UNIT 5 Russia

In Motion Use **StudentWorks™ Plus** or glencoe.com.

Maps, graphs, and diagrams labeled with the In Motion icon have been specially enhanced on the StudentWorks™ Plus CD-ROM and on glencoe.com. These In Motion graphics allow you to interact with layers of displayed data and to listen to audio components.

Entries in blue indicate In Motion graphics.

UNIT 6 North Africa, Southwest Asia, and Central Asia

Maps

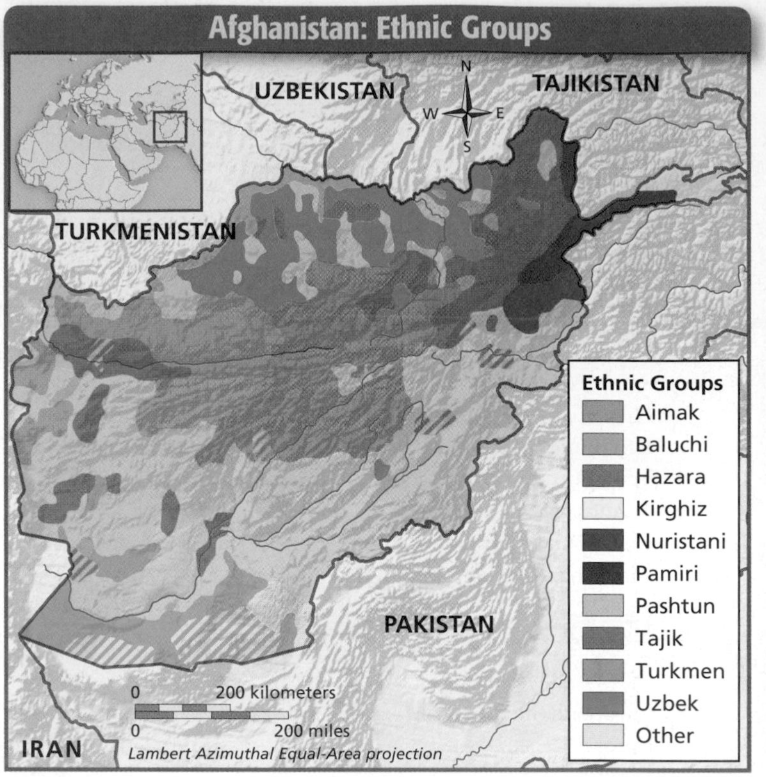

UNIT 10 Southeast Asia

UNIT 11 Australia, Oceania, and Antarctica

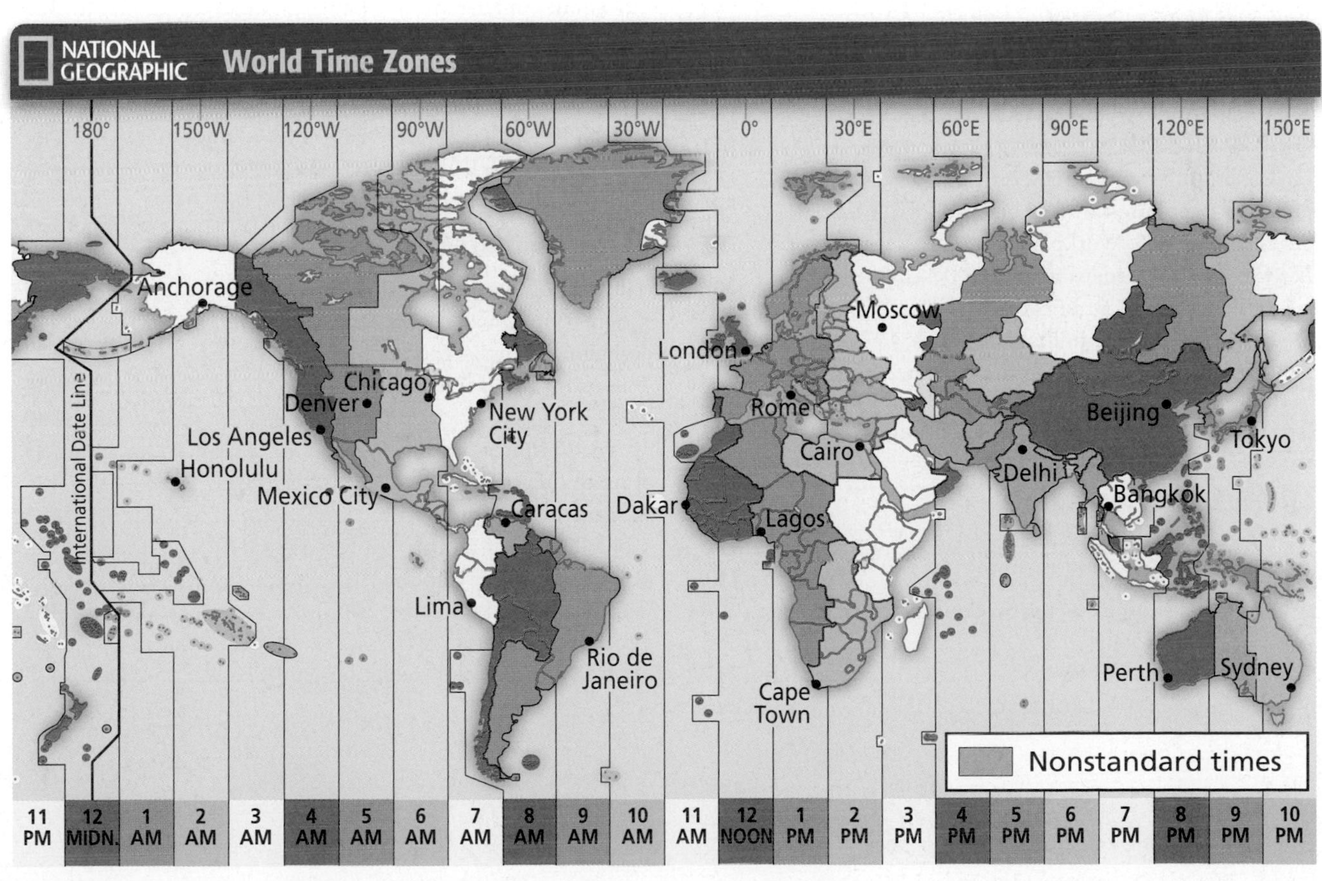

Graphs, Charts, and Diagrams

UNIT 1 The World

UNIT 2 United States and Canada

UNIT 3 Latin America

UNIT 4 Europe

UNIT 5 Russia

UNIT 6 North Africa, Southwest Asia, and Central Asia

UNIT 7 Africa South of the Sahara

UNIT 8 South Asia

UNIT 9 East Asia

UNIT 10 Southeast Asia

UNIT 11 Australia, Oceania, and Antarctica

Primary Sources

NG = National Geographic

Scavenger Hunt

World Geography and Cultures has a wealth of information—the trick is to know where to find it. If you go through this scavenger hunt exercise with your teachers or parents, you will quickly learn how the textbook is organized and how to get the most out of your reading and study time. Let's get started!

1. How many chapters and units are in the book?
2. What world region does Unit 4 cover?
3. Name the place where you can find the Big Ideas for Chapter 12.
4. In what two places can you find the Content Vocabulary for Section 2 of Chapter 12?
5. Where can you find the Foldable summarizing the Cultural Geography of Europe discussed in Chapter 12?
6. How are the Academic Vocabulary words for Section 3 of Chapter 12 highlighted in the narrative?
7. Where can you find listings for latitude and longitude of cities?
8. Where do you look if you want to quickly find all the maps in the book?
9. Each section of a chapter opens with an excerpt from a primary source. Where else can you find primary sources in the textbook?
10. Where can you learn the definition of a physical map, a political map, and a thematic map?

Reference Atlas

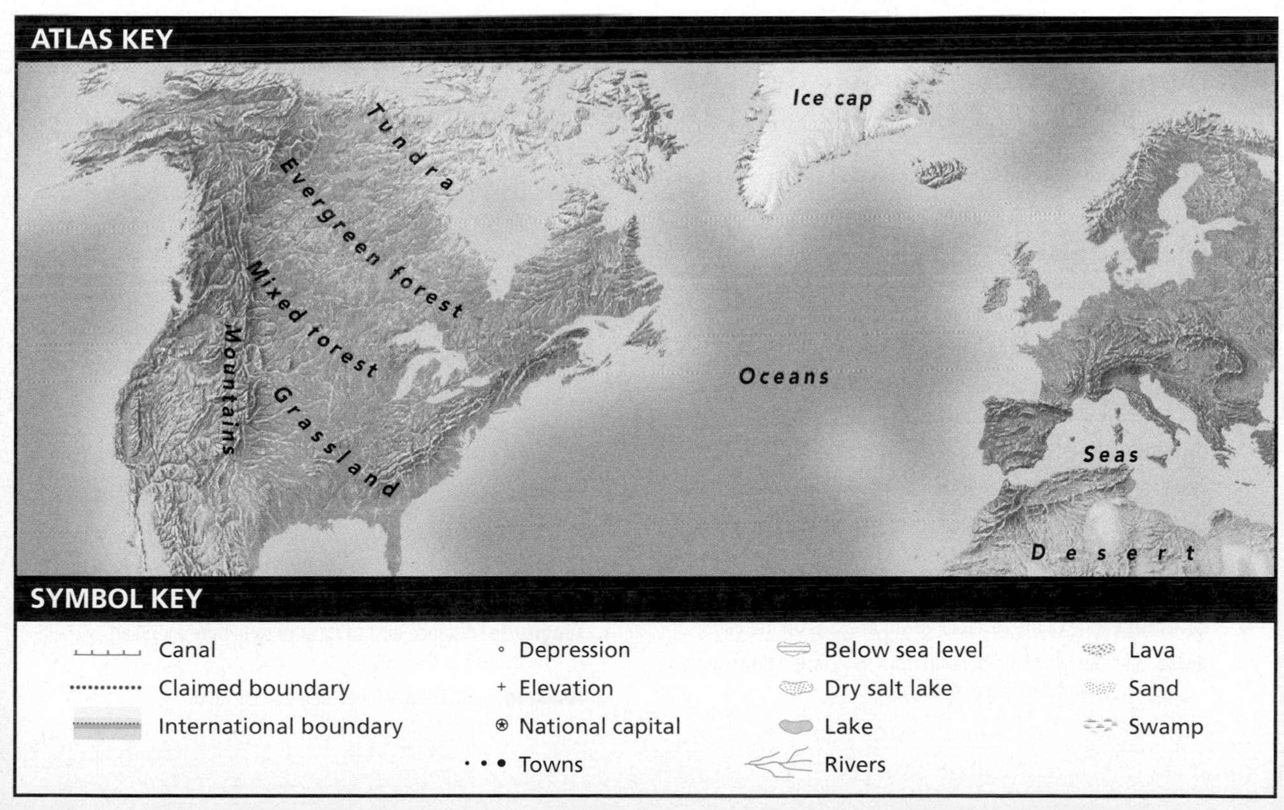

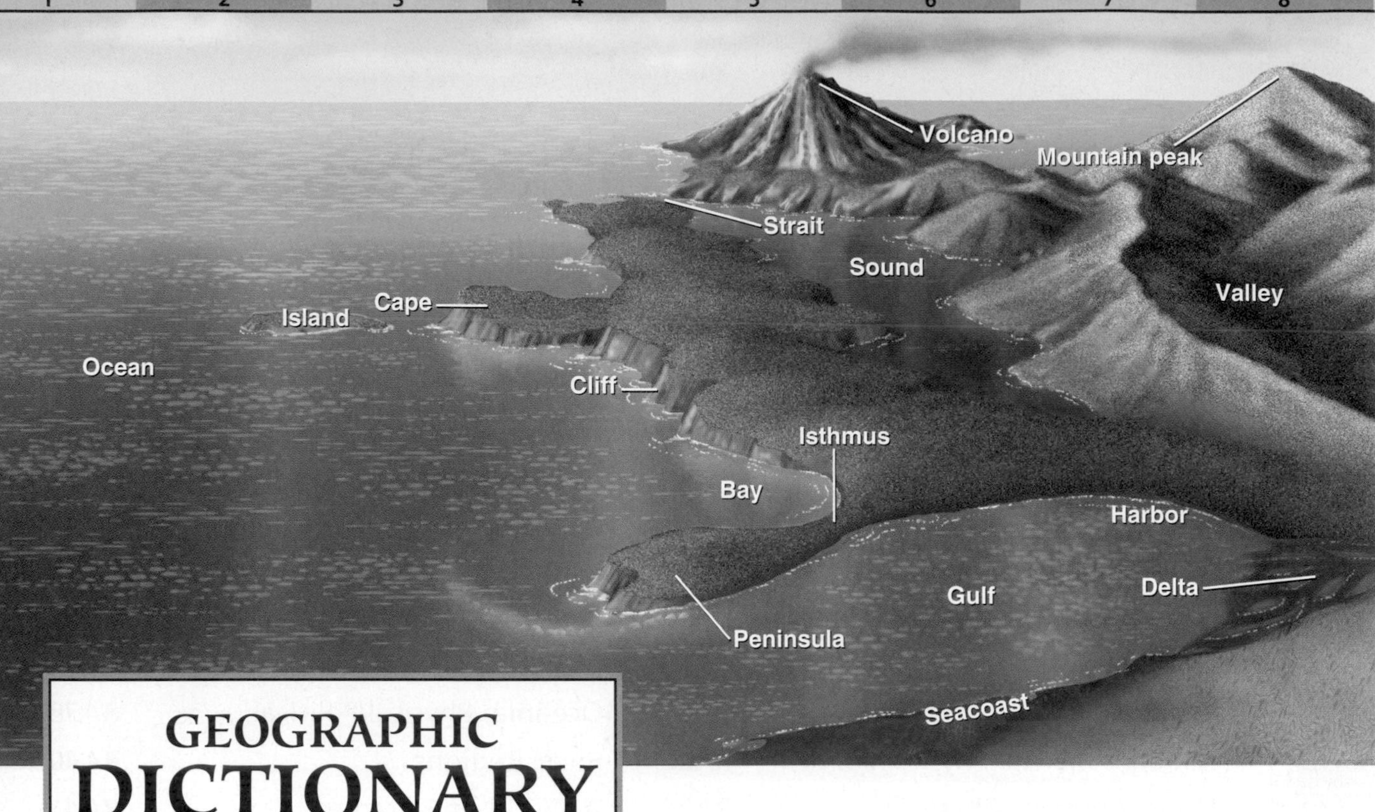

GEOGRAPHIC DICTIONARY

As you read about the world's geography, you will encounter the terms listed below. Many of the terms are pictured in the diagram.

absolute location exact location of a place on the Earth described by global coordinates

basin area of land drained by a given river and its branches; area of land surrounded by lands of higher elevations

bay part of a large body of water that extends into a shoreline, generally smaller than a gulf

canyon deep and narrow valley with steep walls

cape point of land that extends into a river, lake, or ocean

channel wide strait or waterway between two landmasses that lie close to each other; deep part of a river or other waterway

cliff steep, high wall of rock, Earth, or ice

continent one of the seven large landmasses on the Earth

delta flat, low-lying land built up from soil carried downstream by a river and deposited at its mouth

divide stretch of high land that separates river systems

downstream direction in which a river or stream flows from its source to its mouth

elevation height of land above sea level

Equator imaginary line that runs around the Earth halfway between the North and South Poles; used as the starting point to measure degrees of north and south latitude

glacier large, thick body of slowly moving ice

gulf part of a large body of water that extends into a shoreline, generally larger and more deeply indented than a bay

harbor a sheltered place along a shoreline where ships can anchor safely

highland elevated land area such as a hill, mountain, or plateau

hill elevated land with sloping sides and rounded summit; generally smaller than a mountain

island land area, smaller than a continent, completely surrounded by water

isthmus narrow stretch of land connecting two larger land areas

lake a sizable inland body of water

latitude distance north or south of the Equator, measured in degrees

longitude distance east or west of the Prime Meridian, measured in degrees

lowland land, usually level, at a low elevation

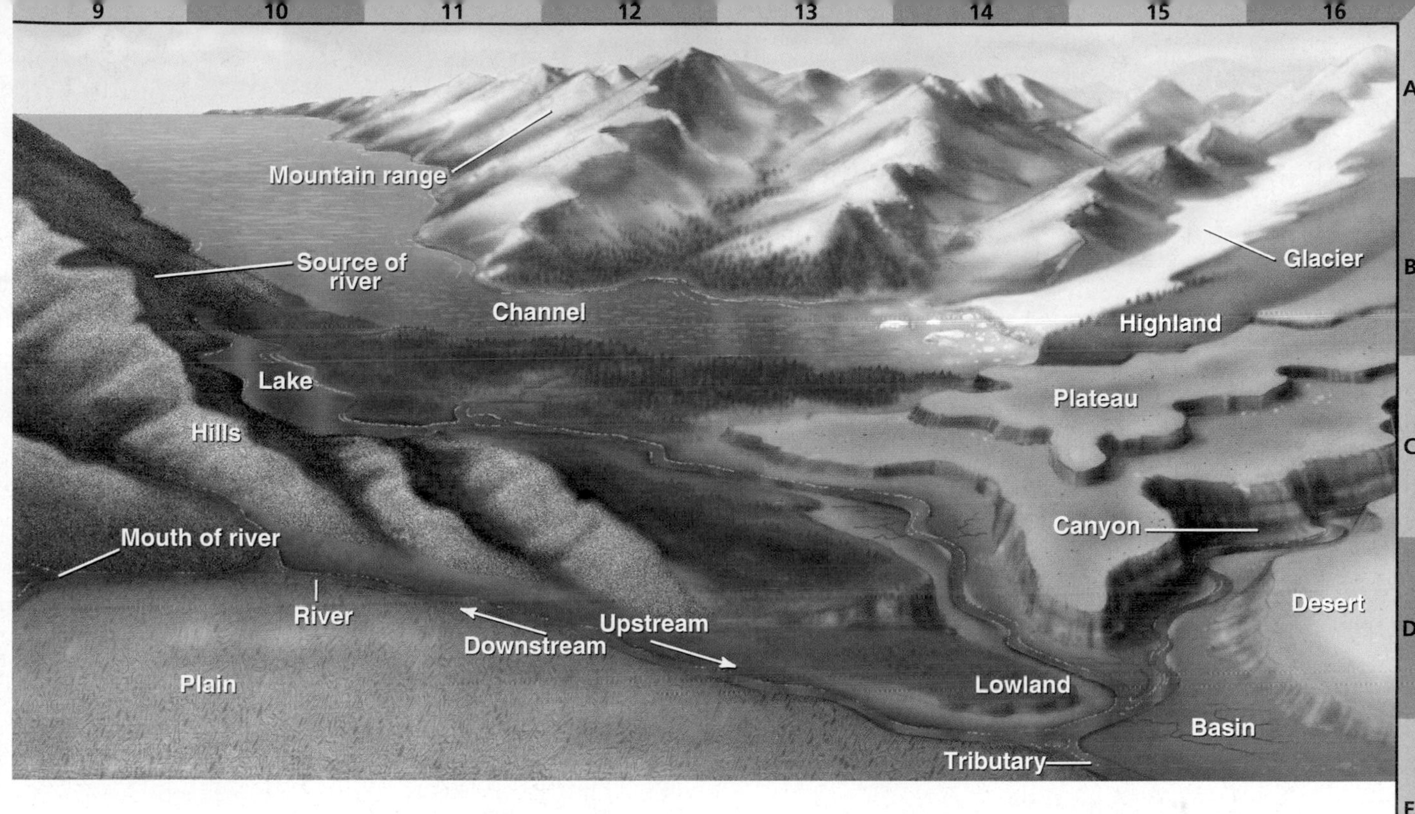

map drawing of the Earth shown on a flat surface

meridian one of many lines on the global grid running from the North Pole to the South Pole; used to measure degrees of longitude

mesa broad, flat-topped landform with steep sides; smaller than a plateau

mountain land with steep sides that rises sharply (1,000 feet or more) from surrounding land; generally larger and more rugged than a hill

mountain peak pointed top of a mountain

mountain range a series of connected mountains

mouth (of a river) place where a stream or river flows into a larger body of water

ocean one of the four major bodies of salt water that surround the continents

ocean current stream of either cold or warm water that moves in a definite direction through an ocean

parallel one of many lines on the global grid that circles the Earth north or south of the Equator; used to measure degrees of latitude

peninsula body of land jutting into a lake or ocean, surrounded on three sides by water

physical feature characteristic of a place occurring naturally, such as a landform, body of water, climate pattern, or resource

plain area of level land, usually at low elevation and often covered with grasses

plateau area of flat or rolling land at a high elevation, about 300 to 3,000 feet (90 to 900 m) high

Prime Meridian line of the global grid running from the North Pole to the South Pole at Greenwich, England; starting point for measuring degrees of east and west longitude

relief changes in elevation over a given area of land

river large natural stream of water that runs through the land

sea large body of water completely or partly surrounded by land

seacoast land lying next to a sea or an ocean

sound broad inland body of water, often between a coastline and one or more islands off the coast

source (of a river) place where a river or stream begins, often in highlands

strait narrow stretch of water joining two larger bodies of water

tributary small river or stream that flows into a large river or stream; a branch of the river

upstream direction opposite the flow of a river; toward the source of a river or stream

valley area of low land usually between hills or mountains

volcano mountain or hill created as liquid rock and ash erupt from inside the Earth

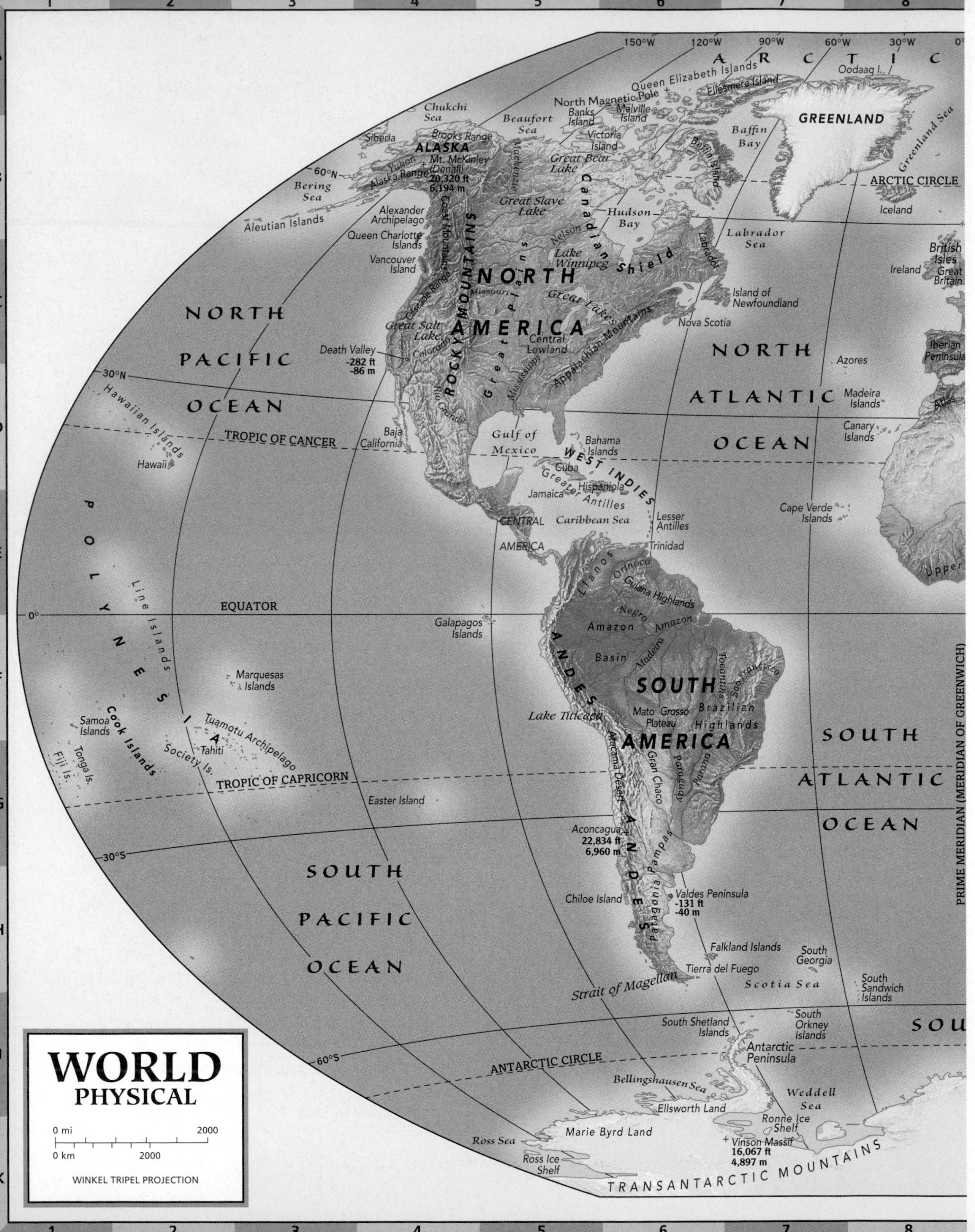

WORLD
PHYSICAL
0 mi
2000
0 km
2000
WINKEL TRIPEL PROJECTION
150°W
120°W
90°W
60°W
30°W
60°N
30°N
0°
30°S
60°S
ARCTIC CIRCLE
TROPIC OF CANCER
EQUATOR
TROPIC OF CAPRICORN
ANTARCTIC CIRCLE
PRIME MERIDIAN (MERIDIAN OF GREENWICH)
A R C T I C
Oodaaq I.
Queen Elizabeth Islands
Ellesmere Island
North Magnetic Pole
Banks Island
Melville Island
Victoria Island
GREENLAND
Greenland Sea
Baffin Bay
Baffin Island
Chukchi Sea
Beaufort Sea
Siberia
Brooks Range
ALASKA
Yukon
Mt. McKinley (Denali)
20,320 ft
6,194 m
Alaska Range
Bering Sea
Aleutian Islands
Mackenzie
Great Bear Lake
Great Slave Lake
Canadian Shield
Hudson Bay
Labrador Sea
Labrador
Iceland
British Isles
Ireland
Great Britain
Alexander Archipelago
Coast Mountains
Queen Charlotte Islands
Vancouver Island
Nelson
Lake Winnipeg
Great Plains
NORTH AMERICA
ROCKY MOUNTAINS
Cascade Range
Missouri
Great Lakes
Island of Newfoundland
Nova Scotia
Appalachian Mountains
Great Salt Lake
Colorado
Central Lowland
Mississippi
Death Valley
-282 ft
-86 m
Rio Grande
NORTH PACIFIC OCEAN
Hawaiian Islands
Hawaii
NORTH ATLANTIC OCEAN
Azores
Iberian Peninsula
Madeira Islands
Atlas
Canary Islands
Baja California
Gulf of Mexico
Bahama Islands
WEST INDIES
Cuba
Greater Antilles
Hispaniola
Jamaica
Caribbean Sea
Lesser Antilles
Cape Verde Islands
CENTRAL AMERICA
Trinidad
Upper
POLYNESIA
Line Islands
Llanos
Orinoco
Guiana Highlands
Negro
Amazon
Amazon Basin
Galapagos Islands
ANDES
Madeira
Tocantins
São Francisco
SOUTH AMERICA
Brazilian Highlands
Mato Grosso Plateau
Lake Titicaca
Marquesas Islands
Samoa Islands
Cook Islands
Tuamotu Archipelago
Tahiti
Society Is.
Tonga Is.
Fiji Is.
Atacama Desert
Gran Chaco
Paraguay
Paraná
SOUTH ATLANTIC OCEAN
Easter Island
Aconcagua
22,834 ft
6,960 m
Pampas
SOUTH PACIFIC OCEAN
Chiloe Island
Patagonia
Valdes Peninsula
-131 ft
-40 m
Falkland Islands
South Georgia
Tierra del Fuego
Strait of Magellan
Scotia Sea
South Sandwich Islands
South Shetland Islands
South Orkney Islands
SOU
Antarctic Peninsula
Bellingshausen Sea
Weddell Sea
Ellsworth Land
Ronne Ice Shelf
Marie Byrd Land
Ross Sea
Vinson Massif
16,067 ft
4,897 m
Ross Ice Shelf
TRANSANTARCTIC MOUNTAINS

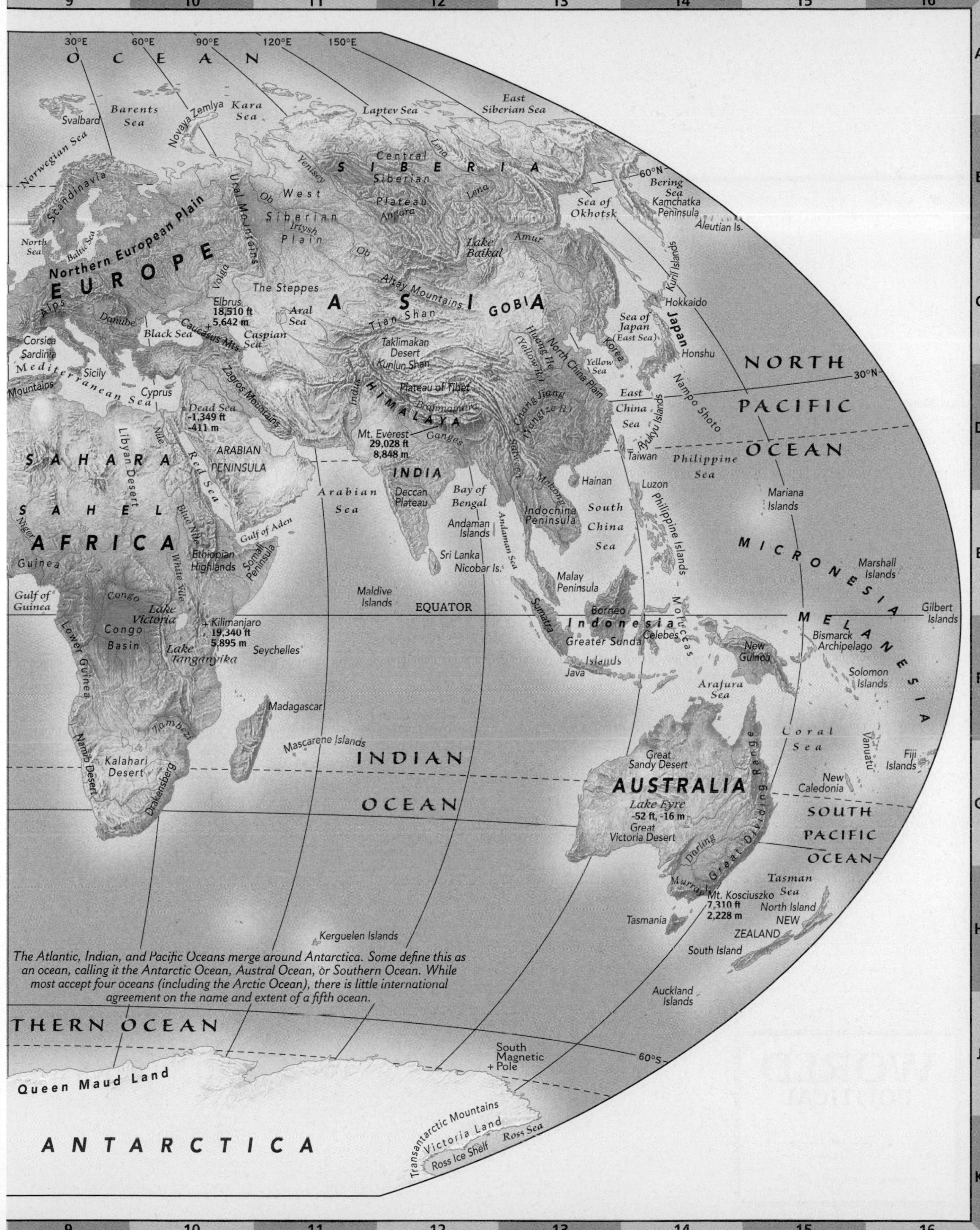

The Atlantic, Indian, and Pacific Oceans merge around Antarctica. Some define this as an ocean, calling it the Antarctic Ocean, Austral Ocean, or Southern Ocean. While most accept four oceans (including the Arctic Ocean), there is little international agreement on the name and extent of a fifth ocean.

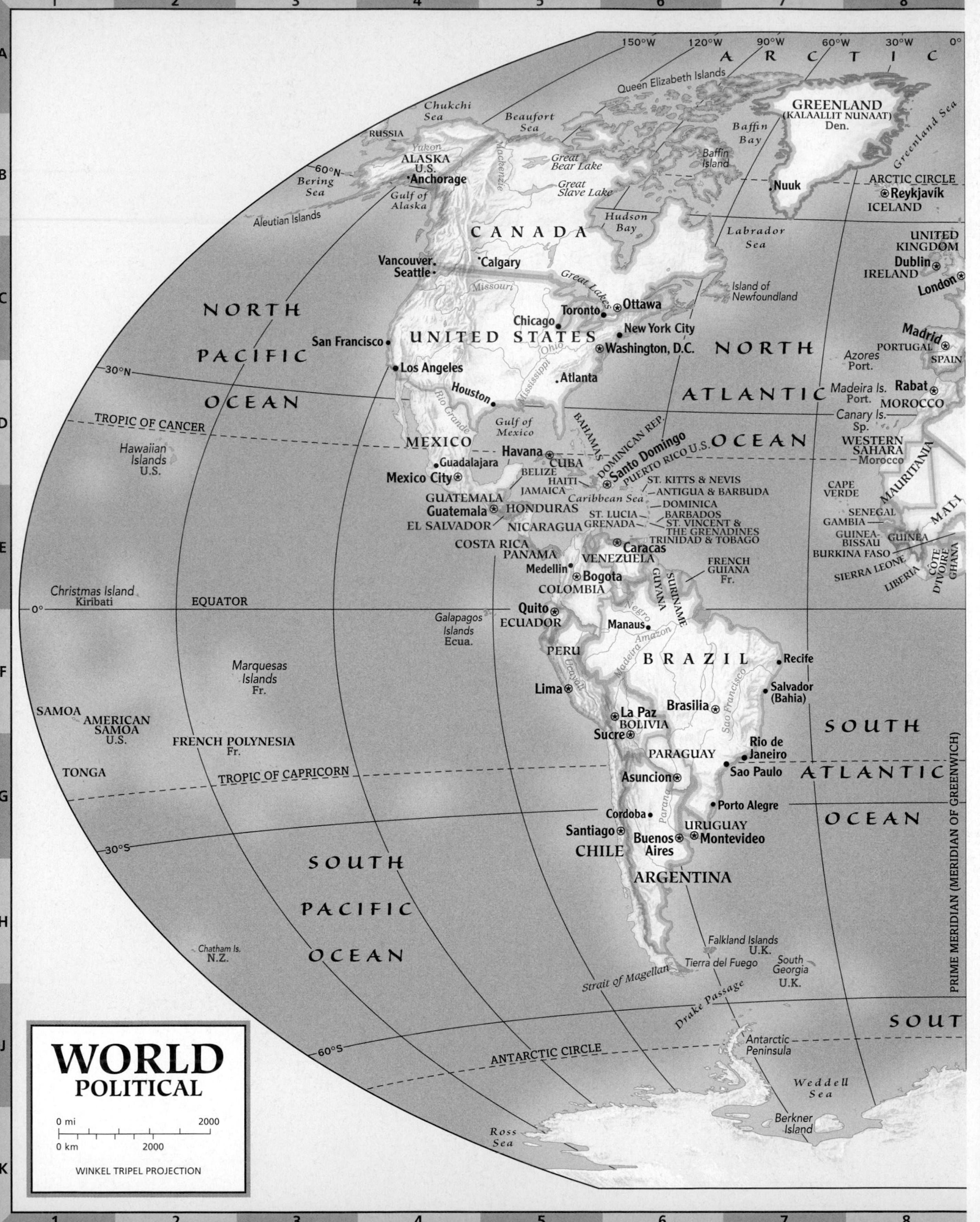

WORLD
POLITICAL
0 mi
2000
0 km
2000
WINKEL TRIPEL PROJECTION
NORTH PACIFIC OCEAN
SOUTH PACIFIC OCEAN
NORTH ATLANTIC OCEAN
SOUTH ATLANTIC OCEAN
ARCTIC
SOUT
CANADA
UNITED STATES
MEXICO
BRAZIL
GREENLAND
(KALAALLIT NUNAAT)
Den.
ALASKA
U.S.
RUSSIA
Anchorage
Vancouver
Seattle
Calgary
San Francisco
Los Angeles
Houston
Chicago
Toronto
Ottawa
New York City
Washington, D.C.
Atlanta
Nuuk
Reykjavík
ICELAND
UNITED KINGDOM
Dublin
IRELAND
London
Madrid
PORTUGAL
SPAIN
Azores
Port.
Madeira Is.
Port.
Rabat
MOROCCO
Canary Is.
Sp.
WESTERN SAHARA
Morocco
MAURITANIA
CAPE VERDE
SENEGAL
GAMBIA
GUINEA-BISSAU
GUINEA
MALI
BURKINA FASO
SIERRA LEONE
LIBERIA
CÔTE D'IVOIRE
GHANA
Chukchi Sea
Beaufort Sea
Bering Sea
Gulf of Alaska
Aleutian Islands
Queen Elizabeth Islands
Baffin Bay
Baffin Island
Greenland Sea
ARCTIC CIRCLE
Great Bear Lake
Great Slave Lake
Hudson Bay
Labrador Sea
Island of Newfoundland
Great Lakes
Yukon
Mackenzie
Missouri
Ohio
Mississippi
Rio Grande
Gulf of Mexico
Hawaiian Islands
U.S.
TROPIC OF CANCER
BAHAMAS
DOMINICAN REP.
Santo Domingo
PUERTO RICO U.S.
Havana
CUBA
Guadalajara
Mexico City
BELIZE
HAITI
JAMAICA
Caribbean Sea
GUATEMALA
Guatemala
HONDURAS
EL SALVADOR
NICARAGUA
COSTA RICA
PANAMA
ST. KITTS & NEVIS
ANTIGUA & BARBUDA
DOMINICA
ST. LUCIA
BARBADOS
GRENADA
ST. VINCENT & THE GRENADINES
TRINIDAD & TOBAGO
Caracas
VENEZUELA
Medellín
Bogota
COLOMBIA
GUYANA
SURINAME
FRENCH GUIANA
Fr.
Christmas Island
Kiribati
EQUATOR
Quito
ECUADOR
Galapagos Islands
Ecua.
Negro
Manaus
Amazon
Madeira
PERU
Ucayali
Lima
Recife
Salvador (Bahia)
Brasilia
Sao Francisco
La Paz
BOLIVIA
Sucre
Marquesas Islands
Fr.
SAMOA
AMERICAN SAMOA
U.S.
FRENCH POLYNESIA
Fr.
TONGA
TROPIC OF CAPRICORN
PARAGUAY
Asuncion
Rio de Janeiro
Sao Paulo
Parana
Porto Alegre
Cordoba
Santiago
CHILE
URUGUAY
Buenos Aires
Montevideo
ARGENTINA
Chatham Is.
N.Z.
Falkland Islands
U.K.
Tierra del Fuego
South Georgia
U.K.
Strait of Magellan
Drake Passage
Antarctic Peninsula
ANTARCTIC CIRCLE
Weddell Sea
Berkner Island
Ross Sea
PRIME MERIDIAN (MERIDIAN OF GREENWICH)
150°W
120°W
90°W
60°W
30°W
0°
60°N
30°N
0°
30°S
60°S
A
B
C
D
E
F
G
H
J
K
1
2
3
4
5
6
7
8

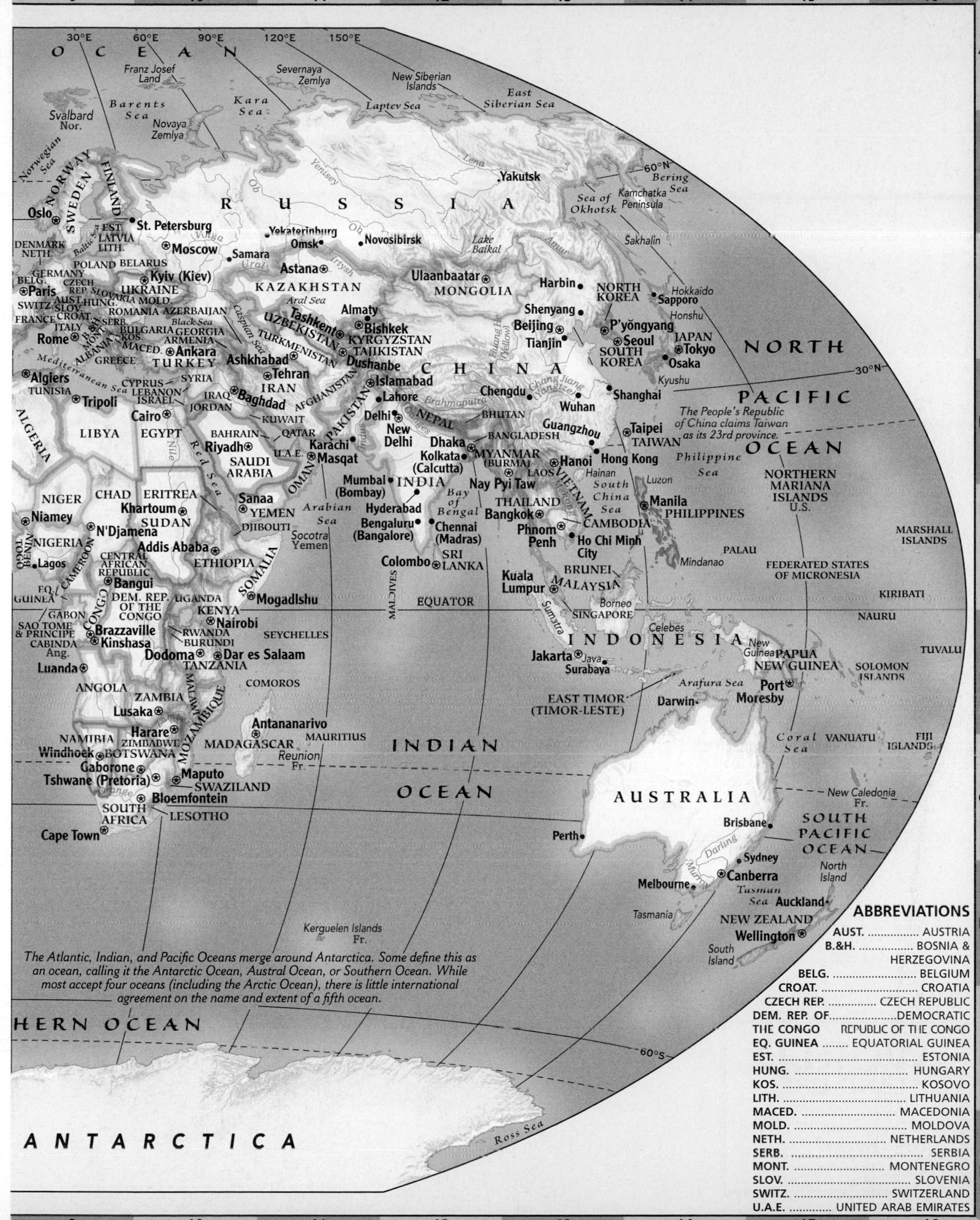

9
10
11
12
13
14
15
16
A
B
C
D
E
F
G
H
J
K
30°E
60°E
90°E
120°E
150°E
O C E A N
Franz Josef Land
Severnaya Zemlya
New Siberian Islands
East Siberian Sea
Laptev Sea
Barents Sea
Kara Sea
Svalbard Nor.
Novaya Zemlya
Norwegian Sea
NORWAY
SWEDEN
FINLAND
Yenisey
Lena
Ob
.Yakutsk
60°N
Bering Sea
Sea of Okhotsk
Kamchatka Peninsula
R U S S I A
Oslo
EST.
LATVIA
LITH.
St. Petersburg
Yekaterinburg
Omsk
Novosibirsk
Lake Baikal
Amur
Sakhalin
DENMARK
NETH.
Baltic Sea
Moscow
Volga
Samara
Ural
Irtysh
POLAND
BELARUS
GERMANY
BELG.
CZECH REP.
Paris
SLOVAKIA
Kyiv (Kiev)
UKRAINE
MOLD.
Astana
KAZAKHSTAN
Aral Sea
Ulaanbaatar
MONGOLIA
Harbin
NORTH KOREA
Hokkaido
Sapporo
SWITZ.
AUST.
HUNG.
SLOV.
FRANCE
CROAT.
ROMANIA
AZERBAIJAN
Almaty
Shenyang
Honshu
ITALY
SERB.
B.&H.
Black Sea
BULGARIA
GEORGIA
Tashkent
UZBEKISTAN
Bishkek
KYRGYZSTAN
Beijing
P'yŏngyang
JAPAN
Rome
MONT.
KOS.
MACED.
ALBANIA
ARMENIA
Caspian Sea
TURKMENISTAN
TAJIKISTAN
Huang He (Yellow)
Tianjin
Seoul
SOUTH KOREA
Tokyo
Osaka
NORTH
Mediterranean Sea
GREECE
Ankara
TURKEY
Ashkhabad
Dushanbe
C H I N A
Algiers
CYPRUS
SYRIA
Tehran
IRAN
AFGHANISTAN
Islamabad
Chang Jiang (Yangtze)
Chengdu
Kyushu
30°N
TUNISIA
LEBANON
ISRAEL
IRAQ
Baghdad
Lahore
Brahmaputra
Shanghai
PACIFIC
Tripoli
JORDAN
PAKISTAN
Delhi
Wuhan
The People's Republic of China claims Taiwan as its 23rd province.
ALGERIA
Cairo
KUWAIT
NEPAL
BHUTAN
Ganges
New Delhi
Taipei
TAIWAN
LIBYA
EGYPT
BAHRAIN
QATAR
Riyadh
Nile
Red Sea
Karachi
Indus
Dhaka
BANGLADESH
Guangzhou
U.A.E.
SAUDI ARABIA
OMAN
Masqat
Kolkata (Calcutta)
MYANMAR (BURMA)
Hanoi
Hong Kong
Philippine Sea
OCEAN
LAOS
Hainan
Luzon
NORTHERN MARIANA ISLANDS U.S.
Mumbai (Bombay)
INDIA
Nay Pyi Taw
VIETNAM
Mekong
South China Sea
NIGER
CHAD
ERITREA
Khartoum
SUDAN
Sanaa
YEMEN
Arabian Sea
Bay of Bengal
Hyderabad
THAILAND
Manila
PHILIPPINES
Niamey
N'Djamena
DJIBOUTI
Bengaluru (Bangalore)
Chennai (Madras)
Bangkok
CAMBODIA
MARSHALL ISLANDS
NIGERIA
BENIN
TOGO
Socotra Yemen
Phnom Penh
Ho Chi Minh City
Addis Ababa
ETHIOPIA
CENTRAL AFRICAN REPUBLIC
SOMALIA
Colombo
SRI LANKA
PALAU
Lagos
CAMEROON
Mindanao
FEDERATED STATES OF MICRONESIA
Bangui
MALDIVES
BRUNEI
Kuala Lumpur
MALAYSIA
EQ. GUINEA
DEM. REP. OF THE CONGO
UGANDA
Mogadishu
EQUATOR
KIRIBATI
CONGO
GABON
KENYA
Nairobi
Borneo
SINGAPORE
Sumatra
NAURU
SAO TOME & PRINCIPE
Brazzaville
RWANDA
BURUNDI
SEYCHELLES
Celebes
CABINDA Ang.
Kinshasa
Dodoma
Dar es Salaam
TANZANIA
I N D O N E S I A
New Guinea
PAPUA NEW GUINEA
TUVALU
Luanda
Jakarta
Java
Surabaya
SOLOMON ISLANDS
ANGOLA
MALAWI
COMOROS
Arafura Sea
Port Moresby
ZAMBIA
EAST TIMOR (TIMOR-LESTE)
Darwin
Lusaka
Zambezi
Antananarivo
MOZAMBIQUE
Harare
ZIMBABWE
MAURITIUS
NAMIBIA
INDIAN
Coral Sea
VANUATU
FIJI ISLANDS
Windhoek
BOTSWANA
MADAGASCAR
Reunion Fr.
Gaborone
Maputo
Tshwane (Pretoria)
SWAZILAND
OCEAN
Orange
Bloemfontein
AUSTRALIA
New Caledonia Fr.
SOUTH AFRICA
LESOTHO
Cape Town
Perth
Brisbane
SOUTH PACIFIC OCEAN
Darling
Sydney
Murray
Canberra
North Island
Melbourne
Tasman Sea
Auckland
Tasmania
NEW ZEALAND
Kerguelen Islands Fr.
Wellington
South Island
The Atlantic, Indian, and Pacific Oceans merge around Antarctica. Some define this as an ocean, calling it the Antarctic Ocean, Austral Ocean, or Southern Ocean. While most accept four oceans (including the Arctic Ocean), there is little international agreement on the name and extent of a fifth ocean.
HERN OCEAN
60°S
Ross Sea
ANTARCTICA
ABBREVIATIONS
AUST. .... AUSTRIA
B.&H. .... BOSNIA & HERZEGOVINA
BELG. .... BELGIUM
CROAT. .... CROATIA
CZECH REP. .... CZECH REPUBLIC
DEM. REP. OF THE CONGO .... DEMOCRATIC REPUBLIC OF THE CONGO
EQ. GUINEA .... EQUATORIAL GUINEA
EST. .... ESTONIA
HUNG. .... HUNGARY
KOS. .... KOSOVO
LITH. .... LITHUANIA
MACED. .... MACEDONIA
MOLD. .... MOLDOVA
NETH. .... NETHERLANDS
SERB. .... SERBIA
MONT. .... MONTENEGRO
SLOV. .... SLOVENIA
SWITZ. .... SWITZERLAND
U.A.E. .... UNITED ARAB EMIRATES

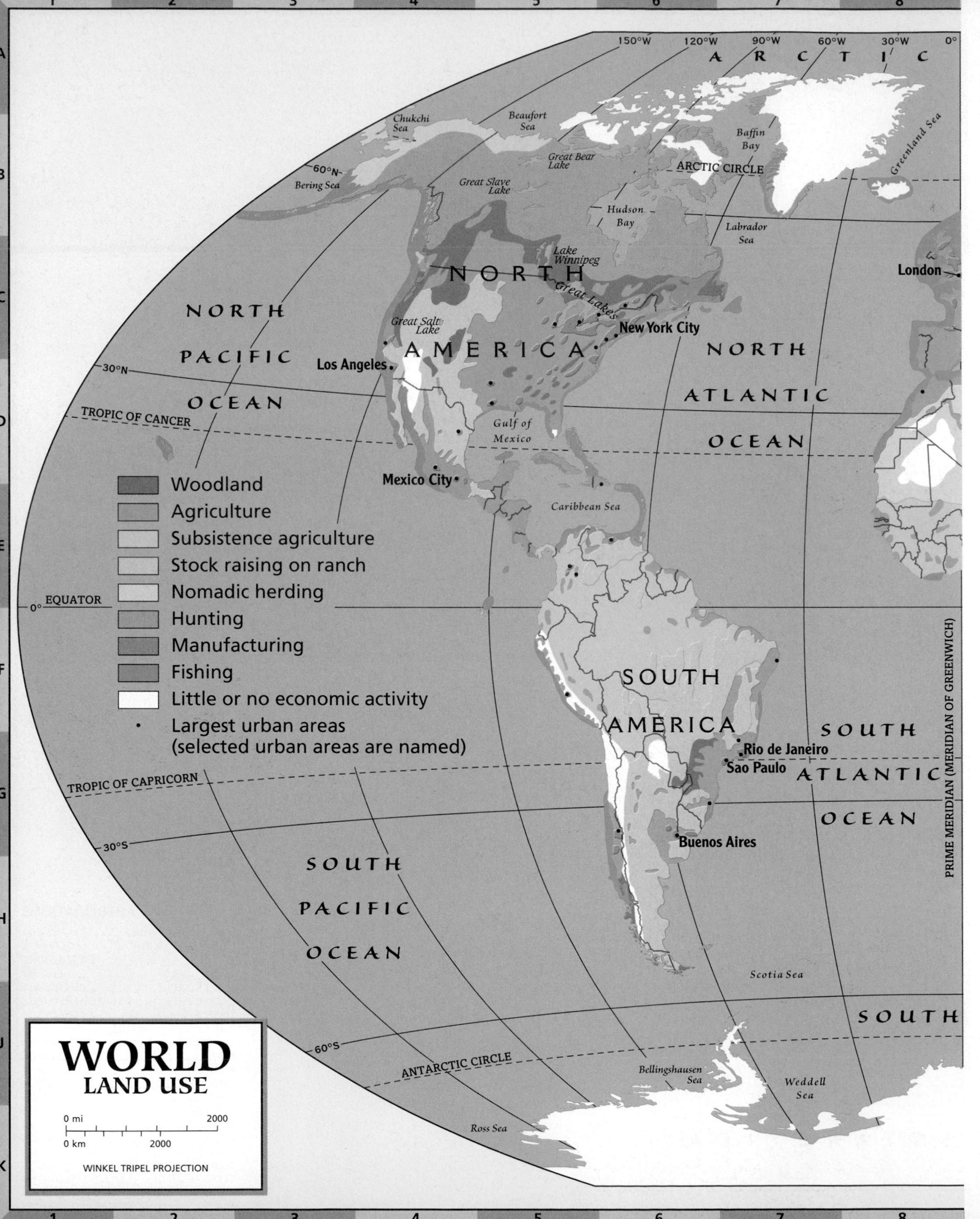
WORLD
LAND USE
0 mi 2000
0 km 2000
WINKEL TRIPEL PROJECTION
Woodland
Agriculture
Subsistence agriculture
Stock raising on ranch
Nomadic herding
Hunting
Manufacturing
Fishing
Little or no economic activity
Largest urban areas
(selected urban areas are named)
ARCTIC
NORTH AMERICA
SOUTH AMERICA
NORTH PACIFIC OCEAN
SOUTH PACIFIC OCEAN
NORTH ATLANTIC OCEAN
SOUTH ATLANTIC OCEAN
SOUTH
Chukchi Sea
Beaufort Sea
Bering Sea
Great Bear Lake
Great Slave Lake
Hudson Bay
Baffin Bay
Greenland Sea
Labrador Sea
Lake Winnipeg
Great Lakes
Great Salt Lake
Gulf of Mexico
Caribbean Sea
Scotia Sea
Bellingshausen Sea
Weddell Sea
Ross Sea
Los Angeles
New York City
Mexico City
London
Rio de Janeiro
Sao Paulo
Buenos Aires
ARCTIC CIRCLE
TROPIC OF CANCER
EQUATOR
TROPIC OF CAPRICORN
ANTARCTIC CIRCLE
PRIME MERIDIAN (MERIDIAN OF GREENWICH)
150°W
120°W
90°W
60°W
30°W
0°
60°N
30°N
0°
30°S
60°S
1 2 3 4 5 6 7 8
A B C D E F G H J K

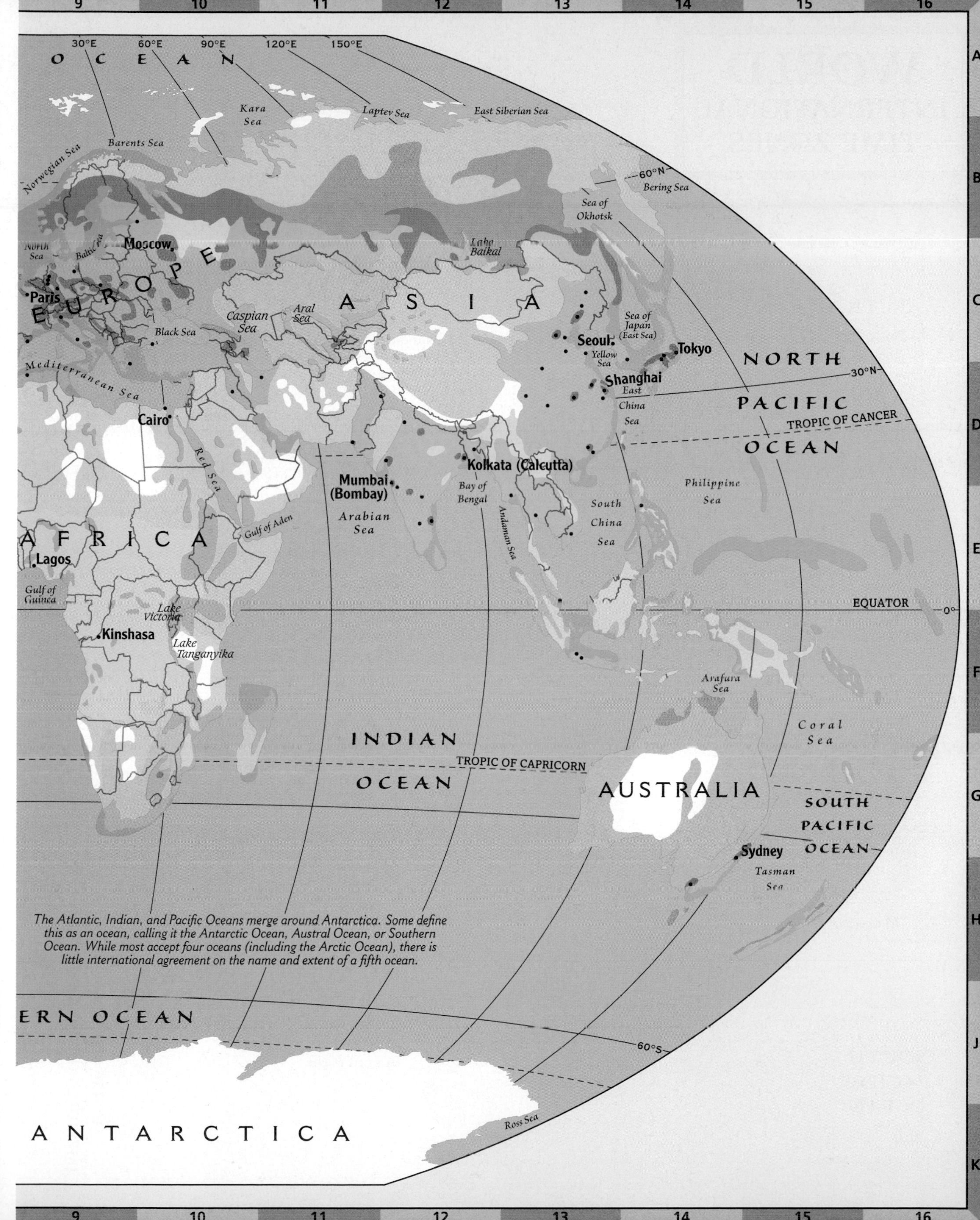

OCEAN
Kara Sea
Laptev Sea
East Siberian Sea
Barents Sea
Norwegian Sea
60°N
Bering Sea
Sea of Okhotsk
Lake Baikal
Moscow
North Sea
Baltic Sea
EUROPE
Paris
ASIA
Caspian Sea
Aral Sea
Sea of Japan (East Sea)
Black Sea
Seoul
Yellow Sea
Tokyo
NORTH PACIFIC OCEAN
Mediterranean Sea
Shanghai
30°N
East China Sea
TROPIC OF CANCER
Cairo
Red Sea
Kolkata (Calcutta)
Mumbai (Bombay)
Bay of Bengal
Philippine Sea
South China Sea
Arabian Sea
Andaman Sea
Gulf of Aden
AFRICA
Lagos
Gulf of Guinea
EQUATOR
Lake Victoria
Kinshasa
Lake Tanganyika
Arafura Sea
Coral Sea
INDIAN OCEAN
TROPIC OF CAPRICORN
AUSTRALIA
SOUTH PACIFIC OCEAN
Sydney
Tasman Sea
The Atlantic, Indian, and Pacific Oceans merge around Antarctica. Some define this as an ocean, calling it the Antarctic Ocean, Austral Ocean, or Southern Ocean. While most accept four oceans (including the Arctic Ocean), there is little international agreement on the name and extent of a fifth ocean.
ERN OCEAN
60°S
Ross Sea
ANTARCTICA

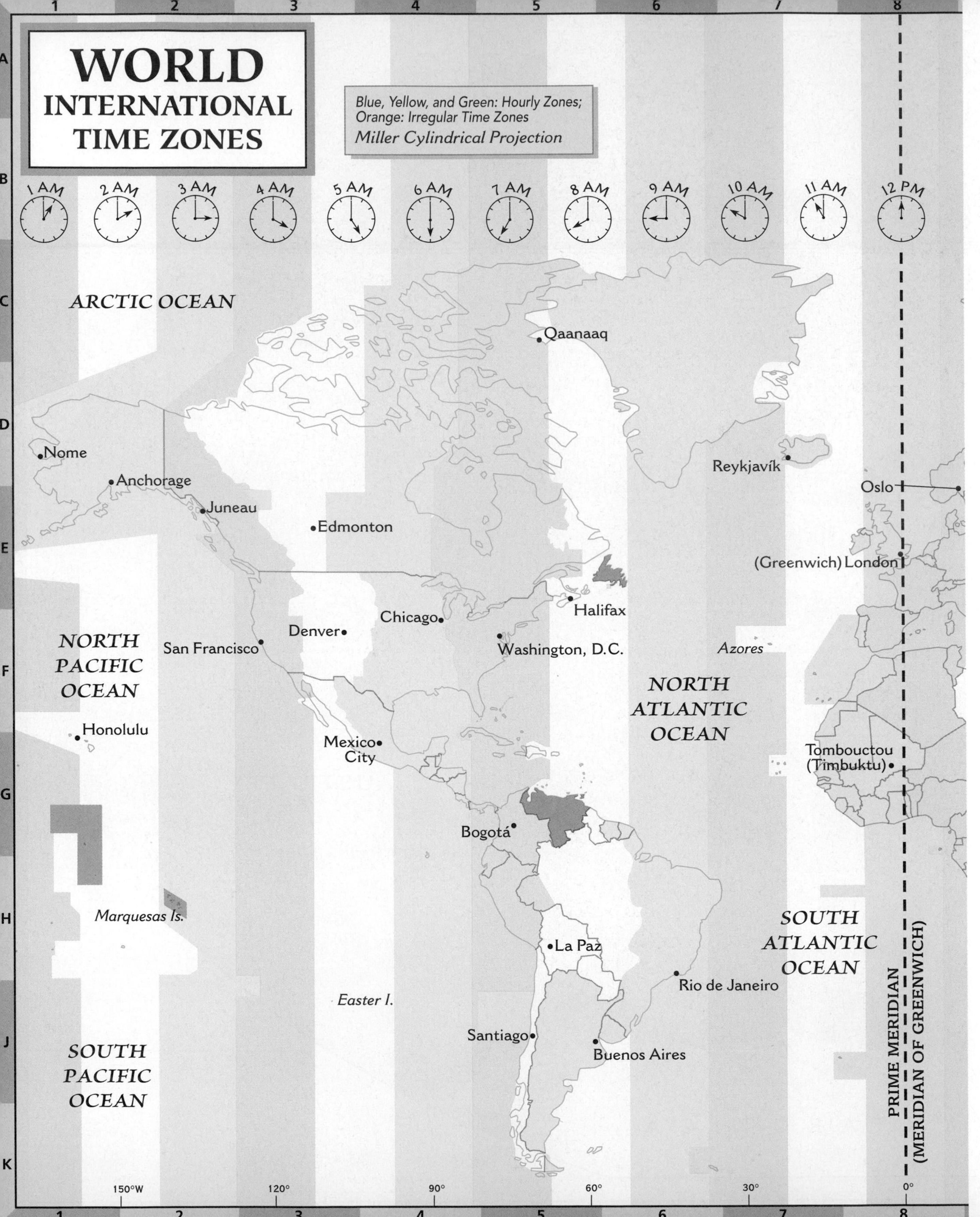

WORLD INTERNATIONAL TIME ZONES
Blue, Yellow, and Green: Hourly Zones; Orange: Irregular Time Zones
Miller Cylindrical Projection
1 AM
2 AM
3 AM
4 AM
5 AM
6 AM
7 AM
8 AM
9 AM
10 AM
11 AM
12 PM
ARCTIC OCEAN
Qaanaaq
Nome
Anchorage
Juneau
Edmonton
Reykjavík
Oslo
(Greenwich) London
Halifax
Chicago
Denver
San Francisco
Washington, D.C.
Azores
NORTH PACIFIC OCEAN
NORTH ATLANTIC OCEAN
Honolulu
Mexico City
Tombouctou (Timbuktu)
Bogotá
Marquesas Is.
SOUTH ATLANTIC OCEAN
La Paz
Rio de Janeiro
Easter I.
Santiago
Buenos Aires
SOUTH PACIFIC OCEAN
PRIME MERIDIAN (MERIDIAN OF GREENWICH)
150°W
120°
90°
60°
30°
0°
1
2
3
4
5
6
7
8
A
B
C
D
E
F
G
H
J
K

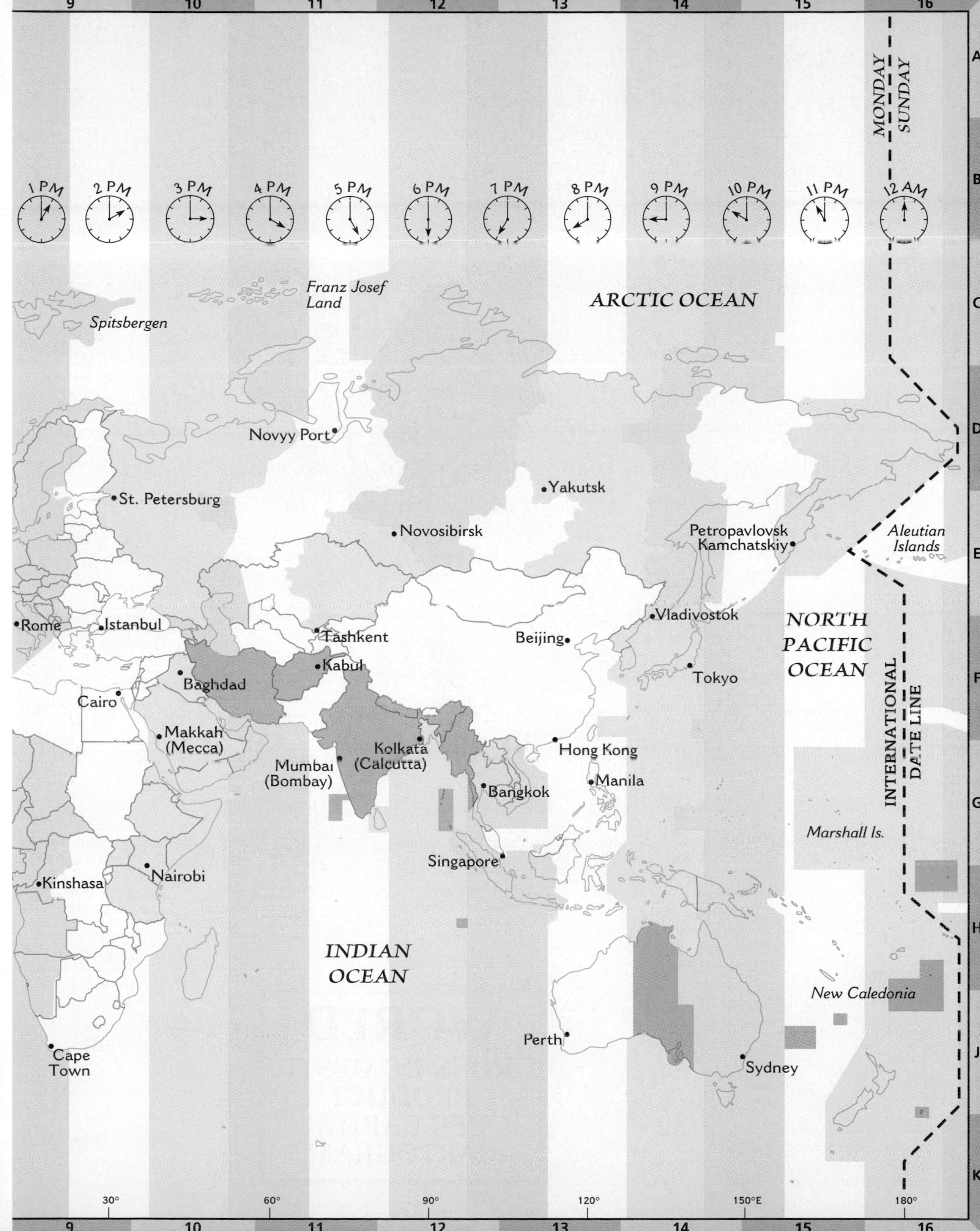
9
10
11
12
13
14
15
16
A
B
C
D
E
F
G
H
J
K
MONDAY
SUNDAY
1 PM
2 PM
3 PM
4 PM
5 PM
6 PM
7 PM
8 PM
9 PM
10 PM
11 PM
12 AM
Franz Josef Land
ARCTIC OCEAN
Spitsbergen
Novyy Port
Yakutsk
St. Petersburg
Novosibirsk
Petropavlovsk Kamchatskiy
Aleutian Islands
Vladivostok
NORTH PACIFIC OCEAN
Rome
Istanbul
Tashkent
Beijing
Kabul
Tokyo
Baghdad
Cairo
INTERNATIONAL DATE LINE
Makkah (Mecca)
Hong Kong
Kolkata (Calcutta)
Mumbai (Bombay)
Manila
Bangkok
Marshall Is.
Singapore
Nairobi
Kinshasa
INDIAN OCEAN
New Caledonia
Perth
Cape Town
Sydney
30°
60°
90°
120°
150°E
180°

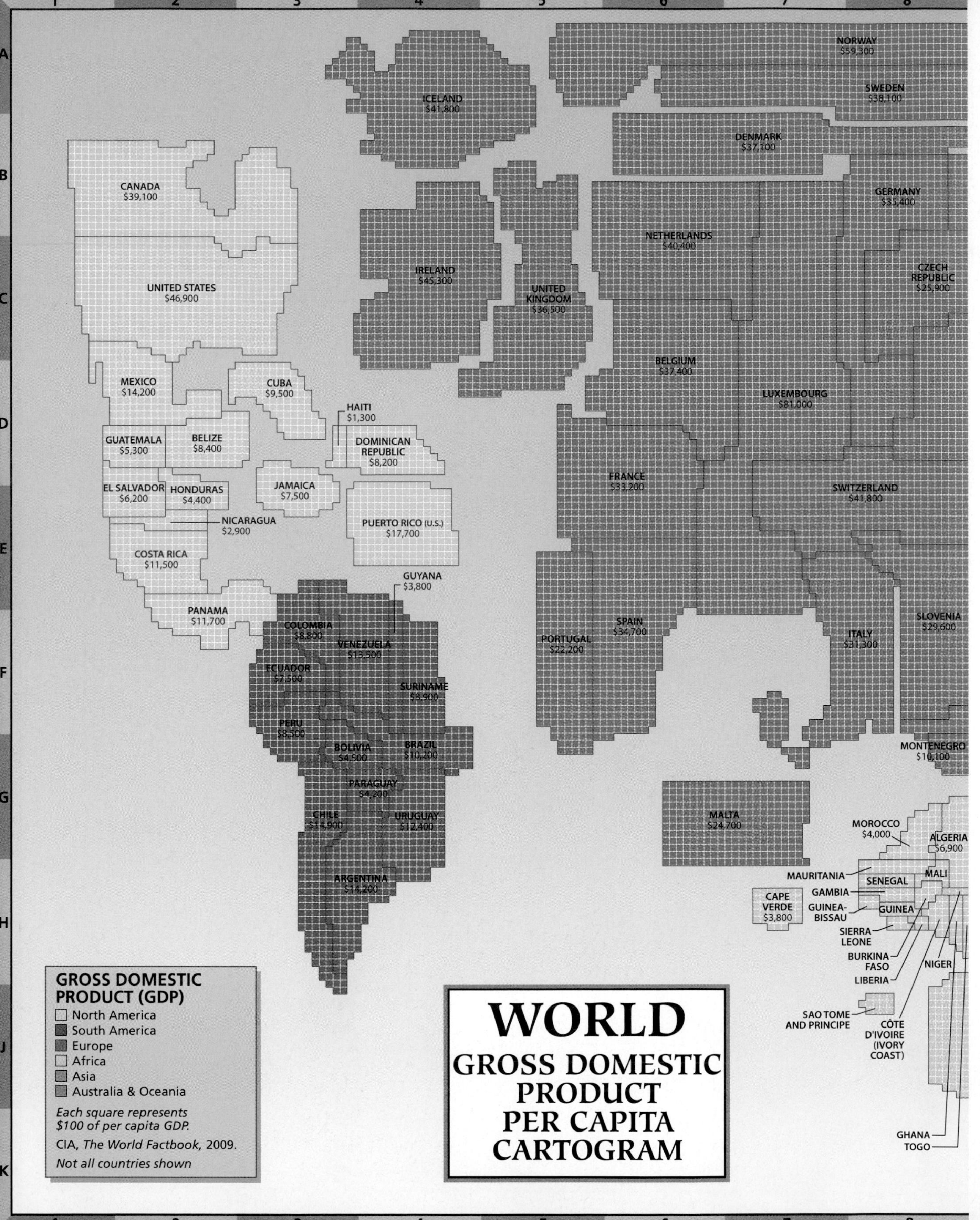

NORWAY
$59,300
SWEDEN
$38,100
ICELAND
$41,800
DENMARK
$37,100
CANADA
$39,100
GERMANY
$35,400
NETHERLANDS
$40,400
IRELAND
$45,300
CZECH REPUBLIC
$25,900
UNITED STATES
$46,900
UNITED KINGDOM
$36,500
BELGIUM
$37,400
MEXICO
$14,200
CUBA
$9,500
LUXEMBOURG
$81,000
HAITI
$1,300
GUATEMALA
$5,300
BELIZE
$8,400
DOMINICAN REPUBLIC
$8,200
FRANCE
$33,200
SWITZERLAND
$41,800
EL SALVADOR
$6,200
HONDURAS
$4,400
JAMAICA
$7,500
NICARAGUA
$2,900
PUERTO RICO (U.S.)
$17,700
COSTA RICA
$11,500
GUYANA
$3,800
PANAMA
$11,700
COLOMBIA
$8,800
SPAIN
$34,700
SLOVENIA
$29,600
PORTUGAL
$22,200
VENEZUELA
$13,500
ITALY
$31,300
ECUADOR
$7,500
SURINAME
$8,900
PERU
$8,500
BOLIVIA
$4,500
BRAZIL
$10,200
MONTENEGRO
$10,100
PARAGUAY
$4,200
CHILE
$14,900
URUGUAY
$12,400
MALTA
$24,700
MOROCCO
$4,000
ALGERIA
$6,900
MAURITANIA
SENEGAL
MALI
ARGENTINA
$14,200
CAPE VERDE
$3,800
GAMBIA
GUINEA-BISSAU
GUINEA
SIERRA LEONE
BURKINA FASO
NIGER
LIBERIA
GROSS DOMESTIC PRODUCT (GDP)
North America
South America
Europe
Africa
Asia
Australia & Oceania
Each square represents $100 of per capita GDP.
CIA, The World Factbook, 2009.
Not all countries shown
WORLD
GROSS DOMESTIC PRODUCT PER CAPITA CARTOGRAM
SAO TOME AND PRINCIPE
CÔTE D'IVOIRE (IVORY COAST)
GHANA
TOGO

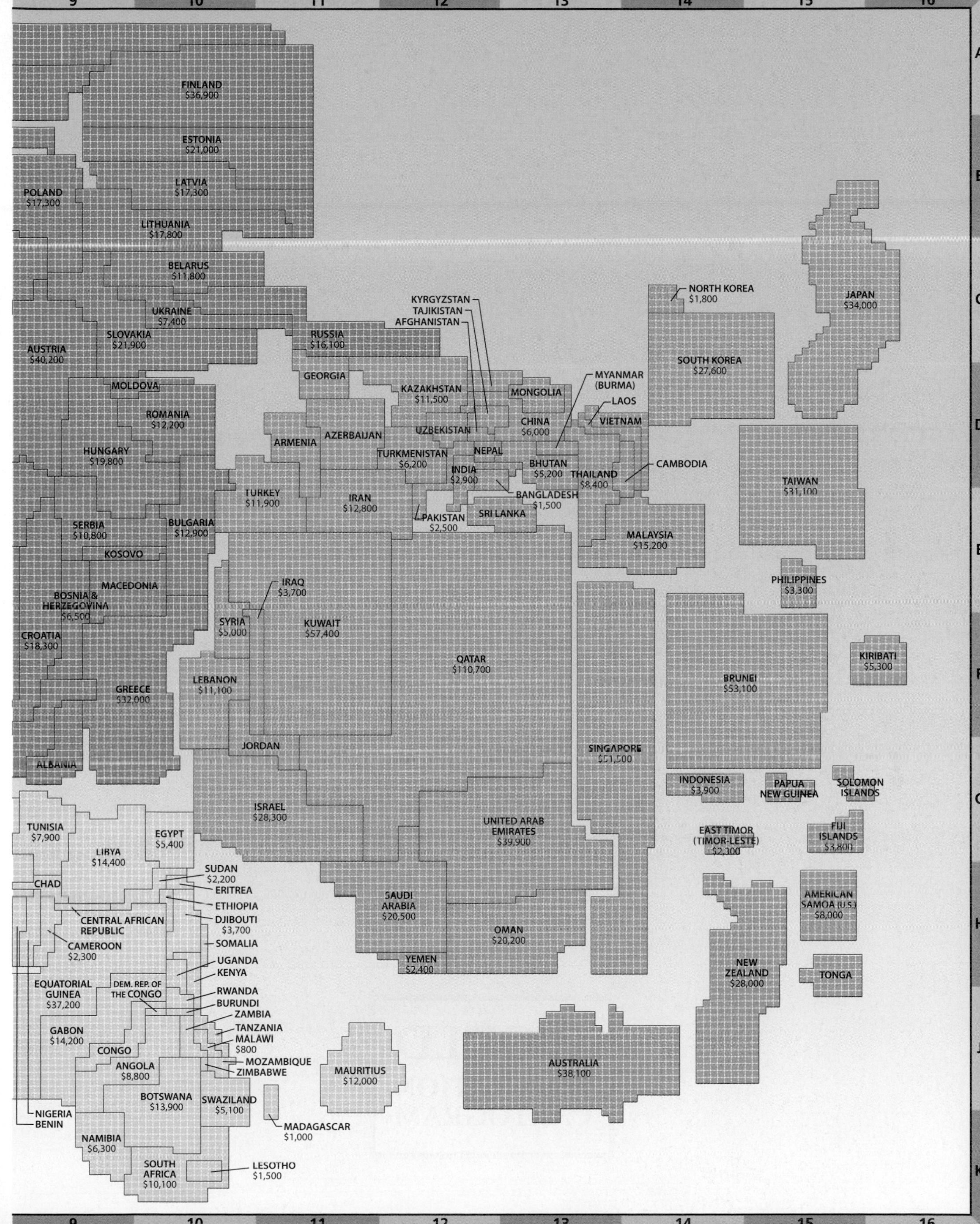

9
10
11
12
13
14
15
16
A
B
C
D
E
F
G
H
J
K
FINLAND $36,900
ESTONIA $21,000
LATVIA $17,300
POLAND $17,300
LITHUANIA $17,800
BELARUS $11,800
UKRAINE $7,400
SLOVAKIA $21,900
AUSTRIA $40,200
MOLDOVA
ROMANIA $12,200
HUNGARY $19,800
SERBIA $10,800
BULGARIA $12,900
KOSOVO
MACEDONIA
BOSNIA & HERZEGOVINA $6,500
CROATIA $18,300
GREECE $32,000
ALBANIA
RUSSIA $16,100
GEORGIA
KYRGYZSTAN
TAJIKISTAN
AFGHANISTAN
KAZAKHSTAN $11,500
MONGOLIA
ARMENIA
AZERBAIJAN
UZBEKISTAN
TURKMENISTAN $6,200
NEPAL
INDIA $2,900
CHINA $6,000
BHUTAN $5,200
MYANMAR (BURMA)
LAOS
VIETNAM
THAILAND $8,400
CAMBODIA
BANGLADESH $1,500
SRI LANKA
PAKISTAN $2,500
TURKEY $11,900
IRAN $12,800
MALAYSIA $15,200
NORTH KOREA $1,800
SOUTH KOREA $27,600
JAPAN $34,000
TAIWAN $31,100
PHILIPPINES $3,300
IRAQ $3,700
KUWAIT $57,400
SYRIA $5,000
LEBANON $11,100
QATAR $110,700
JORDAN
SINGAPORE $51,500
BRUNEI $53,100
KIRIBATI $5,300
ISRAEL $28,300
UNITED ARAB EMIRATES $39,900
INDONESIA $3,900
PAPUA NEW GUINEA
SOLOMON ISLANDS
EAST TIMOR (TIMOR-LESTE) $2,300
FIJI ISLANDS $3,800
SAUDI ARABIA $20,500
OMAN $20,200
YEMEN $2,400
AMERICAN SAMOA (U.S.) $8,000
NEW ZEALAND $28,000
TONGA
TUNISIA $7,900
LIBYA $14,400
EGYPT $5,400
SUDAN $2,200
CHAD
ERITREA
ETHIOPIA
CENTRAL AFRICAN REPUBLIC
DJIBOUTI $3,700
CAMEROON $2,300
SOMALIA
UGANDA
KENYA
EQUATORIAL GUINEA $37,200
DEM. REP. OF THE CONGO
RWANDA
BURUNDI
ZAMBIA
TANZANIA
GABON $14,200
MALAWI $800
CONGO
MOZAMBIQUE
ANGOLA $8,800
ZIMBABWE
MAURITIUS $12,000
AUSTRALIA $38,100
BOTSWANA $13,900
SWAZILAND $5,100
NIGERIA
BENIN
MADAGASCAR $1,000
NAMIBIA $6,300
SOUTH AFRICA $10,100
LESOTHO $1,500

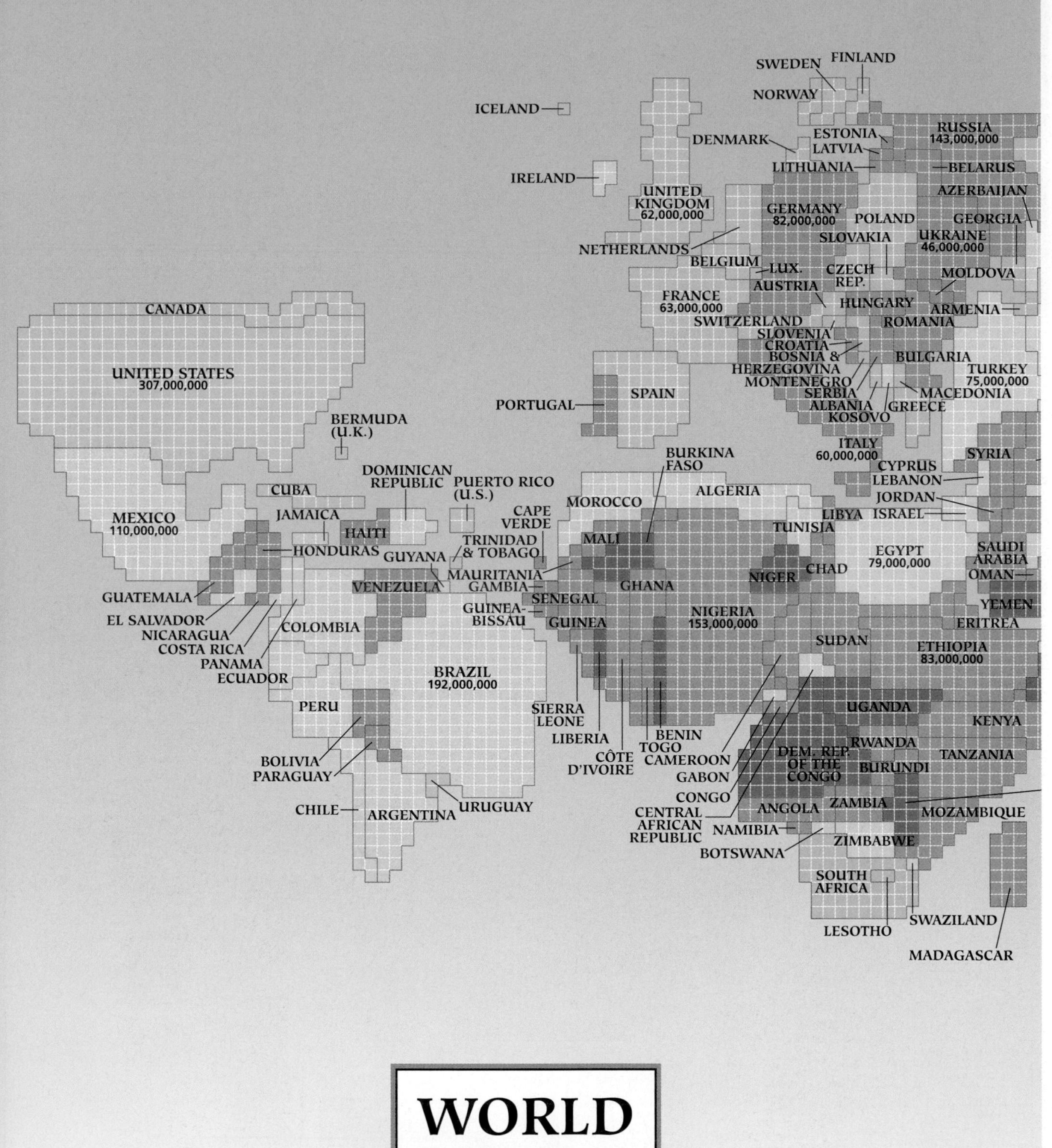

WORLD
POPULATION CARTOGRAM

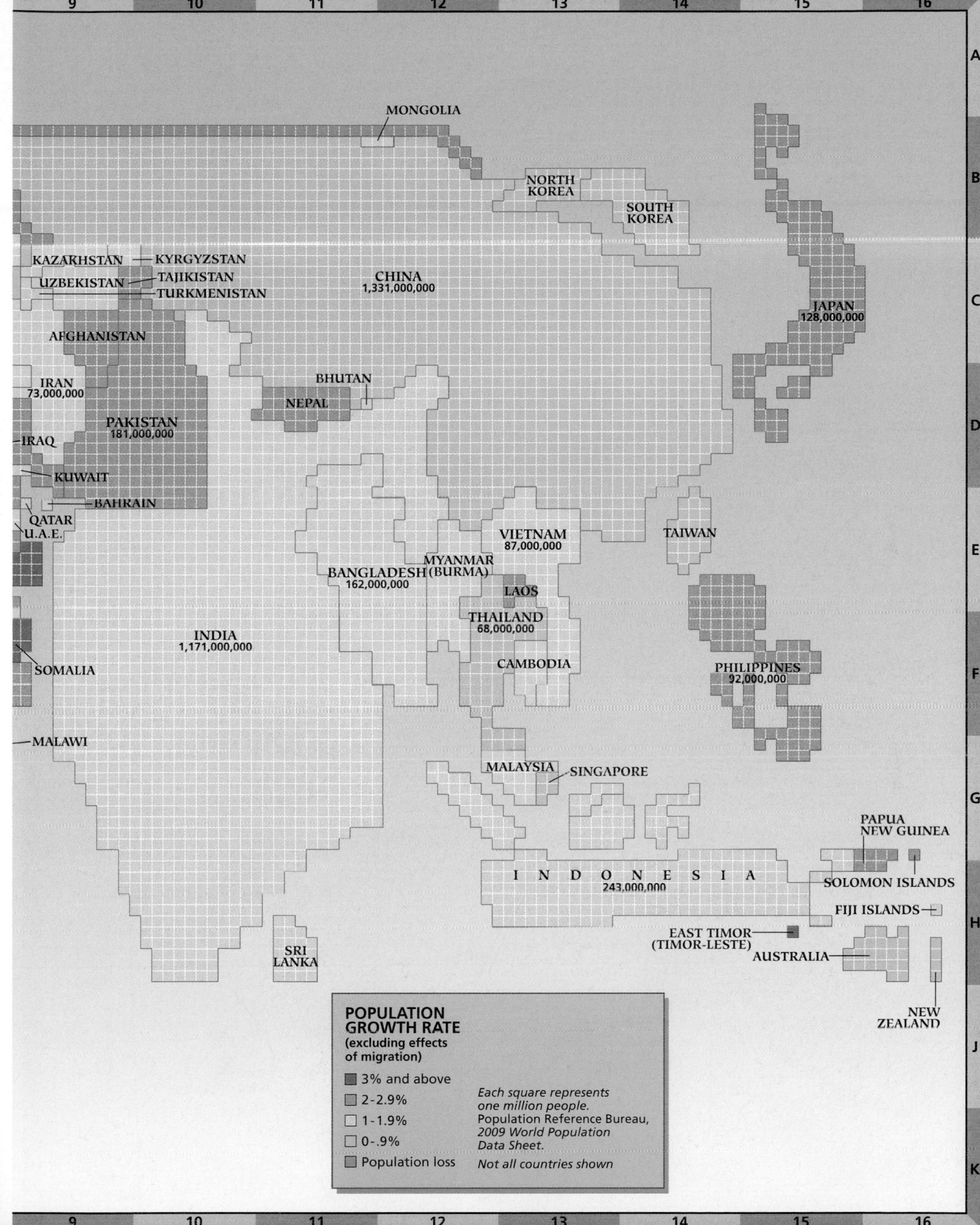
MONGOLIA
NORTH KOREA
SOUTH KOREA
KAZAKHSTAN
KYRGYZSTAN
TAJIKISTAN
UZBEKISTAN
TURKMENISTAN
CHINA
1,331,000,000
JAPAN
128,000,000
AFGHANISTAN
IRAN
73,000,000
BHUTAN
NEPAL
PAKISTAN
181,000,000
IRAQ
KUWAIT
BAHRAIN
QATAR
U.A.E.
VIETNAM
87,000,000
TAIWAN
MYANMAR (BURMA)
BANGLADESH
162,000,000
LAOS
THAILAND
68,000,000
INDIA
1,171,000,000
CAMBODIA
SOMALIA
PHILIPPINES
92,000,000
MALAWI
MALAYSIA
SINGAPORE
PAPUA NEW GUINEA
INDONESIA
243,000,000
SOLOMON ISLANDS
FIJI ISLANDS
EAST TIMOR (TIMOR-LESTE)
AUSTRALIA
SRI LANKA
NEW ZEALAND
POPULATION GROWTH RATE
(excluding effects of migration)
3% and above
2-2.9%
1-1.9%
0-.9%
Population loss
Each square represents one million people.
Population Reference Bureau, 2009 World Population Data Sheet.
Not all countries shown

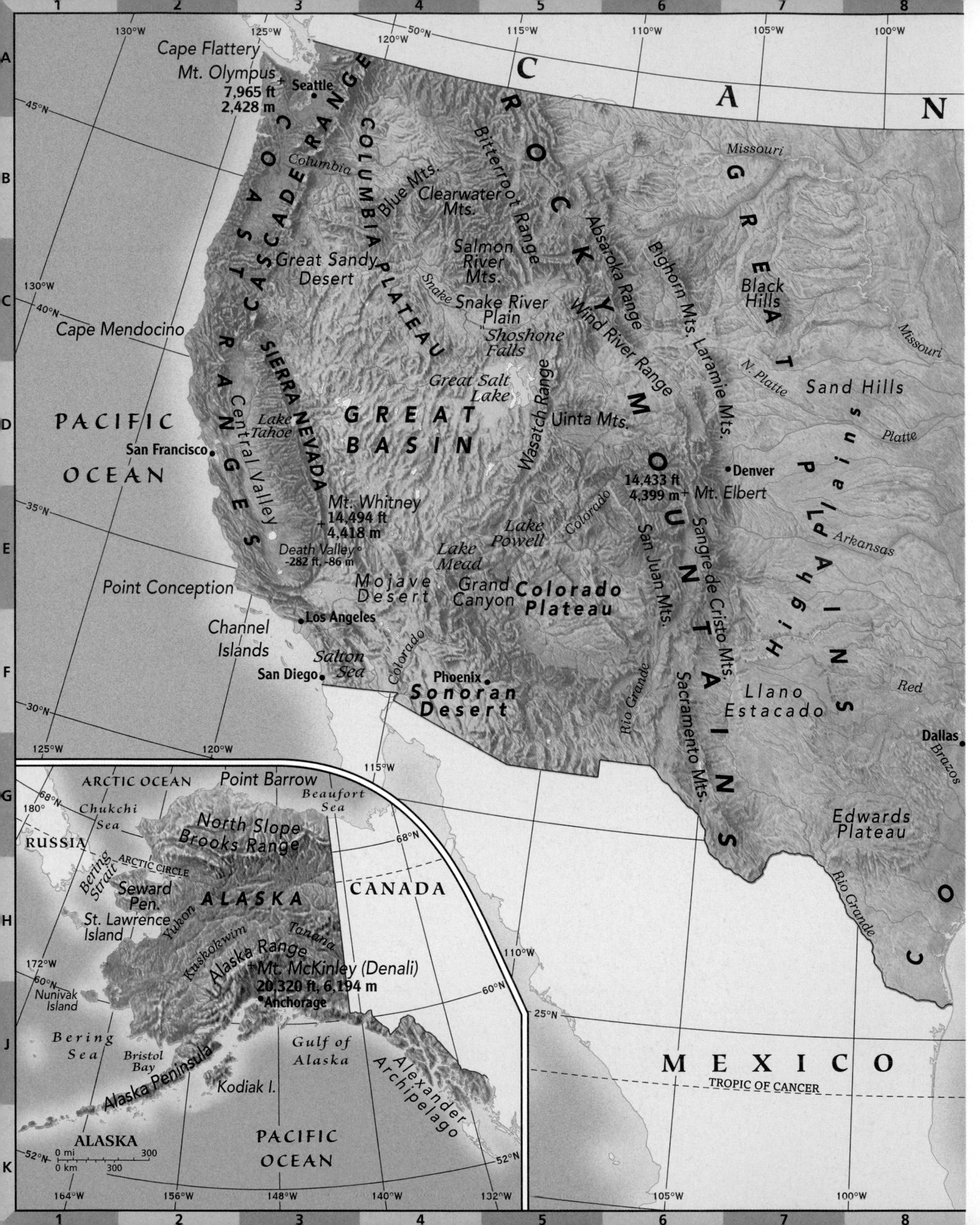

CANADA
PACIFIC OCEAN
Cape Flattery
Mt. Olympus 7,965 ft 2,428 m
Seattle
COAST RANGES
CASCADE RANGE
Columbia
COLUMBIA PLATEAU
Blue Mts.
Clearwater Mts.
Bitterroot Range
Salmon River Mts.
ROCKY MOUNTAINS
Great Sandy Desert
Snake
Snake River Plain
Shoshone Falls
Absaroka Range
Wind River Range
Bighorn Mts.
Laramie Mts.
Missouri
GREAT PLAINS
Black Hills
Cape Mendocino
SIERRA NEVADA
Great Salt Lake
Wasatch Range
Uinta Mts.
N. Platte
Sand Hills
Platte
Lake Tahoe
Central Valley
GREAT BASIN
San Francisco
Denver
14,433 ft 4,399 m Mt. Elbert
Mt. Whitney 14,494 ft 4,418 m
Colorado
Lake Powell
San Juan Mts.
Sangre de Cristo Mts.
High Plains
Arkansas
Death Valley -282 ft, -86 m
Lake Mead
Mojave Desert
Grand Canyon
Colorado Plateau
Point Conception
Los Angeles
Channel Islands
Salton Sea
San Diego
Phoenix
Sonoran Desert
Rio Grande
Sacramento Mts.
Llano Estacado
Red
Dallas
Brazos
Edwards Plateau
MEXICO
TROPIC OF CANCER
ARCTIC OCEAN
Point Barrow
Beaufort Sea
Chukchi Sea
North Slope
Brooks Range
RUSSIA
ARCTIC CIRCLE
Bering Strait
Seward Pen.
ALASKA
CANADA
St. Lawrence Island
Yukon
Tanana
Kuskokwim
Alaska Range
Mt. McKinley (Denali) 20,320 ft, 6,194 m
Anchorage
Nunivak Island
Bering Sea
Bristol Bay
Gulf of Alaska
Alexander Archipelago
Kodiak I.
Alaska Peninsula
ALASKA
PACIFIC OCEAN
0 mi 300
0 km 300

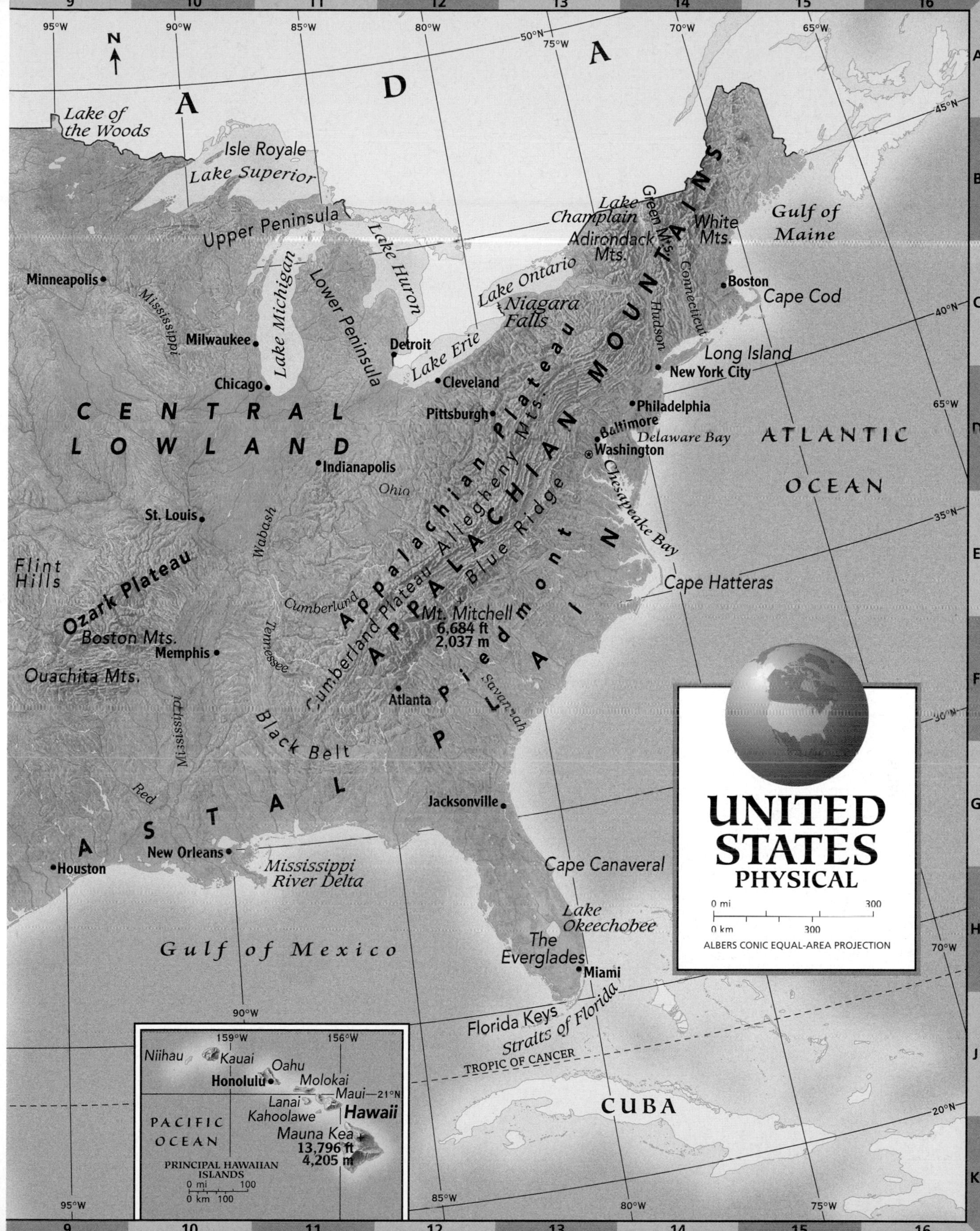
UNITED STATES
PHYSICAL
ALBERS CONIC EQUAL-AREA PROJECTION
Lake of the Woods
Isle Royale
Lake Superior
Upper Peninsula
Lake Michigan
Lower Peninsula
Lake Huron
Lake Ontario
Lake Erie
Niagara Falls
Lake Champlain
Adirondack Mts.
Green Mts.
White Mts.
Gulf of Maine
Boston
Cape Cod
Long Island
New York City
Philadelphia
Baltimore
Washington
Delaware Bay
Chesapeake Bay
Cape Hatteras
ATLANTIC OCEAN
Minneapolis
Milwaukee
Chicago
Detroit
Cleveland
Pittsburgh
Indianapolis
St. Louis
Memphis
Atlanta
Jacksonville
New Orleans
Houston
Miami
CENTRAL LOWLAND
APPALACHIAN MOUNTAINS
Appalachian Plateau
Allegheny Mts.
Blue Ridge
Cumberland Plateau
Piedmont
Mt. Mitchell 6,684 ft 2,037 m
Ozark Plateau
Boston Mts.
Ouachita Mts.
Flint Hills
Black Belt
COASTAL PLAIN
Mississippi River Delta
Gulf of Mexico
Cape Canaveral
Lake Okeechobee
The Everglades
Florida Keys
Straits of Florida
TROPIC OF CANCER
CUBA
PRINCIPAL HAWAIIAN ISLANDS
Niihau
Kauai
Oahu
Honolulu
Molokai
Maui
Lanai
Kahoolawe
Hawaii
Mauna Kea 13,796 ft 4,205 m
PACIFIC OCEAN

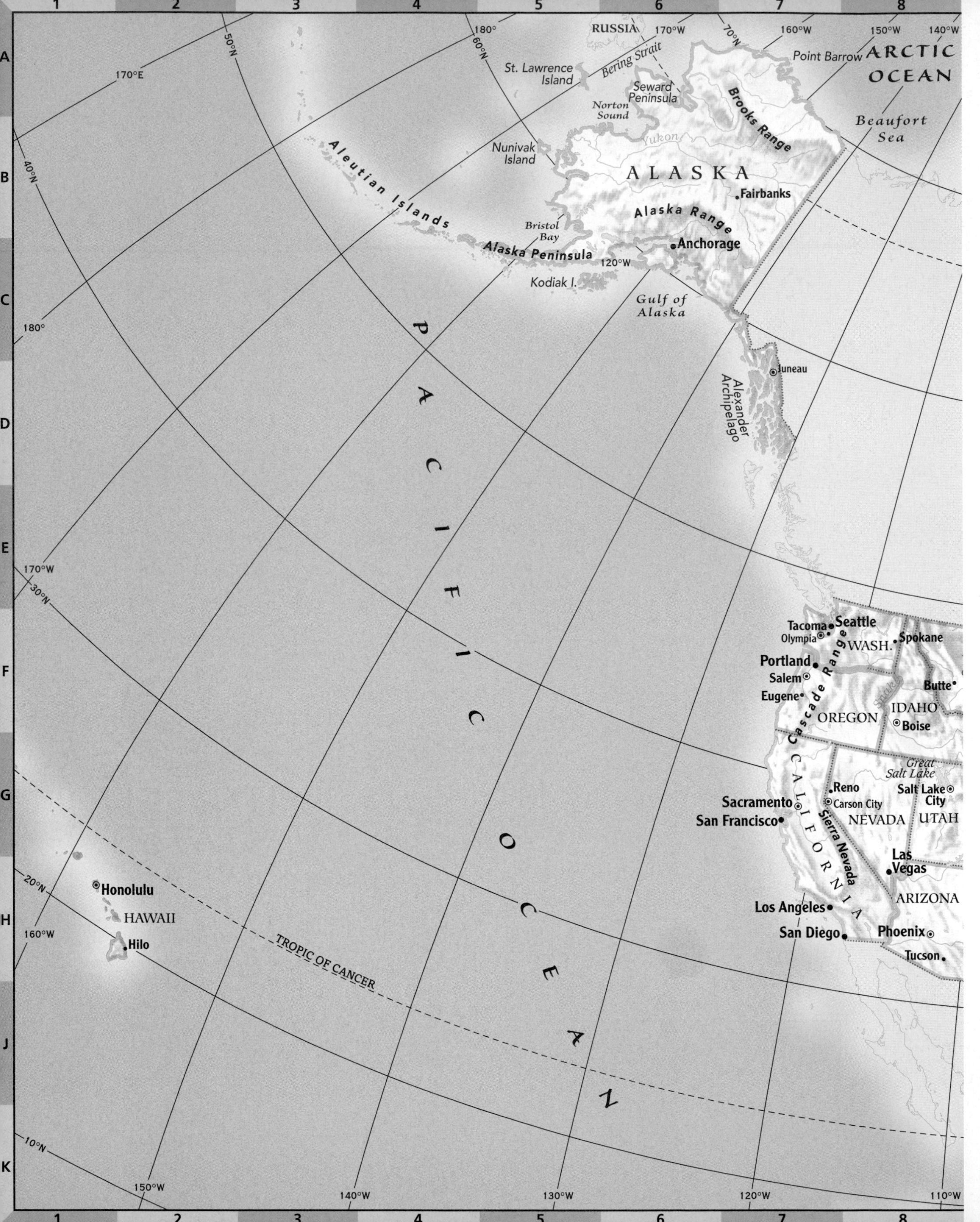

RUSSIA
Bering Strait
St. Lawrence Island
Seward Peninsula
Norton Sound
Point Barrow
ARCTIC OCEAN
Beaufort Sea
Brooks Range
Yukon
Nunivak Island
ALASKA
Fairbanks
Alaska Range
Aleutian Islands
Bristol Bay
Alaska Peninsula
Anchorage
Kodiak I.
Gulf of Alaska
Juneau
Alexander Archipelago
PACIFIC OCEAN
Tacoma
Seattle
Olympia
WASH.
Spokane
Portland
Salem
Eugene
Cascade Range
Butte
IDAHO
OREGON
Boise
Snake
Great Salt Lake
Salt Lake City
Reno
Carson City
Sacramento
San Francisco
NEVADA
UTAH
CALIFORNIA
Sierra Nevada
Las Vegas
ARIZONA
Los Angeles
San Diego
Phoenix
Tucson
Honolulu
HAWAII
Hilo
TROPIC OF CANCER
180°
170°E
170°W
160°W
150°W
140°W
130°W
120°W
110°W
50°N
60°N
70°N
40°N
30°N
20°N
10°N
1
2
3
4
5
6
7
8
A
B
C
D
E
F
G
H
J
K

UNITED STATES
POLITICAL
0 mi 600
0 km 600
OBLIQUE AZIMUTHAL EQUIDISTANT PROJECTION
GREENLAND
(KALAALLIT NUNAAT)
Den.
ARCTIC CIRCLE
CANADA
MONTANA
Helena
Billings
NORTH DAKOTA
Bismarck
MINNESOTA
SOUTH DAKOTA
Pierre
WYOMING
Casper
Cheyenne
NEBRASKA
Lincoln
Omaha
Sioux City
IOWA
Des Moines
Minneapolis
St. Paul
WISCONSIN
Milwaukee
Madison
Lake Superior
MICHIGAN
Lake Michigan
Lake Huron
Lansing
Detroit
L. Erie
Toledo
Cleveland
Chicago
ILLINOIS
IND.
Columbus
OHIO
Indianapolis
Springfield
Cincinnati
Dayton
Pittsburgh
PA.
Harrisburg
Buffalo
Lake Ontario
Albany
NEW YORK
New York City
MAINE
Augusta
Portland
Montpelier
VT.
Concord, N.H.
Boston, MASS.
Providence, R.I.
Hartford, CONN.
Trenton, N.J.
Philadelphia
Dover, DEL.
Baltimore
Annapolis, MD.
Washington, D.C.
W. VA.
Charleston
Richmond
VIRGINIA
Virginia Beach
Frankfort
Louisville
KENTUCKY
Nashville
TENNESSEE
Raleigh
NORTH CAROLINA
Charlotte
SOUTH CAROLINA
Columbia
Charleston
COLORADO
Denver
KANSAS
Kansas City
Topeka
Jefferson City
MISSOURI
St. Louis
Santa Fe
Albuquerque
NEW MEXICO
El Paso
Oklahoma City
OKLAHOMA
Tulsa
ARKANSAS
Little Rock
Memphis
MISS.
Jackson
ALABAMA
Birmingham
Montgomery
Atlanta
GEORGIA
Savannah
Jacksonville
Tallahassee
FLORIDA
Tampa
Miami
TEXAS
Fort Worth
Dallas
Austin
San Antonio
Houston
LOUISIANA
Baton Rouge
New Orleans
ROCKY MOUNTAINS
APPALACHIAN MTS.
Missouri
Mississippi
Arkansas
Rio Grande
MEXICO
Gulf of Mexico
Straits of Florida
BAHAMAS
CUBA
HAITI
DOMINICAN REPUBLIC
JAMAICA
PUERTO RICO U.S.
San Juan
ST. KITTS & NEVIS
ANTIGUA & BARBUDA
DOMINICA
Caribbean Sea
Bermuda Is. U.K.
ATLANTIC OCEAN

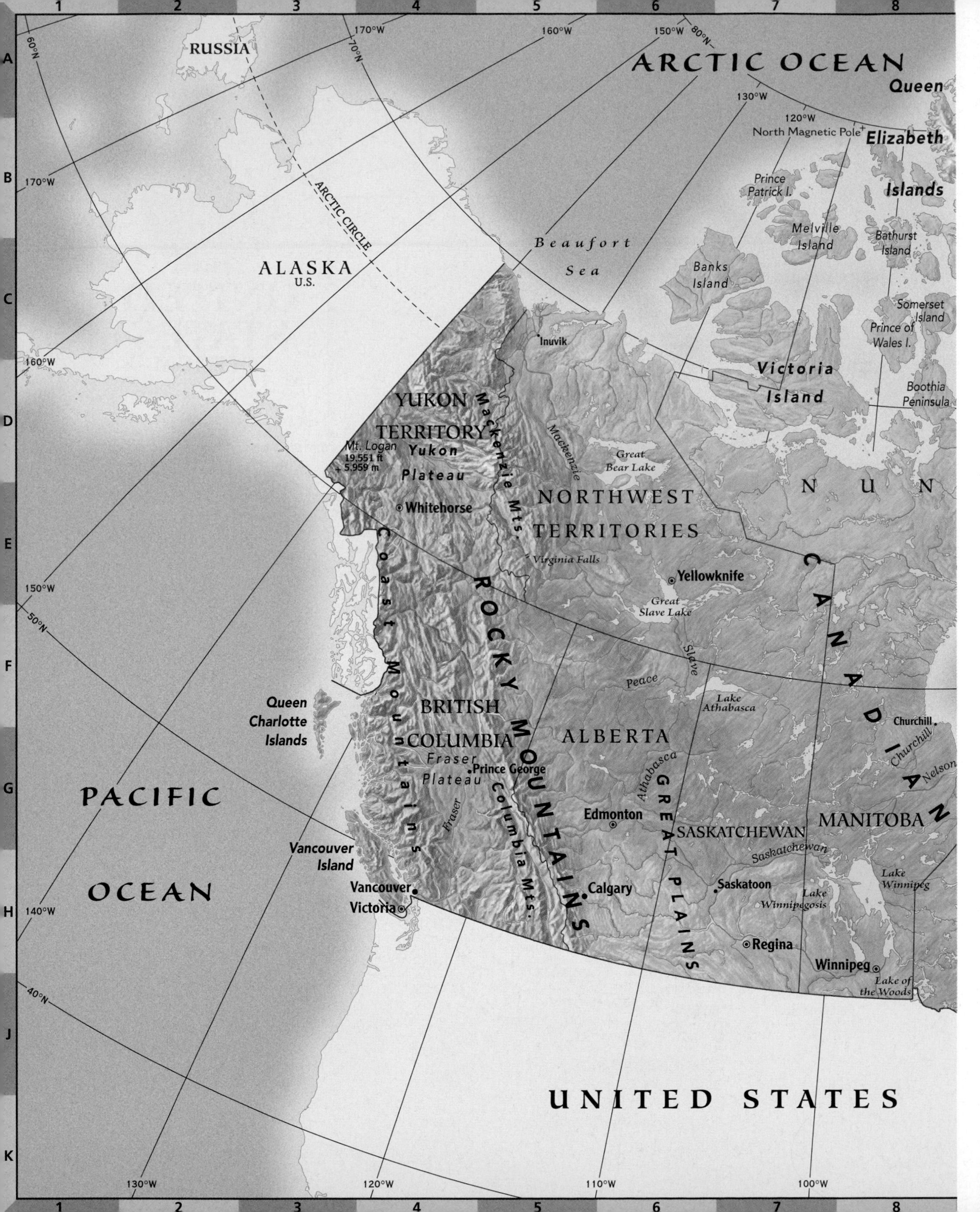

RUSSIA
ARCTIC OCEAN
Queen
Elizabeth
Islands
North Magnetic Pole
Prince Patrick I.
Melville Island
Bathurst Island
Banks Island
Beaufort Sea
ARCTIC CIRCLE
ALASKA
U.S.
Somerset Island
Prince of Wales I.
Inuvik
Victoria Island
Boothia Peninsula
YUKON TERRITORY
Mackenzie Mts.
Mackenzie
Mt. Logan 19,551 ft 5,959 m
Yukon Plateau
Great Bear Lake
N U N
Whitehorse
NORTHWEST TERRITORIES
Virginia Falls
Yellowknife
Great Slave Lake
Coast Mountains
ROCKY MOUNTAINS
CANADIAN
Slave
Peace
Lake Athabasca
Queen Charlotte Islands
BRITISH COLUMBIA
ALBERTA
Churchill
Churchill
Nelson
Fraser Plateau
Prince George
Athabasca
GREAT PLAINS
PACIFIC OCEAN
Columbia Mts.
Fraser
Edmonton
SASKATCHEWAN
MANITOBA
Saskatchewan
Vancouver Island
Lake Winnipeg
Vancouver
Calgary
Saskatoon
Lake Winnipegosis
Victoria
Regina
Winnipeg
Lake of the Woods
UNITED STATES
170°W
160°W
150°W
140°W
130°W
120°W
110°W
100°W
80°N
70°N
60°N
50°N
40°N

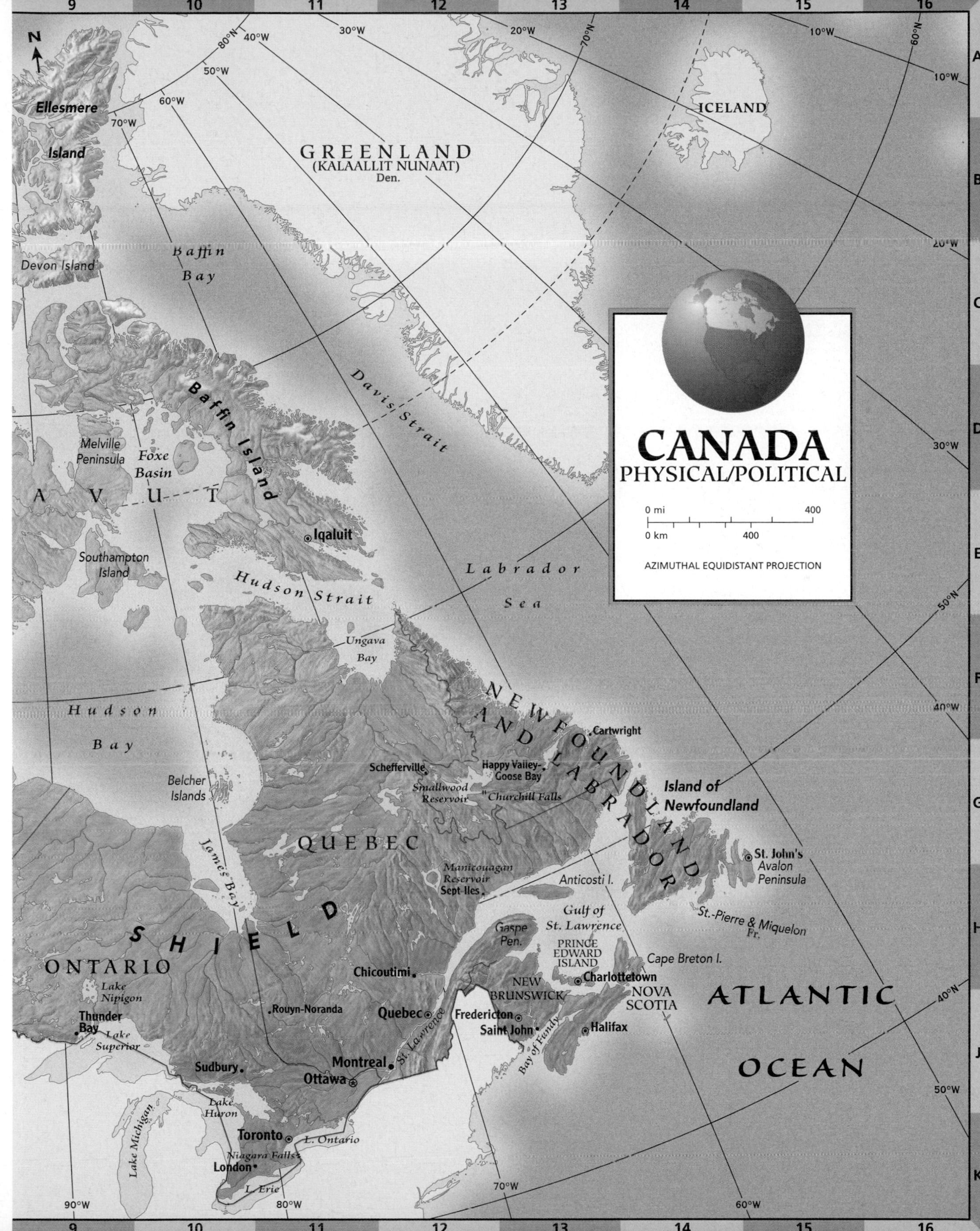
CANADA
PHYSICAL/POLITICAL
0 mi 400
0 km 400
AZIMUTHAL EQUIDISTANT PROJECTION
GREENLAND
(KALAALLIT NUNAAT)
Den.
ICELAND
Ellesmere Island
Devon Island
Baffin Bay
Baffin Island
Davis Strait
Melville Peninsula
Foxe Basin
Southampton Island
Iqaluit
Hudson Strait
Labrador Sea
Ungava Bay
Hudson Bay
Belcher Islands
James Bay
NEWFOUNDLAND AND LABRADOR
Cartwright
Schefferville
Happy Valley-Goose Bay
Smallwood Reservoir
Churchill Falls
Island of Newfoundland
St. John's
Avalon Peninsula
QUEBEC
Manicouagan Reservoir
Sept-Iles
Anticosti I.
Gulf of St. Lawrence
St.-Pierre & Miquelon Fr.
Gaspe Pen.
PRINCE EDWARD ISLAND
Cape Breton I.
Charlottetown
NEW BRUNSWICK
NOVA SCOTIA
Fredericton
Saint John
Halifax
Bay of Fundy
ATLANTIC OCEAN
SHIELD
ONTARIO
Lake Nipigon
Thunder Bay
Lake Superior
Chicoutimi
Rouyn-Noranda
Quebec
St. Lawrence
Montreal
Sudbury
Ottawa
Lake Huron
Lake Michigan
Toronto
L. Ontario
Niagara Falls
London
L. Erie
N
80°N 70°N 60°N 50°N 40°N
90°W 80°W 70°W 60°W 50°W 40°W 30°W 20°W 10°W
A B C D E F G H J K
9 10 11 12 13 14 15 16

MIDDLE AMERICA
PHYSICAL/POLITICAL
0 mi 400
0 km 400
AZIMUTHAL EQUIDISTANT PROJECTION
UNITED STATES
MEXICO
CENTRAL AMERICA
PACIFIC OCEAN
Gulf of Mexico
Gulf of California
Baja California
Sonoran Desert
Sierra Madre Occidental
Sierra Madre Oriental
Sierra Madre del Sur
Sierra Madre
Rio Grande
BAJA CALIFORNIA
BAJA CALIFORNIA SUR
SONORA
CHIHUAHUA
COAHUILA
SINALOA
DURANGO
NUEVO LEON
TAMAULIPAS
ZACATECAS
SAN LUIS POTOSI
NAYARIT
AGUASCALIENTES
QUERETARO
VERACRUZ
TLAXCALA
JALISCO
GUANAJUATO
HIDALGO
COLIMA
MICHOACAN
DISTRITO FEDERAL
MEXICO
MORELOS
PUEBLA
GUERRERO
OAXACA
TABASCO
CHIAPAS
CAMPECHE
YUCATAN
QUINTANA ROO
Tijuana
Mexicali
Ciudad Juarez
Chihuahua
Nuevo Laredo
Monterrey
Matamoros
La Paz
Mazatlan
False Cape
Ciudad Madero
Tampico
San Luis Potosi
Guadalajara
Leon
Mexico City
Orizaba 18,855 ft 5,747 m
Popocatepetl 17,802 ft 5,426 m
Veracruz
Acapulco
Revillagigedo Islands Mex.
Bay of Campeche
Isthmus of Tehuantepec
Gulf of Tehuantepec
Merida
Yucatan Peninsula
Cozumel Island
Belize City
Belmopan
BELIZE
Gulf of Honduras
HON
Tegucigalpa
GUATEMALA
Guatemala
EL SALVADOR
San Salvador
Leon
Cocos Island C.R.
110°W
100°W
90°W
30°N
20°N
10°N
0°
N

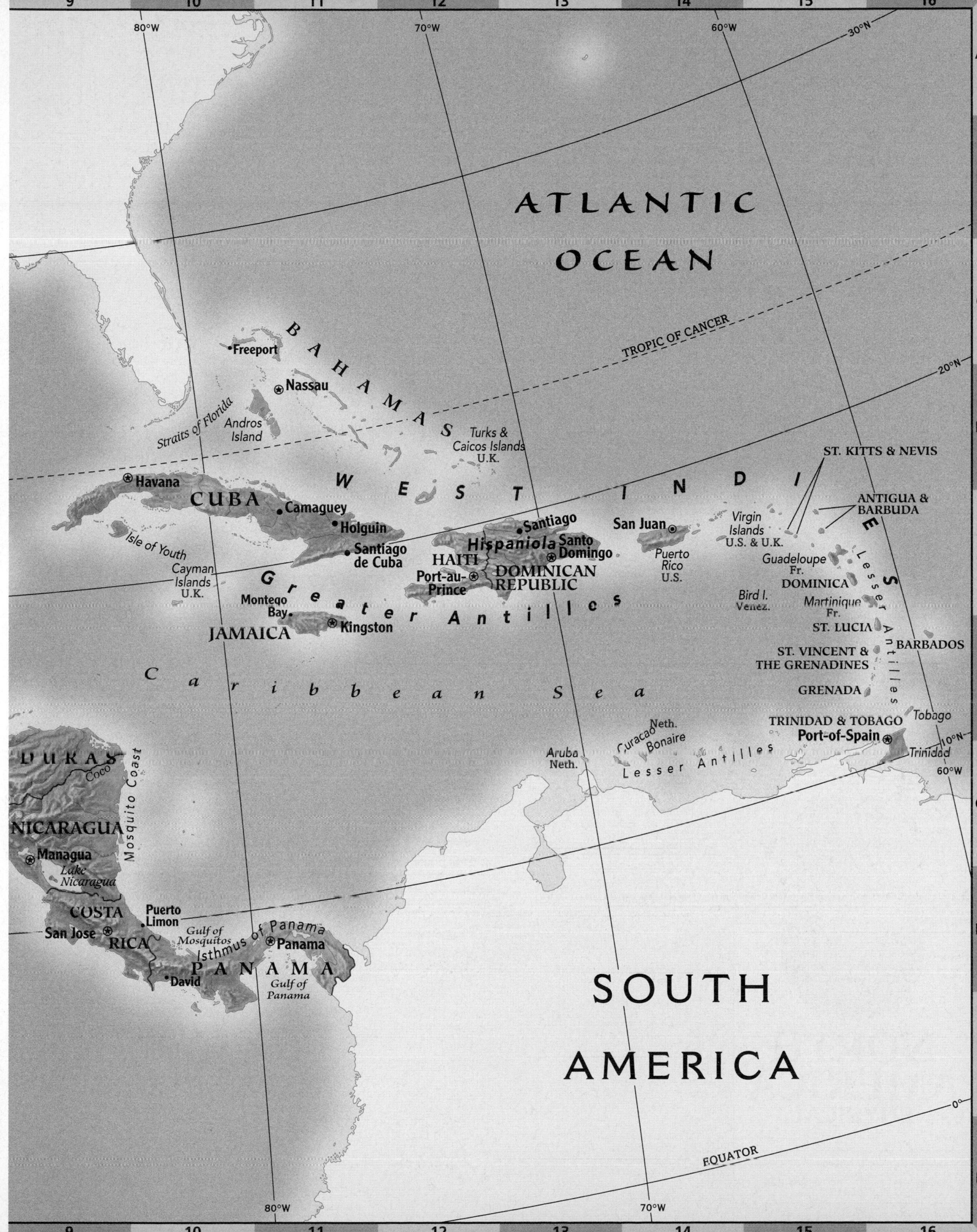

ATLANTIC OCEAN
TROPIC OF CANCER
BAHAMAS
Freeport
Nassau
Andros Island
Straits of Florida
Turks & Caicos Islands U.K.
WEST INDIES
Havana
CUBA
Camaguey
Holguin
Santiago de Cuba
Isle of Youth
Cayman Islands U.K.
Hispaniola
HAITI
Port-au-Prince
Santiago
Santo Domingo
DOMINICAN REPUBLIC
Greater Antilles
Montego Bay
JAMAICA
Kingston
San Juan
Puerto Rico U.S.
Virgin Islands U.S. & U.K.
ST. KITTS & NEVIS
ANTIGUA & BARBUDA
Guadeloupe Fr.
DOMINICA
Martinique Fr.
ST. LUCIA
BARBADOS
ST. VINCENT & THE GRENADINES
GRENADA
Bird I. Venez.
Lesser Antilles
Caribbean Sea
TRINIDAD & TOBAGO
Port-of-Spain
Tobago
Trinidad
Aruba Neth.
Curacao
Neth.
Bonaire
Lesser Antilles
NICARAGUA
Managua
Lake Nicaragua
Mosquito Coast
Coco
COSTA RICA
San Jose
Puerto Limon
Gulf of Mosquitos
Isthmus of Panama
Panama
PANAMA
David
Gulf of Panama
SOUTH AMERICA
EQUATOR
30°N
20°N
10°N
0°
80°W
70°W
60°W

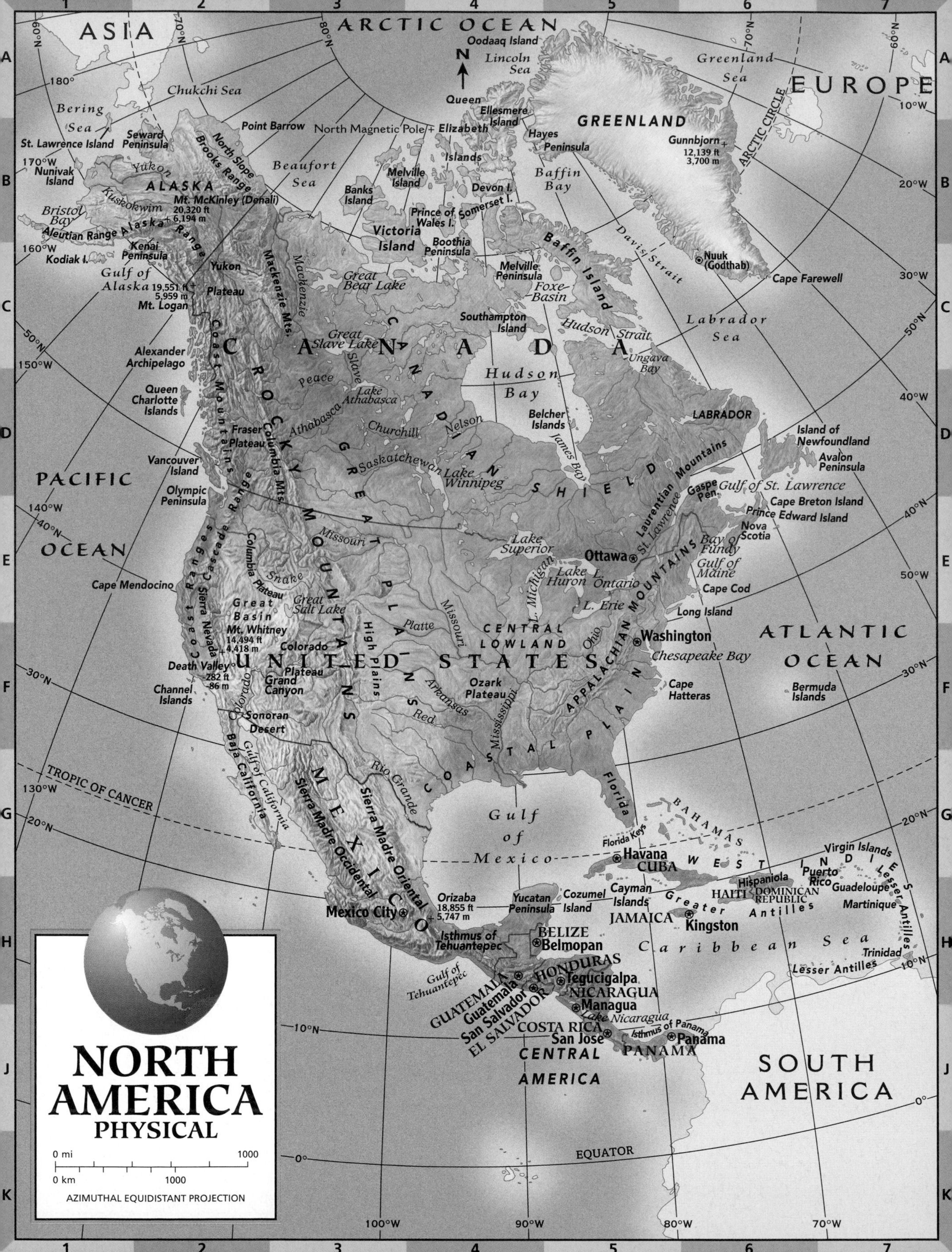

NORTH AMERICA
PHYSICAL
0 mi
1000
0 km
1000
AZIMUTHAL EQUIDISTANT PROJECTION
ASIA
ARCTIC OCEAN
EUROPE
PACIFIC OCEAN
ATLANTIC OCEAN
SOUTH AMERICA
Oodaaq Island
Lincoln Sea
Greenland Sea
GREENLAND
Gunnbjorn
12,139 ft
3,700 m
ARCTIC CIRCLE
Queen Elizabeth Islands
Ellesmere Island
Hayes Peninsula
North Magnetic Pole
Point Barrow
Chukchi Sea
Bering Sea
St. Lawrence Island
Seward Peninsula
Nunivak Island
Yukon
ALASKA
North Slope
Brooks Range
Beaufort Sea
Banks Island
Melville Island
Devon I.
Somerset I.
Baffin Bay
Prince of Wales I.
Victoria Island
Boothia Peninsula
Baffin Island
Davis Strait
Nuuk (Godthab)
Cape Farewell
Mt. McKinley (Denali)
20,320 ft
6,194 m
Kuskokwim
Bristol Bay
Aleutian Range
Alaska Range
Kenai Peninsula
Kodiak I.
Gulf of Alaska
19,551 ft
5,959 m
Mt. Logan
Yukon Plateau
Mackenzie Mts.
Mackenzie
Great Bear Lake
Melville Peninsula
Foxe Basin
Southampton Island
Hudson Strait
Labrador Sea
Ungava Bay
CANADA
Great Slave Lake
Slave
Peace
Lake Athabasca
Athabasca
Churchill
Nelson
Hudson Bay
Belcher Islands
James Bay
LABRADOR
CANADIAN SHIELD
Island of Newfoundland
Avalon Peninsula
Alexander Archipelago
Queen Charlotte Islands
Coast Mountains
ROCKY MOUNTAINS
Fraser Plateau
Columbia Mts.
Vancouver Island
Olympic Peninsula
Cascade Range
Coast Ranges
Saskatchewan
Lake Winnipeg
GREAT PLAINS
Laurentian Mountains
Gaspe Pen.
Gulf of St. Lawrence
Cape Breton Island
Prince Edward Island
Nova Scotia
Bay of Fundy
Gulf of Maine
Cape Cod
Long Island
St. Lawrence
Ottawa
Lake Superior
L. Michigan
Lake Huron
L. Ontario
L. Erie
Missouri
Snake
Columbia Plateau
Cape Mendocino
Sierra Nevada
Great Basin
Great Salt Lake
Mt. Whitney
14,494 ft
4,418 m
Death Valley
-282 ft
-86 m
Colorado Plateau
Grand Canyon
Channel Islands
Colorado
Sonoran Desert
Platte
High Plains
CENTRAL LOWLAND
Ohio
APPALACHIAN MOUNTAINS
Washington
Chesapeake Bay
Cape Hatteras
Bermuda Islands
UNITED STATES
Ozark Plateau
Arkansas
Red
Mississippi
COASTAL PLAIN
Florida
Rio Grande
Baja California
Gulf of California
TROPIC OF CANCER
Sierra Madre Occidental
Sierra Madre Oriental
MEXICO
Gulf of Mexico
Florida Keys
BAHAMAS
Havana
CUBA
WEST INDIES
Hispaniola
HAITI
DOMINICAN REPUBLIC
Puerto Rico
Virgin Islands
Guadeloupe
Martinique
Lesser Antilles
Greater Antilles
Cayman Islands
JAMAICA
Kingston
Cozumel Island
Yucatan Peninsula
Orizaba
18,855 ft
5,747 m
Mexico City
Isthmus of Tehuantepec
Gulf of Tehuantepec
BELIZE
Belmopan
Caribbean Sea
Trinidad
HONDURAS
Tegucigalpa
GUATEMALA
Guatemala
San Salvador
EL SALVADOR
NICARAGUA
Managua
Lake Nicaragua
COSTA RICA
San Jose
Isthmus of Panama
Panama
PANAMA
CENTRAL AMERICA
EQUATOR

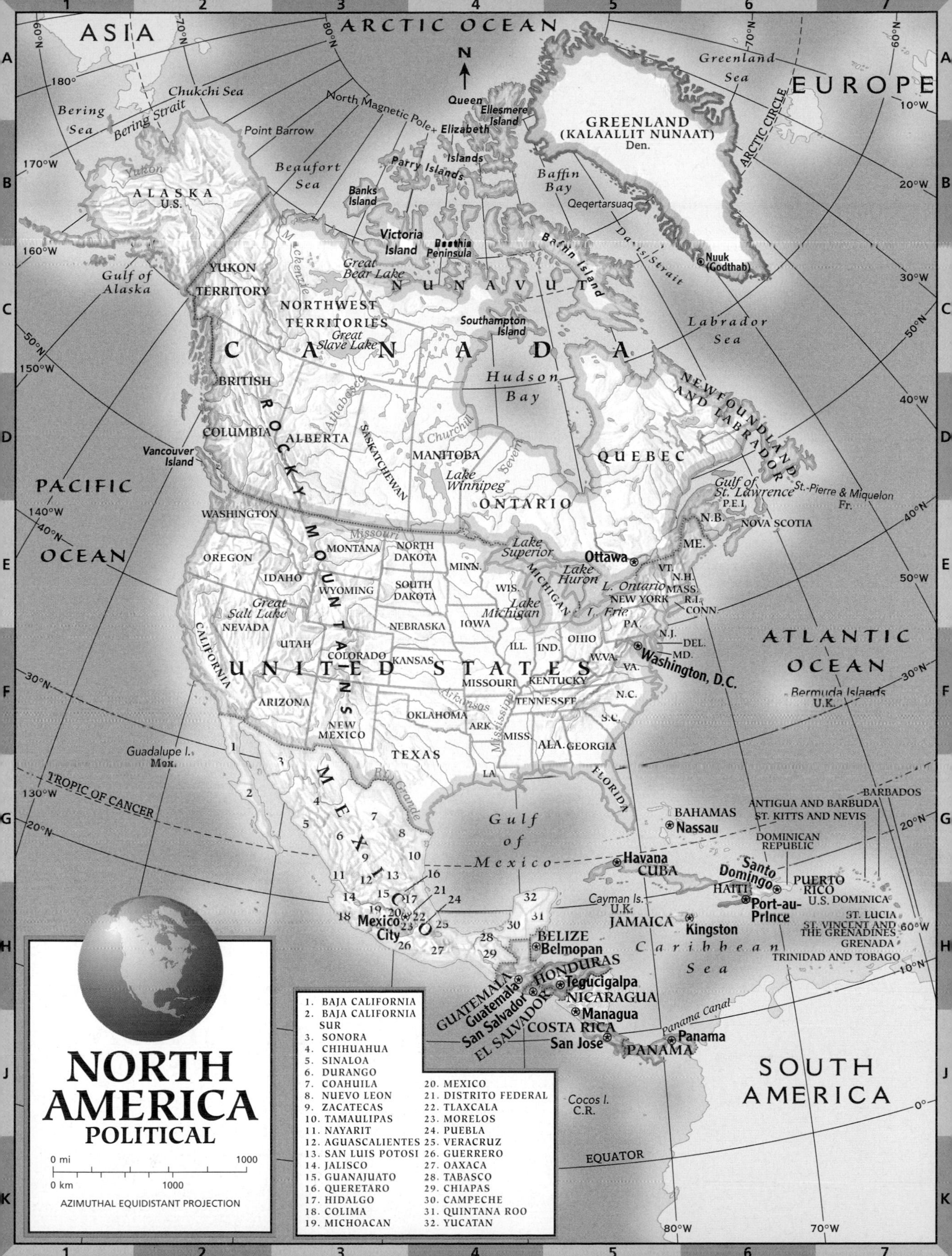

NORTH AMERICA
POLITICAL
0 mi 1000
0 km 1000
AZIMUTHAL EQUIDISTANT PROJECTION
ARCTIC OCEAN
ASIA
EUROPE
PACIFIC OCEAN
ATLANTIC OCEAN
SOUTH AMERICA
N
Chukchi Sea
Bering Sea
Bering Strait
Point Barrow
North Magnetic Pole
Beaufort Sea
Queen Elizabeth Islands
Ellesmere Island
Parry Islands
Banks Island
Victoria Island
Boothia Peninsula
GREENLAND (KALAALLIT NUNAAT) Den.
Greenland Sea
Baffin Bay
Qeqertarsuaq
Baffin Island
Davis Strait
Nuuk (Godthab)
ARCTIC CIRCLE
Labrador Sea
Gulf of Alaska
ALASKA U.S.
Yukon
YUKON TERRITORY
Mackenzie
NORTHWEST TERRITORIES
Great Bear Lake
Great Slave Lake
NUNAVUT
Southampton Island
CANADA
Hudson Bay
BRITISH COLUMBIA
Vancouver Island
ALBERTA
Athabasca
SASKATCHEWAN
MANITOBA
Churchill
Severn
Lake Winnipeg
ONTARIO
QUEBEC
NEWFOUNDLAND AND LABRADOR
Gulf of St. Lawrence
St.-Pierre & Miquelon Fr.
P.E.I.
N.B.
NOVA SCOTIA
Ottawa
ROCKY MOUNTAINS
WASHINGTON
OREGON
IDAHO
MONTANA
Missouri
WYOMING
NORTH DAKOTA
SOUTH DAKOTA
MINN.
WIS.
Lake Superior
Lake Huron
Lake Michigan
MICHIGAN
L. Ontario
L. Erie
NEW YORK
VT.
N.H.
MASS.
R.I.
CONN.
ME.
PA.
N.J.
DEL.
MD.
Washington, D.C.
Great Salt Lake
NEVADA
CALIFORNIA
UTAH
COLORADO
NEBRASKA
IOWA
ILL.
IND.
OHIO
W.VA.
VA.
KANSAS
MISSOURI
KENTUCKY
UNITED STATES
ARIZONA
NEW MEXICO
OKLAHOMA
Arkansas
ARK.
TENNESSEE
Mississippi
N.C.
S.C.
MISS.
ALA.
GEORGIA
TEXAS
LA
FLORIDA
Rio Grande
Bermuda Islands U.K.
Guadalupe I. Mex.
TROPIC OF CANCER
MEXICO
Mexico City
Gulf of Mexico
BAHAMAS
Nassau
Havana
CUBA
Cayman Is. U.K.
JAMAICA
Kingston
HAITI
Port-au-Prince
Santo Domingo
DOMINICAN REPUBLIC
PUERTO RICO U.S.
ANTIGUA AND BARBUDA
ST. KITTS AND NEVIS
BARBADOS
DOMINICA
ST. LUCIA
ST. VINCENT AND THE GRENADINES
GRENADA
TRINIDAD AND TOBAGO
Caribbean Sea
BELIZE
Belmopan
GUATEMALA
Guatemala
HONDURAS
Tegucigalpa
EL SALVADOR
San Salvador
NICARAGUA
Managua
COSTA RICA
San Jose
PANAMA
Panama
Panama Canal
Cocos I. C.R.
EQUATOR
1. BAJA CALIFORNIA
2. BAJA CALIFORNIA SUR
3. SONORA
4. CHIHUAHUA
5. SINALOA
6. DURANGO
7. COAHUILA
8. NUEVO LEON
9. ZACATECAS
10. TAMAULIPAS
11. NAYARIT
12. AGUASCALIENTES
13. SAN LUIS POTOSI
14. JALISCO
15. GUANAJUATO
16. QUERETARO
17. HIDALGO
18. COLIMA
19. MICHOACAN
20. MEXICO
21. DISTRITO FEDERAL
22. TLAXCALA
23. MORELOS
24. PUEBLA
25. VERACRUZ
26. GUERRERO
27. OAXACA
28. TABASCO
29. CHIAPAS
30. CAMPECHE
31. QUINTANA ROO
32. YUCATAN

SOUTH AMERICA
PHYSICAL
0 mi
800
0 km
800
AZIMUTHAL EQUIDISTANT PROJECTION
Caribbean Sea
ATLANTIC OCEAN
PACIFIC OCEAN
Caracas
Lake Maracaibo
VENEZUELA
Orinoco
LLANOS
GUYANA
Georgetown
SURINAME
Paramaribo
Cayenne
FRENCH GUIANA
Angel Falls
Total drop: 3,212 ft 979 m
Bogota
GUIANA HIGHLANDS
COLOMBIA
Malpelo I.
Boundary claimed by Suriname
Quito
ECUADOR
Negro
Amazon
Marajo Island
EQUATOR
AMAZON BASIN
Selvas
Madeira
Tapajos
Xingu
Purus
Ucayali
Teles Pires
Tocantins
Sao Francisco
BRAZIL
PERU
Lima
Machu Picchu
Lake Titicaca
MATO GROSSO PLATEAU
BRAZILIAN HIGHLANDS
Brasilia
BOLIVIA
La Paz
Altiplano
Sucre
Salar de Uyuni
Paraguay
PARAGUAY
GRAN CHACO
Asuncion
Iguazu Falls
TROPIC OF CAPRICORN
San Felix I.
San Ambrosio I.
Parana
Uruguay
ANDES
Aconcagua 22,834 ft 6,960 m
Santiago
Juan Fernandez Is.
Buenos Aires
URUGUAY
Montevideo
Rio de la Plata
PAMPAS
ARGENTINA
CHILE
Negro
PATAGONIA
-131 ft -40 m
Valdes Peninsula
Chiloe Island
Gulf of San Jorge
Taitao Peninsula
Wellington I.
Falkland Islands (Islas Malvinas)
Stanley
Strait of Magellan
Tierra del Fuego
Cape Horn
South Georgia I.
80°W
60°W
50°W
40°W
10°N
0°
10°S
20°S
30°S
40°S
50°S
100°W
90°W
70°W
30°W
20°W
N
1 2 3 4 5 6 7
A B C D E F G H J K

SOUTH AMERICA
POLITICAL
Caribbean Sea
ATLANTIC OCEAN
PACIFIC OCEAN
GUYANA
SURINAME
FRENCH GUIANA
VENEZUELA
COLOMBIA
ECUADOR
PERU
BOLIVIA
BRAZIL
PARAGUAY
URUGUAY
ARGENTINA
AMAZON BASIN
ANDES
Barranquilla
Cartagena
Santa Marta
Maracaibo
Lake Maracaibo
Barquisimeto
Caracas
Valencia
Ciudad Guayana
Georgetown
Paramaribo
Cayenne
Fr.
Boundary claimed by Suriname
Bucaramanga
San Cristobal
Medellin
Bogota
Cali
Malpelo I.
Col.
Boa Vista
Esmeraldas
Quito
Guayaquil
Iquitos
Manaus
Santarem
Belem
Marajo Island
EQUATOR
Fortaleza
Teresina
Natal
Campina Grande
Recife
Porto Velho
Rio Branco
Callao
Lima
Machu Picchu
Cuzco
Ayacucho
Lake Titicaca
Trinidad
La Paz
Arequipa
Salvador (Bahia)
Brasilia
Goiania
Arica
Oruro
Santa Cruz
Sucre
Iquique
Tarija
Uberandia
Uberaba
Campo Grande
Belo Horizonte
Campinas
Nova Iguacu
Rio de Janeiro
Londrina
Sao Paulo
Santos
Curitiba
Antofagasta
Salta
TROPIC OF CAPRICORN
Asuncion
San Miguel de Tucuman
San Felix I.
San Ambrosio I.
Chil.
Uruguaiana
Santa Maria
Porto Alegre
La Serena
Coquimbo
Cordoba
Rosario
Valparaiso
Santiago
Mendoza
Buenos Aires
Montevideo
La Plata
Rio de la Plata
Juan Fernandez Is.
Concepcion
Mar del Plata
Bahia Blanca
Puerto Montt
Comodoro Rivadavia
Falkland Islands (Islas Malvinas)
Stanley
Administered by United Kingdom
(Claimed by Arg.)
Rio Gallegos
Strait of Magellan
Punta Arenas
TIERRA DEL FUEGO
Ushuaia
Cape Horn
South Georgia I.
U.K.
Orinoco
Negro
Amazon
Amazon (Solimoes)
Maranon
Madeira
Tapajos
Xingu
Purus
Ucayali
Teles Pires
Tocantins
Sao Francisco
Paraguay
Parana
Uruguay
0 mi
800
0 km
800
AZIMUTHAL EQUIDISTANT PROJECTION
N
80°W
60°W
50°W
40°W
100°W
90°W
70°W
30°W
20°W
10°N
0°
10°S
20°S
30°S
40°S
50°S
1
2
3
4
5
6
7
A
B
C
D
E
F
G
H
J
K

EUROPE
PHYSICAL
0 mi
400
0 km
400
AZIMUTHAL EQUIDISTANT PROJECTION
ATLANTIC
OCEAN
Reykjavík
ICELAND
ARCTIC CIRCLE
PRIME MERIDIAN (MERIDIAN OF GREENWICH)
Norwegian Sea
Faeroe Islands
Shetland Islands
Orkney Islands
Outer Hebrides
Highlands
British Isles
Edinburgh
Belfast
UNITED KINGDOM
IRELAND
Dublin
Irish Sea
Great Britain
Cardiff
London
North Sea
Jutland
DENMARK
Copenhagen
Zealand
Baltic
NORWAY
SWEDEN
SCANDINAVIA
Oslo
Stockholm
Gulf of
Amsterdam
NETH.
BELGIUM
Brussels
LUX.
Berlin
GERMANY
NORTH
POLAND
Oder
Elbe
Rhine
Prague
CZECH REP.
SLOVAKIA
Bratislava
Vienna
Budapest
HUNGARY
AUSTRIA
LIECH.
Bern
SWITZ.
ALPS
Danube
English Channel
Brittany
Seine
Paris
Loire
FRANCE
Bay of Biscay
Mont Blanc
15,771 ft
4,807 m
Massif Central
Rhone
MONACO
Riviera
SLOVENIA
Ljubljana
Zagreb
CROATIA
Drava
Sava
BOSNIA & HERZEGOVINA
Sarajevo
Po
Adriatic Sea
SAN MARINO
Apennines
ITALY
VATICAN CITY
Rome
MONTENEGRO
Tiranë
ALBANIA
Cantabrian Mountains
Pyrenees
Douro
PORTUGAL
Lisbon
Tagus
IBERIAN PENINSULA
Madrid
SPAIN
Ebro
ANDORRA
Corsica
Sardinia
Tyrrhenian Sea
Ionian Sea
Balearic Islands
Baetic Mountains
GIBRALTAR
Strait of Gibraltar
Mediterranean
Sicily
Etna
10,902 ft
3,323 m
Valletta
MALTA
AFRICA
N
60°N
50°N
40°N
30°N
70°N
40°W
30°W
20°W
10°W
0°
10°E
1 2 3 4 5 6 7 8
A B C D E F G H J K

Barents Sea
North Cape
Kola Peninsula
White Sea
FINLAND
Lake Region
Bothnia
Lake Onega
Lake Ladoga
Helsinki
Gulf of Finland
Tallinn
ESTONIA
Riga
LATVIA
LITHUANIA
Vilnius
RUSSIA
Minsk
BELARUS
Warsaw
RUSSIA
EUROPEAN PLAIN
Northern Dvina
Pechora
Moscow
Volga
Kama
Oka
CENTRAL RUSSIAN UPLAND
Dnieper
Don
URAL MOUNTAINS
Ural
Europe-Asia boundary
ASIA
KAZAKHSTAN
Caspian Depression
Kyiv (Kiev)
UKRAINE
Dniester
Carpathian Mountains
MOLDOVA
Chişinău
Sea of Azov
Crimea
Elbrus 18,510 ft 5,642 m
Caucasus Mountains
GEORGIA
AZERBAIJAN
Baku
Caspian Sea
ROMANIA
Bucharest
Danube
Belgrade
BALKAN PENINSULA
SERBIA
BULGARIA
Balkan Mountains
Sofia
KOSOVO
Skopje
MACED.
Black Sea
Bosporus
TURKEY
Dardanelles
Sea of Marmara
GREECE
Aegean Sea
Athens
Peloponnesus
Crete
Rhodes
Nicosia
CYPRUS
Sea
ASIA
30°E
40°E
50°E
70°N
70°E
80°E
60°N
50°N
40°N
30°N
60°E

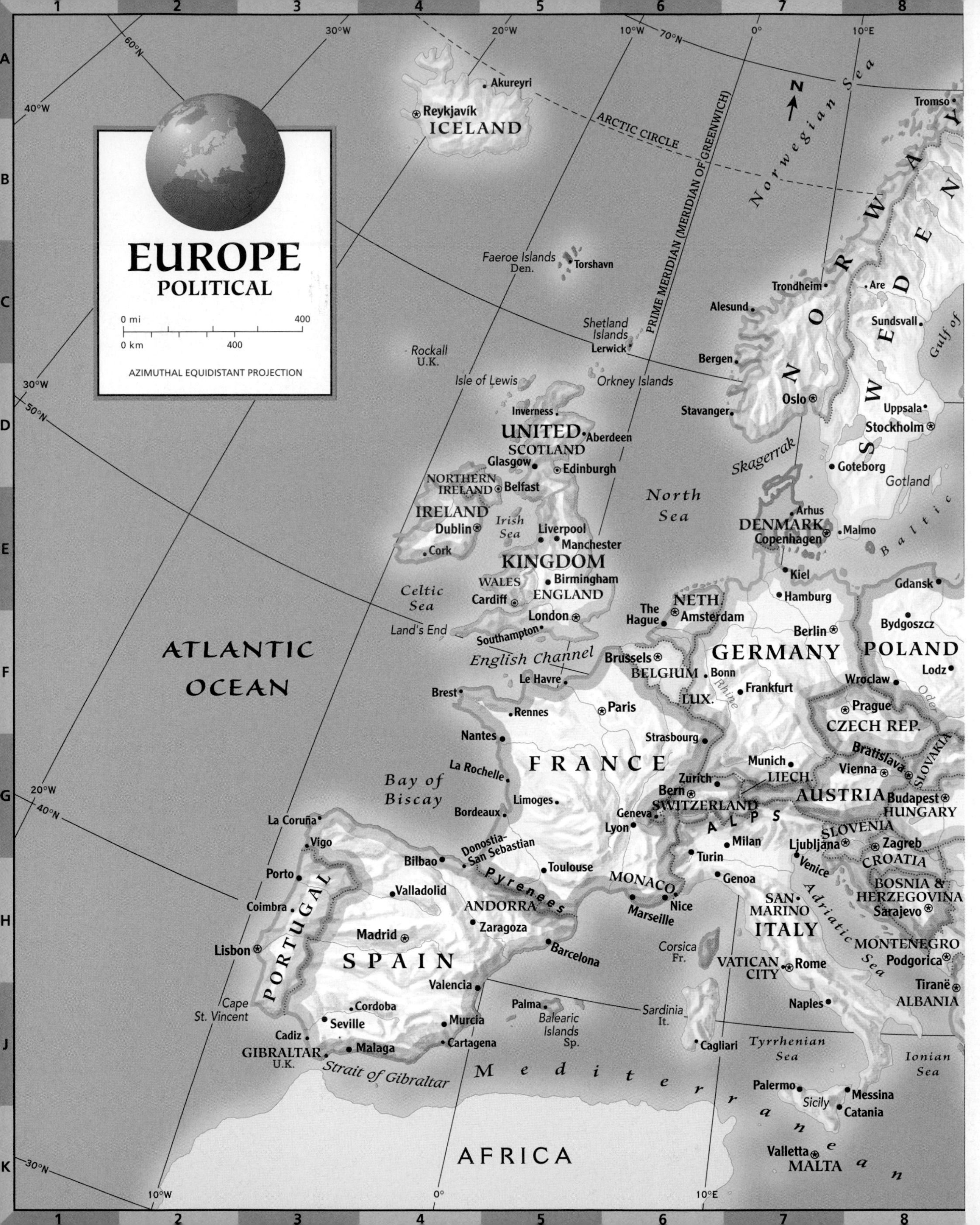
EUROPE
POLITICAL
0 mi
400
0 km
400
AZIMUTHAL EQUIDISTANT PROJECTION
ATLANTIC
OCEAN
ICELAND
Reykjavík
Akureyri
ARCTIC CIRCLE
PRIME MERIDIAN (MERIDIAN OF GREENWICH)
Norwegian Sea
Faeroe Islands
Den.
Torshavn
Shetland
Islands
Lerwick
Orkney Islands
Rockall
U.K.
Isle of Lewis
Inverness
UNITED
KINGDOM
SCOTLAND
Aberdeen
Glasgow
Edinburgh
NORTHERN
IRELAND
Belfast
IRELAND
Dublin
Cork
Irish
Sea
Liverpool
Manchester
WALES
Birmingham
ENGLAND
Cardiff
London
Celtic
Sea
Land's End
Southampton
English Channel
North
Sea
Skagerrak
NORWAY
SWEDEN
Tromso
Trondheim
Are
Alesund
Sundsvall
Gulf of
Bergen
Oslo
Stavanger
Uppsala
Stockholm
Goteborg
Gotland
DENMARK
Arhus
Copenhagen
Malmo
Baltic
Kiel
Hamburg
Gdansk
NETH.
The
Hague
Amsterdam
Berlin
Bydgoszcz
GERMANY
POLAND
Brussels
BELGIUM
Bonn
Rhine
Frankfurt
LUX.
Wroclaw
Lodz
Oder
Prague
CZECH REP.
Le Havre
Brest
Rennes
Paris
Nantes
Strasbourg
FRANCE
Munich
Bratislava
SLOVAKIA
Vienna
La Rochelle
Bay of
Biscay
Zurich
LIECH.
Bern
SWITZERLAND
AUSTRIA
Budapest
HUNGARY
Limoges
Bordeaux
Geneva
ALPS
Lyon
SLOVENIA
Ljubljana
Zagreb
CROATIA
Milan
Turin
Venice
La Coruña
Vigo
Donostia-
San Sebastian
Bilbao
Toulouse
Porto
Pyrenees
MONACO
Genoa
BOSNIA &
HERZEGOVINA
Sarajevo
Valladolid
Coimbra
ANDORRA
Marseille
Nice
SAN
MARINO
ITALY
Adriatic Sea
PORTUGAL
Zaragoza
Madrid
Lisbon
SPAIN
Barcelona
Corsica
Fr.
VATICAN
CITY
Rome
MONTENEGRO
Podgorica
Tiranë
ALBANIA
Valencia
Cape
St. Vincent
Cordoba
Seville
Murcia
Palma
Balearic
Islands
Sp.
Sardinia
It.
Naples
Cadiz
Cartagena
GIBRALTAR
U.K.
Malaga
Cagliari
Tyrrhenian
Sea
Ionian
Sea
Strait of Gibraltar
Mediterranean
Palermo
Sicily
Messina
Catania
Valletta
MALTA
AFRICA
30°W
20°W
10°W
70°N
0°
10°E
60°N
40°W
50°N
40°N
30°N
10°W
0°
10°E
1
2
3
4
5
6
7
8
A
B
C
D
E
F
G
H
J
K

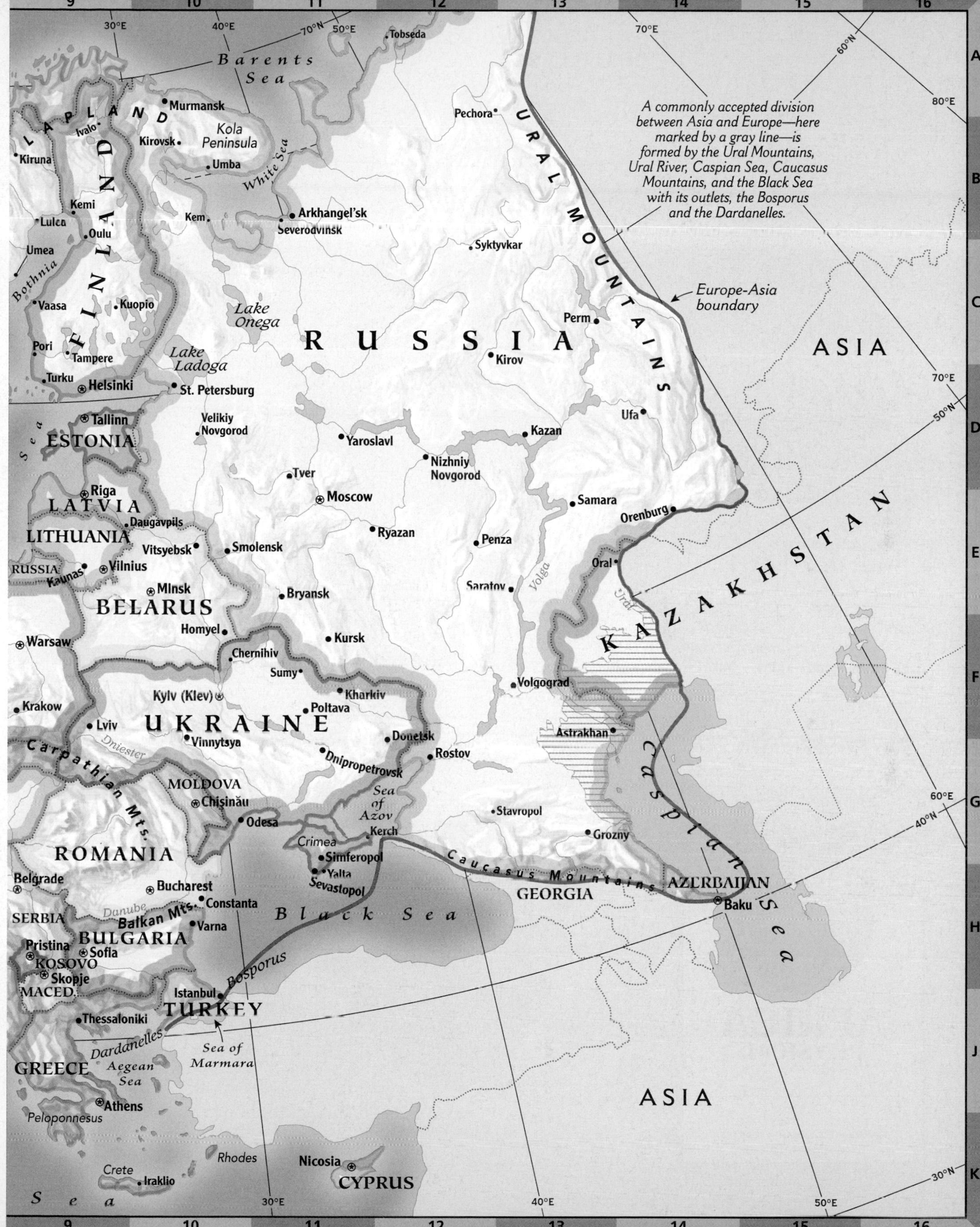

A commonly accepted division between Asia and Europe—here marked by a gray line—is formed by the Ural Mountains, Ural River, Caspian Sea, Caucasus Mountains, and the Black Sea with its outlets, the Bosporus and the Dardanelles.
Europe-Asia boundary
RUSSIA
ASIA
KAZAKHSTAN
URAL MOUNTAINS
Barents Sea
Kola Peninsula
White Sea
LAPLAND
FINLAND
Bothnia
Lake Onega
Lake Ladoga
ESTONIA
LATVIA
LITHUANIA
BELARUS
UKRAINE
MOLDOVA
ROMANIA
SERBIA
BULGARIA
KOSOVO
MACED.
GREECE
TURKEY
GEORGIA
AZERBAIJAN
CYPRUS
Carpathian Mts.
Balkan Mts.
Caucasus Mountains
Black Sea
Caspian Sea
Sea of Azov
Crimea
Bosporus
Dardanelles
Sea of Marmara
Aegean Sea
Peloponnesus
Crete
Rhodes
Dniester
Danube
Volga
Ural
Murmansk
Ivalo
Kirovsk
Umba
Kiruna
Kemi
Lulea
Oulu
Umea
Vaasa
Kuopio
Pori
Tampere
Turku
Helsinki
St. Petersburg
Tallinn
Riga
Daugavpils
Vitsyebsk
Smolensk
Kaunas
Vilnius
Minsk
Warsaw
Krakow
Lviv
Kem
Arkhangel'sk
Severodvinsk
Tobseda
Pechora
Syktyvkar
Perm
Kirov
Ufa
Kazan
Velikiy Novgorod
Yaroslavl
Nizhniy Novgorod
Tver
Moscow
Ryazan
Penza
Samara
Orenburg
Oral
Saratov
Bryansk
Homyel
Kursk
Chernihiv
Sumy
Kharkiv
Kyiv (Kiev)
Poltava
Vinnytsya
Donetsk
Rostov
Dnipropetrovsk
Volgograd
Astrakhan
Stavropol
Grozny
Baku
Chisinău
Odesa
Kerch
Simferopol
Yalta
Sevastopol
Belgrade
Bucharest
Constanta
Varna
Pristina
Sofia
Skopje
Istanbul
Thessaloniki
Athens
Iraklio
Nicosia
RUSSIA
Sea
30°E
40°E
50°E
60°E
70°E
80°E
70°N
60°N
50°N
40°N
30°N
9 10 11 12 13 14 15 16
A B C D E F G H J K

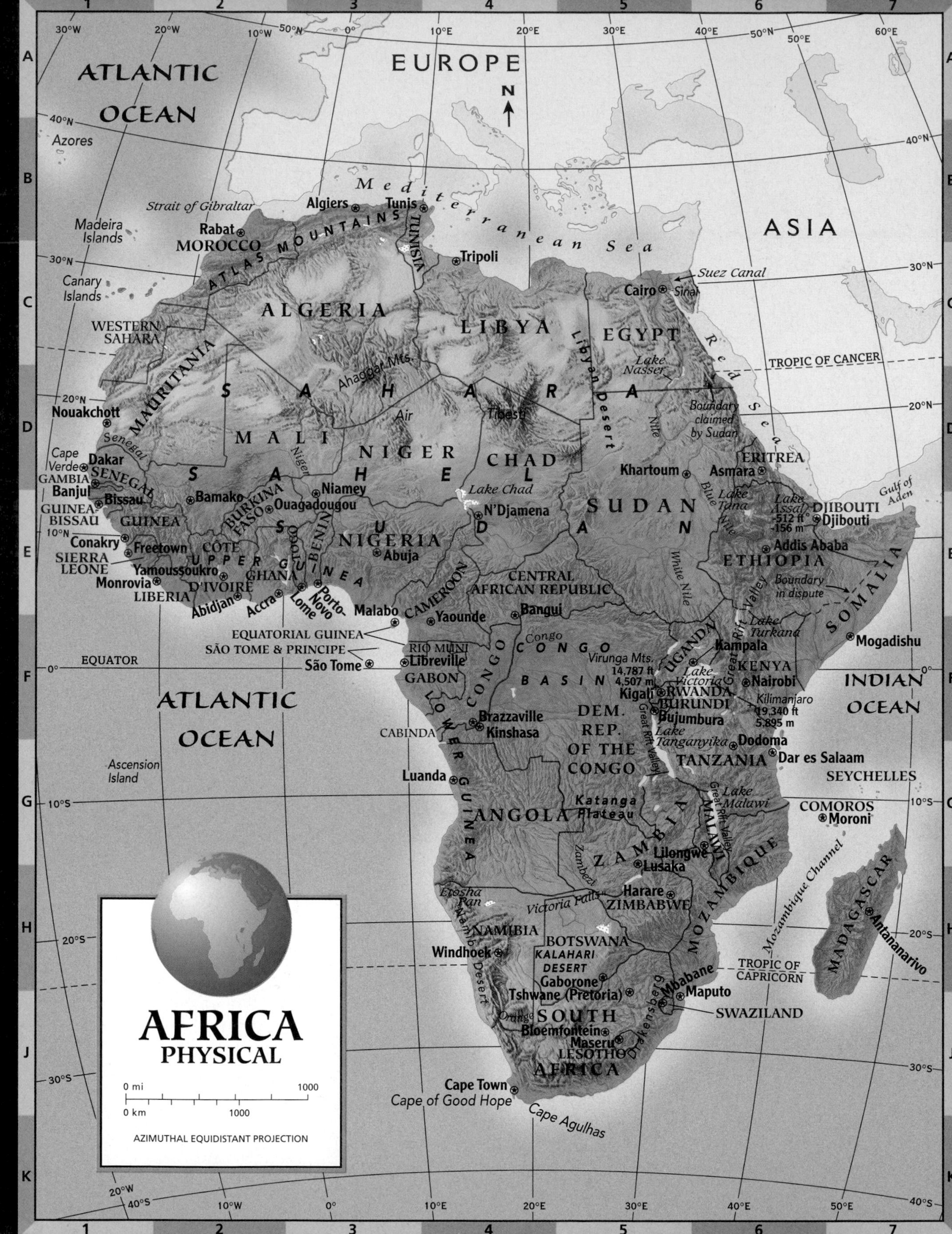
AFRICA
PHYSICAL
AZIMUTHAL EQUIDISTANT PROJECTION
ATLANTIC OCEAN
EUROPE
ASIA
INDIAN OCEAN
Mediterranean Sea
Red Sea
SAHARA
SAHEL
UPPER GUINEA
LOWER GUINEA
CONGO BASIN
ALGERIA
LIBYA
EGYPT
MOROCCO
TUNISIA
WESTERN SAHARA
MAURITANIA
MALI
NIGER
CHAD
SUDAN
ERITREA
DJIBOUTI
ETHIOPIA
SOMALIA
KENYA
UGANDA
RWANDA
BURUNDI
TANZANIA
NIGERIA
CAMEROON
CENTRAL AFRICAN REPUBLIC
DEM. REP. OF THE CONGO
CONGO
GABON
ANGOLA
ZAMBIA
MALAWI
MOZAMBIQUE
ZIMBABWE
NAMIBIA
BOTSWANA
SOUTH AFRICA
LESOTHO
SWAZILAND
MADAGASCAR
SEYCHELLES
COMOROS
Cairo
Khartoum
Addis Ababa
Mogadishu
Nairobi
Kampala
Dodoma
Dar es Salaam
Kinshasa
Brazzaville
Luanda
Lusaka
Harare
Windhoek
Gaborone
Tshwane (Pretoria)
Maputo
Mbabane
Maseru
Bloemfontein
Cape Town
Antananarivo
Moroni
Lilongwe
Algiers
Tunis
Tripoli
Rabat
Nouakchott
Dakar
Banjul
Bissau
Conakry
Freetown
Monrovia
Yamoussoukro
Abidjan
Accra
Lome
Porto-Novo
Abuja
Niamey
Ouagadougou
Bamako
N'Djamena
Yaounde
Malabo
Bangui
Libreville
São Tome
Asmara
Djibouti
Kigali
Bujumbura
Azores
Madeira Islands
Canary Islands
Cape Verde
Ascension Island
Strait of Gibraltar
Atlas Mountains
Ahaggar Mts.
Tibesti
Air
Lake Chad
Libyan Desert
Lake Nasser
Nile
Suez Canal
Sinai
Boundary claimed by Sudan
Boundary in dispute
Lake Tana
Lake Assal -512 ft -156 m
Gulf of Aden
Blue Nile
White Nile
Great Rift Valley
Lake Turkana
Lake Victoria
Virunga Mts. 14,787 ft 4,507 m
Kilimanjaro 19,340 ft 5,895 m
Lake Tanganyika
Lake Malawi
Katanga Plateau
Zambezi
Victoria Falls
Etosha Pan
Namib Desert
KALAHARI DESERT
Orange
Drakensberg
Cape of Good Hope
Cape Agulhas
Mozambique Channel
TROPIC OF CANCER
TROPIC OF CAPRICORN
EQUATOR
EQUATORIAL GUINEA
SÃO TOME & PRINCIPE
RIO MUNI
CABINDA
Senegal
Niger
Congo
GAMBIA
SENEGAL
GUINEA-BISSAU
GUINEA
SIERRA LEONE
LIBERIA
CÔTE D'IVOIRE
GHANA
TOGO
BENIN
BURKINA FASO
0 mi 1000
0 km 1000

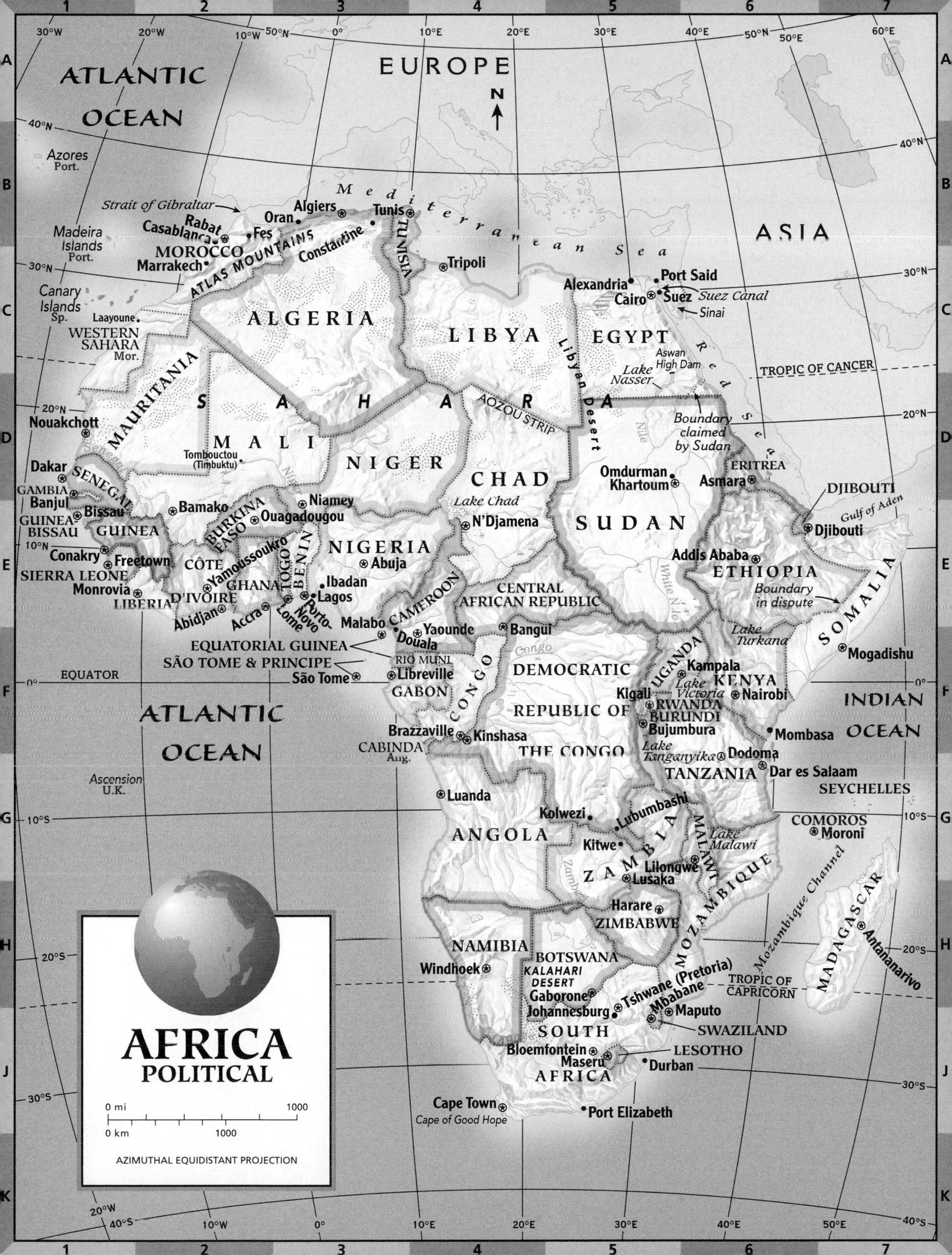
AFRICA
POLITICAL
AZIMUTHAL EQUIDISTANT PROJECTION
0 mi 1000
0 km 1000
EUROPE
ASIA
ATLANTIC OCEAN
ATLANTIC OCEAN
INDIAN OCEAN
Mediterranean Sea
Red Sea
Gulf of Aden
Mozambique Channel
Strait of Gibraltar
Suez Canal
Sinai
Azores Port.
Madeira Islands Port.
Canary Islands Sp.
Ascension U.K.
TROPIC OF CANCER
TROPIC OF CAPRICORN
EQUATOR
ATLAS MOUNTAINS
SAHARA
AOZOU STRIP
Libyan Desert
KALAHARI DESERT
Cape of Good Hope
Boundary claimed by Sudan
Boundary in dispute
Lake Nasser
Aswan High Dam
Lake Chad
Lake Turkana
Lake Victoria
Lake Tanganyika
Lake Malawi
Nile
White Nile
Congo
Niger
Senegal
Zambezi
Orange
MOROCCO
WESTERN SAHARA Mor.
ALGERIA
TUNISIA
LIBYA
EGYPT
MAURITANIA
MALI
NIGER
CHAD
SUDAN
ERITREA
DJIBOUTI
ETHIOPIA
SOMALIA
SENEGAL
GAMBIA
GUINEA-BISSAU
GUINEA
SIERRA LEONE
LIBERIA
CÔTE D'IVOIRE
BURKINA FASO
GHANA
TOGO
BENIN
NIGERIA
CAMEROON
CENTRAL AFRICAN REPUBLIC
EQUATORIAL GUINEA
SÃO TOME & PRINCIPE
RIO MUNI
GABON
CONGO
DEMOCRATIC REPUBLIC OF THE CONGO
CABINDA Ang.
UGANDA
KENYA
RWANDA
BURUNDI
TANZANIA
SEYCHELLES
COMOROS
ANGOLA
ZAMBIA
MALAWI
MOZAMBIQUE
ZIMBABWE
NAMIBIA
BOTSWANA
MADAGASCAR
SOUTH AFRICA
SWAZILAND
LESOTHO
Rabat
Casablanca
Fes
Marrakech
Oran
Algiers
Constantine
Tunis
Tripoli
Alexandria
Port Said
Cairo
Suez
Laayoune
Nouakchott
Tombouctou (Timbuktu)
Dakar
Banjul
Bissau
Conakry
Freetown
Monrovia
Bamako
Ouagadougou
Niamey
Yamoussoukro
Abidjan
Accra
Lome
Porto Novo
Lagos
Ibadan
Abuja
N'Djamena
Omdurman
Khartoum
Asmara
Djibouti
Addis Ababa
Mogadishu
Malabo
Douala
Yaounde
Bangui
São Tome
Libreville
Brazzaville
Kinshasa
Kampala
Kigali
Nairobi
Bujumbura
Mombasa
Dodoma
Dar es Salaam
Moroni
Luanda
Kolwezi
Lubumbashi
Kitwe
Lilongwe
Lusaka
Harare
Windhoek
Gaborone
Tshwane (Pretoria)
Johannesburg
Mbabane
Maputo
Antananarivo
Bloemfontein
Maseru
Durban
Cape Town
Port Elizabeth
N
30°W
20°W
10°W
0°
10°E
20°E
30°E
40°E
50°E
60°E
50°N
40°N
30°N
20°N
10°N
0°
10°S
20°S
30°S
40°S
1
2
3
4
5
6
7
A
B
C
D
E
F
G
H
J
K

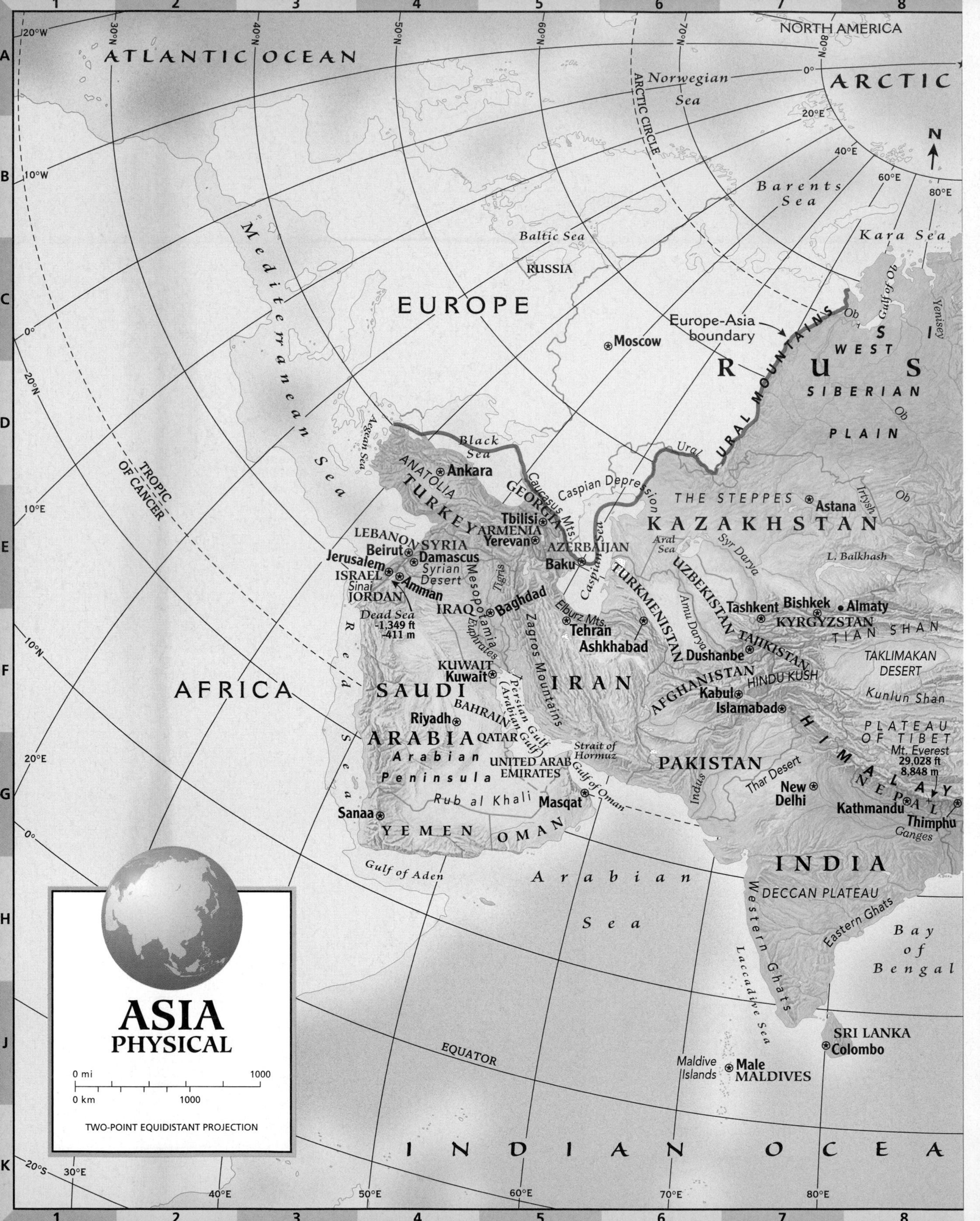
ASIA
PHYSICAL
0 mi
1000
0 km
1000
TWO-POINT EQUIDISTANT PROJECTION
ATLANTIC OCEAN
NORTH AMERICA
ARCTIC
Norwegian Sea
ARCTIC CIRCLE
Barents Sea
Kara Sea
Baltic Sea
RUSSIA
EUROPE
Mediterranean Sea
Moscow
Europe-Asia boundary
URAL MOUNTAINS
Ob
Gulf of Ob
Yenisey
WEST SIBERIAN PLAIN
R U S
Ural
Black Sea
Aegean Sea
Ankara
ANATOLIA
TURKEY
GEORGIA
Caucasus Mts.
Caspian Depression
THE STEPPES
Astana
Irtysh
KAZAKHSTAN
Tbilisi
ARMENIA
Yerevan
AZERBAIJAN
Baku
Caspian Sea
Aral Sea
Syr Darya
L. Balkhash
LEBANON
SYRIA
Beirut
Damascus
Jerusalem
ISRAEL
Syrian Desert
Sinai
JORDAN
Amman
Mesopotamia
Tigris
Euphrates
IRAQ
Baghdad
Dead Sea
-1,349 ft
-411 m
TURKMENISTAN
UZBEKISTAN
Amu Darya
Tashkent
Bishkek
Almaty
KYRGYZSTAN
TIAN SHAN
Elburz Mts.
Tehran
Ashkhabad
Zagros Mountains
TAJIKISTAN
Dushanbe
TAKLIMAKAN DESERT
KUWAIT
Kuwait
IRAN
AFGHANISTAN
HINDU KUSH
Kabul
Islamabad
Kunlun Shan
AFRICA
SAUDI ARABIA
Riyadh
BAHRAIN
QATAR
Persian Gulf (Arabian Gulf)
Strait of Hormuz
PLATEAU OF TIBET
Mt. Everest
29,028 ft
8,848 m
HIMALAYA
Arabian Peninsula
UNITED ARAB EMIRATES
Gulf of Oman
PAKISTAN
Indus
Thar Desert
New Delhi
NEPAL
Kathmandu
Thimphu
Ganges
Red Sea
Rub al Khali
Masqat
Sanaa
YEMEN
OMAN
INDIA
TROPIC OF CANCER
Gulf of Aden
Arabian Sea
DECCAN PLATEAU
Western Ghats
Eastern Ghats
Bay of Bengal
Laccadive Sea
SRI LANKA
Colombo
EQUATOR
Maldive Islands
Male
MALDIVES
INDIAN OCEA

NORTH AMERICA
Bering Strait
Chukchi Sea
Chukchi Peninsula
Gulf of Anadyr
North Pole
OCEAN
Wrangel Island
East Siberian Sea
Bering Sea
Aleutian Islands
North Land
New Siberian Islands
Laptev Sea
Taymyr Peninsula
Kolyma Range
Commander Islands
Cherskiy Range
Kamchatka Peninsula
Verkhoyanski Mountains
CENTRAL SIBERIAN PLATEAU
SIBERIA
Sea of Okhotsk
Kuril Islands
Sakhalin
Yablonovyy Range
Lake Baikal
Greater Khingan Range
Manchurian Plain
Sikhote Alin Range
Hokkaido
Sea of Japan (East Sea)
JAPAN
Tokyo
Honshu
Ulaanbaatar
MONGOLIA
ALTAY MOUNTAINS
GOBI
P'yongyang
NORTH KOREA
Seoul
SOUTH KOREA
Beijing
North China Plain
Shikoku
Kyushu
Nampo Shoto
TROPIC OF CANCER
PACIFIC OCEAN
Yellow Sea
East China Sea
Ryukyu Islands
Qaidam Basin
Huang He (Yellow R.)
CHINA
Sichuan Basin
Chang Jiang (Yangtze R.)
Gongga Shan 24,790 ft 7,556 m
Mariana Islands
Taipei
TAIWAN
Philippine Sea
BHUTAN
BANGLADESH
Dhaka
Hanoi
MYANMAR (BURMA)
Nay Pyi Taw
Vientiane
LAOS
Hainan
South China Sea
Luzon
Manila
PHILIPPINES
PHILIPPINE ISLANDS
Caroline Islands
EQUATOR
THAILAND
Bangkok
CAMBODIA
VIETNAM
Phnom Penh
Andaman Islands
Sulu Sea
Mindanao
Gulf of Thailand
Andaman Sea
Malay Peninsula
Bandar Seri Begawan
BRUNEI
MALAYSIA
Celebes Sea
Moluccas
New Guinea
Nicobar Islands
Kuala Lumpur
SINGAPORE
Borneo
Celebes
Sumatra
INDONESIA
Arafura Sea
GREATER SUNDA ISLANDS
Mentawai Islands
Java Sea
Java
Jakarta
EAST TIMOR (TIMOR-LESTE)
Dili
Timor Sea
AUSTRALIA

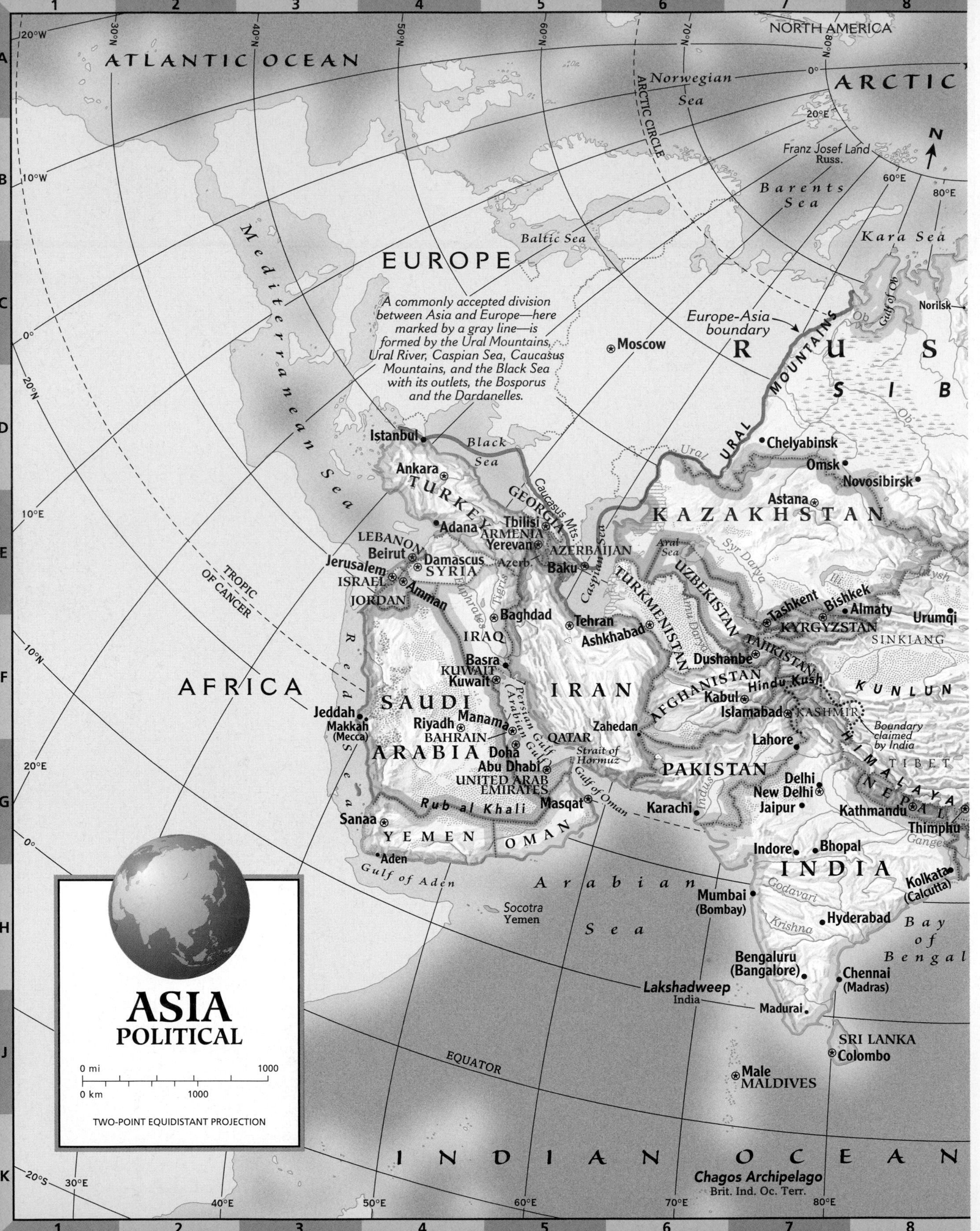
ASIA
POLITICAL
TWO-POINT EQUIDISTANT PROJECTION
0 mi
1000
0 km
1000
ATLANTIC OCEAN
NORTH AMERICA
ARCTIC
Norwegian Sea
ARCTIC CIRCLE
Franz Josef Land
Russ.
Barents Sea
Kara Sea
Baltic Sea
EUROPE
Mediterranean Sea
A commonly accepted division between Asia and Europe—here marked by a gray line—is formed by the Ural Mountains, Ural River, Caspian Sea, Caucasus Mountains, and the Black Sea with its outlets, the Bosporus and the Dardanelles.
Europe-Asia boundary
Moscow
URAL MOUNTAINS
RUS
SIB
Norilsk
Gulf of Ob
Ob
Ural
Chelyabinsk
Omsk
Novosibirsk
Istanbul
Black Sea
Ankara
TURKEY
GEORGIA
Caucasus Mts.
Tbilisi
ARMENIA
Yerevan
AZERBAIJAN
Azerb.
Baku
Caspian Sea
Adana
LEBANON
Beirut
Damascus
SYRIA
Jerusalem
ISRAEL
Amman
JORDAN
Euphrates
Tigris
Baghdad
IRAQ
Tehran
Ashkhabad
TURKMENISTAN
KAZAKHSTAN
Astana
Aral Sea
Syr Darya
Amu Darya
UZBEKISTAN
Tashkent
Bishkek
Almaty
KYRGYZSTAN
Urumqi
SINKIANG
Irtysh
Ili
TAJIKISTAN
Dushanbe
AFGHANISTAN
Hindu Kush
Kabul
Islamabad
KASHMIR
Boundary claimed by India
KUNLUN
TIBET
HIMALAYA
NEPAL
Kathmandu
Thimphu
Ganges
TROPIC OF CANCER
AFRICA
Red Sea
Basra
KUWAIT
Kuwait
IRAN
SAUDI ARABIA
Jeddah
Makkah
(Mecca)
Riyadh
Manama
BAHRAIN
Persian Gulf
(Arabian Gulf)
QATAR
Doha
Abu Dhabi
UNITED ARAB EMIRATES
Strait of Hormuz
Gulf of Oman
Zahedan
PAKISTAN
Indus
Lahore
Karachi
Delhi
New Delhi
Jaipur
Rub al Khali
Masqat
Sanaa
YEMEN
OMAN
Aden
Gulf of Aden
Socotra
Yemen
Arabian Sea
INDIA
Indore
Bhopal
Kolkata
(Calcutta)
Mumbai
(Bombay)
Godavari
Krishna
Hyderabad
Bay of Bengal
Bengaluru
(Bangalore)
Chennai
(Madras)
Lakshadweep
India
Madurai
SRI LANKA
Colombo
Male
MALDIVES
EQUATOR
INDIAN OCEAN
Chagos Archipelago
Brit. Ind. Oc. Terr.
20°W
10°W
0°
10°E
20°E
30°E
40°E
50°E
60°E
70°E
80°E
30°N
40°N
50°N
60°N
70°N
80°N
20°N
10°N
0°
20°S
1
2
3
4
5
6
7
8
A
B
C
D
E
F
G
H
J
K

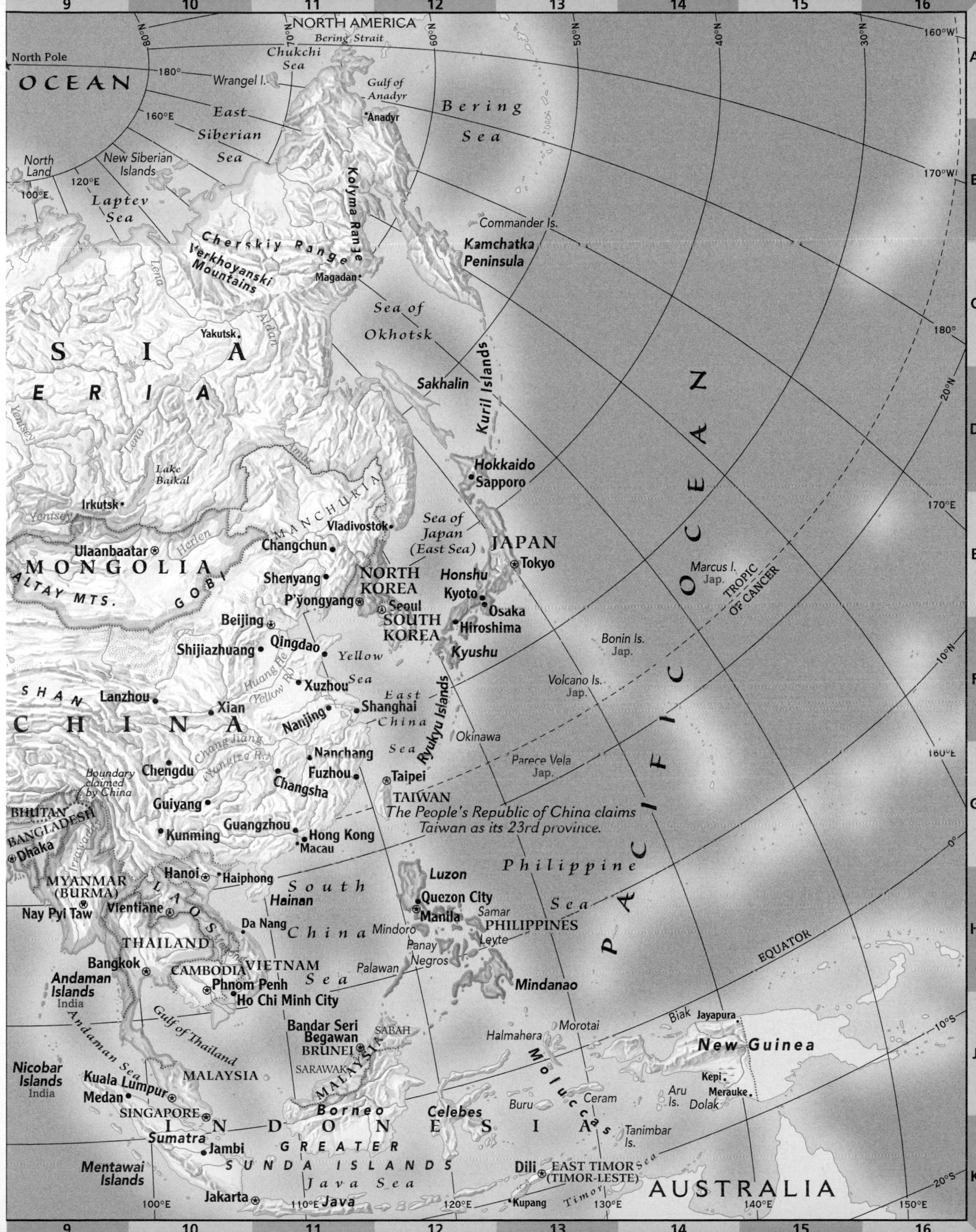

NORTH AMERICA
Bering Strait
Chukchi Sea
North Pole
OCEAN
Wrangel I.
Gulf of Anadyr
Anadyr
Bering Sea
East Siberian Sea
New Siberian Islands
North Land
Laptev Sea
Kolyma Range
Commander Is.
Cherskiy Range
Verkhoyanski Mountains
Kamchatka Peninsula
Magadan
Sea of Okhotsk
Lena
Aldan
Yakutsk
S I A
E R I A
Sakhalin
Kuril Islands
Yenisey
Amur
Lake Baikal
Hokkaido
Sapporo
Irkutsk
MANCHURIA
Vladivostok
Sea of Japan (East Sea)
Herlen
Ulaanbaatar
Changchun
JAPAN
Tokyo
MONGOLIA
ALTAY MTS.
GOBI
Shenyang
NORTH KOREA
Honshu
Marcus I. Jap.
TROPIC OF CANCER
P'yongyang
Seoul
Kyoto
Osaka
Beijing
SOUTH KOREA
Hiroshima
Bonin Is. Jap.
Shijiazhuang
Qingdao
Yellow Sea
Kyushu
Huang He (Yellow R.)
Xuzhou
Volcano Is. Jap.
SHAN
Lanzhou
Xian
East China Sea
Shanghai
Ryukyu Islands
Nanjing
CHINA
Okinawa
Chang Jiang (Yangtze R.)
Nanchang
Parece Vela Jap.
Chengdu
Boundary claimed by China
Fuzhou
Taipei
Changsha
TAIWAN
BHUTAN
Guiyang
The People's Republic of China claims Taiwan as its 23rd province.
BANGLADESH
Kunming
Guangzhou
Hong Kong
Macau
Dhaka
Irrawaddy
PACIFIC OCEAN
Philippine Sea
MYANMAR (BURMA)
Hanoi
Haiphong
Luzon
South China Sea
Hainan
Quezon City
Manila
Samar
Nay Pyi Taw
Vientiane
LAOS
PHILIPPINES
Da Nang
Mindoro
Leyte
THAILAND
Panay
Negros
Mekong
EQUATOR
Bangkok
VIETNAM
CAMBODIA
Palawan
Andaman Islands India
Phnom Penh
Ho Chi Minh City
Mindanao
Andaman Sea
Gulf of Thailand
Bandar Seri Begawan
SABAH
Morotai
Biak
Jayapura
BRUNEI
Halmahera
New Guinea
Nicobar Islands India
MALAYSIA
SARAWAK
Moluccas
Kuala Lumpur
Kepi
Aru Is.
Merauke
Medan
Borneo
Celebes
Ceram
Dolak
SINGAPORE
Buru
INDONESIA
Sumatra
Tanimbar Is.
Jambi
GREATER SUNDA ISLANDS
Mentawai Islands
Dili
EAST TIMOR (TIMOR-LESTE)
Java Sea
AUSTRALIA
Jakarta
Java
Kupang
Timor Sea
80°N
70°N
60°N
50°N
40°N
30°N
20°N
10°N
0°
10°S
20°S
180°
160°E
120°E
100°E
110°E
130°E
140°E
150°E
170°E
160°E
160°W
170°W
9 10 11 12 13 14 15 16
A B C D E F G H J K

NORTH PACIFIC OCEAN
TROPIC OF CANCER
NORTHERN MARIANA ISLANDS U.S.
Saipan
GUAM U.S.
Hagatna
Wake Island U.S.
Bikini Atoll
MARSHALL ISLANDS
Ralik Chain
Ratak Chain
Majuro
MICRONESIA
ASIA
PALAU
Melekeok
Yap Islands
Truk Islands
Caroline Islands
Palikir
Pohnpei (Ponape)
FEDERATED STATES OF MICRONESIA
Tarawa (Bairiki)
Gilbert Islands
EQUATOR
Yaren
NAURU
New Guinea
Mt. Wilhelm 14,793 ft 4,509 m
New Britain
PAPUA NEW GUINEA
Port Moresby
Torres Strait
MELANESIA
Solomon Is.
SOLOMON ISLANDS
Honiara
TUVALU
Funafuti
Santa Cruz Is.
Darwin
Gulf of Carpentaria
Coral Sea
CORAL SEA ISLANDS TERRITORY Austral.
VANUATU
Port-Vila
Suva
FIJI ISLANDS
NEW CALEDONIA Fr.
Noumea
Kimberley Plateau
NORTHERN TERRITORY
AUSTRALIA
GREAT DIVIDING RANGE
TROPIC OF CAPRICORN
WESTERN AUSTRALIA
Macdonnell Ranges
QUEENSLAND
Brisbane
GREAT VICTORIA DESERT
SOUTH AUSTRALIA
Lake Eyre -52 ft -16 m
Norfolk Island Austral.
Darling
NEW SOUTH WALES
Lord Howe Island Austral.
Perth
Great Australian Bight
Sydney
Adelaide
Murray
AUSTRALIAN CAPITAL TERRITORY
Canberra
Mt. Kosciuszko 7,310 ft 2,228 m
VICTORIA
Melbourne
North Island
Auckland
Tasman Sea
NEW ZEALAND
Wellington
INDIAN OCEAN
TASMANIA
Hobart
Mt. Cook 12,316 ft 3,754 m
Christchurch
South Island
Stewart Island
120°E 130°E 140°E 150°E 160°E 170°E
20°N 10°N 0° 10°S 20°S 30°S 40°S 50°S

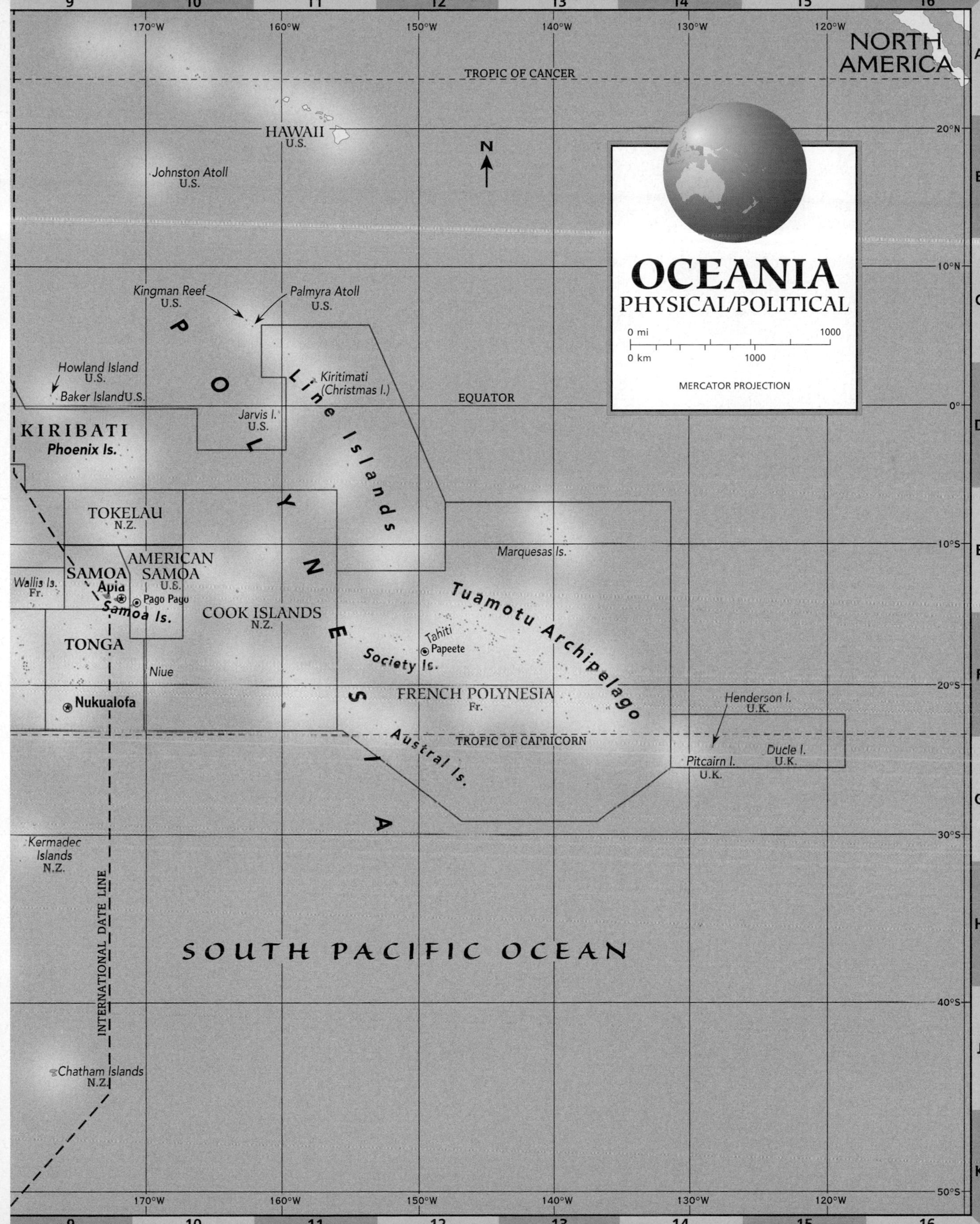

OCEANIA
PHYSICAL/POLITICAL
0 mi
1000
0 km
1000
MERCATOR PROJECTION
NORTH AMERICA
TROPIC OF CANCER
HAWAII
U.S.
Johnston Atoll
U.S.
N
Kingman Reef
U.S.
Palmyra Atoll
U.S.
POLYNESIA
Howland Island
U.S.
Baker Island U.S.
Kiritimati
(Christmas I.)
Line Islands
EQUATOR
Jarvis I.
U.S.
KIRIBATI
Phoenix Is.
TOKELAU
N.Z.
AMERICAN SAMOA
U.S.
SAMOA
Apia
Pago Pago
Samoa Is.
Wallis Is.
Fr.
TONGA
Nukualofa
Niue
COOK ISLANDS
N.Z.
Marquesas Is.
Tuamotu Archipelago
Tahiti
Papeete
Society Is.
FRENCH POLYNESIA
Fr.
Austral Is.
TROPIC OF CAPRICORN
Henderson I.
U.K.
Ducie I.
U.K.
Pitcairn I.
U.K.
Kermadec
Islands
N.Z.
INTERNATIONAL DATE LINE
SOUTH PACIFIC OCEAN
Chatham Islands
N.Z.
170°W
160°W
150°W
140°W
130°W
120°W
20°N
10°N
0°
10°S
20°S
30°S
40°S
50°S
9
10
11
12
13
14
15
16
A
B
C
D
E
F
G
H
J
K

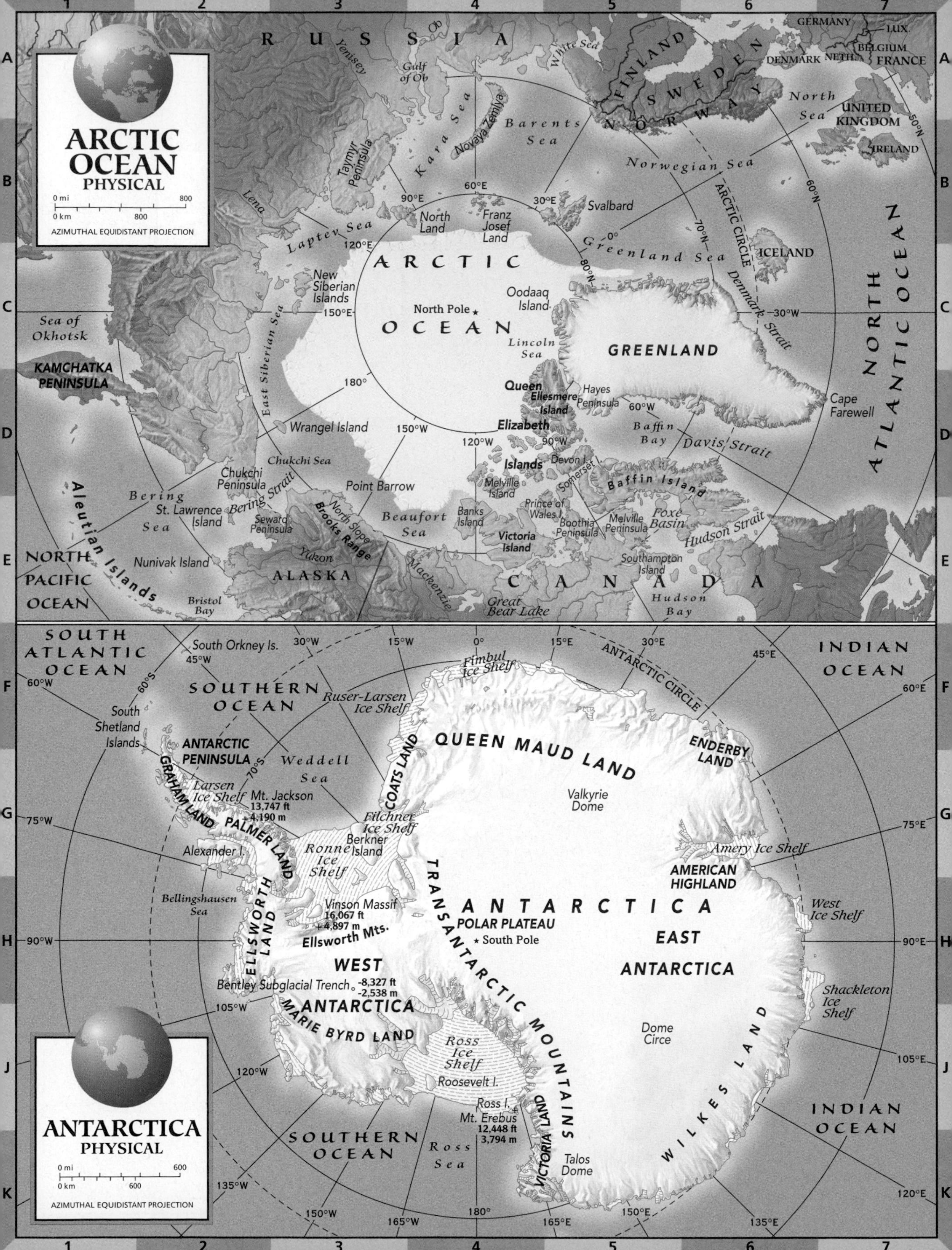
ARCTIC OCEAN
PHYSICAL
0 mi 800
0 km 800
AZIMUTHAL EQUIDISTANT PROJECTION
RUSSIA
FINLAND
SWEDEN
NORWAY
GERMANY
LUX.
BELGIUM
DENMARK
NETH.
FRANCE
UNITED KINGDOM
IRELAND
North Sea
Barents Sea
Kara Sea
Norwegian Sea
Greenland Sea
Laptev Sea
East Siberian Sea
Chukchi Sea
Bering Sea
Beaufort Sea
Lincoln Sea
Baffin Bay
Davis Strait
Denmark Strait
Hudson Strait
Hudson Bay
Sea of Okhotsk
ARCTIC OCEAN
North Pole
NORTH ATLANTIC OCEAN
NORTH PACIFIC OCEAN
GREENLAND
ICELAND
Svalbard
Franz Josef Land
North Land
Novaya Zemlya
New Siberian Islands
Wrangel Island
Taymyr Peninsula
Gulf of Ob
White Sea
Yenisey
Ob
Lena
KAMCHATKA PENINSULA
Chukchi Peninsula
Bering Strait
St. Lawrence Island
Seward Peninsula
Nunivak Island
Aleutian Islands
Bristol Bay
ALASKA
Point Barrow
North Slope
Brooks Range
Yukon
Mackenzie
Great Bear Lake
CANADA
Oodaaq Island
Queen Elizabeth Islands
Ellesmere Island
Hayes Peninsula
Devon I.
Somerset I.
Baffin Island
Melville Island
Banks Island
Prince of Wales I.
Victoria Island
Boothia Peninsula
Melville Peninsula
Foxe Basin
Southampton Island
Cape Farewell
ARCTIC CIRCLE
ANTARCTICA
PHYSICAL
0 mi 600
0 km 600
AZIMUTHAL EQUIDISTANT PROJECTION
SOUTH ATLANTIC OCEAN
SOUTHERN OCEAN
INDIAN OCEAN
South Orkney Is.
South Shetland Islands
ANTARCTIC PENINSULA
GRAHAM LAND
PALMER LAND
Larsen Ice Shelf
Mt. Jackson 13,747 ft 4,190 m
Weddell Sea
Filchner Ice Shelf
Berkner Island
Ronne Ice Shelf
Alexander I.
Bellingshausen Sea
ELLSWORTH LAND
Vinson Massif 16,067 ft 4,897 m
Ellsworth Mts.
Bentley Subglacial Trench -8,327 ft -2,538 m
WEST ANTARCTICA
MARIE BYRD LAND
Ross Ice Shelf
Roosevelt I.
Ross I.
Mt. Erebus 12,448 ft 3,794 m
Ross Sea
VICTORIA LAND
Talos Dome
TRANSANTARCTIC MOUNTAINS
ANTARCTICA
POLAR PLATEAU
South Pole
EAST ANTARCTICA
Dome Circe
WILKES LAND
Shackleton Ice Shelf
West Ice Shelf
AMERICAN HIGHLAND
Amery Ice Shelf
Valkyrie Dome
QUEEN MAUD LAND
ENDERBY LAND
COATS LAND
Ruser-Larsen Ice Shelf
Fimbul Ice Shelf
ANTARCTIC CIRCLE

A WORLD OF EXTREMES

1 The largest continent is Asia with an area of 12,262,691 sq. miles (31,758,898 sq. km).

2 The smallest continent is Australia with an area of 2,988,888 sq. miles (7,741,184 sq. km).

3 The largest country is Russia with an area of 6,592,819 sq. miles (17,075,322 sq. km).

4 The smallest country is Vatican City with an area of 1 sq. mile (2.6 sq. km).

5 The longest river is the Nile River with a length of 4,160 miles (6,695 km).

6 The deepest lake is Lake Baikal with a maximum depth of 5,715 feet (1,742 m).

7 The highest waterfall is Angel Falls with a height of 3,212 feet (979 m).

8 The highest mountain is Mount Everest with a height of 29,028 feet (8,848 m) above sea level.

9 The largest desert is the Sahara with an area of 3,500,000 sq. miles (9,065,000 sq. km).

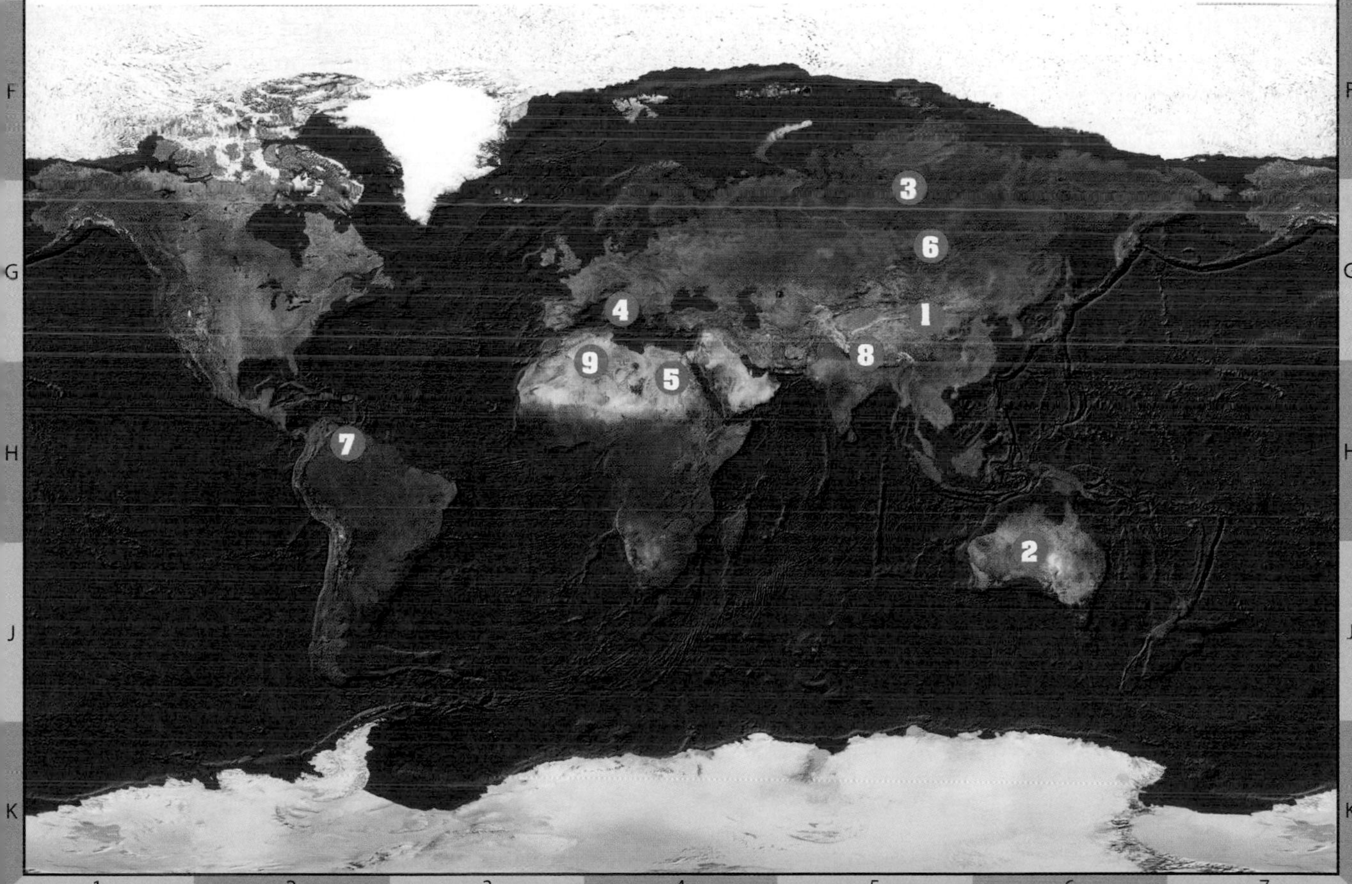

The Big Ideas in Geography

As you read *World Geography and Cultures*, you will be given help in sorting out all the information you encounter. This textbook organizes geographic concepts around Big Ideas in social studies to help you better understand how geography affects you, your family, and your community today and in the future.

Geography is used to interpret the past, understand the present, and plan for the future.

Knowing about the human and physical geography of a place informs us of patterns and relationships that help us interpret, understand, and plan the world around us.

Physical processes shape Earth's surface.

Natural forces such as earthquakes, volcanic activity, erosion, and climate patterns affect the surface of the Earth by creating many landforms and environments like mountains, valleys, lakes, deserts, and swamps.

Geographers study how people, places, and environments are distributed on Earth's surface.

The study of how and where people, places, and natural environments are located on Earth helps us understand the patterns and relationships that affect our lives.

Countries are affected by their relationships with each other.

Interactions among countries influence their economies, politics, culture, and environment. Sometimes countries even have to decide how much they want to interact with others.

Geography and the environment play an important role in shaping a society over time.

The physical environment of a place influences human culture. For example, the types of economic activities a society is involved in, the types of houses they build, and where they build them can all be affected by the physical geography and surrounding environment.

Certain processes, patterns, and functions help determine where people settle.

Physical processes and patterns such as tectonic activity, erosion, and climate can influence where people live. In addition, the function, or purpose of human settlement—economic, religious, or political—also affects where people choose to live.

Places reflect the relationship between humans and the physical environment.

A place—a town, a state, a country—is defined by the interaction between humans and their surroundings. For example, abundant oil reserves near the port city of Houston, Texas, have resulted in the growth of oil refining plants and created a city known for its energy industry.

Technology impacts people and economies and can change a people's way of life.

Using the BIG Ideas

You will find a Big Idea at the beginning of each chapter of your text. Each section opens with an Essential Question that helps you put it all together to better understand how geographic concepts are connected—and to see why geography is important to you.

The characteristics and distribution of human populations affect human and physical systems.

People and the activities in which they are involved influence human, or cultural, characteristics such as government, economic systems, belief systems, and language. Human populations also affect the natural environment by such actions as clearing land for agriculture and building roads and cities.

Cultures are held together by shared beliefs and common practices and values.

Elements of culture like language, religion, ethnicity, and political and economic systems that are shared by a group of people help the group maintain a sense of unity.

The movement of people, goods, and ideas causes societies to change over time.

The exchange of goods and ideas between people has occurred for thousands of years as a result of migration and conquest. Coming into contact with other people, goods, and ideas influences the way people live.

Disputes over ideas, resources, values, and politics can lead to change.

Not every person or group of people agrees on everything all the time. When people or groups disagree over ideas or resources, conflicts and their results—whether they are agreed-upon solutions or violent takeovers—can lead to change.

Economic systems shape relationships in society.

All societies develop economic systems to provide for the wants and needs of their citizens. Each society's economic system influences interactions among individuals, businesses, institutions, and other societies.

The characteristics and distribution of ecosystems help people understand environmental issues.

An understanding of Earth's ecosystems—communities of plants and animals that depend upon one another and their surroundings for survival—provides people with the information they need to be aware of environmental issues affecting their lives.

Recycling is a choice people make to protect Earth's physical environment.

How Do I Study Geography?

By Richard G. Boehm, Ph.D.

Geographers have tried to understand the best way to teach and learn about geography. In order to do this, geographers created the *Five Themes of Geography*. The themes acted as a guide for teaching the basic ideas about geography to students like yourself.

People who teach and study geography, though, thought that the Five Themes were too broad. In 1994, geographers created 18 national geography standards. These standards were more detailed about what should be taught and learned. The Six Essential Elements act as a bridge connecting the Five Themes with the standards.

These pages show you how the Five Themes are related to the Six Essential Elements and the 18 standards.

5 Themes of Geography

1 Location

Location describes where something is. Absolute location describes a place's exact position on the Earth's surface. Relative location expresses where a place is in relation to another place.

2 Place

Place describes the physical and human characteristics that make a location unique.

3 Regions

Regions are areas that share common characteristics.

4 Movement

Movement explains how and why people and things move and are connected.

5 Human-Environment Interaction

Human-Environment Interaction describes the relationship between people and their environment.

6 Essential Elements

18 Geography Standards

I. The World in Spatial Terms
Geographers look to see where a place is located. Location acts as a starting point to answer "Where Is It?" The location of a place helps you orient yourself as to where you are.

1 How to use maps and other tools
2 How to use mental maps to organize information
3 How to analyze the spatial organization of people, places, and environments

II. Places and Regions
Place describes physical characteristics such as landforms, climate, and plant or animal life. It might also describe human characteristics, including language and way of life. Places can also be organized into regions. **Regions** are places united by one or more characteristics.

4 The physical and human characteristics of places
5 How people create regions to interpret Earth's complexity
6 How culture and experience influence people's perceptions of places and regions

III. Physical Systems
Geographers study how physical systems, such as hurricanes, volcanoes, and glaciers, shape the surface of the Earth. They also look at how plants and animals depend upon one another and their surroundings for their survival.

7 The physical processes that shape Earth's surface
8 The distribution of ecosystems on Earth's surface

IV. Human Systems
People shape the world in which they live. They settle in certain places but not in others. An ongoing theme in geography is the movement of people, ideas, and goods.

9 The characteristics, distribution, and migration of human populations
10 The complexity of Earth's cultural mosaics
11 The patterns and networks of economic interdependence
12 The patterns of human settlement
13 The forces of cooperation and conflict

V. Environment and Society
How does the relationship between people and their natural surroundings influence the way people live? Geographers study how people use the environment and how their actions affect the environment.

14 How human actions modify the physical environment
15 How physical systems affect human systems
16 The meaning, use, and distribution of resources

VI. The Uses of Geography
Knowledge of geography helps us understand the relationships among people, places, and environments over time. Applying geographic skills helps you understand the past and prepare for the future.

17 How to apply geography to interpret the past
18 How to apply geography to interpret the present and plan for the future

UNIT 1 The World

Why It Matters

The world today in the twenty-first century is a much smaller place than it was at the time of your grandparents. Advances in technology, communications, and transportation have narrowed vast distances and made neighbors of the world's people. The Internet, for example, now puts you in immediate touch with people in other parts of the world. In the years to come, you and your generation—here and elsewhere—will be challenged to use this and other technology to make the world a better place for everyone.

Japanese tourist and Indian vendor, Goa, India

NATIONAL GEOGRAPHIC
NGS ONLINE To learn more about the basics of geography visit www.nationalgeographic.com/education.

CHAPTER 1

How Geographers Look at the World

BiG Idea

Geography is used to interpret the past, understand the present, and plan for the future. News coverage of local, national, and international events is part of daily life in the twenty-first century. An understanding of geography will help you learn how such events affect your life.

Essential Questions

Section 1:
Geography Skills Handbook
What tools do geography skills provide?

Section 2:
The Geographer's Craft
What are the elements of geography?

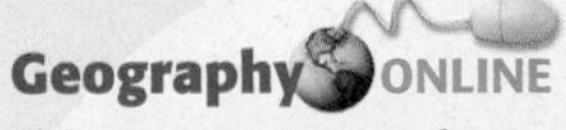

Geography ONLINE
Visit glencoe.com and enter ***QuickPass***™ code WGC9952C1 for Chapter 1 resources.

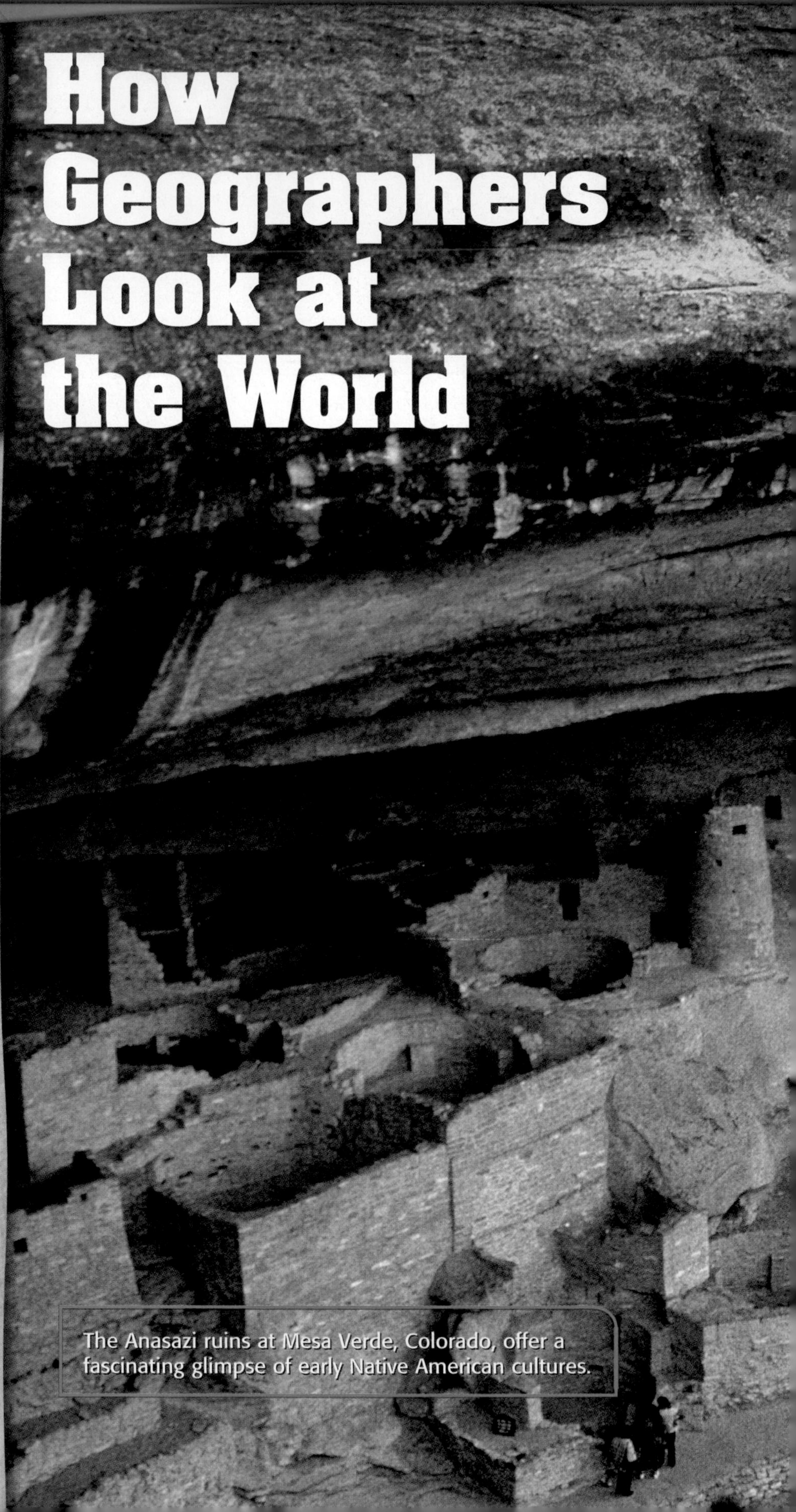

The Anasazi ruins at Mesa Verde, Colorado, offer a fascinating glimpse of early Native American cultures.

FOLDABLES™ Study Organizer

Organizing Information Make a Top-Tab Book to help you understand how geographers look at the world.

Elements of Geography | Research Methods | Geography and Other Subjects | Geography as a Career

How Geographers Look at the World

Reading and Writing As you read the chapter, write at least five facts, generalizations, or observations for each of the following headings: Elements of Geography, Research Methods, Geography and Other Subjects, and Geography as a Career.

section audio

spotlight video

SECTION 1

Geography Skills Handbook

Contents

Essential Question

What tools do geography skills provide?

Geography skills provide the tools and methods for us to understand the relationships between people, places, and environments. We use geographic skills when we make daily personal decisions—where to buy a home; where to get a job; how to get to the shopping mall; where to go on vacation. Community decisions, such as where to locate a new school or how to solve problems of air and water pollution, also require the skillful use of geographic information.

Geographers use a wide array of tools and technologies—from basic globes to high-tech global positioning systems—to understand the Earth. These help us collect and analyze a great deal of information. However, the study of geography is more than knowing a lot of facts about places. Rather, it has more to do with asking questions about the Earth, pursuing their answers, and solving problems. Thus, one of the most important geographic tools is inside your head: the ability to think geographically.

—Dr. Richard Boehm,
September 2006

World Geography and Cultures
author Dr. Richard Boehm

Globes and Maps

A **globe** is a scale model of the Earth. Because Earth is round, a globe presents the most accurate depiction of geographic information such as area, distance, and direction. However, globes show little close-up detail. A printed **map** is a symbolic representation of all or part of the planet. Unlike globes, maps can show small areas in great detail.

From 3-D to 2-D

Think about the surface of the Earth as the peel of an orange. To flatten the peel, you have to cut it like the globe shown here. To create maps that are not interrupted, mapmakers, or **cartographers,** use mathematical formulas to transfer information from the three-dimensional globe to the two-dimensional map. However, when the curves of a globe become straight lines on a map, distortion of size, shape, distance, or area occurs.

Great Circle Routes

A straight line of true direction—one that runs directly from west to east, for example—is not always the shortest distance between two points on Earth. This is due to the curvature of the Earth. To find the shortest distance between any two places, stretch a piece of string around a globe from one point to the other. The string will form part of a *great circle,* an imaginary line that follows the curve of the Earth. Traveling along a great circle is called following a **great circle route.** Ship captains and airline pilots use great circle routes to reduce travel time and conserve fuel.

The idea of a great circle route is an important difference between globes and maps. A round globe accurately shows a great circle route, as indicated on the map below. However, as shown on the flat map, the great circle distance (dotted line) between Tokyo and Los Angeles appears to be far longer than the true direction distance (solid line). In fact, the great circle distance is about 315 miles (506 km) shorter.

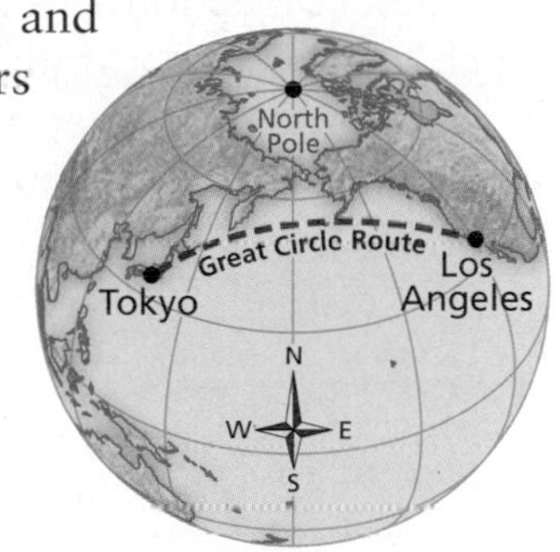

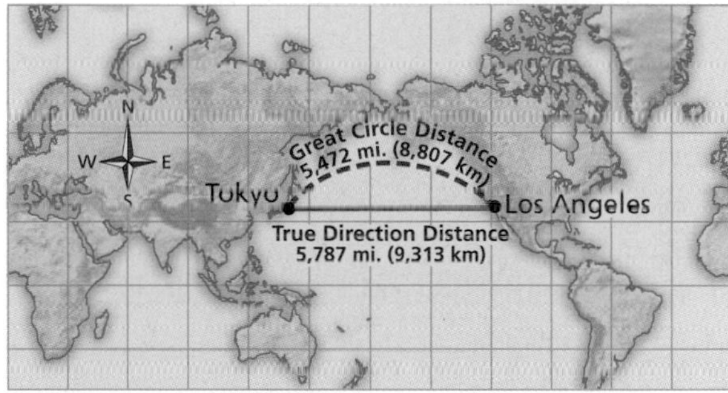

PRACTICING THE SKILL

1. **Explain** the significance of: globe, map, cartographer, great circle route.
2. **Describe** the problems that arise when the curves of a globe become straight lines on a map.
3. **Answering the Essential Question** Use a Venn diagram like the one below to identify the similarities and differences between globes and maps.

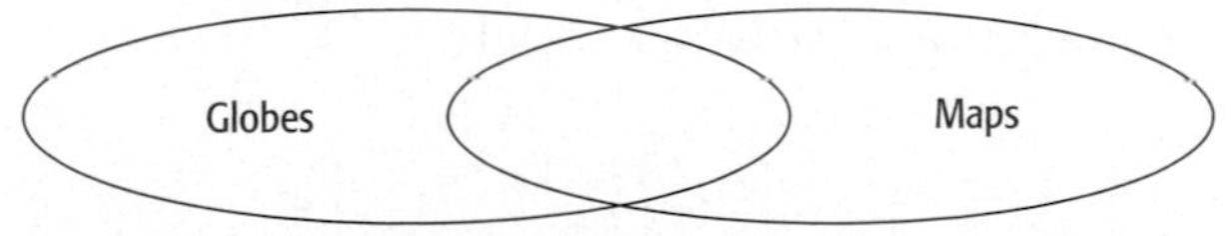

THE WORLD

Projections

To create maps, cartographers project the round Earth onto a flat surface—making a **map projection.** Distance, shape, direction, or size may be distorted by a projection. As a result, the purpose of the map usually dictates which projection is used. There are many kinds of map projections, some with general names and some named for the cartographers who developed them. Three basic categories of map projections are shown here: **planar, cylindrical,** and **conic.**

Planar Projection

A planar projection shows the Earth centered in such a way that a straight line coming from the center to any other point represents the shortest distance. Also known as an azimuthal projection, it is most accurate at its center. As a result, it is often used for maps of the Poles.

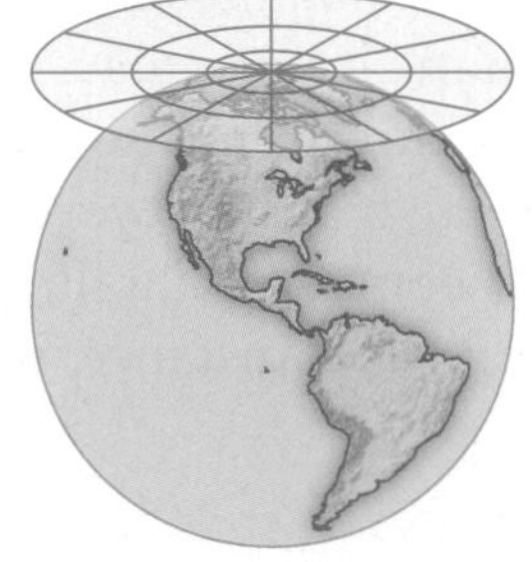

Cylindrical Projection

A cylindrical projection is based on the projection of the globe onto a cylinder. This projection is most accurate near the Equator, but shapes and distances are distorted near the Poles.

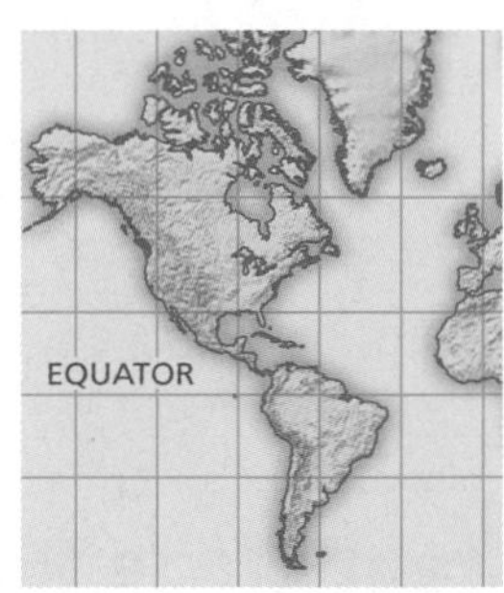

Conic Projection

A conic projection comes from placing a cone over part of a globe. Conic projections are best suited for showing limited east-west areas that are not too far from the Equator. For these uses, a conic projection can indicate distances and directions fairly accurately.

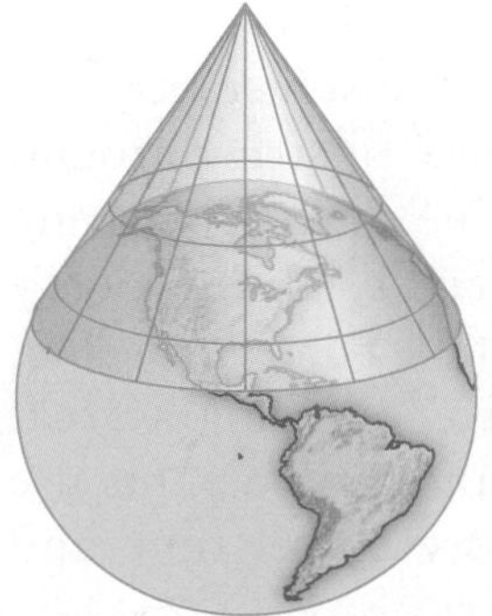

Common Map Projections

Each type of map projection has advantages and some degree of inaccuracy. Four of the most common projections are shown here.

Winkel Tripel Projection

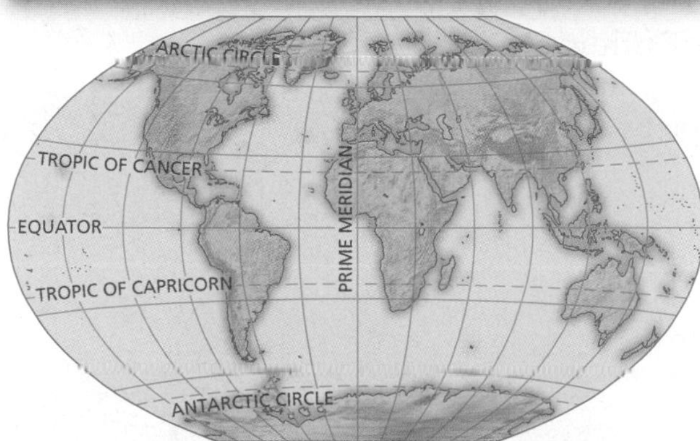

Most general reference world maps are the Winkel Tripel projection. It provides a good balance between the size and shape of land areas as they are shown on the map. Even the polar areas are depicted with little distortion of size and shape.

Goode's Interrupted Equal-Area Projection

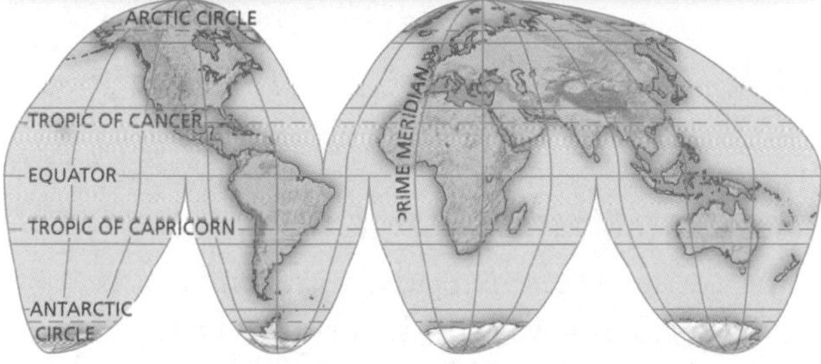

An **interrupted projection** resembles a globe that has been cut apart and laid flat. Goode's Interrupted Equal-Area projection shows the true size and shape of Earth's landmasses, but distances are generally distorted.

Robinson Projection

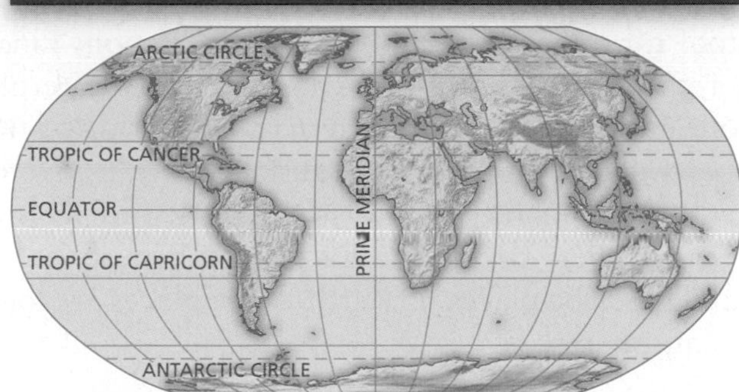

The Robinson projection has minor distortions. The sizes and shapes near the eastern and western edges of the map are accurate, and outlines of the continents appear much as they do on the globe. However, the polar areas are flattened.

Mercator Projection

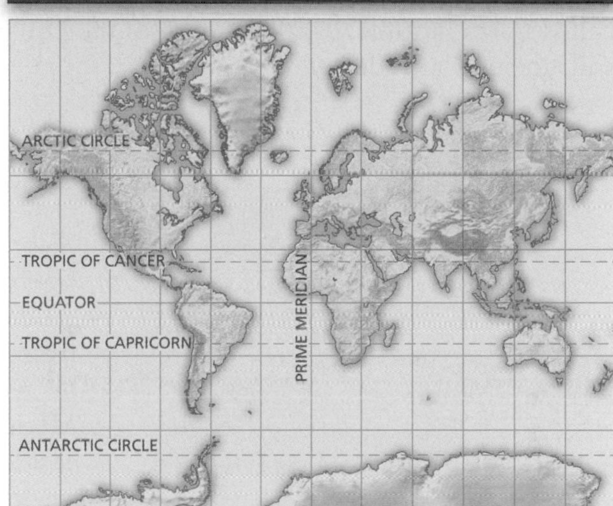

The Mercator projection increasingly distorts size and distance as it moves away from the Equator. However, Mercator projections do accurately show true directions and the shapes of landmasses, making these maps useful for sea travel.

PRACTICING THE SKILL

1. **Explain** the significance of: map projection, planar, cylindrical, conic, interrupted projection.
2. **Which** of the four common projections described above is the best one to use when showing the entire world? Why?
3. **Draw** a map of the world from memory, labeling continents, oceans, and as many countries as you can. Then trade maps with a partner and look for similarities and differences between your maps. Discuss how each person's spatial perspective is reflected in his or her map.
4. **Answering the Essential Question** Use a Venn diagram like the one below to identify the similarities and differences between the Winkel Tripel and Mercator projections.

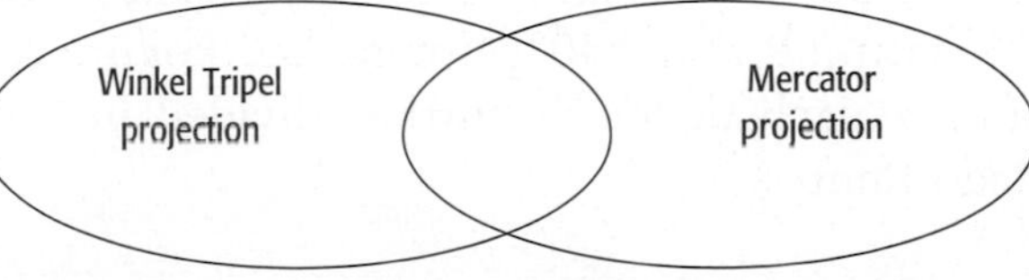

THE WORLD

Determining Location

Geography is often said to begin with the question *Where?* The basic tool for answering the question is **location.** Lines on globes and maps provide information that can help you locate places. These lines cross one another forming a pattern called a **grid system,** which helps you find exact places on the Earth's surface.

A **hemisphere** is one of the halves into which the Earth is divided. Geographers divide the Earth into hemispheres to help them classify and describe places on Earth. Most places are located in two of the four hemispheres.

Latitude

Lines of **latitude,** or parallels, circle the Earth parallel to the Equator and measure the distance north or south of the Equator in degrees. The **Equator** is measured at 0° latitude, while the Poles lie at latitudes 90°N (north) and 90°S (south). Parallels north of the Equator are called north latitude. Parallels south of the Equator are called south latitude.

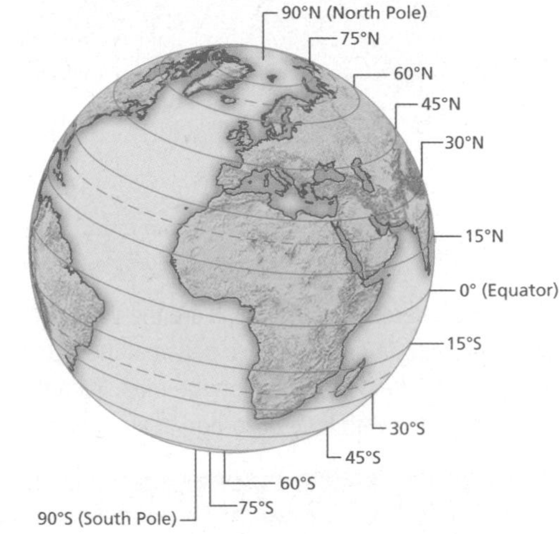

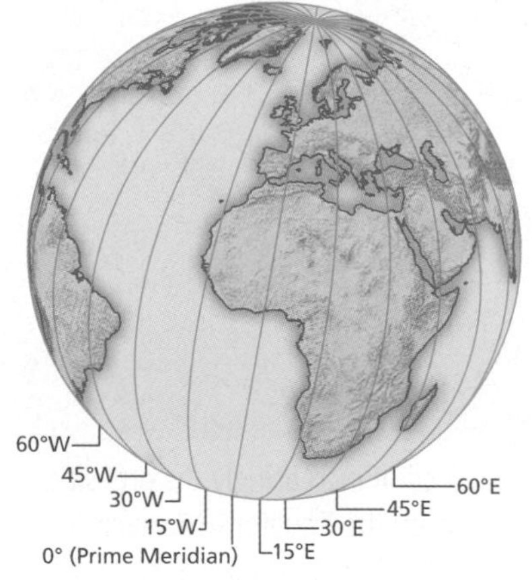

Longitude

Lines of **longitude,** or meridians, circle the Earth from Pole to Pole. These lines measure distance east or west of the **Prime Meridian** at 0° longitude. Meridians east of the Prime Meridian are known as east longitude. Meridians west of the Prime Meridian are known as west longitude. The 180° meridian on the opposite side of the Earth is called the International Date Line.

The Global Grid

Every place has a global address, or **absolute location.** You can identify the absolute location of a place by naming the latitude and longitude lines that cross exactly at that place. For example, Tokyo, Japan, is located at 36°N latitude and 140°E longitude. For more precise readings, each degree is further divided into 60 units called minutes.

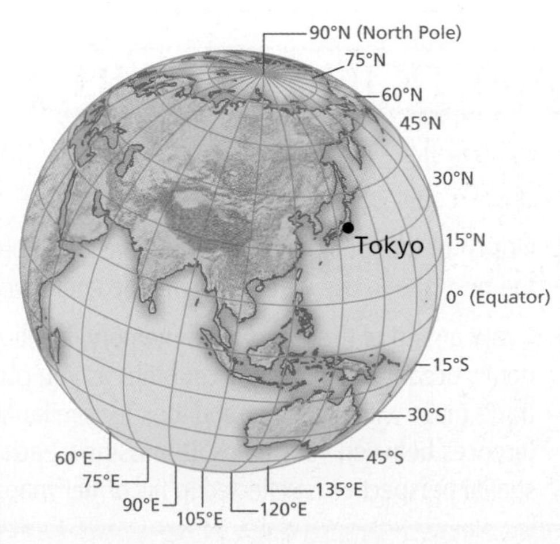

Northern and Southern Hemispheres

The diagram below shows that the Equator divides the Earth into the Northern and Southern Hemispheres. Everything north of the Equator is in the **Northern Hemisphere.** Everything south of the Equator is in the **Southern Hemisphere.**

Northern Hemisphere

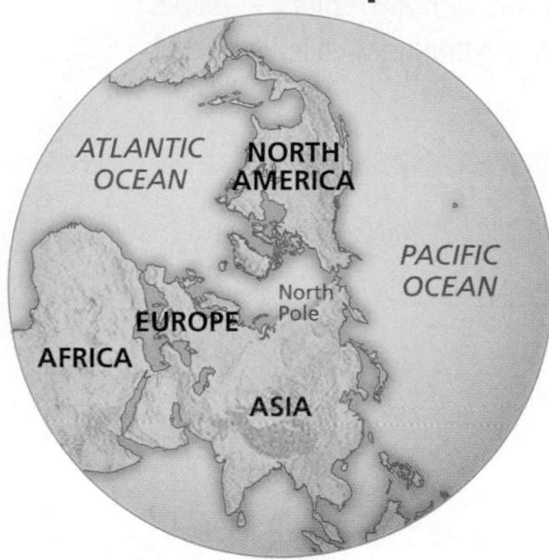

Southern Hemisphere

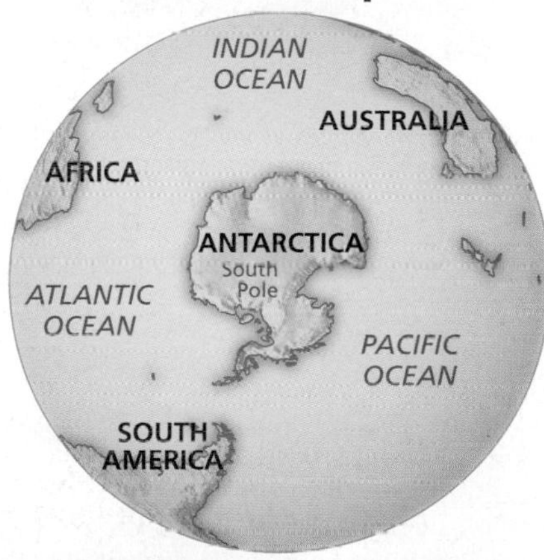

Eastern and Western Hemispheres

The Prime Meridian and the International Date Line divide the Earth into the Eastern and Western Hemispheres. Everything east of the Prime Meridian for 180° is in the **Eastern Hemisphere.** Everything west of the Prime Meridian for 180° is in the **Western Hemisphere.**

Eastern Hemisphere

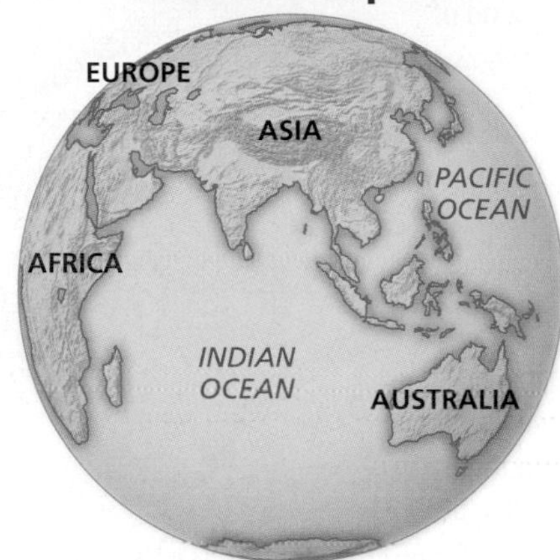

Western Hemisphere

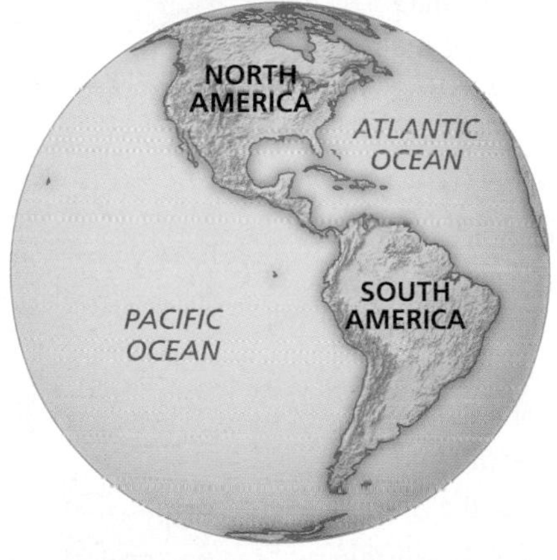

PRACTICING THE SKILL

1. **Explain** the significance of: location, grid system, hemisphere, latitude, Equator, longitude, Prime Meridian, absolute location, Northern Hemisphere, Southern Hemisphere, Eastern Hemisphere, Western Hemisphere.
2. **Which** lines of latitude and longitude divide the Earth into hemispheres?
3. **Use** the Reference Atlas maps to create a chart listing the latitude and longitude of three world cities. Have a partner try to identify the cities.
4. **Answering the Essential Question** Use a chart like the one below to identify the continents in each hemisphere. Continents will appear in more than one hemisphere.

Hemisphere	Continents
Northern	
Southern	
Eastern	
Western	

THE WORLD

Reading a Map

In addition to latitude and longitude, maps feature other important tools to help you interpret the information they contain. Learning to use these map tools will help you read the symbolic language of maps more easily.

Title

The title tells you what kind of information the map is showing.

Key

The **key** lists and explains the symbols, colors, and lines used on the map. The key is sometimes called a legend.

Scale Bar

The **scale bar** shows the relationship between map measurements and actual distances on the Earth. By laying a ruler along the scale bar, you can calculate how many miles or kilometers are represented per inch or centimeter. The map projection used to create the map is often listed near the scale bar.

Compass Rose

The **compass rose** indicates directions. The four **cardinal directions**—north, south, east, and west—are usually indicated with arrows or the points of a star. The **intermediate directions**—northeast, northwest, southeast, and southwest—may also be shown.

Cities

Cities are represented by a dot. Sometimes the relative sizes of cities are shown using dots of different sizes.

Capitals

National capitals are often represented by a star within a circle.

Boundary Lines

On political maps of large areas, boundary lines highlight the borders between different countries and states.

Using Scale

All maps are drawn to a certain scale. **Scale** is a consistent, proportional relationship between the measurements shown on the map and the measurement of the Earth's surface.

Small-Scale Maps A small-scale map, like this political map of France, can show a large area but little detail. Note that the scale bar on this map indicates that about 1 inch is equal to 200 miles.

Large-Scale Maps A large-scale map, like this map of Paris, can show a small area with a great amount of detail. Study the scale bar. Note that the map measurements correspond to much smaller distances than on the map of France.

France: Political

U.K.
English Channel
Lille
BELG.
LUX.
GERMANY
Paris
Nantes
Orléans
ATLANTIC OCEAN
FRANCE
SWITZ.
Bay of Biscay
Lyon
ITALY
Bordeaux
Toulouse
Nice
Corsica
Marseille
ANDORRA
SPAIN
National boundary
Regional boundary
National capital
Major city
0 200 kilometers
0 200 miles
Lambert Azimuthal Equal-Area projection

The City of Paris

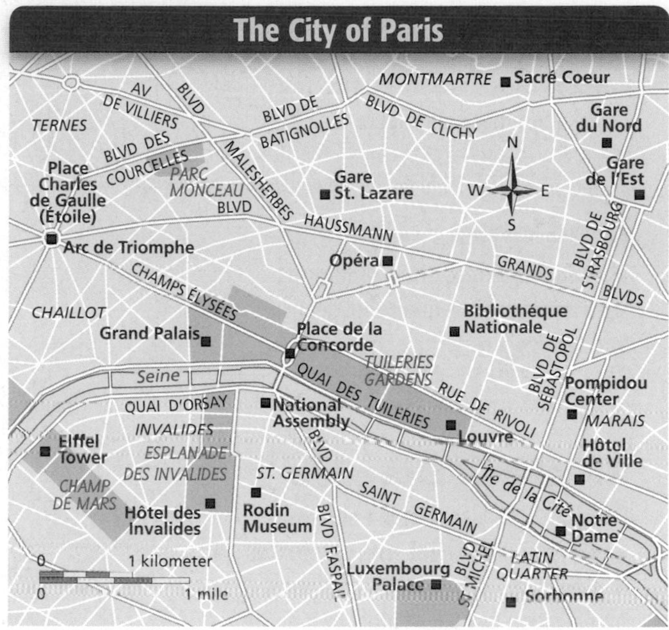

Absolute and Relative Location

As you learned on page 8, absolute location is the exact point where a line of latitude crosses a line of longitude. Another way to indicate location is by **relative location,** or the location of one place in relation to another. To find relative location, find a reference point—a location you already know—on a map. Then look in the appropriate direction for the new location. For example, locate Paris (your reference point) on the map of France above. The relative location of Lyon can be described as southeast of Paris.

PRACTICING THE SKILL

1. **Explain** the significance of: key, compass rose, cardinal directions, intermediate directions, scale bar, scale, relative location.
2. **Describe** the elements of a map that help you interpret the information displayed on the map.
3. **How** does the scale bar help you determine distances on the Earth's surface?
4. **Describe** the relative location of your school in two ways.
5. **Answering the Essential Question** Use a Venn diagram to identify the similarities and differences between small-scale maps and large-scale maps.

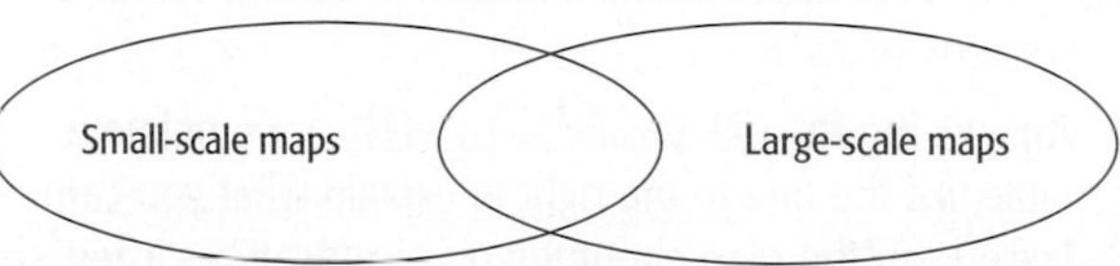

THE WORLD

Physical Maps

A **physical map** shows the location and the **topography,** or shape, of the Earth's physical features. A study of a country's physical features often helps to explain the historical development of the country. For example, mountains may be barriers to transportation, and rivers and streams can provide access into the interior of a country.

Water Features

Physical maps show rivers, streams, lakes, and other water features.

Landforms

Physical maps may show landforms such as mountains, plains, plateaus, and valleys.

Relief and Elevation

Physical maps use shading and texture to show general **relief**—the differences in **elevation,** or height, of landforms. An elevation key uses colors to indicate specific measured differences in elevation above sea level.

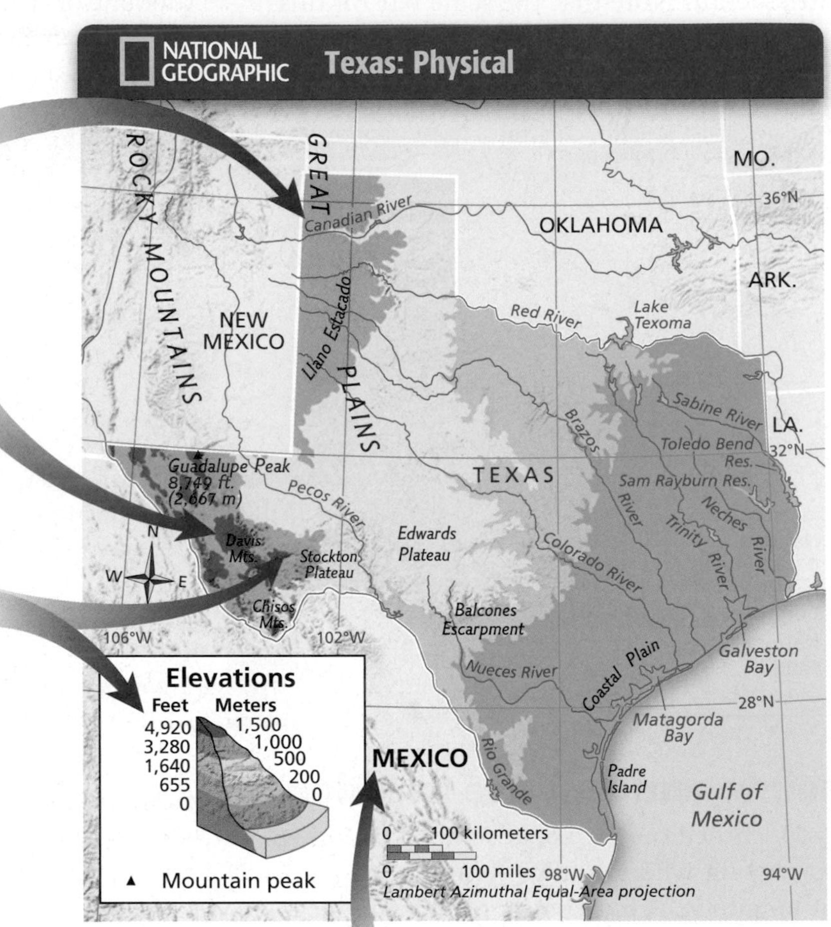

Political Features

Some physical maps also show political features such as boundary lines, countries, and states.

PRACTICING THE SKILL

1. **Explain** the significance of: physical map, topography, relief, elevation.
2. **What** is the approximate elevation of central Texas? Of western Texas?
3. **Answering the Essential Question** Complete a table like the one to the right to explain what you can learn from the map about the physical features listed.

Physical Feature	What You Can Learn from the Map
Davis Mountains	
Red River	
Gulf Coastal Plains	

Political Maps

A **political map** shows the boundaries and locations of political units such as countries, states, counties, cities, and towns. Many features depicted on a political map are **human-made,** or determined by humans rather than by nature. Political maps can show the networks and links that exist within and between political units.

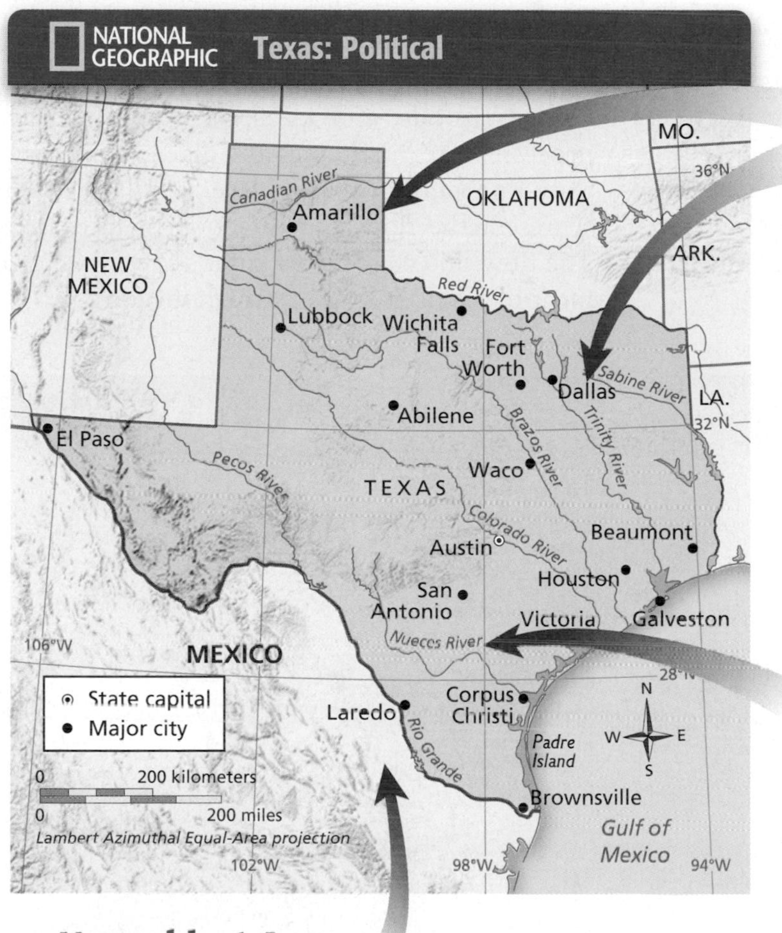

Human-Made Features

Political maps show human-made features such as boundaries, capitals, cities, roads, highways, and railroads.

Physical Features

Political maps may show some physical features such as relief, rivers, and mountains.

Nonsubject Area

Areas surrounding the subject area of the map are usually a different color to set them apart. They are labeled to give you a context for the area you are studying.

PRACTICING THE SKILL

1. **Explain** the significance of: political map, human-made.
2. **What** types of information would you find on a political map that would not appear on a physical map?
3. **Answering the Essential Question** Complete a table like the one to the right to explain what you can learn from the map about the human-made features listed.

Human-Made Feature	What You Can Learn from the Map
Austin	
El Paso	
Texas state boundary	

THE WORLD

Thematic Maps

Maps that emphasize a single idea or a particular kind of information about an area are called **thematic maps.** There are many kinds of thematic maps, each designed to serve a different need. This textbook includes thematic maps that show climate, natural vegetation, population density, and economic activities.

Qualitative Maps

Maps that use colors, symbols, lines, or dots to show information related to a specific idea are called **qualitative maps.** Such maps are often used to depict historical information. For example, the qualitative map below left shows resources and exports in Latin America over time.

Flow-Line Maps

Maps that illustrate the movement of people, animals, goods, and ideas, as well as physical processes like hurricanes and glaciers, are called **flow-line maps.** Arrows are usually used to represent the flow and direction of movement. The flow-line map below right shows the movement of Slavic peoples throughout Europe.

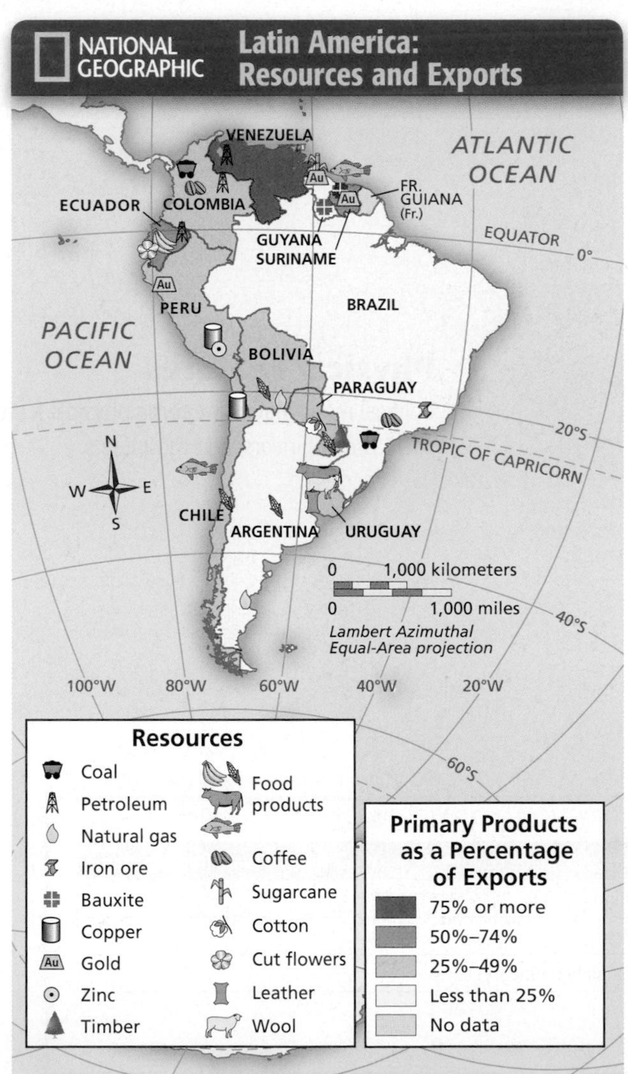

Geographic Information Systems

Modern technology has changed the way maps are made. Most cartographers use computers with software programs called geographic information systems (GIS). A GIS is designed to accept data from different sources—maps, satellite images, printed text, and statistics. The GIS converts the data into a digital code, which arranges it in a database. Cartographers then program the GIS to process the data and produce maps. With GIS, each kind of information on a map is saved as a separate electronic layer. This modern technolgy allows cartographers to make maps—and change them—quickly and easily.

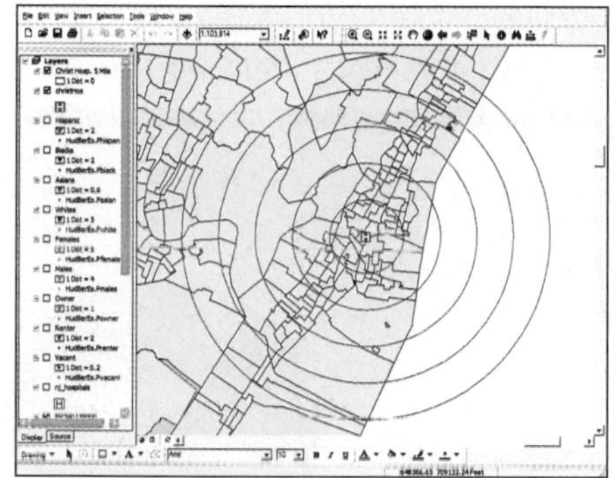

1 *The first layer of information in a GIS pinpoints the area of interest. This allows the user to see, in detail, the area he or she needs to study. In this case, the area of study is a 5-mile (8-km) radius around Christ Hospital in Jersey City, New Jersey.*

2 *Additional layers of information are added based on the problem or issue being studied. In this case, hospital administrators want to find out about the population living near the hospital so they can offer the community the services it needs. A second layer showing African Americans who live within the 5-mile (8-km) radius has been added to the GIS.*

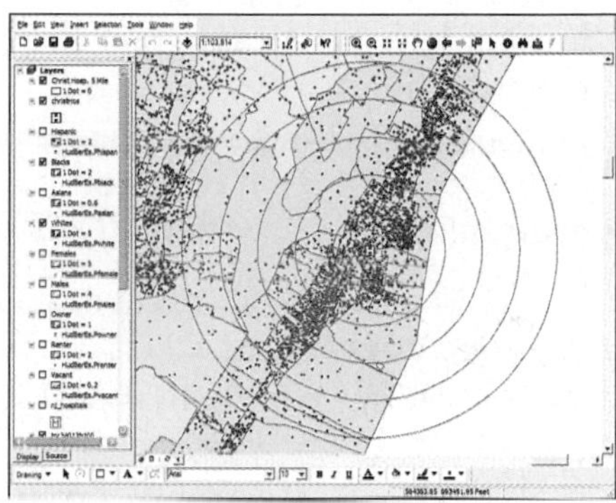

3 *Complex information can be presented using more than one layer. For example, the hospital's surrounding neighborhoods include other groups in addition to African Americans. A third layer showing whites who live within the 5-mile (8-km) radius has been added to the GIS. Administrators can now use this information to help them make decisions about staffing and services associated with the hospital.*

THE WORLD

PRACTICING THE SKILL

1. **Explain** the significance of: thematic maps, qualitative maps, flow-line maps.
2. **Which** type of thematic map would best show natural vegetation regions in Europe?
3. **Which** type of thematic map would best show trade routes between the United States, Canada, and Mexico?
4. **How** does GIS allow cartographers to create maps and make changes to maps quickly and easily?
5. **Answering the Essential Question** Complete a chart like the one below by identifying three examples of each type of thematic map found in this textbook. Note the page numbers of each.

Qualitative Maps	Flow-Line Maps

SECTION 2

 section audio  spotlight video

The Geographer's Craft

Guide to Reading

Essential Question
What are the elements of geography?

Content Vocabulary
- site *(p. 18)*
- situation *(p. 18)*
- place *(p. 18)*
- region *(p. 18)*
- formal region *(p. 18)*
- functional region *(p. 18)*
- perceptual region *(p. 18)*
- ecosystem *(p. 19)*
- movement *(p. 19)*
- human-environment interaction *(p. 19)*
- cartography *(p. 20)*
- geographic information systems (GIS) *(p. 21)*

Academic Vocabulary
- occur *(p. 17)*
- traditional *(p. 18)*
- aspect *(p. 19)*
- obtain *(p. 21)*
- alter *(p. 22)*
- assist *(p. 24)*

Reading Strategy
Organizing As you read about the work of geographers, complete a graphic organizer similar to the one below by listing the specialized research methods geographers use.

Research Methods

Geography is more than just learning place-names. It also has practical uses. For example, ecologist J. Michael Fay conducted a flyover to identify the physical changes and human impact on the African continent.

NATIONAL GEOGRAPHIC **Voices Around the World**

"The next morning we were airborne over Kruger National Park. This was the kind of thing every boy dreams of, cruising at low level over an African game park in your own little Cessna. . . . The land was covered with animal trails and water holes were heavily trodden by elephants and buffalo. It was very clear that nature was intact here. . . .

As we neared the western boundary, I could see a line along the border of the park. . . . Elephants, rhinos, and lions ruled one side while humans dominated the other. . . . [A]s human populations grow, they tend to move closer to the artificial boundaries of protected areas until one day somebody has to put up a fence."

—J. Michael Fay,
Africa Megaflyover: Air Dispatches,
National Geographic (online),
June 14, 2004

J. Michael Fay

The Elements of Geography

MAIN Idea Geographers study the location of people and places on Earth's surface and the patterns in which they are arranged.

GEOGRAPHY AND YOU Is your community located near a river or in the mountains? Why do you think this is so? Read to learn how the study of geography can help you understand the world around you.

Geographers study the Earth's physical and human features and the interactions of people, places, and environments. They search for patterns in these features and interactions, seeking to explain how and why they exist or occur. In their work, geographers consider six elements: the world in spatial terms, places and regions, physical systems, human systems, environment and society, and the uses of geography.

The World in Spatial Terms

Spatial relationships are the links people and places have to one another because of their locations. For geographers, location, or a specific place on the Earth, is a reference point in the same way that dates are reference points for historians.

One way of locating a place is by describing its absolute location the exact spot at which the place is found on the Earth. To determine absolute location, geographers use a network of imaginary lines around the Earth.

Remember that the Equator, the Prime Meridian, and other lines of latitude and longitude cross one another to form a grid system. Using the grid, you can name the absolute location of any place on Earth. This location is stated in terms of *latitude*, degrees north or south of the Equator, and *longitude*, degrees east or west of the Prime Meridian. For example, Dallas, Texas, is located at latitude 32° N (north) and longitude 96° W (west).

Although absolute location is useful, most people locate a place in relation to other places, known as its relative location. For example, New Orleans is located near the mouth of the Mississippi River. Knowing the relative location of a place helps you create mental maps to orient yourself in space and to develop an awareness of the world around you.

Learn how to find absolute location and relative location by following the steps below.

Absolute Location

1. To find the absolute location of Indianapolis, first identify the line of latitude that runs near the city. This is 40° N.
2. Then identify the line of longitude that runs near Indianapolis. This is 86° W.
3. Finally, write the location of Indianapolis using latitude and longitude. This is latitude 40° N, longitude 86° W.

Relative Location

4. To find the relative location of Lansing, identify places such as cities, lakes, rivers, and states near Lansing. Unlike absolute location, relative location can be described in many ways.
5. For example, Lansing is northwest of Ann Arbor.
6. Lansing is also east of Lake Michigan.

1. **Location** What is the absolute location of Chicago? How did you determine it?
2. **Place** Describe the relative location of Chicago in two different ways.
3. **Location** What is the absolute location of Springfield? The relative location?
4. **Place** Describe the relative location of Madison using physical features.

Maps in MOtion Use **StudentWorks™ Plus** or glencoe.com.

THE WORLD

Using the concepts of absolute location and relative location, geographers make a distinction between the site and situation of a place. **Site** refers to the specific location of a place, including its physical setting. For example, the site of San Francisco is its location at the end of a peninsula, surrounded by the Pacific Ocean and San Francisco Bay. **Situation** is an expression of relative location. It refers to the geographic position of a place in relation to other places and its connections to other regions. San Francisco's situation is as a port city on the Pacific coast, close to California's agricultural lands.

Places and Regions

A **place** is a particular space with physical and human meaning. Every place on Earth has its own unique characteristics, determined by the surrounding environment and the people who live there. One task of geographers is to understand and explain how places are similar to and different from one another. To interpret the Earth's complexity, geographers often group places into **regions,** or areas with similar characteristics. The defining characteristics of a region may be physical, such as climate, landforms, soils, vegetation, and animal life. A region may also be defined by human characteristics. These may include language, religion, political systems, economic systems, and population distribution.

Geographers identify three types of regions: formal, functional, and perceptual. A **formal region** is defined by a common characteristic, such as a product produced there. The Corn Belt—a band of farmland from Ohio to Nebraska in the United States—is a formal region because corn is its major crop. A **functional region** is a central place and the surrounding area linked to it. Metropolitan areas, as well as smaller cities and towns, are functional regions. A **perceptual region** is defined by popular feelings and images rather than by objective data. For example, the term "heartland" refers to a central area in which **traditional** values are believed to predominate.

NATIONAL GEOGRAPHIC *Some traditional Hawaiian beliefs are used to explain the flow of fiery lava from a volcano.*

Human-Environment Interaction How does volcanic activity shape the Earth's surface and impact human activities?

Physical Systems and Human Systems

Geography covers a broad range of topics. To make their work easier, geographers divide their subject area into different branches. The two major branches are physical geography and human geography. *Physical geography* focuses on the study of the Earth's physical features. It looks at climate, land, water, plants, and animal life in terms of their relationships to one another and to humans. *Human geography,* or cultural geography, is the study of human activities and their relationship to the cultural and physical environments. It focuses on political, economic, social, and cultural factors, such as population growth, urban development, and economic production and consumption.

Physical geography and human geography are further divided into smaller subject areas. For example, climatology is the study of weather, climate, long-term conditions of the atmosphere, and their impact on ecology and society. Historical geography is the study of places and human activities over time and the geographic factors that have shaped them.

Geographers analyze how certain natural phenomena, such as volcanoes, hurricanes, and floods, shape the Earth's surface. A Hawaiian uses traditional beliefs to describe the fascinating force of a volcano:

> *"We don't see her work as destruction but as cleansing. She's a creator. When she comes through, she wipes the land clean and leaves us new fertile ground. We don't get mad. It is all hers to begin with."*
>
> —Jennifer S. Holland, "Red Hot Hawaii," *National Geographic,* October 2004

Geographers study how physical features interact with plants and animals to create, support, or change ecosystems. An **ecosystem** is a community of plants and animals that depend upon one another, and their surroundings, for survival.

Geographers also examine how people shape the world—how they settle the Earth, form societies, and create permanent features. A recurring theme in geography is the ongoing **movement** of people, goods, and ideas. For example, new people entering a long-established society usually bring different ideas and practices that may transform that society's existing culture. In studying human systems, geographers look at how people compete or cooperate to change or control **aspects** of the Earth to meet their needs.

NATIONAL GEOGRAPHIC *Farmers in many countries alter the natural environment by creating terraced fields.*

Human-Environment Interaction How have people changed the physical environment in your community?

Environment and Society

Human-environment interaction, or the study of the interrelationship between people and their physical environment, is another theme of geography. Geographers examine the ways people use their environment, how and why they change it, and what consequences result from these changes. In some cases the physical environment affects human activities. For example, mountains and deserts often pose barriers to human movement. In other instances human activities, such as building a dam, cause changes to the physical environment. By understanding how the Earth's physical features and processes shape and are shaped by human activity, geographers help societies make informed decisions.

The Uses of Geography

Geography provides insight into how physical features and living things developed in the past. It also interprets current trends to plan for future needs. Governments, businesses, and individuals use geographic information in planning and decision making. Data on physical features and processes can determine whether a site is suitable for human habitation or has resources worth developing. Geographic information on human activities, such as population growth and migration, can help planners decide whether to build new schools or highways in a particular place. As geographers learn more about the relationships among people, places, and the environment, their knowledge helps us plan and build a better future.

READING Check **Location** How is absolute location different from relative location?

NATIONAL GEOGRAPHIC *Through direct observation, a scientist studies ice cores to learn about patterns of global warming and cooling.*

Location Why is direct observation in areas like Antarctica important?

Research Methods

MAIN Idea Geographers use different research methods to conduct their work.

GEOGRAPHY AND YOU How do you prepare to write a research report? Read to learn how geographers organize and study geography.

Geographers use specialized research methods in their work. These methods include direct observation, mapping, interviewing, statistics, and the use of technology.

Direct Observation

Geographers use direct observation to study the Earth and the patterns of human activities that take place on its surface. They will often visit a place to gather specific information about it and its geographic features. Geographers also employ remote sensing to study the Earth, using aerial photographs and satellite images. For example, aerial photographs or satellite images can be used to locate mineral deposits or to determine the size of freshwater sources.

Mapping

Maps are essential to geographers. Specialists who make and design maps are known as cartographers. Their area of work, known as **cartography,** involves designing and making maps.

Many findings from geographic research can be shown on maps better than they can be explained in written text. Cartographers select complex pieces of information about an area and present them in a more understandable form on a map. In this way they show the location, features, patterns, and relationships of people, places, and things. In addition, maps allow a visual comparison between places and regions. For example, a geographer might compare population density maps of two counties in order to determine where to build new schools.

Interviewing

To answer a geographic question, geographers must often go beyond observation. In many cases, they want to find out how people think or feel about certain places. They also may want to examine the ways in which people's beliefs and attitudes have affected the physical environment.

NATIONAL GEOGRAPHIC **Skills for Thinking Like a Geographer**

Skill	Examples	Tools and Technologies
Asking Geographic Questions–helps you pose questions about your surroundings	• Why has traffic increased along this road? • What should be considered when building a new community sports facility?	• Maps • Globes • Internet • Remote sensing • News media
Acquiring Geographic Information–helps you answer geographic questions	• Compare aerial photographs of a region over time. • Design a survey to determine who might use a community facility.	• Direct observation • Interviews • Reference books • Satellite images • Historical records
Organizing Geographic Information–helps you analyze and interpret information you have collected	• Compile a map showing the spread of housing development over time. • Summarize information obtained from interviews.	• Field maps • Databases • Statistical tables • Graphs • Diagrams • Summaries
Analyzing Geographic Information–helps you look for patterns, relationships, and connections	• Draw conclusions about the effects of road construction on traffic patterns. • Compare information from different maps that show available land and zoning districts.	• Maps • Charts • Graphs • GIS • Spreadsheets
Answering Geographic Questions–helps you apply information to real-life situations and problem solving	• Present a report conveying the results of a case study. • Suggest locations for a new facility based on geographic data gathered.	• Sketch maps • Reports • Research papers • Oral or multimedia presentations

CHART STUDY

One of the most important geographic tools is the ability to think geographically. The five skills identified above are key to geographic understanding.

1. **Place** What types of information can you gain from a map that would help you ask questions about why traffic has increased along a specific road?
2. **Regions** Why are the news media and the Internet important tools for geographers?

Such information is **obtained** by interviewing. Geographers choose a particular group of people for study. Instead of contacting everyone in that group, however, geographers talk to a carefully chosen sample whose answers represent the whole group.

Analyzing Statistics

Some of the information geographers use is numerical. Temperature and rainfall data indicate a region's climate, for example. Geographers use computers to organize and present this information. They also analyze the data to find patterns and trends. For example, census data can be studied to learn about the age, ethnic, and gender makeup of the population. After identifying these patterns and trends, geographers use statistical tests to see whether their ideas are valid.

Using Technology

As noted in the chart above, geographers often use scientific instruments in their work. They especially depend on advanced technological tools, such as satellites and computers. Satellites orbiting the Earth carry remote sensors, high-tech cameras, and radar that gather data and images related to the Earth's environment, weather, human settlement patterns, and vegetation. **Geographic information systems** (**GIS**) are computer tools that process and organize data and satellite images with other types of information gathered by geographers and other scientists. GIS technology can be used for many purposes. For example, urban planners use it to help determine where to build roads. Biologists use it to monitor wildlife populations in a specific area. Public safety officials use it to pinpoint safe and efficient evacuation routes from hurricane paths.

The development of computer technology has also transformed the process of mapmaking. Today, most cartographers rely on computers and computer software to make maps. Each type of data on a map is kept as a separate "layer" in the map's digital files. This method allows cartographers to make and change maps quickly and easily.

READING **Check** **Location** How has technology changed the way maps are created?

Geography and Other Subjects

MAIN Idea Geography is related to other subject areas such as history and economics.

GEOGRAPHY AND YOU Do you use math when figuring out a science problem? Read to learn how geographers use knowledge from other subject areas to understand the world around them.

Geographers use geographic tools and methods to understand historical patterns, politics and political patterns, societies and culture and their impact on landscape, and economies.

Past Environments and Politics

Geographers use historical perspectives to understand what places could have looked like in the past. For example, geographers might want to know about the changes that have occurred in Boston, Massachusetts, over the past two centuries. They may begin by gathering information about different time periods in the city's history. This information could be used to answer such geographic questions as: How have human activities changed the natural vegetation? Are the waterways different today than they were in the past? Answers to such questions help people make informed decisions about the present and better plans for the future.

Geographers study political patterns to see how people in different places are governed. They look at how political boundaries have formed and changed. Geographers are also interested in how the natural environment has influenced political decisions and how governments change natural environments. For example, in the 1960s the Egyptian government built the massive Aswān High Dam on the Nile River to help irrigate the land. The dam **altered** the Nile River valley in profound ways and has had a significant impact on the region's people.

Society and Culture

Human geographers, or cultural geographers, use the tools of sociology and anthropology to understand cultures around the world. They study the relationships between physical environment and social structures. They examine people's ways of life in different parts of the world. Human geographers also seek to understand how the activities of different groups affect physical systems and how the physical systems affect human systems differently.

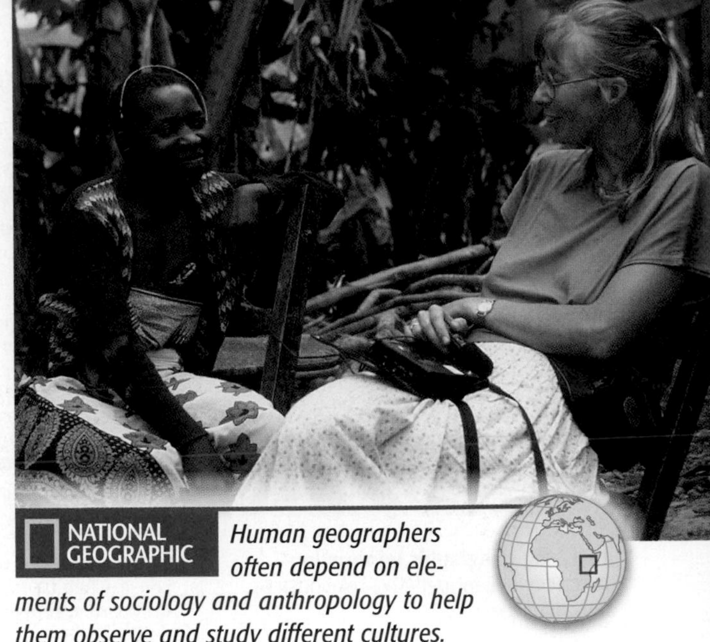

NATIONAL GEOGRAPHIC *Human geographers often depend on elements of sociology and anthropology to help them observe and study different cultures.*

Human-Environment Interaction How can discussion with an area's inhabitants evolve into a greater understanding of a region's people?

Economies

Geographers study economies to understand how the locations of resources affect the ways people make, transport, and use goods, and how and where services are provided. Geographers are interested in how locations are chosen for various economic activities, such as farming, mining, manufacturing, and trade. A desirable location usually includes plentiful resources and good transportation routes.

Geographers are also interested in the interdependence of people's economic activities around the world. New developments in communications and transportation—such as the Internet, cellular phones, and overnight air delivery—make the movement of information and goods faster and more efficient than ever before. With such ever-changing technology, a modern business can operate globally without depending on any one place to fill all of its needs.

READING Check **Human-Environment Interaction** Why do geographers study economies?

Geography as a Career

MAIN Idea Geography skills are useful in a variety of jobs and work environments.

GEOGRAPHY AND YOU What skills do you use when learning how to get to the library? Read to learn how geography is used in the workplace.

Although people trained in geography are in great demand in the workforce, many of them do not have *geographer* as a job title. Geography skills are useful in so many different situations that geographers have more than a hundred different job titles. Geographers work in a variety of jobs in government, business, and education. They often combine the study of geography with other areas of study. For example, an ecologist must know the geographic characteristics of a place or region in which he or she studies living organisms. Also, a travel agent must have knowledge of geography in order to plan trips for clients.

One broad cluster of career opportunities in geography is teaching and education. Teaching opportunities exist at all levels—from elementary school to high school to university. Teachers with some background and training in geography are in demand in elementary and high schools in the United States. At the university level, teaching and research focused on specific branches of geography have been established for decades. Students with formal geographic training from a university have found work in many different industries.

Because geography itself has many specialized fields, there are many different kinds of geographers. Those with knowledge of physical geography work as climate and weather experts and soil scientists. Recently, geographers with training in environmental studies have been in demand for work as environmental managers and technicians. Such work includes assessing the environmental impact of proposed development projects on air, water quality, and wildlife. They also prepare the environmental impact report that is often required before construction can begin.

Human geographers find work in many areas, including health care, transportation, population

NATIONAL GEOGRAPHIC Storm chasers use technology to record, measure, and document the intensity of storms.

Human-Environment Interaction Aside from dramatic images, what other information can the work of storm chasers provide?

studies, economic development, and international economics. Some human geographers with a background in urban planning are hired as planners in local and state government agencies. They focus on housing and community development, park and recreation planning, and urban and regional planning. Planners map and analyze land use and transportation systems, and monitor urban land development.

Geographers who specialize in a specific branch of geography—such as economic geography or regional geography—also find jobs outside of the university setting. For example, an economic geographer examines human economic activities and their relationship to the environment. He or she may work at such tasks as market analysis and site selection for stores, factories, and restaurants. A regional geographer studies the features of a particular region and may **assist** government and businesses in making decisions about land use. Geographers also find employment as writers and editors for publishers of textbooks, maps, atlases, and news and travel magazines.

READING Check **Place** Why are there many different types of geographers?

NATIONAL GEOGRAPHIC *Geography skills are used to survey the land near a construction site.*

Human-Environment Interaction In what other types of situations may geography skills be necessary?

Geography ONLINE

Student Web Activity Visit glencoe.com, select the *World Geography and Cultures* Web site, and click on Student Web Activities—Chapter 1 for an activity about careers in geography.

SECTION 2 REVIEW

Vocabulary

1. Explain the significance of: site, situation, place, region, formal region, functional region, perceptual region, ecosystem, movement, human-environment interaction, cartography, geographic information systems (GIS).

Main Ideas

2. Describe the research methods geographers use.
3. What other subject areas is geography related to?
4. List examples of jobs and work environments in which geography skills are useful.
5. Use a table like the one below to describe the elements geographers use to study people and places on Earth's surface.

The Elements of Geography	
The World in Spatial Terms	
Places and Regions	
Physical Systems	
Human Systems	
Environment and Society	
The Uses of Geography	

Critical Thinking

6. **Answering the Essential Question** How does the study of human systems help geographers in their work as countries become increasingly interdependent?
7. **Comparing and Contrasting** Think about the physical and human characteristics that constitute a region. Identify the differences and similarities among formal, functional, and perceptual regions.
8. **Analyzing Visuals** Study the physical map of the United States on pages RA16–RA17. What kinds of information can you learn from this map? How does the information on this map differ from the political map on pages RA18–RA19?

Writing About Geography

9. **Expository Writing** As a geographer working on a plan for a new community center, what research methods would you use? Explain your choices in a paragraph.

Geography ONLINE

Study Central To review this section, go to glencoe.com and click on Study Central.

What Is Geography?

- Geography is the study of the location of people and places and the patterns in which they are arranged on Earth.
- Physical geography focuses on the physical features and processes of Earth.
- Human geography focuses on the political, economic, or cultural characteristics of human populations.
- An important element in geography is the interaction between people and their environment. Geographers try to understand how Earth's physical environment shapes and is shaped by human activities.

A researcher takes ice core samples to study climate change.

The Geographer's Tools

- To understand geography, you first need to understand how maps work.
- Three-dimensional globes are the most accurate depiction of Earth. For example, the shortest distance between two places is not always a straight line but a great circle. This is due to the curvature of the Earth.
- Flat maps use one of several types of projections that distort Earth's features in some way.
- It is important to understand how each projection distorts Earth so you can pick the map projection that best suits your needs.

Types of Maps

- Maps can be used to show many different types of information.
- Most maps show location. The two most common types of maps show the location and physical features of a place, or the location and political boundaries.
- Some other types of maps are qualitative maps and flow-line maps. These are useful when dealing with historical information or when trying to show movement.

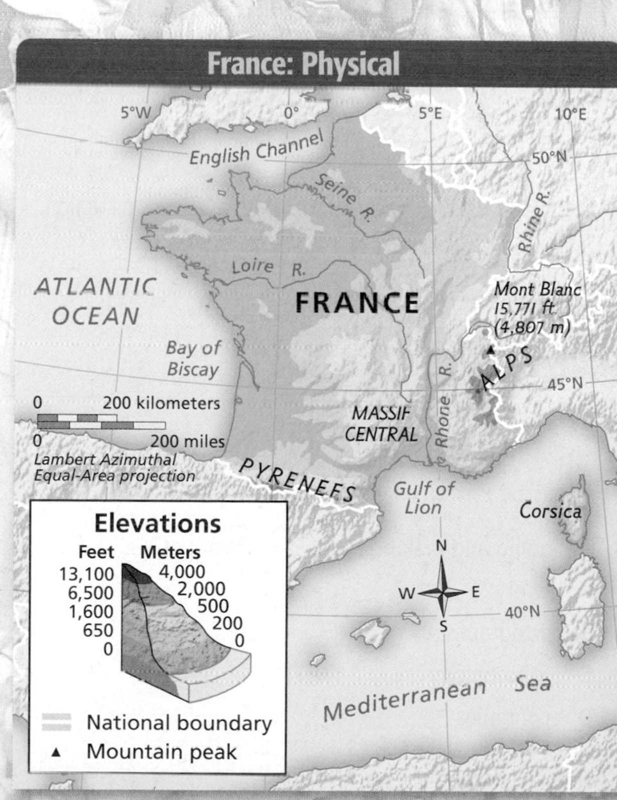

CHAPTER 1

STANDARDIZED TEST PRACTICE

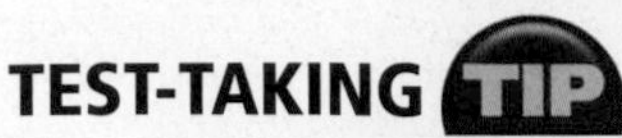

Be sure to read all the choices before you answer a question so that you are sure that you have picked the *best* choice.

Reviewing Vocabulary

Directions: Choose the word or words that best complete the sentence.

1. To draw a map of the round Earth on a flat surface, mapmakers use ________.

A absolute locations
B relative locations
C map projections
D hemispheres

2. Lines of ________ are drawn on maps from the North Pole to the South Pole to measure distance east and west.

A latitude
B parallels
C location
D longitude

3. A city and its surroundings make up a ________ region.

A functional
B perceptual
C formal
D virtual

4. Designing and making maps is the work of ________.

A history
B cartography
C anthropology
D economies

Reviewing Main Ideas

Directions: Choose the best answers to complete the sentences or to answer the following questions.

Section 1 *(pp. 4–15)*

5. The shortest distance between two places on the Earth follows a ________.

A great circle
B straight line on a map
C scenic route
D map projection

6. ________ location is a place's global address.

A Relative
B Global
C Absolute
D Cartographic

Section 2 *(pp. 16–24)*

7. To study and interpret the Earth's complexity, geographers often divide it into ________.

A projections
B regions
C places
D situations

8. In studying the Earth and its people, geographers use ________.

A direct observation
B interviews
C technology
D all of these

GO ON

Critical Thinking

Directions: Choose the best answers to complete the sentences or to answer the following questions.

9. How has the geographer's craft changed over the last 100 years?

A The world has literally become smaller.

B Maps are no longer as important as they once were.

C Technology has provided computers and views from the air and from space.

D Interviewing is used less often to learn about and compare cultures.

Base your answer to question 10 on the map and on your knowledge of Chapter 1.

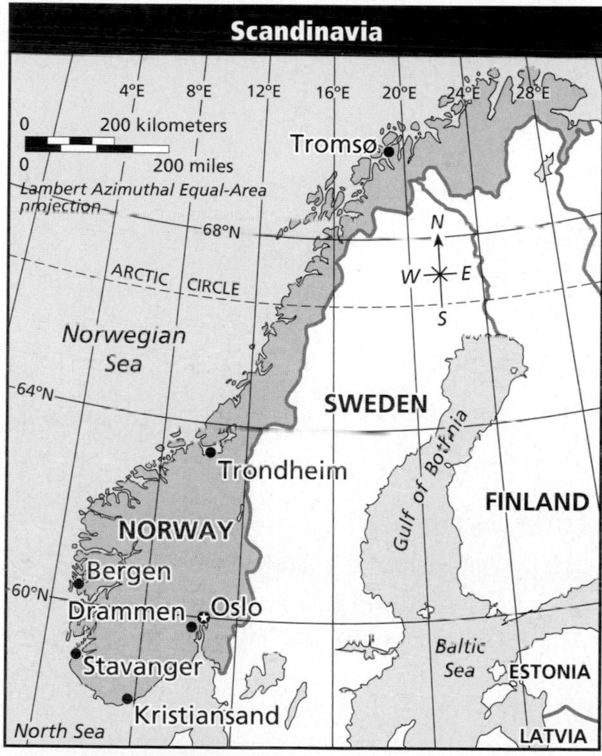

10. The absolute location of Oslo is ________.

A 60° south and 10° west

B 60° west and 10° north

C 60° north and 10° east

D 10° east and 60° south

Document-Based Questions

Directions: Analyze the document and answer the short-answer questions that follow the document.

Geography is increasingly important because our ties to the rest of the world increase every day. Here is what geographer George J. Demko said about geography:

> *Geography—real-world geography—is the art and science of location, or place. It is about spatial patterns and spatial processes. It is about which way the wind blows from Chernobyl, the Pacific "ring of fire," AIDS, terrorists, and refugees. It is about acid rain, El Niño, ocean dumping, cultural censorship, droughts and famines. . . .*
>
> *Real-world geography also explores things in locations: why something is where it is and what processes change its distribution. Geography is the why of where of an ever-changing universe. Its surpassing objective is to discover the processes that move over space and connect places and continually transform the location and character of everything.*
>
> —George J. Demko,
> *Why in the World: Adventures in Geography*

11. What are some of the world issues that Dr. Demko lists as concerns of geography?

12. According to the excerpt, what is the main objective of geography?

Extended Response

13. Why is it important for geographers to use a variety of methods to study the Earth?

14. Exploring the BIG Idea
How are each of the six essential elements used in the study of geography?

For additional test practice, use Self-Check Quizzes—Chapter 1 on glencoe.com.

Need Extra Help?

If you missed questions. . .	1	2	3	4	5	6	7	8	9	10	11	12	13	14
Go to page. . .	6	8	18	20	5	8	18	20	10	8	20	21	22	17

CHAPTER 2

The Physical World

BIG Idea

Physical processes shape Earth's surface. Understanding that Earth is part of a larger physical system called the solar system helps us understand how life on our planet is possible. Earth's physical systems are affected by natural forces such as earthquakes and volcanoes that can influence human activities on the planet.

Essential Questions

Section 1: Planet Earth
As a physical system, what makes Earth suitable for plant and animal life?

Section 2: Forces of Change
How have internal and external forces shaped Earth's surface?

Section 3: Earth's Water
What physical process keeps Earth's water constant?

Visit glencoe.com and enter QuickPass™ code WGC9952C2 for Chapter 2 resources.

Kilauea, in Hawaii, is one of the most active volcanoes in the world.

FOLDABLES™ Study Organizer

Organizing Information Make a Three-Pocket Book to help you organize information about the physical systems and processes that affect life on Earth.

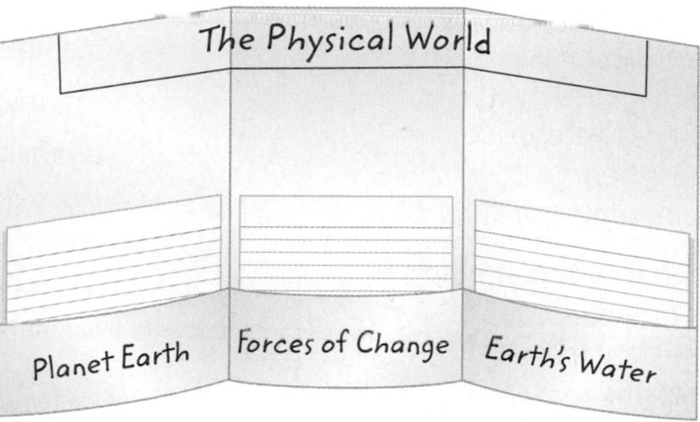

Reading and Writing As you read this chapter, write information on note cards about the structure of planet Earth and its place in the solar system, the natural forces that affect Earth's physical systems, and Earth's water. Place the cards in the correct pocket of your Foldable.

SECTION 1

section audio

spotlight video

Guide to Reading

Essential Question

As a physical system, what makes Earth suitable for plant and animal life?

Content Vocabulary
- hydrosphere *(p. 32)*
- lithosphere *(p. 32)*
- atmosphere *(p. 32)*
- biosphere *(p. 32)*
- continental shelf *(p. 33)*

Academic Vocabulary
- approach *(p. 31)*
- assistance *(p. 32)*
- features *(p. 33)*

Places to Locate
- Isthmus of Panama *(p. 33)*
- Sinai Peninsula *(p. 33)*
- Mount Everest *(p. 33)*
- Dead Sea *(p. 33)*
- Mariana Trench *(p. 33)*

Reading Strategy

Categorizing As you read about Earth, complete a graphic organizer similar to the one below by describing the four components of Earth.

Component	Description
Hydrosphere	
Lithosphere	
Atmosphere	
Biosphere	

Planet Earth

An astronaut, seeing Earth from the blackness of space, described it as "piercingly beautiful." From the vantage point of space, the Earth's great beauty resembles a blue and white marble, with contrasts of water and land beneath huge swirls of white clouds.

NATIONAL GEOGRAPHIC Voices Around the World

"To the ancient Egyptians the heavens were almost close enough to touch—a benign canopy of light and dark held up by mountain peaks. But modern science has exploded that ancient, peaceful mirage, replacing it with . . . change, and processes that sometimes defy human understanding. Guided by leaps of imagination and armed with potent new technologies . . . scientists have . . . claimed the universe itself as a titanic laboratory."

—Kathy Sawyer,
"Unveiling the Universe,"
National Geographic, October 1999

An astronomer with radio telescopes

Our Solar System

MAIN Idea Earth is part of a larger physical system that contains other planets, moons, and stars.

GEOGRAPHY AND YOU Have you ever seen a movie or read a book about outer space? Read to learn how Earth fits into the ever-changing solar system.

Earth is part of our solar system, which includes the sun and the objects that revolve around it. At the center of the solar system is the sun—a star, or ball of burning gases. The sun's enormous mass, or the amount of matter it contains, creates a strong pull of gravity. This basic physical force keeps the Earth and the other objects revolving around the sun.

The Planets

Except for the sun, spheres called planets are the largest objects in the solar system. At least eight planets exist, and each is in its own orbit around the sun. Mercury, Venus, Earth, and Mars are the inner planets, or those nearest the sun. Earth, the third planet from the sun, is about 93 million miles (about 150 million km) away. Farthest from the sun are the outer planets—Jupiter, Saturn, Uranus, and Neptune.

The planets vary in size. Jupiter is the largest. Earth ranks fifth in size, and Mercury is the smallest. All of the planets except Mercury and Venus have moons, smaller spheres or satellites that orbit them. Earth has 1 moon, and Saturn has at least 18 moons. Ceres and Pluto are dwarf planets. *Dwarf planets* are small round bodies that orbit the sun, but have not cleared the area around their orbits of other orbiting bodies.

All of the planets are grouped into two types—terrestrial planets and gas giant planets. Mercury, Venus, Earth, and Mars are called *terrestrial planets* because they have solid, rocky crusts. Mercury and Venus are scalding hot, and Mars is a cold, barren desert. Only Earth has liquid water at the surface and can support varieties of life.

Farther from the sun are the *gas giant planets*—Jupiter, Saturn, Uranus, and Neptune. They are more gaseous and less dense than the terrestrial planets, even though they are larger in diameter. Each gas giant planet is like a miniature solar system, with orbiting moons and thin, encircling rings. Only Saturn's rings, however, are easily seen from Earth by telescope.

Asteroids, Comets, and Meteoroids

Thousands of smaller objects—including asteroids, comets, and meteoroids—revolve around the sun. Asteroids are small, irregularly shaped, planet-like objects. They are found mainly between Mars and Jupiter in the *asteroid belt*. A few asteroids follow paths that cross Earth's orbit.

Comets, made of icy dust particles and frozen gases, look like bright balls with long, feathery tails. Their orbits are inclined at every possible angle to Earth's orbit. They may **approach** from any direction.

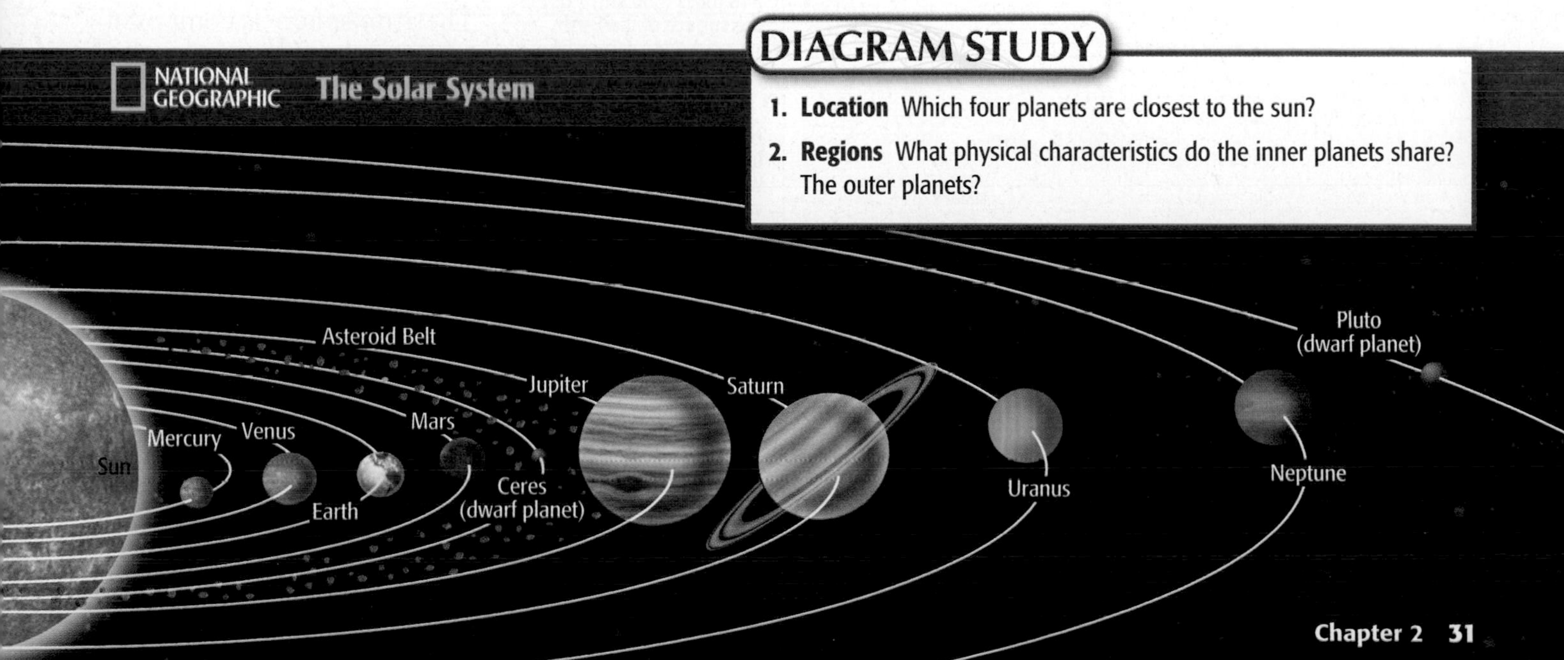

DIAGRAM STUDY

1. **Location** Which four planets are closest to the sun?
2. **Regions** What physical characteristics do the inner planets share? The outer planets?

Meteoroids are pieces of space debris—chunks of rock and iron. When they occasionally enter Earth's atmosphere, friction usually burns them up before they reach the Earth's surface. Those that collide with Earth are called meteorites. Meteorite strikes, though rare, can significantly affect the landscape, leaving craters and causing other devastation. In 1908 a huge area of forest in the remote Russian region of Siberia was flattened and burned by a "mysterious fireball." Scientists theorize it was a meteorite or comet. A writer describes the effects:

> *"The heat incinerated herds of reindeer and charred tens of thousands of evergreens across hundreds of square miles. For days, and for thousands of miles around, the sky remained bright with an eerie orange glow—as far away as western Europe people were able to read newspapers at night without a lamp."*
>
> —Richard Stone, "The Last Great Impact on Earth," *Discover,* September 1996

READING Check **Movement** Besides the planets, what other things revolve around the sun?

Getting to Know Earth

MAIN Idea Earth's surface is a complex mix of landforms and water systems.

GEOGRAPHY AND YOU What do you know about Earth's water, land, and air? Read to learn how these features support life on Earth.

The Earth is a rounded object wider around the center than from top to bottom. Earth has a larger diameter at the Equator—about 7,930 miles (12,760 km)—than from Pole to Pole, but the difference is less than 1 percent. With a circumference of about 24,900 miles (40,060 km), Earth is the largest of the inner planets.

Water, Land, and Air

The surface of the Earth is made up of water and land. About 70 percent of our planet's surface is water. Oceans, lakes, rivers, and other bodies of water make up a part of the Earth called the **hydrosphere.**

About 30 percent of the Earth's surface is land, including continents and islands. Land makes up a part of the Earth called the **lithosphere,** the Earth's crust. The lithosphere also includes the ocean basins, or the land beneath the oceans.

The air we breathe is part of Earth's **atmosphere,** a layer of gases extending above the planet's surface. The atmosphere is composed of 78 percent nitrogen, 21 percent oxygen, and small amounts of argon and other gases.

All people, animals, and plants live on or close to the Earth's surface or in the atmosphere. The part of the Earth that supports life is the **biosphere.** Life outside the biosphere, such as on a space station orbiting Earth, exists only with the **assistance** of mechanical life-support systems.

NATIONAL GEOGRAPHIC **Water, Land, and Air**

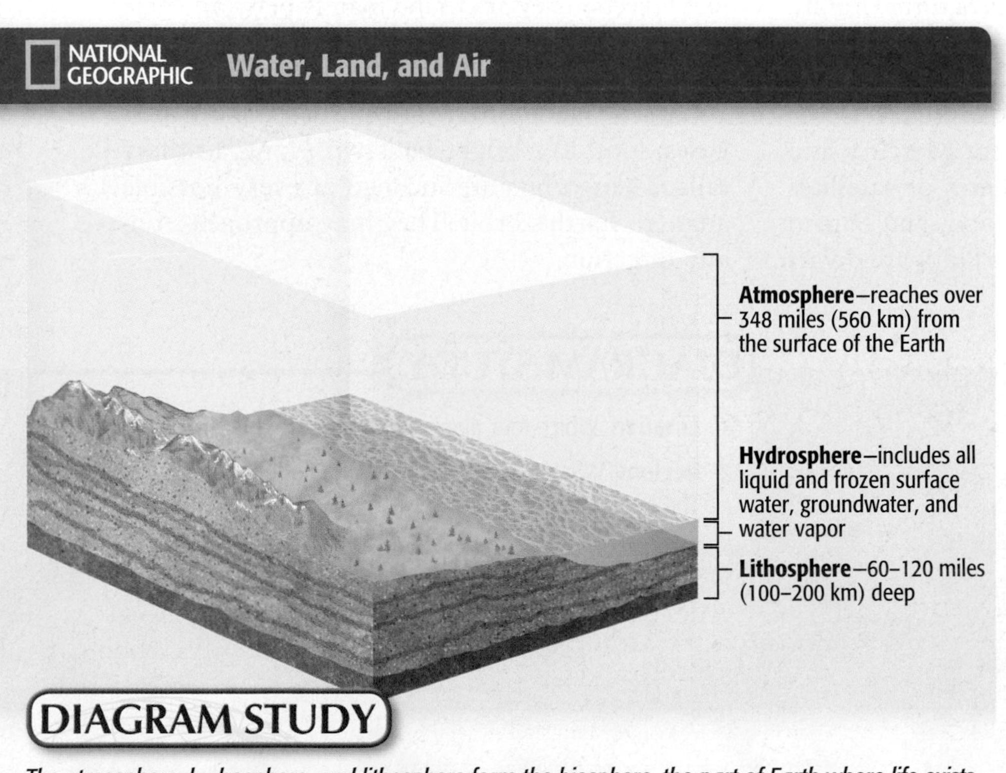

DIAGRAM STUDY

The atmosphere, hydrosphere, and lithosphere form the biosphere, the part of Earth where life exists.

1. **Place** What are Earth's water systems called?
2. **Human-Environment Interaction** How does human activity impact the biosphere?

Landforms

Landforms are the natural **features** of the Earth's surface. So are bodies of water. The diagram on pages RA2–RA3 shows many of the Earth's landforms, which have a particular shape or elevation. Landforms often contain rivers, lakes, and streams.

Underwater landforms are as diverse as those found on dry land. In some places the ocean floor is a flat plain. Other parts feature mountain ranges, cliffs, valleys, and deep trenches.

Seen from space, Earth's most visible landforms are the seven large landmasses called continents. Australia and Antarctica stand alone, while the others are joined in some way. Europe and Asia are parts of one landmass called Eurasia. A narrow strip of land called the **Isthmus of Panama** links North America and South America. At the **Sinai Peninsula,** the human-made Suez Canal separates Africa and Asia.

The **continental shelf** is an underwater extension of the coastal plain. Continental shelves slope out from land for as much as 800 miles (1,287 km) and descend gradually to a depth of about 660 feet (200 m), where a sharp drop marks the beginning of the continental slope. This area drops more sharply to the ocean floor.

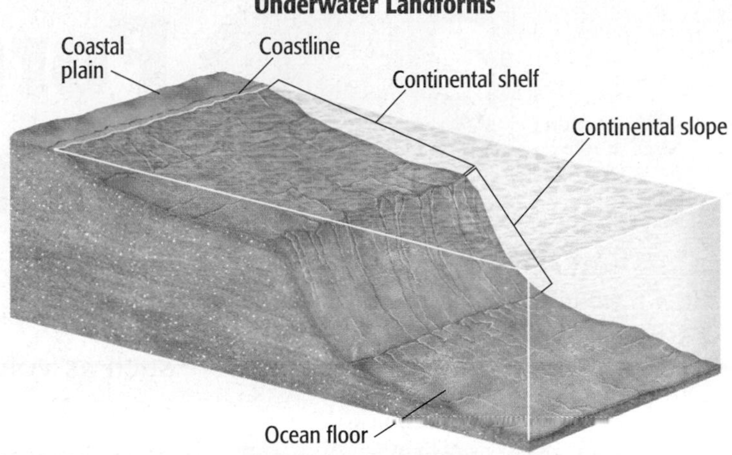

Earth's Heights and Depths

Great contrasts exist in the heights and depths of the Earth's surface. The highest point on Earth is in South Asia at the top of **Mount Everest,** which is 29,028 feet (8,848 m) above sea level. The lowest dry land point, at 1,349 feet (411 m) below sea level, is the shore of the **Dead Sea** in Southwest Asia. Earth's deepest known depression lies under the Pacific Ocean southwest of Guam in the **Mariana Trench,** a narrow, underwater canyon about 35,827 feet (10,920 m) deep.

READING Check **Human-Environment Interaction** How does the biosphere support life on Earth?

THE WORLD

SECTION 1 REVIEW

Vocabulary

1. Explain the significance of: hydrosphere, lithosphere, atmosphere, biosphere, continental shelf.

Main Ideas

2. List examples of Earth's landforms and water systems. How do these features help support life on our planet?
3. How are terrestrial planets and gas giant planets similar? How are they different?
4. Use a diagram like the one below to describe Earth's place in the larger physical system that includes other planets, moons, and stars.

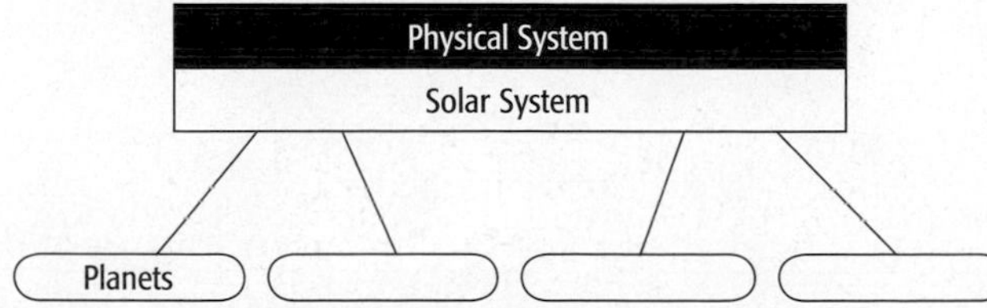

Critical Thinking

5. **Answering the Essential Question** How do the three parts of the biosphere help support life on Earth?
6. **Comparing and Contrasting** How are Earth's underwater landforms similar to the landforms found on dry land? How are they different?
7. **Analyzing Visuals** Study the diagram of the solar system on page 31. How is the size of a planet's orbit influenced by its distance from the sun?

Writing About Geography

8. **Descriptive Writing** Consider the ratio of water and land on Earth. Describe how Earth's landforms and bodies of water would be different if the proportions were reversed.

SECTION 2

 section audio  spotlight video

Guide to Reading

Essential Question

How have internal and external forces shaped Earth's surface?

Content Vocabulary

- core *(p. 35)*
- mantle *(p. 35)*
- crust *(p. 35)*
- continental drift *(p. 35)*
- plate tectonics *(p. 35)*
- magma *(p. 35)*
- subduction *(p. 37)*
- accretion *(p. 37)*
- spreading *(p. 37)*
- fold *(p. 37)*
- fault *(p. 37)*
- faulting *(p. 38)*
- weathering *(p. 39)*
- erosion *(p. 39)*
- glacier *(p. 39)*
- moraine *(p. 39)*

Academic Vocabulary

- releasing *(p. 35)*
- constantly *(p. 35)*
- tension *(p. 38)*

Places to Locate

- Himalaya *(p. 37)*
- San Andreas Fault *(p. 37)*
- Kōbe *(p. 38)*
- San Francisco *(p. 38)*
- Ring of Fire *(p. 38)*
- Greenland *(p. 40)*
- Antarctica *(p. 40)*

Reading Strategy

Taking Notes As you read about the forces that change Earth, use the headings of the section to create an outline like the one below.

Forces of Change
I. Earth's Structure
A.
B.
II. Internal Forces of Change
A.
B.

Forces of Change

The center of Earth is filled with intense heat and pressure. These natural forces drive numerous changes such as volcanoes and earthquakes that renew and enrich Earth's surface. These physical processes can also disrupt, and often destroy, human life. As a result, scientists are working to learn how to predict them.

NATIONAL GEOGRAPHIC Voices Around the World

"[S]cientists are doing everything they can to solve the mysteries of earthquakes. They break rocks in laboratories, studying how stone behaves under stress. They hike through ghost forests where dead trees tell of long-ago tsunamis. They make maps of precarious, balanced rocks to see where the ground has shaken in the past, and how hard. They dig trenches across faults, searching for the active trace. They have wired up fault zones with so many sensors it's as though the Earth is a patient in intensive care."

—Joel Achenbach, "The Next Big One," *National Geographic,* April 2006

A geologist studying earthquake activity

Earth's Structure

MAIN Idea The Earth's internal and external structure, including the tectonic plates, is responsible for the creation of the continents, oceans, and mountain ranges.

GEOGRAPHY AND YOU Are there mountains near where you live? Read to learn how Earth's mountains are formed.

For hundreds of millions of years, the surface of the Earth has been in slow but constant motion. Some forces that change the Earth, such as wind and water, occur on the surface. Others, such as volcanic eruptions and earthquakes, originate deep in the Earth's interior.

A Layered Planet

The diagram at the right shows that Earth is composed of three main layers—the core, the mantle, and the crust. At the very center of the planet is a super-hot but solid inner **core.** Scientists believe that the inner core is made up of iron and nickel under enormous pressure. Surrounding the inner core is a band of melted iron and nickel called the liquid outer core.

Next to the outer core is a thick layer of hot, dense rock called the **mantle.** The mantle consists of silicon, aluminum, iron, magnesium, oxygen, and other elements. This mixture continually rises, cools, sinks, warms up, and rises again, **releasing** 80 percent of the heat generated from the Earth's interior.

The outer layer is the **crust,** a rocky shell forming the Earth's surface. This relatively thin layer of rock ranges from about 2 miles (3.2 km) thick under oceans to about 75 miles (121 km) thick under mountains. The crust is broken into more than a dozen great slabs of rock called plates that rest—or more accurately, float—on a partially melted layer in the upper portion of the mantle. The plates carry the Earth's oceans and continents.

Plate Movement

If you had seen the Earth from space 500 million years ago, the planet probably would not have looked at all like it does today. Many scientists believe that most of the landmasses forming our present-day continents were once part of one gigantic supercontinent called *Pangaea* (pan•JEE•uh). The maps on the next page show that over millions of years, this supercontinent has broken apart into smaller continents. These continents in turn have drifted and, in some places, recombined. The theory that the continents were once joined and then slowly drifted apart is called **continental drift.**

The term **plate tectonics** refers to all of the physical processes that create many of the Earth's physical features. Many scientists theorize that plates moving slowly around the globe have produced Earth's largest features—not only continents, but also oceans and mountain ranges. Most of the time, plate movement is so gradual—only about 1 inch (2 to 3 cm) a year—that it cannot be felt. As they move, the plates may crash into each other, pull apart, or grind and slide past each other. Whatever their actions, plates are **constantly** changing the face of the planet. They push up mountains, create volcanoes, and produce earthquakes. When the plates spread apart, **magma,** or molten rock, is pushed up from the mantle, and ridges are formed. When plates bump together, one may slide under another, forming a trench.

NATIONAL GEOGRAPHIC **Inside the Earth**

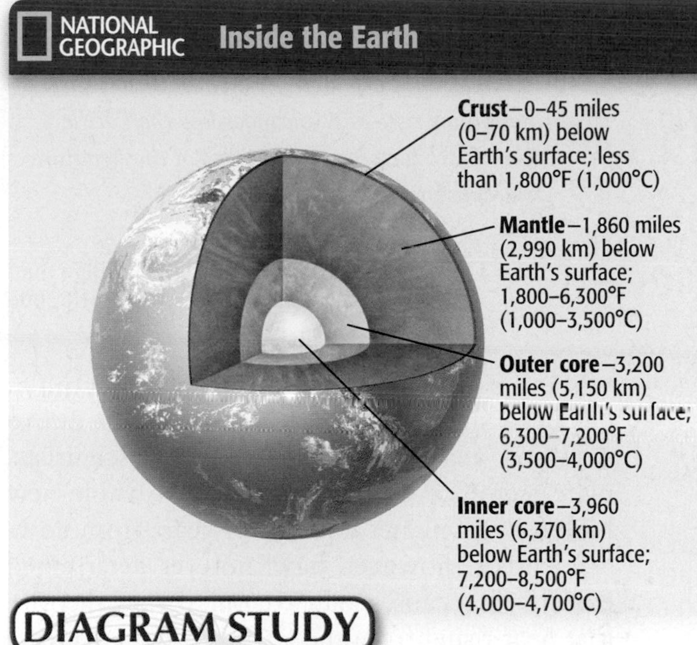

DIAGRAM STUDY

1. **Location** Which of Earth's layers is between the crust and the outer core?
2. **Location** How much does the temperature change from the inner core to the outer core? From the outer core to the mantle?

Concepts In Motion Use **StudentWorks™ Plus** or glencoe.com.

THE WORLD

> *[W]e have a highly successful theory, called plate tectonics, that explains . . . why continents drift, mountains rise, and volcanoes line the Pacific Rim. Plate tectonics may be one of the signature triumphs of the human mind. . . .*
>
> Joel Achenbach, "The Next Big One," *National Geographic,* April 2006

Many scientists estimate that plate tectonics has been shaping the Earth's surface for 2.5 to 4 billion years. According to some scientists, plate tectonics will have sculpted a whole new look for our planet millions of years from now.

Scientists, however, have not yet determined exactly what causes plate tectonics. They theorize that heat rising from the Earth's core may create slow-moving currents within the mantle. Over millions of years, these currents of molten rock may shift the plates around, but the movements are extremely slow and difficult to detect.

READING Check **Movement** What is the theory of continental drift?

Internal Forces of Change

MAIN Idea Plate tectonics is responsible for folding, lifting, bending, and breaking parts of the Earth's surface.

GEOGRAPHY AND YOU Have you seen news coverage about earthquakes or volcanoes in different parts of the world? Read to learn how the internal forces of plate tectonics can cause such natural disasters.

Earth's surface has changed greatly over time. Scientists believe that some of these changes come from forces associated with plate tectonics. One of these forces relates to the movement of magma within the Earth. Others involve movements that can fold, lift, bend, or break the solid rock at the Earth's crust.

MAP STUDY

1. **Regions** How does the first map of Pangaea compare to the last map? The map of plate movement?
2. **Movement** Which plates are moving toward each other? Away from each other?

Maps In MOtion Use StudentWorks™ Plus or glencoe.com.

NATIONAL GEOGRAPHIC Continental Drift

225 million years ago

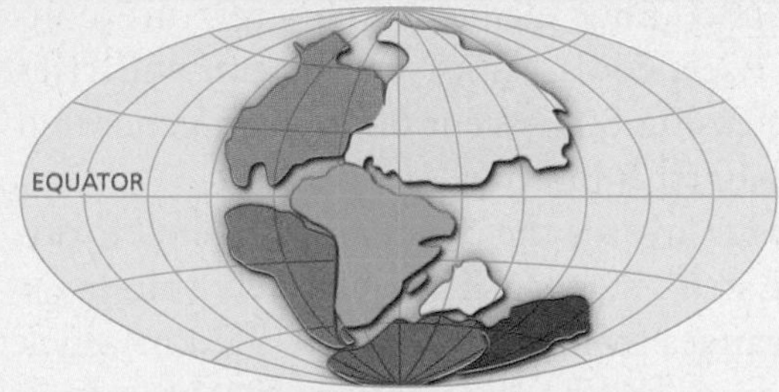

200 million years ago

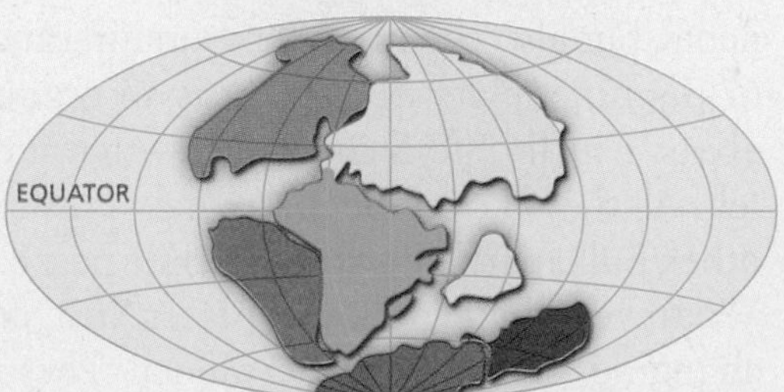

135 million years ago

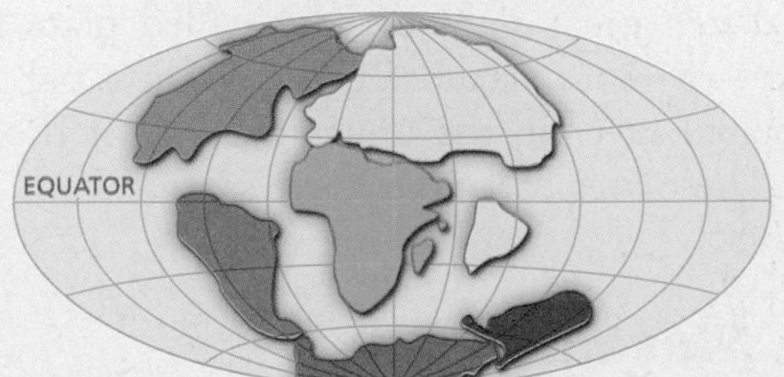

65 million years ago

Africa | Australia | Antarctica
Eurasia | North America | South America

Colliding and Spreading Plates

Mountains are formed in areas where giant continental plates collide. For example, the **Himalaya** mountain ranges in South Asia were thrust upward when the Indian landmass drifted against Eurasia. Himalayan peaks are getting higher as the Indian landmass continues to move northward.

Mountains are also created when a sea plate collides with a continental plate. The diagram on page 38 shows how in a process called **subduction** (suhb•DUHK•shuhn) the heavier sea plate dives beneath the lighter continental plate. Plunging into the Earth's interior, the sea plate becomes molten material. Then, as magma, it bursts through the crust to form volcanic mountains. The Andes, a mountain system in South America, was formed over millions of years as a result of subduction when the Nazca Plate slid beneath the South American Plate.

In other cases where continental and sea plates meet, a different process, known as accretion, occurs. During **accretion** (uh•KREE•shuhn), pieces of the Earth's crust come together slowly as the sea plate slides under the continental plate. This movement levels off seamounts, underwater mountains with steep sides and sharp peaks, and piles up the resulting debris in trenches. This buildup can cause continents to grow outward. Most scientists believe that much of western North America expanded outward into the Pacific Ocean over more than 200 million years as a result of accretion.

New land can also form when two sea plates converge. In this process, one plate moves under the other, often forming an island chain at the boundary. Sea plates also can pull apart in a process known as **spreading.** The resulting rift, or deep crack, allows magma from within the Earth to well up between the plates. The magma hardens to build undersea volcanic mountains or ridges and some islands. This spreading activity occurs down the middle of the Atlantic Ocean's floor, pushing Europe and North America away from each other.

Folds and Faults

Moving plates sometimes squeeze the Earth's surface until it buckles. This activity forms **folds,** or bends, in layers of rock. In other cases, plates may grind or slide past each other, creating cracks in the Earth's crust called **faults.** One famous fault is the **San Andreas Fault** in California.

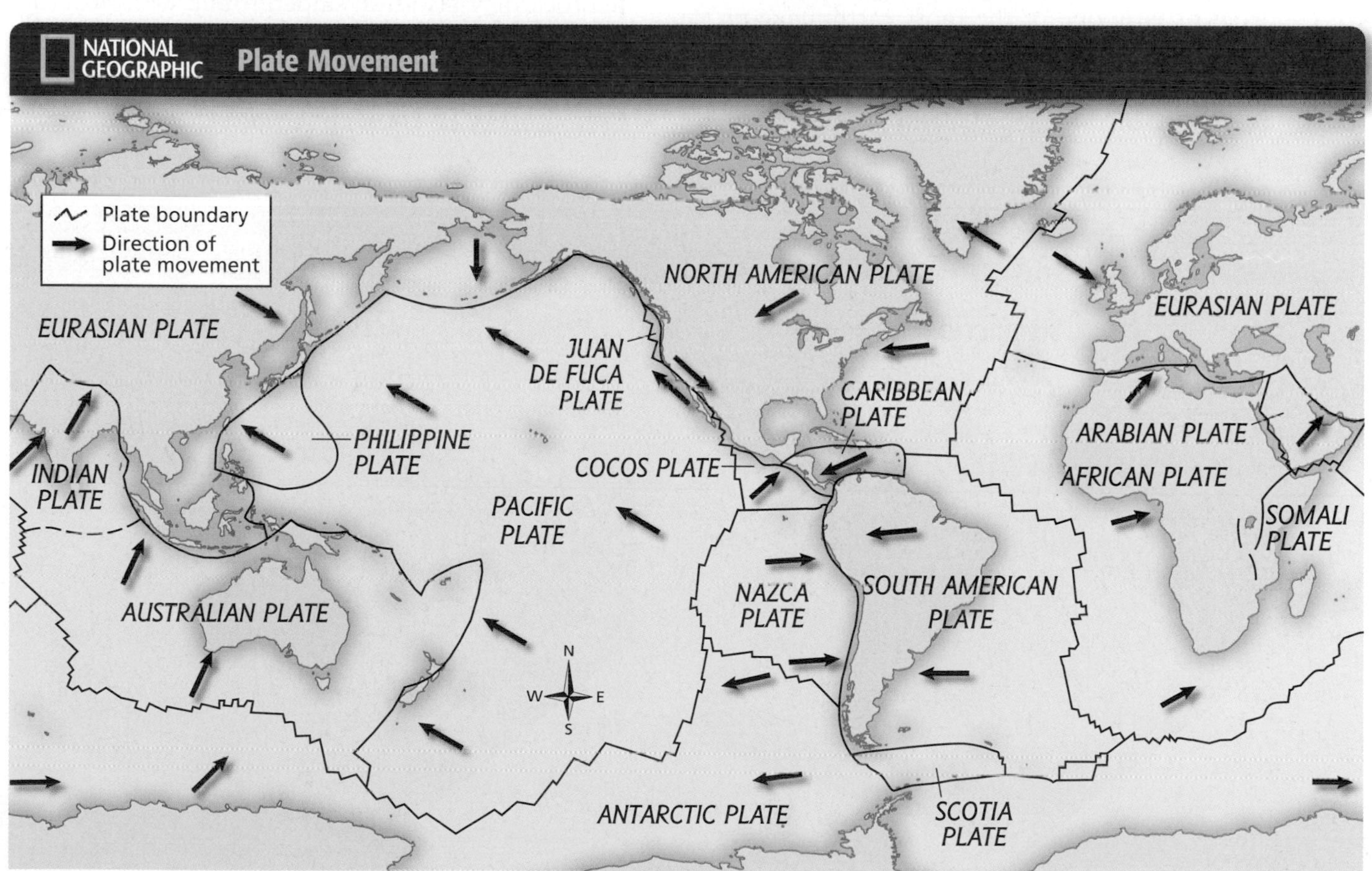

The process of **faulting** occurs when the folded land cannot be bent any further. Then the Earth's crust cracks and breaks into huge blocks. The blocks move along the faults in different directions, grinding against each other. The resulting **tension** may release a series of small jumps, felt as minor tremors on the Earth's surface.

Earthquakes

Sudden, violent movements of tectonic plates along a fault line are known as earthquakes. These shaking activities dramatically change the surface of the land and the floor of the ocean. During a severe earthquake in Alaska in 1964, a portion of the ground lurched upward 38 feet (11.6 m).

Earthquakes often occur where plates meet. Tension builds up along fault lines as the plates stick. The strain eventually becomes so intense that the rocks suddenly snap and shift. This movement releases stored-up energy along the fault. The ground then trembles and shakes as shock waves surge through it moving away from the area where the rocks first snapped apart.

In recent years disastrous earthquakes have occurred in **Kōbe,** Japan, and in Los Angeles and **San Francisco.** These cities are located along the **Ring of Fire,** one of the most earthquake-prone areas on the planet. The Ring of Fire is a zone of earthquake and volcanic activity around the perimeter of the Pacific Ocean. Here the plates that cradle the Pacific meet the plates that hold the continents surrounding the Pacific. North America, South America, Asia, and Australia are affected by their location on the Ring of Fire.

Volcanic Eruptions

Volcanoes are mountains formed by lava or by magma that breaks through the Earth's crust. Volcanoes often rise along plate boundaries where one plate plunges beneath another, as along the Ring of Fire. In such a process, the rocky plate melts as it dives downward into the hot mantle. If the molten rock is too thick, its flow is blocked and pressure builds. A cloud of ash and gas may then spew forth, creating a funnel through which the red-hot magma rushes to the surface. There the lava flow may eventually form a large volcanic cone topped by a crater, a bowl-shaped depression at a volcano's mouth.

Volcanoes also arise in areas away from plate boundaries. Some areas deep in the Earth are hotter than others, and magma often blasts through the surface as volcanoes. As a moving plate passes over these hot spots, molten rock flowing out of the Earth may create volcanic island chains, such as the Hawaiian Islands. At some hot spots, molten rock may also heat underground water, causing hot springs or geysers like Old Faithful in Yellowstone National Park.

READING **Check** **Location** Where is the Ring of Fire?

NATIONAL GEOGRAPHIC **Forces of Change**

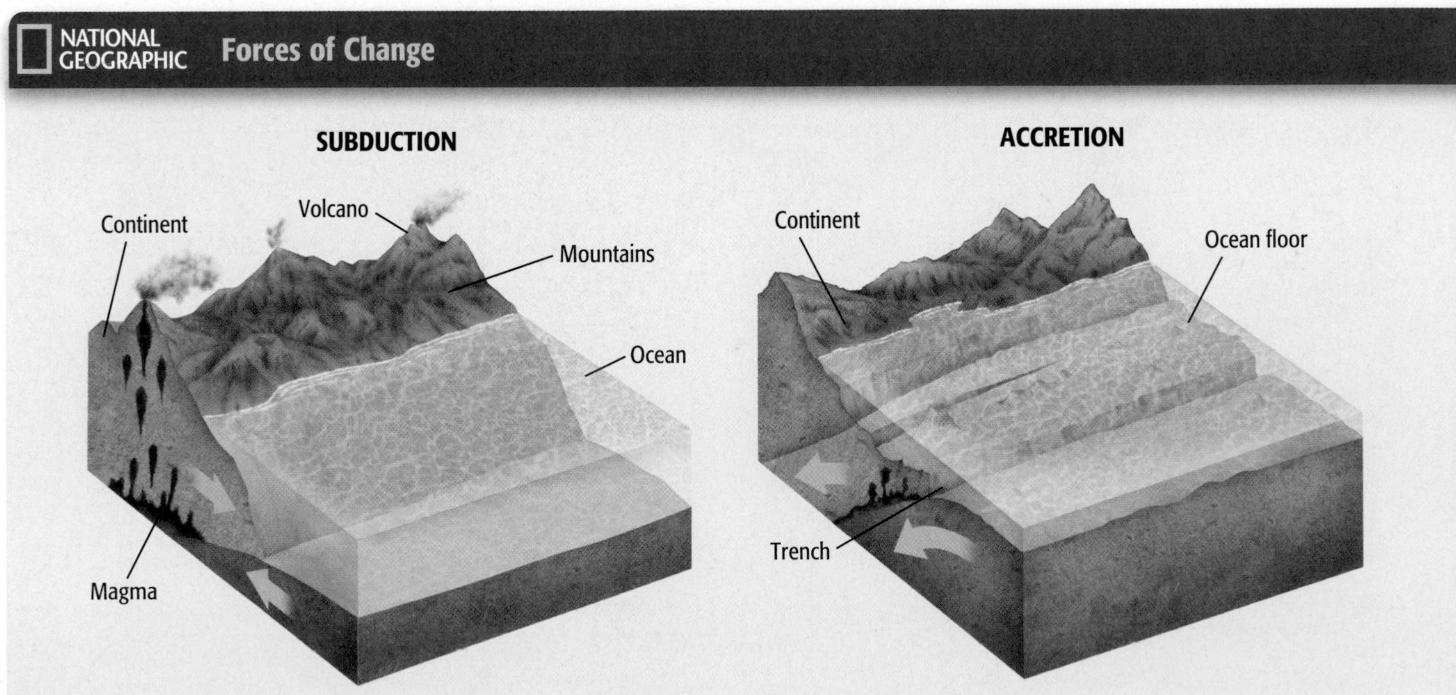

External Forces of Change

MAIN Idea External forces such as weathering and erosion also shape the surface of the Earth.

GEOGRAPHY AND YOU Have you ever seen soil washed over the road after a heavy rain? Read to learn how wind and rain can shape Earth's surface.

External forces, such as wind and water, also change the Earth's surface. Wind and water movements involve two processes. **Weathering** breaks down rocks, and **erosion** wears away the Earth's surface by wind, glaciers, and moving water.

Weathering

The Earth is changed by two basic kinds of weathering. Physical weathering occurs when large masses of rock are physically broken down into smaller pieces. For example, water seeps into the cracks in a rock and freezes, then expands and causes the rock to split. Chemical weathering changes the chemical makeup of rocks. For example, rainwater that contains carbon dioxide from the air easily dissolves certain rocks, such as limestone. Many of the world's caves have been and continue to be formed by this process.

Wind Erosion

Wind erosion involves the movement of dust, sand, and soil from one place to another. Plants help protect the land from wind erosion; however, in dry places where people have cut down trees and plants, winds pick up large amounts of soil and blow it away. Wind erosion can provide some benefits. The dust carried by wind often forms large deposits of mineral-rich soil.

Glacial Erosion

Another cause of erosion is **glaciers,** or large bodies of ice that move across the Earth's surface. Glaciers form over time as layers of snow press together and turn to ice. Their great weight causes them to move slowly downhill or spread outward. As they move, glaciers pick up rocks and soil in their paths, changing the landscape. They can destroy forests, carve out valleys, alter the courses of rivers, and wear down mountaintops.

When glaciers melt and recede, in some places they leave behind large piles of rocks and debris called **moraines.** Some moraines form long ridges of land, while others form dams that hold water back and create glacial lakes.

THE WORLD

DIAGRAM STUDY

1. **Place** How does accretion create deep trenches?
2. **Human-Environment Interaction** How have human settlements been affected by the process of faulting?

Concepts In Motion Use **StudentWorks™ Plus** or glencoe.com.

SPREADING

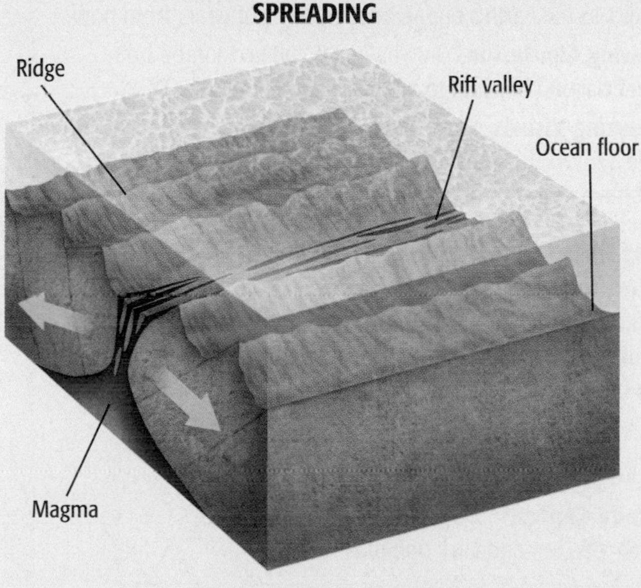

FAULTING

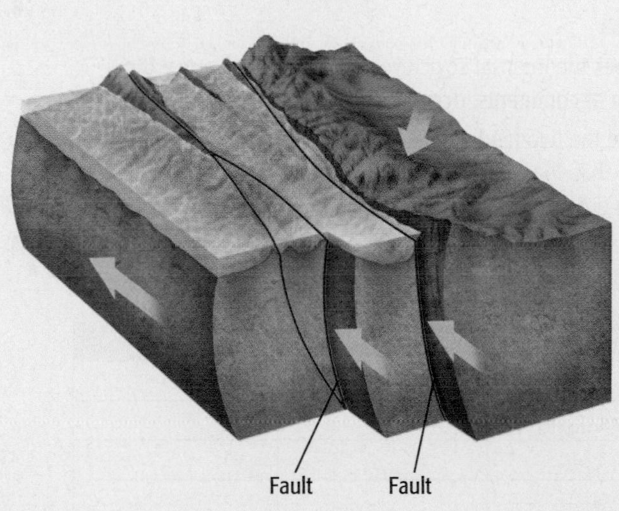

There are two types of glaciers. Sheet glaciers are flat, broad sheets of ice. Today sheet glaciers cover most of **Greenland** and all of **Antarctica.** They advance a few feet each winter and recede in the summer. Large blocks of ice often break off from the coastal edges of sheet glaciers to become icebergs floating in the ocean. More common mountain glaciers, located in high mountain valleys where the climate is cold, gouge out round, U-shaped valleys as they move downhill. As these mountain glaciers melt, rock and soil are deposited in new locations.

Water Erosion

Water erosion begins when springwater and rainwater flow downhill in streams, cutting into the land, and wearing away the soil and rock. The resulting sediment grinds away the surface of rocks along the stream's path. Over time, the eroding action of water forms first a gully and then a V-shaped valley. Sometimes valleys are eroded even further to form canyons. The Grand Canyon is an example of the eroding power of water.

Oceans also play an important role in water erosion. Pounding waves continually erode coastal cliffs, wear rocks into sandy beaches, and move sand away to other coastal areas.

Soil Building

Soil is the product of thousands of years of weathering and biological activity. The process of soil development begins when weathering breaks down solid rock into smaller pieces. Worms and other organisms help break down organic matter (dead plant and animal material) that comes to rest on these particles. Living organisms also add nutrients to the soil and create passages for air and water.

Five factors influence soil formation, with *climate* being the most significant. Wind, temperature, and rainfall determine the type of soil that can develop. *Topography*—the shape and position of Earth's physical features—affects surface runoff of water, drainage, and rate of erosion. *Geology* determines the parent material (original rock), which influences depth, texture, drainage, and nutrient content of soil. *Biology*, living and dead plants and animals, adds organic matter to the soil. The length of *time* the other four factors have been interacting also affects soil formation. These factors combine to produce different types of soils from region to region.

READING **Check** **Regions** How have many of the world's caves been formed?

SECTION 2 REVIEW

Vocabulary

1. Explain the significance of: core, mantle, crust, continental drift, plate tectonics, magma, subduction, accretion, spreading, fold, fault, faulting, weathering, erosion, glacier, moraine.

Main Ideas

2. How does the internal structure of the Earth influence the creation of continents, oceans, and mountain ranges?
3. Describe the two kinds of weathering and the three kinds of erosion that shape the surface of the Earth. How do weathering and erosion help create soil?
4. Use a chart like the one below to explain how plate tectonics folds, lifts, bends, and breaks parts of Earth's surface.

Forces of Change		
Process	How It Works	Example
Subduction		

Critical Thinking

5. **Answering the Essential Question** Based on your understanding of plate tectonics, what changes would you predict to the Earth's appearance millions of years from now?
6. **Drawing Conclusions** In what ways can erosion be both beneficial and harmful to agricultural communities?
7. **Analyzing Visuals** Study the map of plate movement on page 37. Which plates are responsible for the earthquakes that have occurred in California?

Writing About Geography

8. **Descriptive Writing** Review how internal forces shape the surface of the Earth. Now imagine that the mantle ceased to circulate molten rock. Write a description of how land formation on the surface of the Earth would be different.

Study Central™ To review this section, go to glencoe.com and click on Study Central.

SECTION 3

 section audio  spotlight video

Guide to Reading

Essential Question

What physical process keeps Earth's water constant?

Content Vocabulary
- water cycle *(p. 42)*
- evaporation *(p. 42)*
- condensation *(p. 42)*
- precipitation *(p. 42)*
- desalination *(p. 43)*
- groundwater *(p. 44)*
- aquifer *(p. 44)*

Academic Vocabulary
- area *(p. 43)*
- focus *(p. 43)*
- source *(p. 44)*

Places to Locate
- Pacific Ocean *(p. 43)*
- Atlantic Ocean *(p. 43)*
- Indian Ocean *(p. 43)*
- Arctic Ocean *(p. 43)*
- Mediterranean Sea *(p. 43)*
- Gulf of Mexico *(p. 43)*

Reading Strategy

Organizing As you read, complete a graphic organizer similar to the one below by listing the processes that contribute to the water cycle.

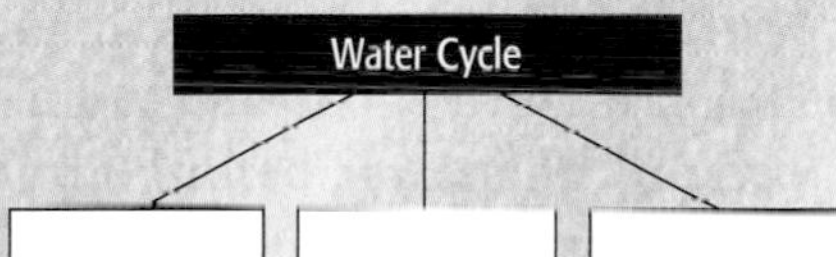

Earth's Water

A submarine crew investigating the Arctic Ocean can still experience the thrill of exploring uncharted territory—one of Earth's last frontiers. Although humans live mostly on land, water is important to our lives, and all living things need water to survive.

NATIONAL GEOGRAPHIC VOICES AROUND THE WORLD

"In a world that's been almost completely mapped, it's easy to forget why cartographers used to put monsters in the blank spots. Today we got a reminder. The submarine captain had warned us that we were in uncharted waters. . . . Yet the first days of our cruise through this ice-covered ocean, Earth's least explored frontier, were . . . smooth. . . . Even when we passed over a mile-high mountain that no one on the planet knew existed, the reaction was one of quiet enthusiasm—'Neat.'"

—Glenn Hodges, "The New Cold War," *National Geographic*, March 2000

A pilot aboard a submarine research vessel

CHAPTER 3

Climates of the Earth

BIG Idea

Geographers study how people, places, and environments are distributed on Earth's surface. Climate affects where and how people live. An understanding of Earth's climates and the factors that influence them adds to a more complete view of life on Earth.

Essential Questions

Section 1: Earth-Sun Relationships
How does Earth's position in relation to the sun affect life on Earth?

Section 2: Factors Affecting Climate
What factors can affect how climates are distributed on Earth's surface?

Section 3: World Climate Patterns
How do geographers classify Earth's climate and vegetation?

Geography ONLINE
Visit glencoe.com and enter *QuickPass*™ code WGC9952C3 for Chapter 3 resources.

Monaco Glacier, Svalbard, Norway, is a popular hunting ground for polar bears.

Identifying Make a Vocabulary Book to help you organize and learn the content vocabulary terms introduced as you read about the climates of the Earth. You may need to make more than one Vocabulary Book.

Reading and Writing As you read the chapter, write a term on each tab and its definition underneath. You may want to illustrate visual terms such as *rain shadow.*

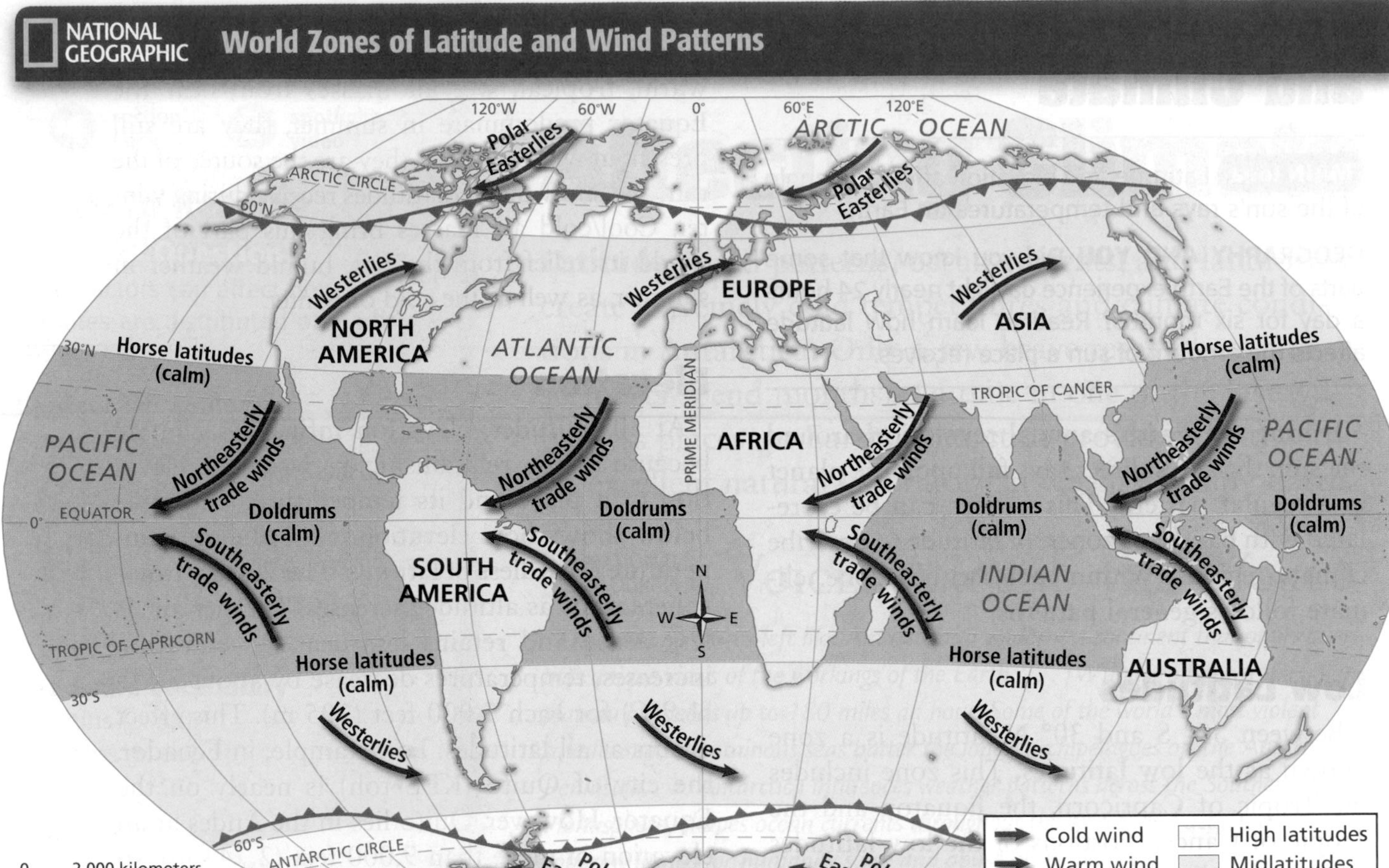

Winds and Ocean Currents

MAIN Idea Wind and water combine with the effects of the sun to influence Earth's weather and climate.

GEOGRAPHY AND YOU How does wind affect the weather in your community? Does it bring cold air or warm air? Read to learn about the patterns of global winds and ocean currents that affect climate.

Air moving across the surface of the Earth is called wind. Winds occur because sunlight heats the Earth's atmosphere and surface unevenly. Rising warm air **creates** areas of low pressure, and sinking cool air causes areas of high pressure. The cool air then flows in to replace the warm rising air. These movements over the Earth's surface cause winds, which distribute the sun's energy around the planet.

Wind Patterns

Winds blow because of temperature differences on Earth's surface, with tropical air moving toward the Poles and polar air moving toward the Equator. Global winds blow in fairly constant patterns called **prevailing winds,** as shown on the map above. The direction of prevailing winds is determined by latitude and is affected by the Earth's movement. Because Earth rotates to the east, the global winds are displaced clockwise in the Northern Hemisphere and counterclockwise in the Southern Hemisphere. This phenomenon, called the **Coriolis effect,** causes prevailing winds to blow diagonally rather than along strict north-south or east-west directions.

Winds are often named for the direction from which they blow, but they sometimes were given names from the early days of sailing. Named for their ability to move trading ships through the region, the prevailing winds of the low latitudes are called *trade winds.* They blow from the north-

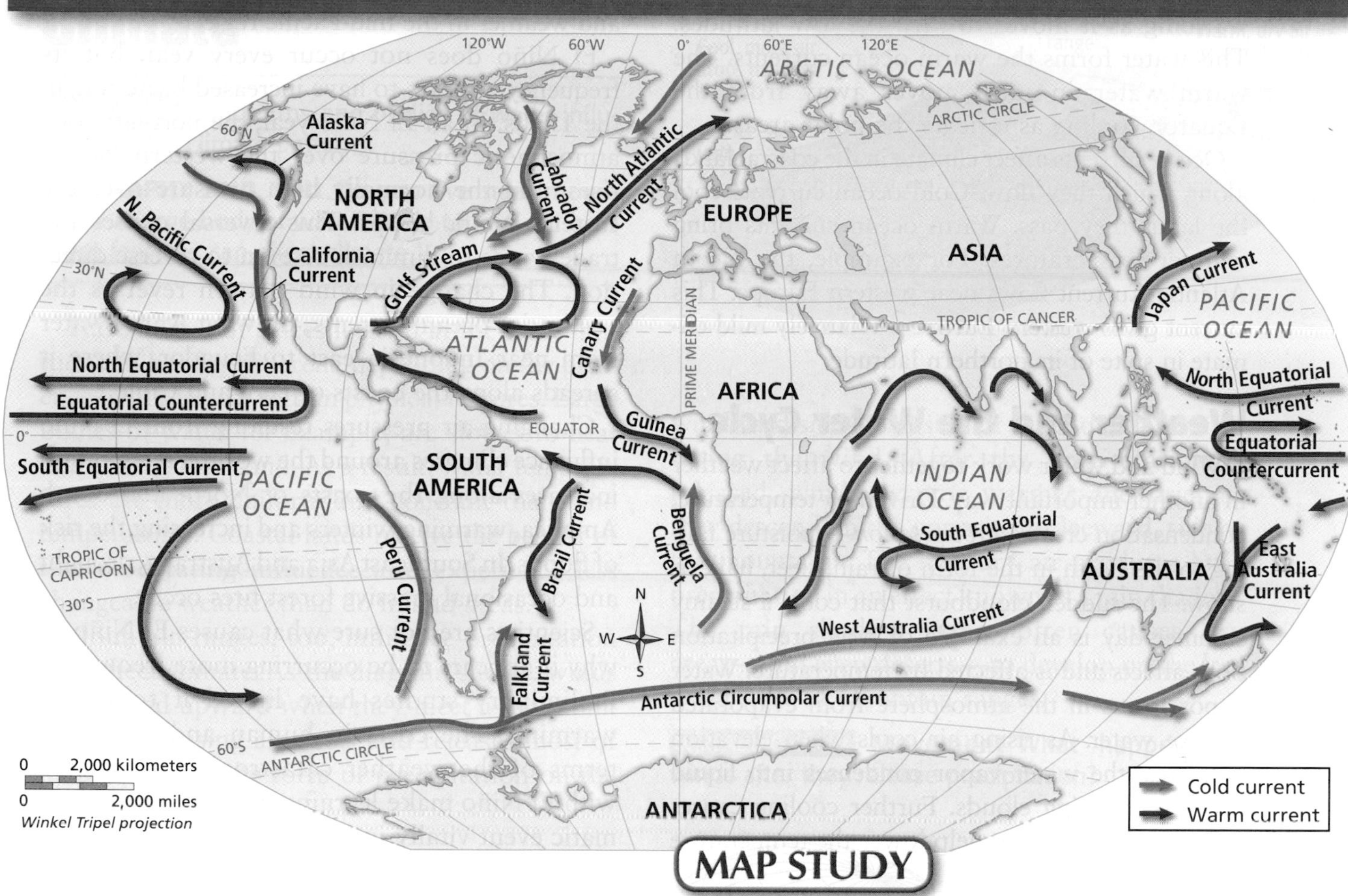

MAP STUDY

1. **Regions** What air currents flow over the midlatitudes?
2. **Movement** How do ocean currents develop? How is this process similar to the development of wind patterns?

Maps In MOtion Use **StudentWorks™ Plus** or glencoe.com.

east toward the Equator from about latitude 30° N and from the southeast toward the Equator from about latitude 30° S. *Westerlies* are the prevailing winds in the midlatitudes, blowing diagonally west to east between about 30° N and 60° N and between about 30° S and 60° S. In the high latitudes, the *polar easterlies* blow diagonally east to west, pushing cold air toward the midlatitudes.

The Horse Latitudes

At the Equator, global winds are diverted north and south, leaving a narrow, generally windless band called the **doldrums.** Two other narrow bands of calm air encircle the globe just north of the Tropic of Cancer and just south of the Tropic of Capricorn. In the days of wind-powered sailing ships, crews feared being stranded in these windless areas. With no moving air to lift the sails, ships were stranded for weeks in the hot, still weather. Food supplies dwindled, and perishable cargoes spoiled as the ships sat.

To lighten the load so the ships could take advantage of the slightest breeze, sailors would toss excess cargo and supplies overboard, including livestock being carried to colonial settlements. This practice gave rise to the name by which the calm areas at the edges of the Tropics are known—the *horse latitudes.*

Ocean Currents

Just as winds move in patterns, cold and warm streams of water, known as **currents,** move through the oceans. Ocean currents are caused by many of the same **factors** that cause winds, including the Earth's rotation, changes in air pressure, and differences in water temperature. The Coriolis effect is also observed in ocean currents.

Climate Changes

MAIN Idea Climate changes over time. Although the causes of change are unclear, evidence suggests that human activity has influenced some of the changes.

GEOGRAPHY AND YOU Does your community have problems with smog? Read to learn how human activity can influence climate change.

Climates change gradually over time, although the causes of these changes are unclear. Scientists search for answers by studying the interrelationships among ocean temperatures, greenhouse gases, wind patterns, and cloud cover.

During the last 1 to 2 million years, for example, the Earth passed through four ice ages, eras when glaciers covered large areas of the planet's surface. One **hypothesis,** or scientific explanation, for these ice ages is that the Earth absorbed less solar energy because of variations in the sun's output of energy or because of variations in the Earth's orbit. Another hypothesis suggests that dust clouds from volcanic activity reflected sunlight back into space, cooling the atmosphere and lowering surface temperatures.

Human interaction with the environment also affects climate. Burning fossil fuels releases gases that mix with water in the air, forming acids that fall in rain and snow. Acid rain can destroy forests. Fewer forests may result in climatic change. The exhaust released from burning fossil fuels in automobile engines and factories is heated in the atmosphere by the sun's ultraviolet rays, forming **smog,** a visible chemical haze in the atmosphere that endangers people's health. Other human-driven changes result from dams and river diversions. These projects, intended to supply water to dry areas, may cause new areas to flood or to dry out and may affect climate over time.

READING Check **Human-Environment Interaction** How does the burning of fossil fuels create smog?

Student Web Activity Visit glencoe.com, select the *World Geography and Cultures* Web site, and click on Student Web Activities—Chapter 3 for an activity about global climate change.

SECTION 3 REVIEW

Vocabulary

1. Explain the significance of: natural vegetation, oasis, coniferous, deciduous, mixed forest, prairie, permafrost, hypothesis, smog.

Main Ideas

2. Describe one hypothesis for climate change. Then list examples of human activities that affect climate.
3. Create a table like the one below to show how geographers divide the Earth into regions with similar climates. Add information and a brief description about each of the world's climate regions.

Earth's Climates	
Climate Region	Features

Critical Thinking

4. **Answering the Essential Question** How are the five major climate regions related to the three zones of latitude?
5. **Comparing and Contrasting** What factors account for the similarities and differences between the subdivisions in tropical climate zones?
6. **Drawing Conclusions** What are the two main categories of factors causing climate change?
7. **Analyzing Visuals** Study the map of world vegetation regions on page 63. What vegetation type dominates Russia? Canada?

Writing About Geography

8. **Summary Writing** On the map of world climate regions on page 62, locate the climate regions for Tashkent, Cape Town, Lima, Chicago, London, and Jakarta. Then write a paragraph summarizing the relationship between climate and settlement.

Study Central™ To review this section, go to glencoe.com and click on Study Central.

CHAPTER 3 VISUAL SUMMARY

Study anywhere, anytime by downloading quizzes and flashcards to your PDA from glencoe.com.

Earth-Sun Relationships

- The relationship of the Earth to the sun affects climate patterns around the world.
- The Earth's tilt and revolution cause the seasons by changing the relationship of the Earth's surface to the sun.
- When the sun is directly over the Tropic of Cancer, it is summer in the Northern Hemisphere. When it is directly over the Tropic of Capricorn, it is winter in the Northern Hemisphere.

Inuit boy with sled dogs during midnight sun, Canada

Latitude and Temperature

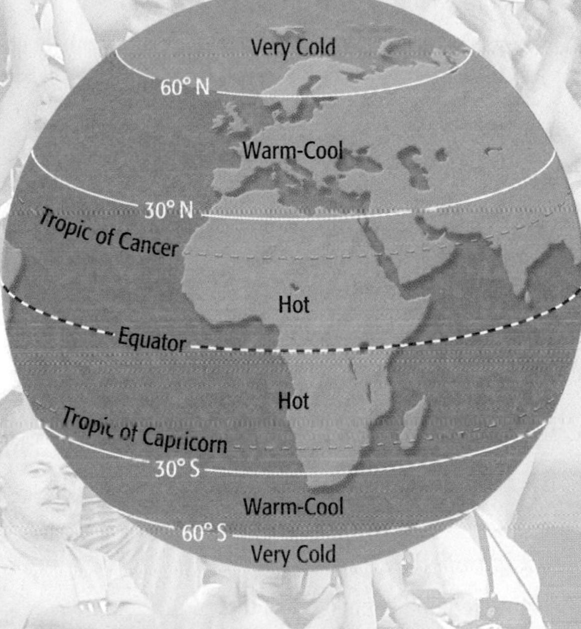

Factors Affecting Climate

- Latitude plays a major role in climate. The farther one gets from the Equator, the cooler the climate.
- High elevations are generally cooler than the surrounding landscape.
- Other factors that help determine climate are wind and water currents, recurring phenomena such as El Niño, and large landforms.

THE WORLD

World Climate Patterns

- Geographers divide the world into major climate regions.
- The major climate regions are tropical, dry, midlatitude, high latitude, and highland climates. Each of these can be broken down into smaller categories.
- Each climate region has its own characteristic natural vegetation.
- Climate patterns change over time as a result of both natural processes and human activity.

Viedma glacier breaking into lake, Argentina

STANDARDIZED TEST PRACTICE

If you find a question that is more difficult for you, skip it and go on to answer the other questions. Then return to the more difficult question.

Reviewing Vocabulary

Directions: Choose the word or words that best complete the sentence.

1. Day-to-day conditions of the atmosphere make up ________.

A weather
B climate
C axis
D equinox

2. Long-term atmospheric conditions that people can expect to be generally true make up ________.

A weather
B climate
C solstice
D tilt

3. Because of ________, winds tend to blow diagonally rather than from due north, south, east, or west.

A weather
B rain shadow
C the Coriolis effect
D El Niño

4. Grasslands in midlatitude climates are called ________.

A savannas
B mixed vegetation
C prairies
D tundras

Reviewing Main Ideas

Directions: Choose the best answers to complete the sentences or to answer the following questions.

Section 1 *(pp. 50–53)*

5. Why is the greenhouse effect necessary to life?

A People's activities increase the greenhouse effect.
B Without the greenhouse effect, Earth would be too cold for life.
C Farmers grow crops in greenhouses.
D Scientists are afraid that climate change will harm society.

Section 2 *(pp. 54–59)*

6. During El Niño years, weather in western South America becomes ________.

A hotter and drier
B warmer and rainier
C colder and rainier
D colder and drier

Section 3 *(pp. 60–64)*

7. Each climate region has its own kind of ________.

A rock structure
B tectonic plate
C natural vegetation
D Coriolis effect

GO ON

Critical Thinking

Directions: Choose the best answers to complete the sentences or to answer the following questions.

8. How do human activities impact climate?

A The sun may give off varying amounts of solar energy.

B Volcanic eruptions may increase cloud cover.

C People add gases to the atmosphere by burning fossil fuels.

D Scientists do research to learn what causes climate change.

Base your answer to question 9 on the map and on your knowledge of Chapter 3.

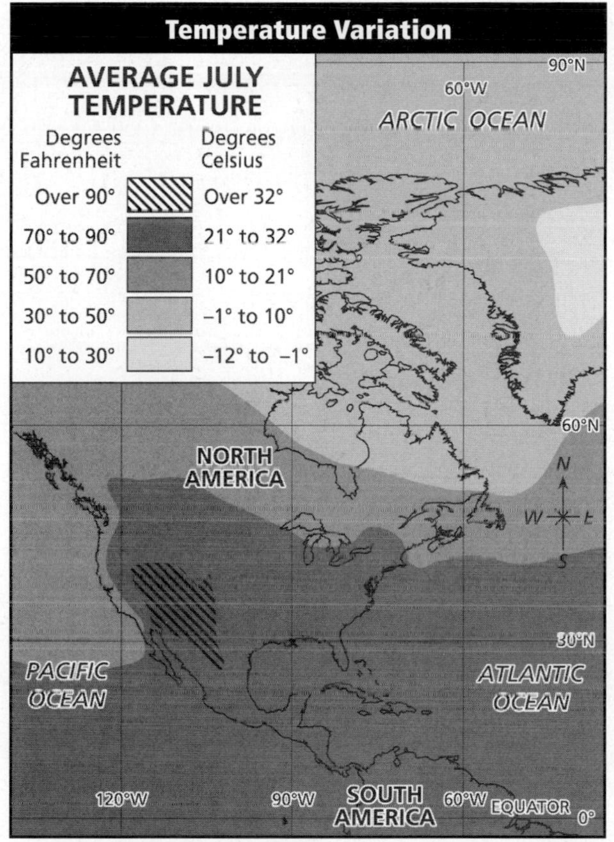

9. Going from the Equator to the North Pole, average temperatures _______.

A become progressively warmer

B become warmer, then colder

C become colder, then warmer

D become progressively colder

Document-Based Questions

Directions: Analyze the document and answer the short-answer questions that follow the document.

The United Nations Framework Convention on Climate Change went into force in 1994. The excerpt below explains why its framers believed it was necessary.

> *The Parties to this Convention,*
>
> Acknowledging *that change in the Earth's climate and its adverse effects are a common concern of humankind,*
>
> Concerned *that human activities have been substantially increasing the atmospheric concentrations of greenhouse gases, that these increases enhance the natural greenhouse effect, and that this will result on average in an additional warming of the Earth's surface and atmosphere and may adversely affect natural ecosystems and humankind,*
>
> Acknowledging *that the global nature of climate change calls for the widest possible cooperation by all countries and their participation in an effective and appropriate international response, in accordance with their common but differentiated responsibilities and respective capabilities and their social and economic conditions, . . .*
>
> Affirming *that responses to climate change should be coordinated with social and economic development in an integrated manner with a view to avoiding adverse impacts on the latter, taking into full account the legitimate priority needs of developing countries for the achievement of sustained economic growth and the eradication of poverty. . . .*

10. Why were the writers of the framework concerned about climate?

11. What attitude does the framework take regarding social and economic development?

Extended Response

12. Exploring the BIG Idea

In which climate region do you live? What factors influence the climate there?

For additional test practice, use Self-Check Quizzes—Chapter 3 on glencoe.com.

Need Extra Help?

If you missed questions. . .	1	2	3	4	5	6	7	8	9	10	11	12
Go to page. . .	51	51	56	62	52	58	61	64	55	64	64	61–63

CHAPTER 4

The Human World

BiG Idea

The characteristics and distribution of human populations affect human and physical systems. A study of the human world—population, culture, political and economic systems, and resources—will help you understand the world around you.

Essential Questions

Section 1: World Population
What factors influence population growth in a given area or region?

Section 2: Global Cultures
How does the spatial interaction of cultures affect human systems?

Section 3: Political and Economic Systems
What types of human systems provide the power for groups of people to control Earth's surface?

Section 4: Resources, Trade, and the Environment
How does the availability and use of natural resources affect economic activities and the environment?

Geography ONLINE
Visit glencoe.com and enter QuickPass™ code WGC9952C4 for Chapter 4 resources.

A woman sells fresh seafood in a Korean market.

Know-Want-Learn Create a Trifold Book to help you keep track of what you know, what you want to learn, and what you learn about the geography of the human world.

KNOW

WANT

LEARN

Reading and Writing Before you read, make a list on your Foldable of what you know about world population, cultures, political and economic systems, and resources, trade, and the environment. Then make a list of what you want to learn about these topics. When you have finished reading the chapter, write down what you learned.

THE WORLD

SECTION 1

 section audio  spotlight video

Guide to Reading

Essential Question
What factors influence population growth in a given area or region?

Content Vocabulary
- birthrate *(p. 71)*
- death rate *(p. 71)*
- natural increase *(p. 71)*
- migration *(p. 71)*
- demographic transition *(p. 71)*
- doubling time *(p. 71)*
- population distribution *(p. 72)*
- population density *(p. 72)*

Academic Vocabulary
- trend *(p. 71)*
- community *(p. 72)*
- negative *(p. 72)*

Places to Locate
- Hungary *(p. 72)*
- Germany *(p. 72)*
- Canada *(p. 73)*
- Bangladesh *(p. 73)*
- Mexico City *(p. 74)*

Reading Strategy
Organizing As you read about changes in world population, create a web diagram like the one below by listing the challenges created by population growth.

Challenges of Population Growth

World Population

Explorer and conservationist J. Michael Fay's flight across Africa showed that the effects of rapid population growth reflect the global challenge humans face today. How can people maintain conditions favorable to human life without endangering those very conditions through overpopulation?

NATIONAL GEOGRAPHIC VOICES AROUND THE WORLD

"As we flew outside of the park, we saw African settlement areas where the land had been overused for mass cultivation and cattle grazing. There were wall to wall fields and masses of people on the ground. Little of the land could still be identified as wild. Farther on, we hit white farm areas, where the land had been irrigated well. Although it looked like a giant oasis, I wondered what the cost was in terms of fossil fuel and ground water. Even farther out, we saw a great hole in the ground. This was Phalaborwa Mine, one of the biggest open pit mines in the world and extreme example of eliminating nature from the planet."

—J. Michael Fay,
Africa Megaflyover:
Air Dispatches, *National Geographic* (online),
June 14, 2004

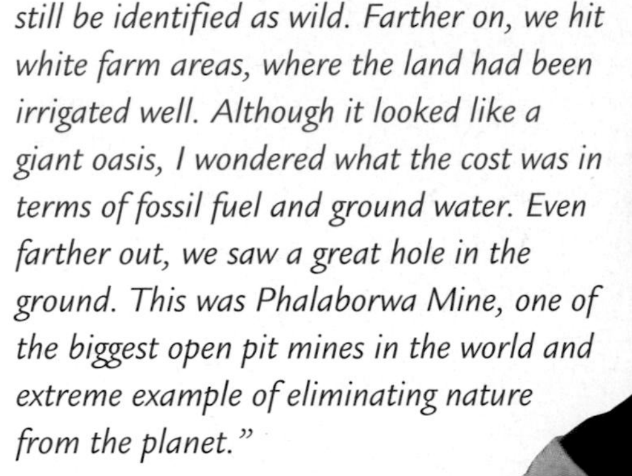

A Sudanese refugee

Population Growth

MAIN Idea Population growth varies from country to country and is influenced by cultural ideas, migration, and level of development.

GEOGRAPHY AND YOU Do you have any siblings? If so, how many? Read to learn how changes in family size create challenges to population growth.

More than 6.8 billion people now live on Earth, inhabiting about 30 percent of the planet's land. Global population is growing rapidly and is expected to reach 9 billion by the year 2050. Such rapid growth was not always the case. From the year 1000 until 1800, the world's population increased slowly. Then the number of people on Earth more than doubled between 1800 and 1950. By 2006 the world's population had soared to more than 6 billion.

The Demographic Transition

Scientists in the field of *demography,* the study of populations, use statistics to learn about population growth. The **birthrate** is the number of births per year for every 1,000 people. The **death rate** is the number of deaths per year for every 1,000 people. **Natural increase,** or the growth rate, of a population is the difference between an area's birthrate and its death rate. **Migration,** or the movement of people from place to place, must also be considered when examining population growth.

The **demographic transition** model uses birthrates and death rates to show changes in the population **trends** of a country or region. The model was first used to show the relationship of declining birthrates and death rates to industrialization in Western Europe. Falling death rates are due to more abundant and reliable food supplies, improved health care, access to medicine and technology, and better living conditions. Birthrates decline more slowly as cultural traditions change.

Today, most of the world's industrialized and technologically developed countries have experienced the transition from high birthrates and death rates to low birthrates and death rates. These countries have reached what is known as *zero population growth,* in which the birthrate and death rate are equal. When this balance occurs, a country's population does not grow.

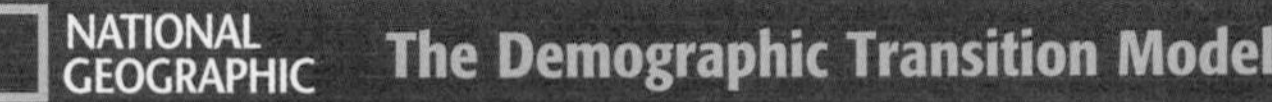

The Demographic Transition Model

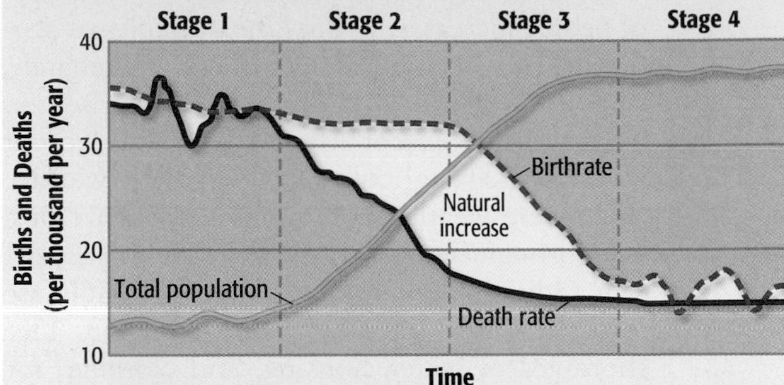

Stage 1 A balance between birthrates and death rates characterizes this stage. Death rates are very high due to a lack of medical knowledge, infectious diseases, and food shortages.

Stage 2 This stage is marked by a high birthrate and a decline in the death rate. As a result, population begins to increase. The declining death rate is due to technology, new farming techniques, and improved health care.

Stage 3 Death rates continue to decline and birthrates begin to decline. This is due to social changes, including urbanization and an increase in opportunities for women.

Stage 4 This stage includes both low birthrates and low death rates. The population of a country in this stage begins to decline and grow older. In some cases, birthrates drop below replacement level.

GRAPH STUDY

1. **Place** How does total population change from Stage 1 to Stage 4?
2. **Place** What happens to birthrates between Stage 2 and Stage 3? What influences this change?

Graphs In MOtion Use **StudentWorks™ Plus** or glencoe.com.

Although birthrates have fallen significantly in many countries in Asia, Africa, and Latin America over the past 40 years, they are still higher than in the industrialized world. Families in these regions traditionally are large because of cultural beliefs about marriage, family, and the value of children. For example, a husband and wife in a rural agricultural area may choose to have several children who will help farm the land. A high number of births combine with low death rates to greatly increase population growth in these regions. As a result, the **doubling time,** or the number of years it takes a population to double in size, has been reduced to only 25 years in some parts of Asia, Africa, and Latin America. In contrast, the average doubling time of a developed country can be more than 300 years.

Challenges of Growth

Rapid population growth presents many challenges to the global **community.** As the number of people increases, so does the difficulty of producing enough food to feed them. Fortunately, since 1950 world food production has risen on all continents except Africa. In Africa, lack of investment in agriculture, along with warfare and severe weather conditions that ruin crops, have brought hunger to this region.

In addition, populations that grow rapidly use resources more quickly. Some countries face shortages of water, housing, and clothing, for instance. Rapid population growth strains these limited resources. Another concern is that the world's population is unevenly distributed by age, with the majority of some countries' populations being infants and young children who cannot contribute to food production.

While some experts are pessimistic about the long-term effects of rapid population growth, others are optimistic that, as the number of humans increases, the levels of technology and creativity will also rise. For example, scientists continue to study and develop ways to boost agricultural productivity. Fertilizers can improve crop yields. Irrigation systems can help increase the amount of land available for farming. New varieties of crops have been created to withstand severe conditions and yield more food.

Negative Population Growth

In the late 1900s, some countries in Europe began to experience *negative population growth,* in which the annual death rate exceeds the annual birthrate. **Hungary** and **Germany,** for example, show change rates of –0.3 and –0.2, respectively. This situation has economic consequences different from, but just as serious as, those caused by high growth rates. In countries with negative population growth, it is difficult to find enough workers to keep the economy going. Labor must be recruited from other countries, often by encouraging immigration or granting temporary work permits. Although the use of foreign labor has helped countries with **negative** change rates maintain their levels of economic activity, it also has created tensions between the "host" population and the communities of newcomers.

READING Check **Place** Where was the demographic transition model first used?

Population Distribution

MAIN Idea World population distribution is uneven and is influenced by migration and the Earth's physical geography.

GEOGRAPHY AND YOU Do you live in a crowded city or an open rural area? Read to learn how population distribution varies and how it is measured.

Not only do population growth rates vary among the Earth's regions, but the planet's **population distribution,** the pattern of human settlement, is uneven as well. Population distribution is related to the Earth's physical geography. Only about 30 percent of the Earth's surface is made up of land, and much of that land is inhospitable. High mountain peaks, barren deserts, and frozen tundra make human activity very difficult. As the world population density map on page 73 shows, almost everyone on Earth lives on a relatively small portion of the planet's land—a little less than one-third. Most people live where fertile soil, available water, and a climate without harsh extremes make human life possible.

Of all the continents, Europe and Asia are the most densely populated. Asia alone contains more than 60 percent of the world's people. Throughout the world, where populations are highly concentrated, many people live in *metropolitan areas*—cities and their surrounding urbanized areas. Today most people in Europe, North America, South America, and Australia live in or around urban areas.

Population Density

Geographers determine how crowded a country or region is by measuring **population density**—the number of people living on a square mile or square kilometer of land. To determine population density in a country, geographers divide the total population of the country by its total land area.

Student Web Activity Visit glencoe.com, select the *World Geography and Cultures* Web site, and click on Student Web Activities–Chapter 4 for an activity on world population.

Population density varies widely from country to country. **Canada,** with a low population density of about 10 people per square mile (4 people per sq. km), offers wide-open spaces and the choice of living in thriving cities or quiet rural areas. In contrast, **Bangladesh** has one of the highest population densities in the world—about 3,227 people per square mile (1,246 people per sq. km).

Countries with populations of about the same size do not necessarily have similar population densities. For example, both Bolivia and the Dominican Republic have about 10 million people. With a smaller land area, the Dominican Republic has 541 people per square mile (209 people per sq. km). However, Bolivia has only 24 people per square mile (9 people per sq. km).

Because the measure of population density includes all the land area of a country, it does not account for uneven population distribution within a country. In Egypt, for example, overall population density is 205 people per square mile (79 people per sq. km). In reality, over 90 percent of Egypt's people live along the Nile River. The rest of Egypt is desert. Thus, some geographers describe a country's population density in terms of land that can be used to support the population rather than total land area. When Egypt's population density is measured this way, it is about 6,962 people per square mile (2,688 people per sq. km)!

Population Movement

The Earth's population is moving in great numbers. People are moving from city to city or from city to suburbs. Others are migrating from rural villages to cities. The resulting growth of city populations brought about by such migration and the changes that come with this increase are called *urbanization.*

MAP STUDY

1. **Regions** Which areas of the Northern Hemisphere are the most densely populated?
2. **Human-Environment Interaction** Study the population densities for South America. What conclusions can you draw about its physical geography?

Maps In MOtion Use **StudentWorks™ Plus** or glencoe.com.

NATIONAL GEOGRAPHIC **World Population Density**

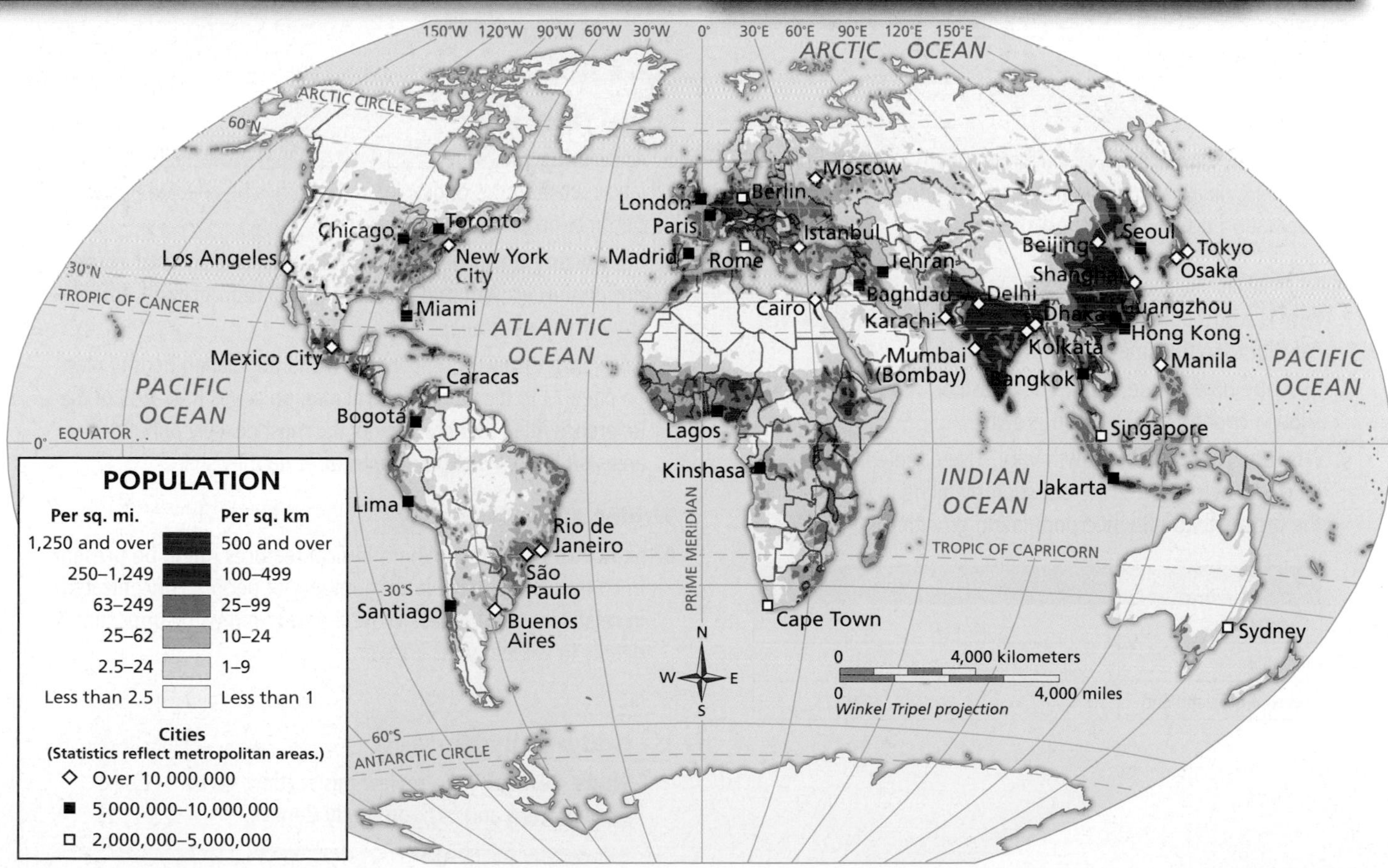

The primary cause of urbanization is the desire of rural people to find jobs and a better life in more prosperous urban areas. Rural populations certainly have grown, but the amount of farmland has not increased to meet the growing number of people. As a result, many rural migrants find urban jobs in manufacturing and service industries.

About half of the world's people live in cities. Between 1960 and 2010, the population of metropolitan **Mexico City** rose from about 5 million to more than 19 million. The graph at the right shows that other cities in Latin America, as well as in Asia and Africa, have seen similar growth. Some of these cities contain a large part of their country's entire population. For example, about one-third of Argentina's people live in Buenos Aires.

Population movement also occurs between countries. Some people emigrate from the country of their birth. They are known as emigrants in their homeland and immigrants in their new country. In the past 40 years, economic pull factors have drawn millions of people from Africa, Asia, and Latin America to the wealthier countries of Europe, North America, and Australia. Some were forced to flee their country because of the push factors of wars, food shortages, or other problems. They are *refugees,* or people who flee to escape persecution or disaster.

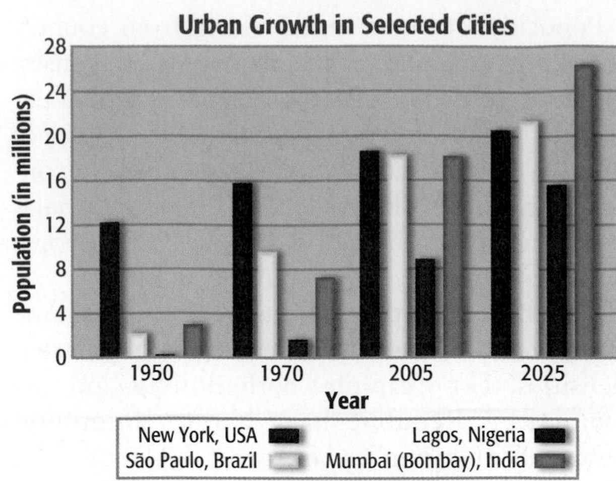

Source: United Nations, *World Urbanization Prospects: The 2007 Revision.*

READING Check **Movement** What influences the migration of people from one country to another?

SECTION 1 REVIEW

Vocabulary

1. Explain the significance of: birthrate, death rate, natural increase, migration, demographic transition, doubling time, population distribution, population density.

Main Ideas

2. Why is world population distribution uneven? What factors contribute to this uneven distribution?
3. Describe how the demographic transition model is used to show a country's population growth.
4. Why does population growth vary? Use a table like the one below to help answer the question by filling in examples of the factors that influence population growth.

Influence	Example
Cultural ideas	
Migration	
Level of development	

Critical Thinking

5. **Answering the Essential Question** How might the population growth rates of developing countries be affected as they become increasingly industrialized?
6. **Comparing and Contrasting** How do the effects of zero population growth and negative population growth differ? How are they similar?
7. **Analyzing Visuals** Compare the world population density map on page 73 to the world physical map on pages RA4–RA5 of the Reference Atlas. Identify three of the most densely populated areas on Earth. What physical features do they share?

Writing About Geography

8. **Expository Writing** What physical features might be present in countries that have large numbers of people concentrated in relatively small areas? Write a paragraph with supporting details to explain your answer.

Study Central™ To review this section, go to glencoe.com and click on Study Central.

SECTION 2

 section audio

 spotlight video

Guide to Reading

Essential Question

How does the spatial interaction of cultures affect human systems?

Content Vocabulary

- culture *(p. 76)*
- language family *(p. 76)*
- ethnic group *(p. 77)*
- culture region *(p. 77)*
- cultural diffusion *(p. 78)*
- culture hearth *(p. 78)*

Academic Vocabulary

- similar *(p. 76)*
- major *(p. 78)*

Places to Locate

- Egypt *(p. 78)*
- Iraq *(p. 78)*
- Pakistan *(p. 78)*
- China *(p. 78)*
- Mexico *(p. 78)*

Reading Strategy

Organizing As you read about global cultures, complete a graphic organizer like the one below by listing the world culture regions.

Global Cultures

The world's people organize communities, develop their ways of life, and adjust to the differences and similarities they experience. Many people struggle to maintain some elements of their traditional cultures while establishing ties with the global community.

NATIONAL GEOGRAPHIC VOICES AROUND THE WORLD

". . . Berber struggle for cultural recognition has grown stronger over time. The urban Berbers leading this revival movement are intellectuals who use French, a language they associate with culture and human rights, rather than Arabic, which they despise as the language of their oppressors. But the language they're really pushing is Tamazight, or Berber. During the last decade of Hassan II's rule (which ended with the monarch's death in 1999), they founded Berber language and cultural associations, set up websites and newspapers, and, in 1994, won the right to broadcast news in Berber on national television."

—Jeffrey Tayler, "Among the Berbers," *National Geographic,* January 2005

A Tuareg man belonging to the Berber ethnic group

Elements of Culture

MAIN Idea Geographers divide the Earth into culture regions, which are defined by the presence of common elements such as language and religion.

GEOGRAPHY AND YOU What language do most people in your community speak? Read to learn how language is important to a culture's development.

Geographers study **culture,** the way of life of a group of people who share **similar** beliefs and customs. A particular culture can be understood by looking at language, religion, daily life, history, art, government, and the economy.

Language

Language is a key element in a culture's development. Through language, people communicate information and experiences and pass on cultural values and traditions. Even within a culture, however, there are language differences. Some people may speak a dialect, or a local form of a language that differs from the main language. These differences may include variations in the pronunciation and meaning of words.

Linguists, scientists who study languages, organize the world's languages into **language families**—large groups of languages having similar roots. Seemingly diverse languages may belong to the same language family. For example, English, Spanish, and Russian are all members of the Indo-European language family.

Religion

Religious beliefs vary significantly around the world, and struggles over religious differences can be a source of conflict. In many cultures, however, religion enables people to find a sense of identity. It also influences aspects of daily life, from the practice of moral values to the celebration of holidays and festivals.

Throughout history, religious symbols and stories have shaped cultural expressions such as painting, architecture, and music. The feature on pages 80–99 discusses world religions.

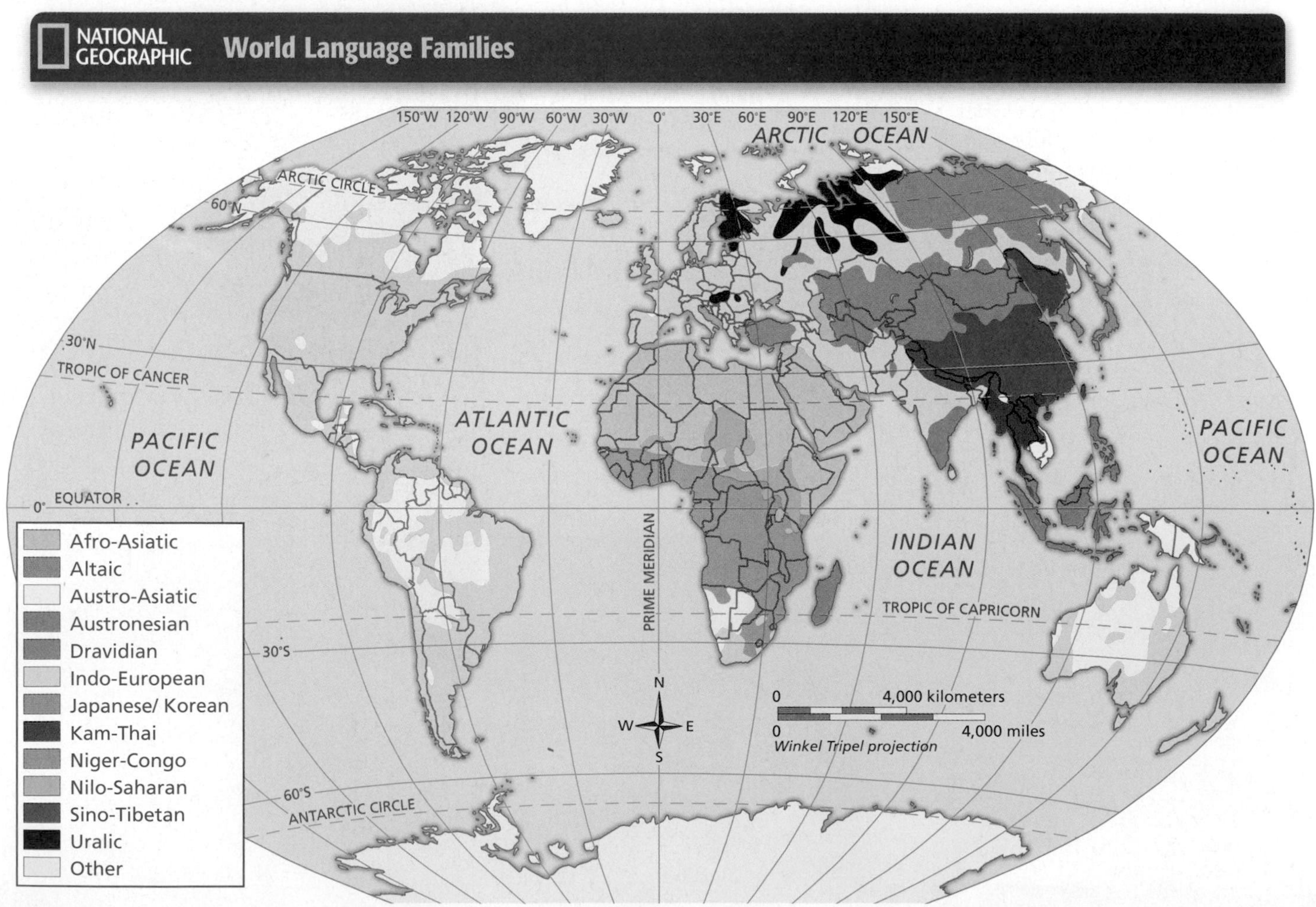

Social Groups

A social system develops to help the members of a culture work together to meet basic needs. In all cultures the family is the most important group. Most cultures are also made up of *social classes,* groups of people ranked according to ancestry, wealth, education, or other criteria. Moreover, cultures may include people who belong to different ethnic groups. An **ethnic group** is made up of people who share a common language, history, place of origin, or a combination of these elements.

Government and Economy

Governments of the world share certain features. For example, each maintains order within the country, provides protection from outside dangers, and supplies other services to its people. Governments are organized by levels of power—national, regional, and local—and by type of authority—a single ruler, a small group of leaders, or a body of citizens and their representatives.

When examining cultures, geographers look at economic activities. They study how a culture utilizes its natural resources to meet human needs. They also analyze the ways in which people produce, obtain, use, and sell goods and services.

Culture Regions

To organize their understanding of cultural development, geographers divide the Earth into culture regions. Each **culture region** includes countries that have certain traits in common. They may share similar economic systems, forms of government, and social groups. Their histories, religions, and art forms may share similar influences. The map below shows the culture regions you will study in this textbook.

READING Check **Place** Why are social groups important to the development of a culture?

MAP STUDY

1. **Regions** In which culture regions do Indo-European languages predominate?
2. **Place** Do people within the same culture region necessarily speak the same language? Explain.

NATIONAL GEOGRAPHIC **World Culture Regions**

Cultural Change

MAIN Idea Internal and external factors change cultures over time.

GEOGRAPHY AND YOU Have you ever moved to a new town or made friends with someone who has just moved into your community? Read to learn how the movement of people can change cultures.

Internal factors—new ideas, lifestyles, and inventions—create change within cultures. Change can also come through spatial interaction such as trade, migration, and war. The process of spreading new knowledge from one culture to another is called **cultural diffusion.**

The Agricultural Revolution

Cultural diffusion has been a **major** factor in cultural development since the dawn of human history. The earliest humans were *nomads,* groups of hunters and herders who had no fixed home but moved from place to place in search of food, water, and grazing land. As the Earth's climate warmed about 10,000 years ago, many of these nomads settled first in hilly areas and later in river valleys and on fertile plains. They became farmers who lived in permanent villages and grew crops on the same land every year. This shift from hunting and gathering food to producing food is known as the Agricultural Revolution.

By about 3500 B.C., some of these early farming villages had evolved into *civilizations,* highly organized, city-based societies with an advanced knowledge of farming, trade, government, art, and science.

Culture Hearths The world's first civilizations arose in **culture hearths,** early centers of civilization whose ideas and practices spread to surrounding areas. The map below shows that the most influential culture hearths developed in areas that make up the modern countries of **Egypt, Iraq, Pakistan, China,** and **Mexico.**

These five culture hearths had certain geographic features in common. They all emerged from farming settlements in areas with a mild climate and fertile land. In addition, they were located near a major river or source of water. The peoples made use of these favorable environments. They dug canals and ditches to irrigate the land. All of these factors enabled people to grow surplus crops.

Surplus food set the stage for the rise of cities and civilizations. With more food available, there was less need for everyone in a settlement to farm the land. People developed other ways of making a living. They created new technology and carried out specialized economic activities, such as metalworking and shipbuilding, that spurred the development of long-distance trade.

In turn, the increased wealth from trade led to the rise of cities and complex social systems. The people of a city needed a well-organized government to coordinate harvests, plan building projects, and manage an army for defense. Officials and merchants created writing systems to record and transmit information.

NATIONAL GEOGRAPHIC **World Culture Hearths**

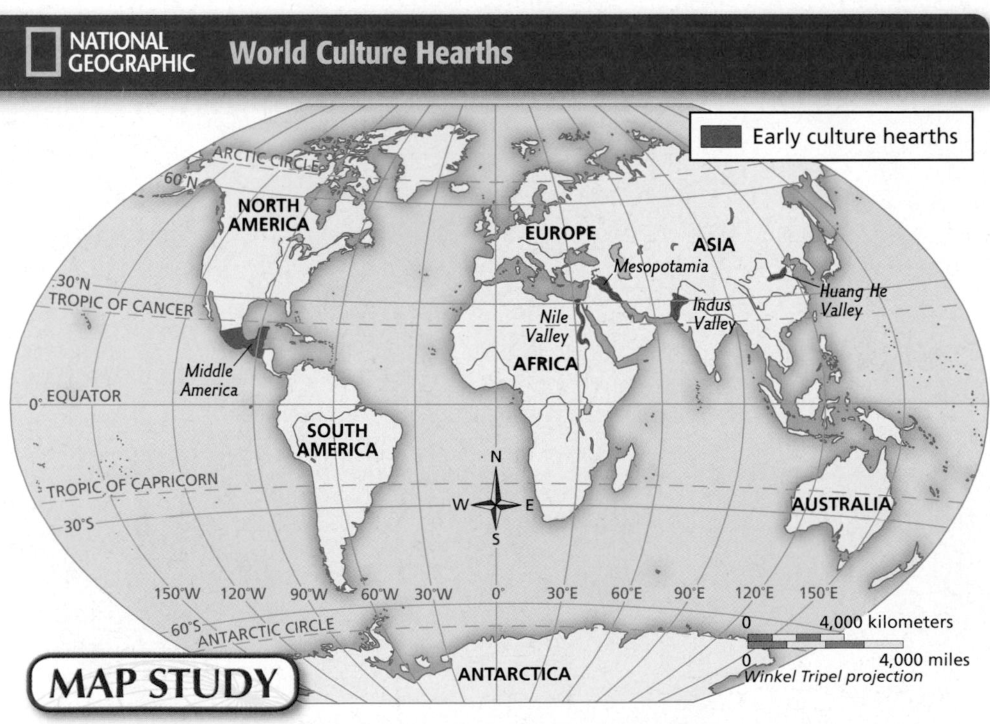

MAP STUDY

1. **Location** Where in Asia were the first major settlements located?
2. **Human-Environment Interaction** What kind of body of water do most of the culture hearths have in common?

Cultural Contacts Cultural contact among different civilizations promoted cultural change as ideas and practices spread through trade and travel. Migration has also fostered cultural diffusion. People migrate to avoid wars, persecution, and famines. In some instances, as in the case of enslaved Africans brought to the Americas, mass migrations have been forced. Conversely, positive factors—better economic opportunities and religious or political freedoms—may draw people from one place to another. Migrants carry their cultures with them, and their ideas and practices often blend with those of the people already living in the migrants' adopted countries.

Industrial and Information Revolutions

Cultural diffusion has increased rapidly during the last 250 years. In the 1700s and 1800s, some countries began to industrialize, using power-driven machines and factories to mass-produce goods. With new production methods, these countries produced goods quickly and cheaply, and their economies changed dramatically. This development, known as the Industrial Revolution, also led to social changes. As people left farms for jobs in factories and mills, cities grew larger.

NATIONAL GEOGRAPHIC *Although geisha are part of traditional Japanese culture, their daily lives are connected to the modern world.*

Movement How did the information revolution help connect cultures around the world?

At the end of the 1900s, the world experienced a new turning point—the information revolution. Computers now make it possible to store huge amounts of information and instantly send it all over the world, thus linking the cultures of the world more closely than ever before.

READING Check **Location** Where were the five earliest world culture hearths located?

THE WORLD

SECTION 2 REVIEW

Vocabulary

1. Explain the significance of: culture, language family, ethnic group, culture region, cultural diffusion, culture hearth.

Main Ideas

2. Describe the elements of culture geographers use to organize the world into culture regions.
3. What are the internal and external factors that change cultures over time? Use a web diagram like the one below to list factors that influence cultural change.

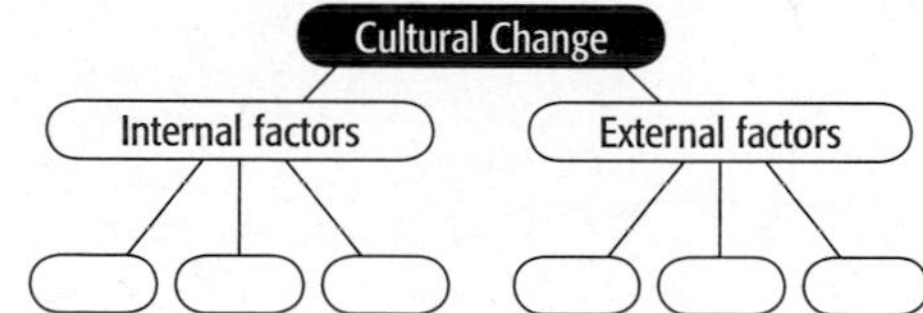

Critical Thinking

4. **Answering the Essential Question** How did cultural diffusion influence the Agricultural Revolution?
5. **Making Generalizations** Explain the factors that influence a country's ability to control territory.
6. **Identifying Cause and Effect** What cultural changes have resulted from the information revolution?
7. **Analyzing Visuals** Study the map of culture hearths on page 78. What factors influenced their location?

Writing About Geography

8. **Descriptive Writing** Use the Internet to find information about how various cultures view particular places or features. Then write an essay describing the similarities and differences you found.

Study Central™ To review this section, go to glencoe.com and click on Study Central.

World Religions

Geography of Religion

A *religion* is a set of beliefs in an ultimate reality and a set of practices used to express those beliefs. Religion is a key component of culture.

Each religion has its own special celebrations and worship styles. Most religions also have their own sacred texts, symbols, and sites. All of these aspects of religion help to unite followers regardless of where in the world they live.

TERMS

animism–belief that spirits inhabit natural objects and forces of nature

atheism–disbelief in the existence of any god

monotheism–belief in one God

polytheism–belief in more than one god

secularism–belief that life's questions can be answered apart from religious belief

sect–a subdivision within a religion that has its own distinctive beliefs and/or practices

NATIONAL GEOGRAPHIC **World Religions Today**

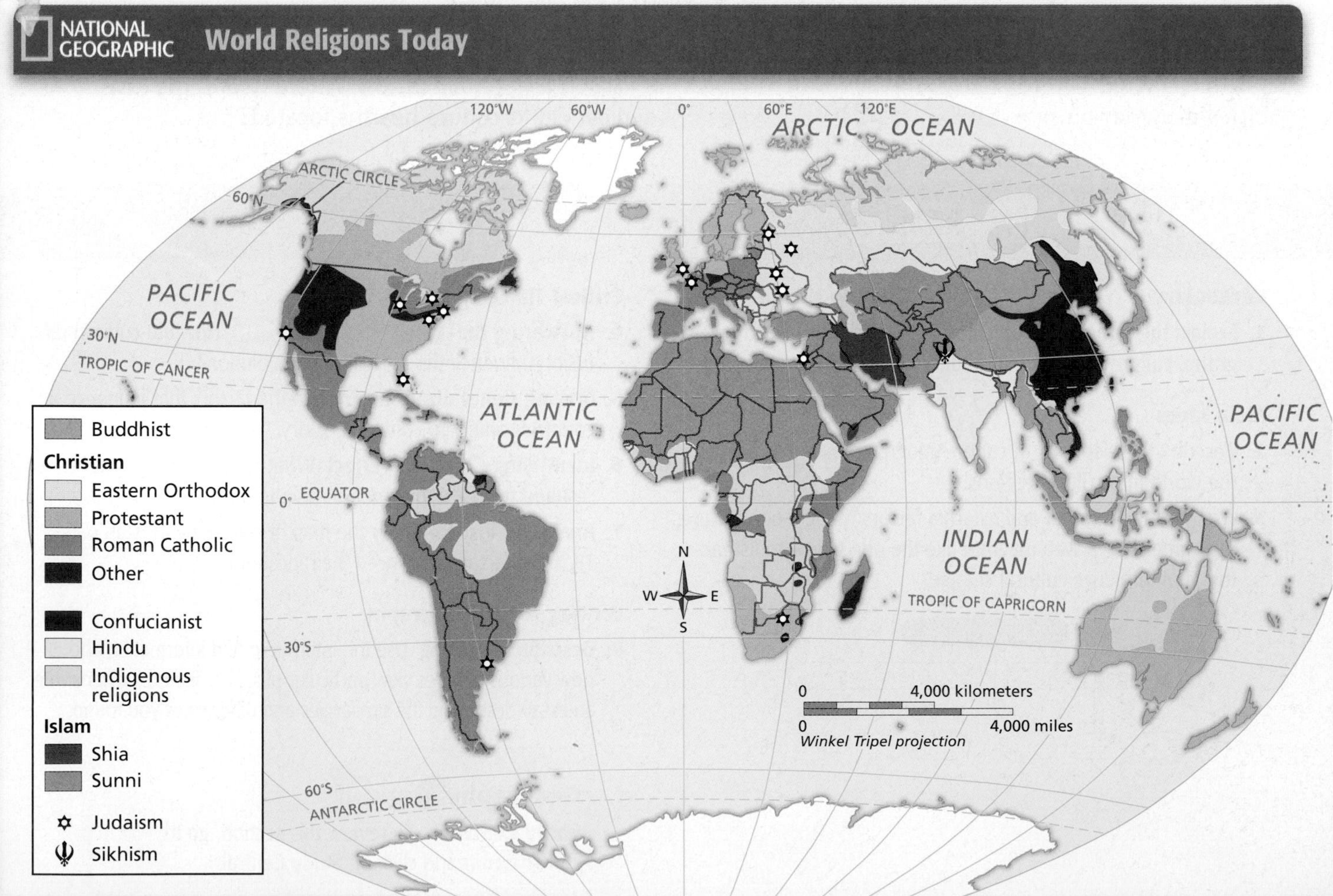

We study religion because it is an important component of culture, shaping how people interact with one another, dress, and eat. Religion is at the core of the belief system of a region's culture.

The diffusion of religion throughout the world has been caused by a variety of factors including migration, missionary work, trade, and war. Buddhism, Christianity, and Islam are the three major religions that spread their religion through missionary activities. Religions such as Hinduism, Sikhism, and Judaism are associated with a particular culture group. Followers are usually born into these religions. Sometimes close contact and differences in beliefs have resulted in conflict between religious groups.

Percentage of World Population

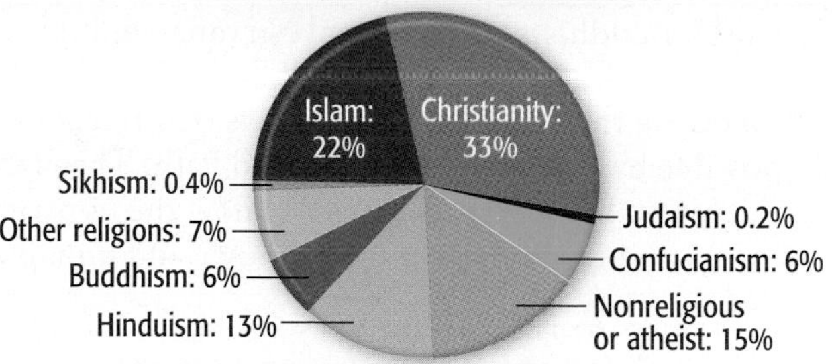

Note: Total exceeds 100% because numbers were rounded.
Sources: www.cia.gov, The World Factbook 2008; www.adherents.com.

NATIONAL GEOGRAPHIC Early Diffusion of Major World Religions

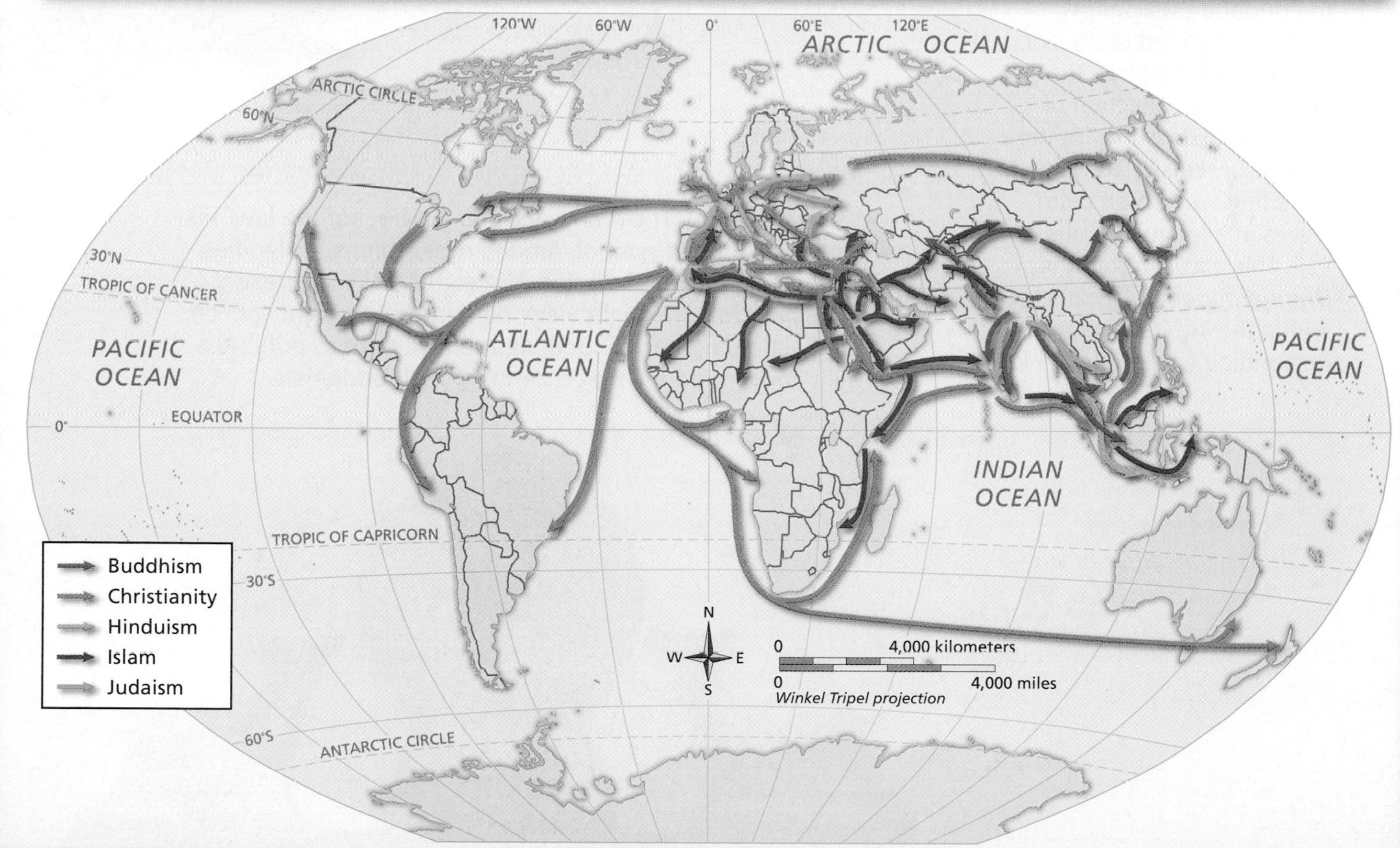

- Buddhism
- Christianity
- Hinduism
- Islam
- Judaism

Winkel Tripel projection

BUDDHISM

Siddhartha Gautama, known as the Buddha ("the Awakened") after his enlightenment at the age of 35, was born some 2,500 years ago in what is now Nepal. The Buddha's followers adhere to his teachings (dharma, meaning "divine law"), which aim to end suffering in the world. Buddhists call this goal Nirvana; and they believe that it can be achieved only by understanding the Four Noble Truths and by following the 4th Truth, which says that freedom from suffering is possible by practicing the Eightfold Path. Through the Buddha's teachings, his followers come to know the impermanence of all things and reach the end of ignorance and unhappiness.

Over time, as Buddhism spread throughout Asia, several branches emerged. The largest of these are Theravada Buddhism, the monk-centered Buddhism which is dominant in Sri Lanka, Burma, Thailand, Laos, and Cambodia; and Mahayana, a complex, more liberal variety of Buddhism that has traditionally been dominant in Tibet, Central Asia, Korea, China, and Japan.

Statue of the Buddha, Kamakura, Japan

Sacred Text For centuries the Buddha's teachings were transmitted orally. For Theravada Buddhists, the authoritative collection of Buddhist texts is the Tripitaka ("three baskets"). These texts were first written on palm leaves in a language called Pali. This excerpt from the *Dhammapada,* a famous text within the Tripitaka, urges responding to hatred with love:

> *"Never in this world is hate*
> *Appeased by hatred.*
> *It is only appeased by love—*
> *This is an eternal law."*
>
> —*Dhammapada I.5*

Sacred Symbol The *dharmachakra* ("wheel of the law") is a major Buddhist symbol. Among other things, it signifies the overcoming of obstacles. The eight spokes represent the Eightfold Path—right view, right intention, right speech, right action, right livelihood, right effort, right mindfulness, right concentration—that is central for all Buddhists.

Sacred Site Buddhists believe that Siddhartha Gautama achieved enlightenment beneath the Bodhi Tree in Bodh Gayā, India. Today, Buddhists from around the world flock to Bodh Gayā in search of their own spiritual awakening.

Worship and Celebration The ultimate goal of Buddhists is to achieve Nirvana, the enlightened state in which individuals are free from ignorance, greed, and suffering. Theravada Buddhists believe that monks are most likely to reach Nirvana because of their lifestyle of renunciation, moral virtue, study, and meditation.

Christianity

Christianity claims more members than any of the other world religions. It dates its beginning to the death of Jesus in A.D. 33 in what is now Israel. It is based on the belief in one God and on the life and teachings of Jesus. Christians believe that Jesus, who was born a Jew, is the son of God and is fully divine and human. Christians regard Jesus as the Messiah (Christ), or savior, who died for humanity's sins. Christians feel that people are saved and achieve eternal life by faith in Jesus.

The major forms of Christianity are Roman Catholicism, Eastern Orthodoxy, and Protestantism. All three are united in their belief in Jesus as savior, but have developed their own individual theologies.

Stained glass window depicting Jesus

Sacred Text The Christian Bible is the spiritual text for all Christians and is considered to be inspired by God. This excerpt, from Matthew 5:3-12, is from Jesus' Sermon on the Mount.

> *"Blessed are the poor in spirit, for theirs is the kingdom of heaven.*
> *Blessed are those who mourn, for they shall be comforted.*
> *Blessed are the meek, for they shall inherit the earth.*
> *Blessed are those who hunger and thirst for righteousness, for they shall be satisfied.*
> *Blessed are the merciful, for they shall obtain mercy.*
> *Blessed are the pure in heart, for they shall see God.*
> *Blessed are the peacemakers, for they shall be called sons of God.*
> *Blessed are those who are persecuted for righteousness' sake, for theirs is the kingdom of heaven.*
> *Blessed are you when men revile you and persecute you and utter all kinds of evil against you falsely on my account.*
> *Rejoice and be glad, for your reward is great in heaven, for so men persecuted the prophets who were before you."*

Sacred Symbol Christians believe that Jesus died for their sins. His death redeemed those who follow his teachings. The statue *Christ the Redeemer,* located in Rio de Janeiro, Brazil, symbolizes this fundamental belief.

Sacred Site The Gospels affirm that Bethlehem was the birthplace of Jesus. Consequently, it holds great importance to Christians. The Church of the Nativity is located in the heart of Bethlehem. It houses the spot where Christians believe Jesus was born.

Worship and Celebration Christians celebrate many events commemorating the life and death of Jesus. Among the most widely known and observed are Christmas, Good Friday, and Easter. Christmas is often commemorated by attending church services to celebrate the birth of Jesus. As part of the celebration, followers often light candles.

CONFUCIANISM

Confucianism began more than 2,500 years ago in China. Although considered a religion, it is actually a philosophy. It is based upon the teachings of Confucius, which are grounded in ethical behavior and good government.

The teachings of Confucius focused on three areas: social philosophy, political philosophy, and education. Confucius taught that relationships are based on rank. Persons of higher rank are responsible for caring for those of lower rank. Those of lower rank should respect and obey those of higher rank. Eventually his teachings spread from China to other East Asian societies.

Students studying Confucianism, Chunghak-dong, South Korea

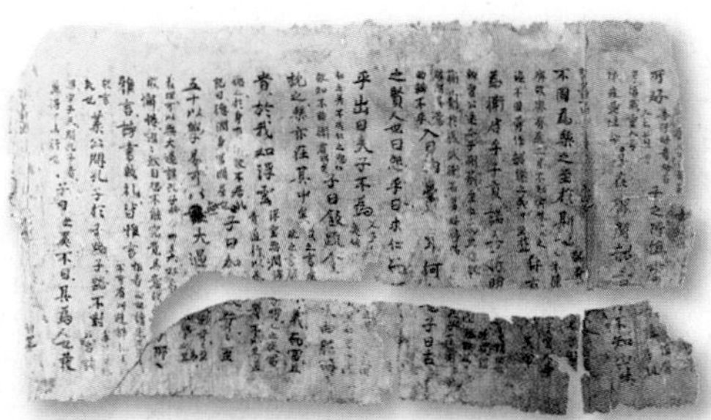

The Analects

Sacred Text Confucius was famous for his sayings and proverbs. These teachings were gathered into a book called the *Analects* after Confucius's death. Below is an example of Confucius's teachings.

Confucius said:

> *"To learn and to practice what is learned time and again is pleasure, is it not? To have friends come from afar is happiness, is it not? To be unperturbed when not appreciated by others is gentlemanly, is it not?"*
>
> —*The* Analects

Sacred Symbol Yin-yang, associated with both Confucianism and Daoism, symbolizes the harmony offered by the philosophies. The light half represents *yang,* the creative, firm, strong elements in all things. The dark half represents *yin,* the receptive, yielding, weak elements. The two act together to balance one another.

Sacred Site The temple at Qufu is a group of buildings dedicated to Confucius. It is located on Confucius's ancestral land. It is one of the largest ancient architectural complexes in China. Every year followers gather at Qufu to celebrate the birthday of Confucius.

Worship and Celebration Confucianism does not have a god or clergy, but there are temples dedicated to Confucius, the spiritual leader. Those who follow his teachings see Confucianism as a way of life and a guide to ethical behavior and good government.

Hinduism

Hinduism is the oldest of the world's major living religions. It developed among the cultures in India as they spread out over the plains and forests of the subcontinent. It has no single founder or founding date. Hinduism is complex: it has numerous sects and many different divinities are honored. Among the more famous Hindu gods are Brahma, Vishnu, and Shiva, who represent respectively the creative, sustaining, and destructive forces in the universe. Major Hindu beliefs are reincarnation, karma, and dharma.

Hindus believe the universe contains several heavens and hells. According to the concept of rebirth or reincarnation, which is central to their beliefs, souls are continually reborn. In what form one is reborn is determined by the good and evil actions performed in his or her past lives. Those acts are karma. A soul continues in the cycle of rebirth until release is achieved.

Sacred Text The Vedas consist of hymns, prayers, and speculations composed in ancient Sanskrit. They are the oldest religious texts in an Indo-European language. The Rig Veda, Sama Veda, Yajur Veda, and Atharva Veda are the four great Vedic collections. Together, they make up one of the most significant and authoritative Hindu religious texts.

Statue of Vishnu

"Now, whether they perform a cremation for
such a person or not,
people like him pass into the flame,
from the flame into the day,
from the day into the fortnight of the waxing moon
from the fortnight of the waxing moon into the six
months when the sun moves north,
from these months into the year,
from the year into the sun,
from the sun into the moon, and from
the moon into the lightning.
Then a person who is not human—
he leads them to Brahman.
This is the path to the gods, the path to Brahman.
Those who proceed along this path do not
return to this human condition."

—The Chandogya Upanishad 4:15.5

Sacred Symbol One important symbol of Hinduism is actually a symbol for a sound. "Om" is a sound that Hindus often chant during prayer, mantras, and rituals.

Sacred Site Hindus believe that when a person dies his or her soul is reborn. This is known as reincarnation. Many Hindus bathe in the Ganges and other sacred rivers to purify their soul and to be released from rebirth.

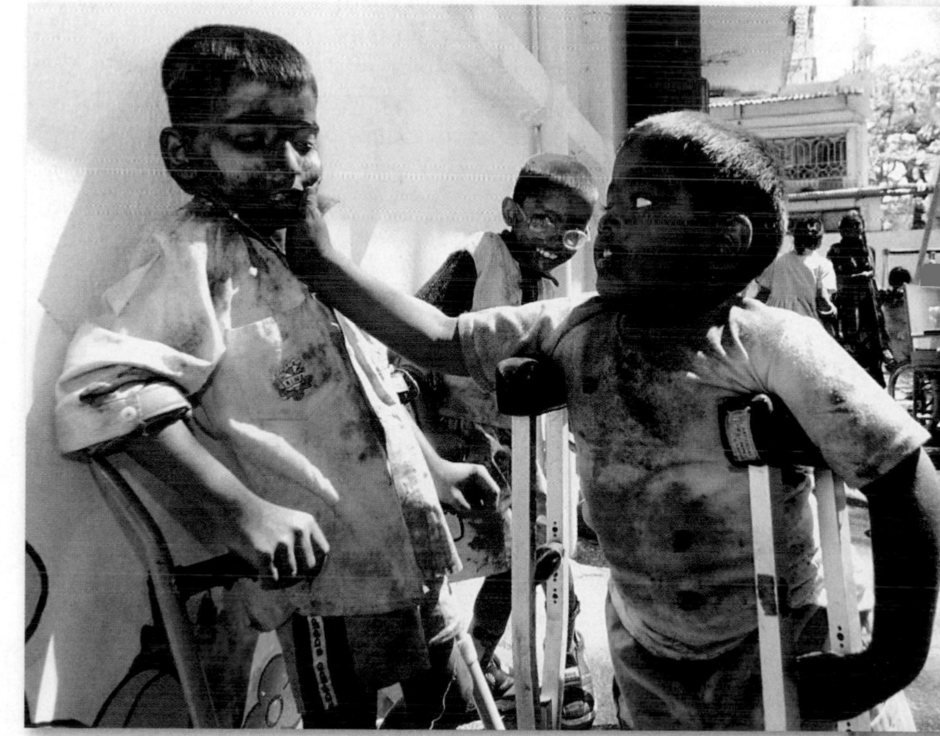

Worship and Celebration Holi is a significant North Indian Hindu festival celebrating the triumph of good over evil. As part of the celebration, men, women, and children splash colored powders and water on each other. In addition to its religious significance, Holi also celebrates the beginning of spring.

Islam

The Dome of the Rock, Jerusalem

Followers of Islam, known as Muslims, believe in one God, whom they call Allah. The word *Allah* is Arabic for "the god." The spiritual founder of Islam, Muhammad, began his teachings in Makkah (Mecca) in A.D. 610. Eventually the religion spread throughout much of Asia, including parts of India to the borders of China, and a substantial portion of Africa. According to Muslims, the Quran, their holy book, contains the direct word of God, revealed to their prophet Muhammad sometime between A.D. 610 and A.D. 632. Muslims believe that God created nature and without his intervention, there would be nothingness. God serves four functions: creation, sustenance, guidance, and judgment.

Central to Islamic beliefs are the Five Pillars. These are affirmation of the belief in Allah and Muhammad as his prophet; group prayer; tithing, or the giving of money to charity; fasting during Ramadan; and a pilgrimage to Makkah once in a lifetime if physically and financially able. Within Islam, there are two main branches, the Sunni and the Shia. The differences between the two are based on the history of the Muslim state. The Shia believed that the rulers should descend from Muhammad. The Sunni believed that the rulers need only be followers of Muhammad. Most Muslims are Sunni.

The Quran

Sacred Text The sacred text of Islam is the Quran. Preferably, it is written and read only in Arabic, but translations have been made into many languages. The excerpt below is a verse repeated by all Muslims during their five daily prayers.

> *"In the Name of Allah, the Compassionate,*
> *the Merciful,*
> *Praise be to Allah, the Lord of the World,*
> *The Compassionate, the Merciful,*
> *Master of the Day of Judgment,*
> *Only You do we worship, and only You*
> *Do we implore for help.*
> *Lead us to the right path,*
> *The path of those you have favored*
> *Not those who have incurred*
> *Your wrath or*
> *Have gone astray."*
>
> —*The Quran*

Sacred Symbol Islam is often symbolized by the crescent moon. It is an important part of Muslim rituals, which are based on the lunar calendar.

Sacred Site Makkah is a sacred site for all Muslims. One of the Five Pillars of Islam states that all those who are physically and financially able must make a hajj, or pilgrimage, to the holy city once in their life. Practicing Muslims are also required to pray facing Makkah five times a day.

Worship and Celebration Ramadan is a month-long celebration commemorating the time during which Muhammad received the Quran from Allah. It is customary for Muslims to fast from dawn until sunset all month long. Muslims believe that fasting helps followers focus on spiritual rather than bodily matters and creates empathy for one's fellow men and women. Ramadan ends with a feast known as Eid-al-Fitr, or Feast of the Fast.

Judaism

Judaism is a monotheistic religion. In fact, Judaism was the first major religion to believe in one God. Jews trace their national and religious origins back to God's call to Abraham. Jews have a covenant with God. They believe that God, who expects them to pursue justice and live ethical lives, will one day usher in an era of universal peace.

Over time Judaism has separated into branches, including Orthodox, Reform, Conservative, and Reconstructionist. Orthodox Jews are the most traditional of all the branches.

El Ghriba Synagogue, Jerba, Tunisia

Sacred Text The Torah is the five books of Moses, which tell the story of the origins of the Jews and explain Jewish laws. The remainder of the Hebrew Bible contains the writings of the prophets, Psalms, and ethical and historical works.

> *"I am the Lord your God, who brought you out of the land of Egypt, out of the house of slavery; you shall have no other gods before me."*
>
> —*Exodus 20:2*

The Torah scroll

Sacred Symbol The menorah is used in the celebration of Hanukkah, commemorating the rededication of the Temple of Jerusalem following the Maccabees' victory over the Syrian Greeks.

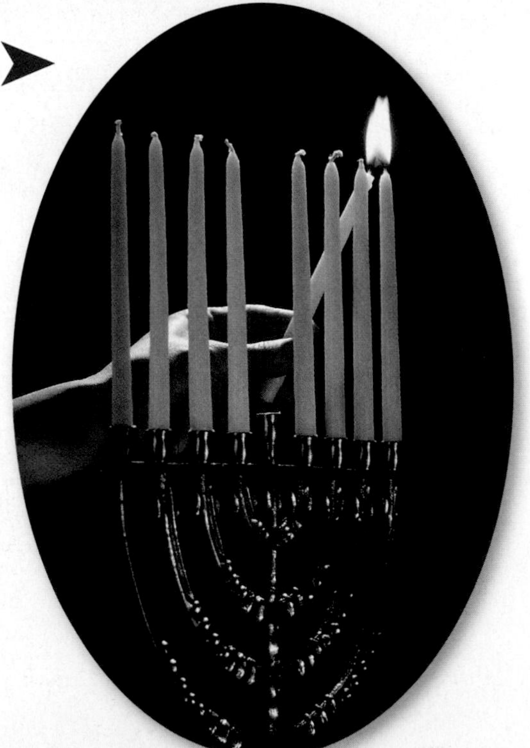

Sacred Site The Western Wall is what remains of the structure surrounding the Second Jerusalem Temple, built after the Jews' return from the Babylonian captivity. It is considered a sacred spot in Jewish religious tradition. Prayers are offered at the wall morning, afternoon, and evening.

Worship and Celebration The day-long Yom Kippur service ends with the blowing of the ram's horn (shofar). Yom Kippur is the holiest day in the Jewish calendar. During Yom Kippur, Jews do not eat or drink for 25 hours. The purpose is to reflect on the past year and gain forgiveness from God for one's sins. It falls in September or October, ten days after Rosh Hashanah, the Jewish New Year.

Sikhism

Sikhism emerged in the mid-1500s in the Punjab, in northwest India, rising from the religious experience and teachings of Guru Nanak. The religion exhibits influences from Islam and Hinduism, but it is distinct from both. Sikh traditions teach that Nanak encountered God directly and was commissioned by Him to be His servant.

Sikhs ("students, disciples") believe in one almighty god who is formless and without qualities (*nirguna*) but can be known through meditation and heard directly. Sikhism forbids discrimination on the basis of class, color, religion, caste, or gender. While over 80 percent of the world's 23 million Sikhs live in the Punjab, Sikhism has spread widely as many Sikhs have migrated to new homes around the world.

Sikh man, Chapeltown, England

The Adi Granth

Sacred Text The great authoritative sacred text for Sikhs is the Adi Granth ("Principal Book," also known as the Guru Granth Sahib). Compiled from the mid-1500s through the 1600s, it includes contributions from Sikh gurus and from some persons also claimed as saints by Hindus and Muslims, such as Namdev, Ravidas, and Kabir.

> *"Enshrine the Lord's Name within your heart. The Word of the Guru's Bani prevails throughout the world, through this Bani, the Lord's Name is obtained."*
>
> —*Guru Amar Das, page 1066*

Sacred Symbol The sacred symbol of the Sikhs is the *khanda.* It is composed of four traditional Sikh weapons: the *khanda* or double-edged sword (in the center), from which the symbol takes its name; the *cakkar* (disk), and two curved daggers (*kirpan*) representing temporal and spiritual power, respectively Piri and Miri.

Sacred Sites Amritsar is the spiritual capital of Sikhism. The Golden Temple (*Harimandir Sahib*) in Amritsar is the most sacred of Sikh shrines.

Worship and Celebration Vaisakhi is a significant Punjabi and Sikh festival in April celebrating the new year and the beginning of the harvest season. Celebrations often take place along riverbanks with participants dancing and wearing brightly colored clothes.

Indigenous Religions

There are many varieties of religious belief that are limited to particular ethnic groups. These local religions are found in Africa as well as isolated parts of Japan, Australia, and the Americas.

Most local religions reflect a close relationship with the environment. Some groups teach that people are a part of nature, not separate from it. Animism is characteristic of many indigenous religions. Natural features are sacred, and stories about how nature came to be are an important part of religious heritage. Although many of these stories have been written down in modern times, they were originally transmitted orally.

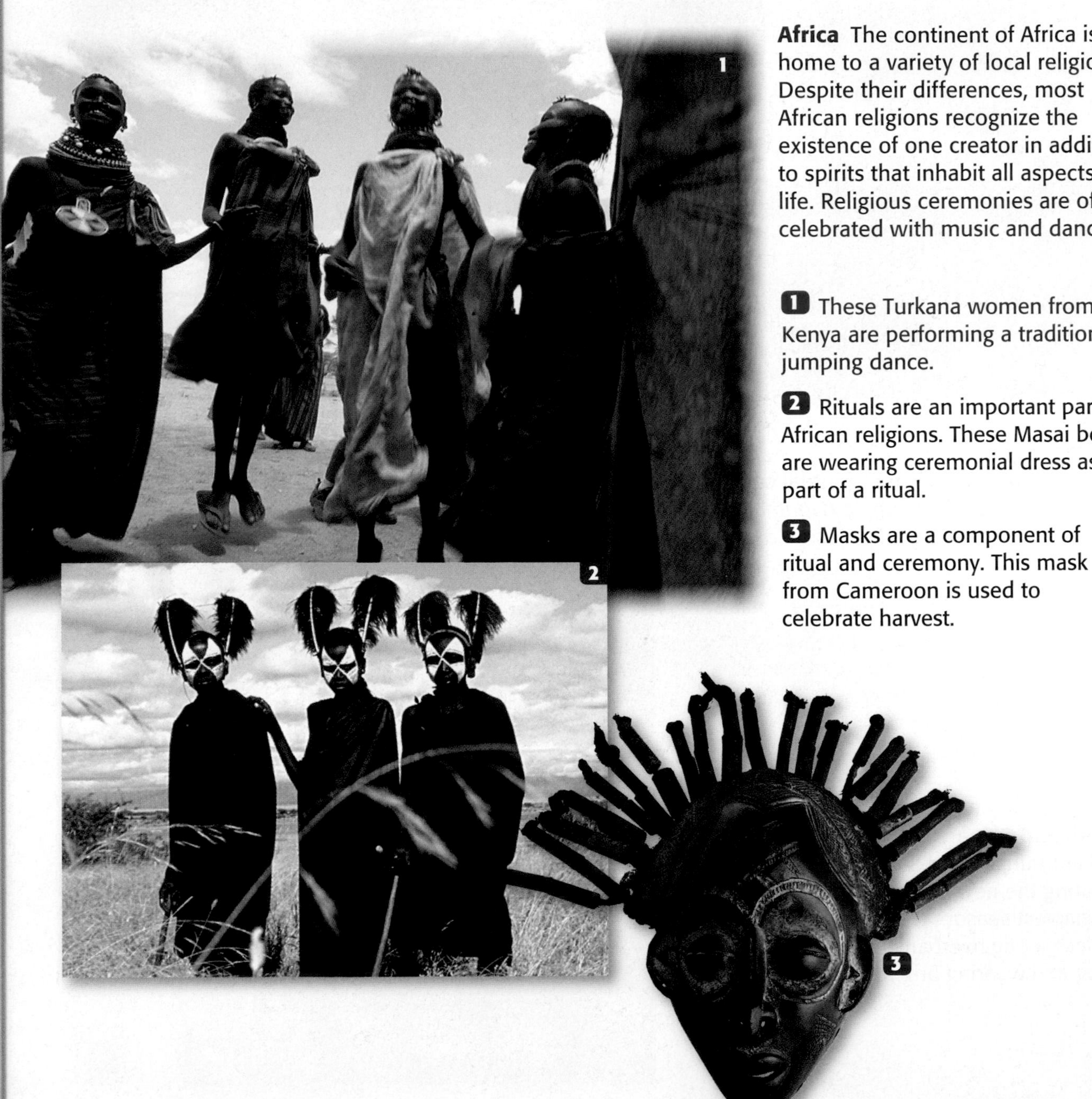

Africa The continent of Africa is home to a variety of local religions. Despite their differences, most African religions recognize the existence of one creator in addition to spirits that inhabit all aspects of life. Religious ceremonies are often celebrated with music and dance.

1 These Turkana women from Kenya are performing a traditional jumping dance.

2 Rituals are an important part of African religions. These Masai boys are wearing ceremonial dress as part of a ritual.

3 Masks are a component of ritual and ceremony. This mask from Cameroon is used to celebrate harvest.

Japan Shinto, founded in Japan, is the largest indigenous religion. It dates back to prehistoric times and has no formal doctrine. The gods are known as kami. Ancestors are also revered and worshiped. Its four million followers often practice Buddhism in addition to practicing Shinto.

4 This Shinto priest is presiding over a ritual at a Japanese temple. These priests often live on shrine grounds.

5 Shinto shrines, like this one, are usually built in places of great natural beauty to emphasize the relationship between people and nature.

Australia The Australian Aboriginal religion has no deities. It is based upon a belief known as the Dreaming, or Dreamtime. Followers believe that ancestors sprang from the Earth and created all people, plant, and animal life. They also believe that these ancestors continue to control the natural world.

6 These Aborigine women are blessing a newborn with smoke during a traditional ritual intended to ensure the child's health and good fortune.

7 Aborigines, like these young girls, often paint their faces with the symbols of their clan or family group.

Indigenous Religions

Native Americans The beliefs of most Native Americans center on the spirit world; however, the rituals and practices of individual groups vary. Most Native Americans believe in a Great Spirit who, along with other spirits, influences all aspects of life. These spirits make their presence known primarily through acts of nature.

The rituals, prayers, and ceremonies of Native Americans are often centered on health and good harvest and hunting. Rituals used to mark the passage through stages of life, including birth, adulthood, and death, are passed down as tribal traditions. Religious ceremonies often focus on important points in the agricultural and hunting seasons. Prayers, which are offered in song and dance, also concentrate on agriculture and hunting themes as well as health and well-being.

1 Rituals are passed down from generation to generation. These Native Americans are performing a ritual dance in Utah.

2 There are many different Native American groups throughout the United States and Canada. This Pawnee is wearing traditional dress during a celebration in Oklahoma.

3 Totem poles, like this one in Alaska, were popular among the Native American peoples of the Northwest Coast. They were often decorated with mythical beings, family crests, or other figures. They were placed outside homes.

Assessment

Reviewing Vocabulary

Match the following terms with their definition.

1. sect
2. monotheism
3. polytheism
4. animism
5. atheism

a. belief that spirits inhabit natural objects and forces of nature
b. belief in one God
c. a subdivision within a religion that has its own distinctive belief and/or practices
d. belief in more than one god
e. disbelief in the existence of any god

Reviewing the Main Ideas

World Religions (pp. 80–81)

6. Which religion has the most followers worldwide? Which has the second-largest group of followers?
7. **Analyzing Visuals** Compare the maps at the bottom of pages 80 and 81. Which religions have spread throughout the world? What factors may have contributed to this spread?
8. On a separate sheet of paper, make a table of the major world religions. Use the chart below to get you started.

Name	Founder	Geographic distribution	Sacred sites
Buddhism			
Christianity			
Confucianism			
Hinduism			
Islam			
Judaism			
Sikhism			
Indigenous			

Buddhism (pp. 82–83)

9. According to Buddhism, how can the end of suffering in the world be achieved?
10. What is Nirvana? According to Buddhists, who is most likely to achieve Nirvana and why?

Christianity (pp. 84–85)

11. In what religion was Jesus raised?
12. Why do Christians regard Jesus as their savior?

Confucianism (pp. 86–87)

13. What is Confucianism based on? Why might some not consider it a religion?
14. What does yin-yang symbolize?

Hinduism (pp. 88–89)

15. What type of religion is Hinduism? Where did it develop?
16. Describe reincarnation. What role do Hindus believe karma plays in this process?

Islam (pp. 90–91)

17. What are the two branches of Islam? What is the main difference between the two groups?
18. What role does Makkah play in the Islamic faith?

Judaism (pp. 92–93)

19. What is the Torah?
20. What is the purpose of Yom Kippur?

Sikhism (pp. 94–95)

21. Where do most Sikhs live? Why?
22. What other religions have contributed to the Adi Granth?

Indigenous Religions (pp. 96–98)

23. Many of the sacred stories in local religions explain the creation of people, animals, and plant life. Why would religions feature such stories?
24. Which of the indigenous religions has the largest membership?

Critical Thinking

25. **Drawing Conclusions** How are major religions similar? How are they different?
26. **Analyzing Information** How do people's religious beliefs affect what people eat and how they dress?
27. **Making Inferences** How do religious beliefs influence a society's laws?

Problem-Solving Activity

28. **Research Project** Use library and Internet sources to research the role of food and food customs in one of the world's major religions. Create a presentation to report your findings to the class.

SECTION 3

 section audio  spotlight video

Guide to Reading

Essential Question

What types of human systems provide the power for groups of people to control Earth's surface?

Content Vocabulary
- unitary system (p. 101)
- federal system (p. 101)
- autocracy (p. 101)
- monarchy (p. 102)
- oligarchy (p. 102)
- democracy (p. 102)
- traditional economy (p. 103)
- market economy (p. 103)
- mixed economy (p. 103)
- command economy (p. 103)

Academic Vocabulary
- unique (p. 101)
- authority (p. 101)
- assembly (p. 102)

Places to Locate
- United States (p. 101)
- Saudi Arabia (p. 102)
- United Kingdom (p. 102)
- China (p. 104)
- Vietnam (p. 104)

Reading Strategy

Taking Notes Use the major headings of the section to create an outline similar to the one below.

Political and Economic Systems
I. Features of Government
A.
B.

Political and Economic Systems

Governments and economies of countries around the world are becoming increasingly interconnected. Some countries or groups of countries, such as the European Union, have strong economies that allow them to help improve standards of living in other countries.

NATIONAL GEOGRAPHIC VOICES AROUND THE WORLD

"I drove east on the highway that connects the capital of Tallinn with Narva, on the Russian border. Nearly everywhere I looked I saw the handiwork of the European Union, starting with the road itself. The EU has already invested millions of euros to improve the highway, which serves as the main link to St. Petersburg, Russia. This highway passes the town of Sillamäe, once a "closed" city run by the Soviet military, which enriched uranium for weapons programs in a huge factory overlooking the sea. The EU is here, too, kicking in more than a million dollars to help prevent the radioactive waste from leaching into the Baltic Sea."

—Don Belt, "Europe's Big Gamble," *National Geographic,* May 2004

The night express at Tallinn, Estonia

Features of Government

MAIN Idea Territory, population, and sovereignty influence levels and types of governments in countries around the world.

GEOGRAPHY AND YOU What do you know about the powers of government in the U.S. federal system? Read to learn about the powers of government in other places.

Today the world includes nearly 200 independent countries that vary in size, military might, natural resources, and world influence. Each country is defined by characteristics such as territory, population, and *sovereignty,* or freedom from outside control. These elements are brought together under a government. A government must make and enforce policies and laws that are binding on all people living within its territory.

NATIONAL GEOGRAPHIC *Ellen Johnson-Sirleaf, Africa's first elected female head of state, is president of the unitary government of Liberia.*

Place How does the unitary system of government differ from that of a federal system?

Levels of Government

The government of each country has **unique** characteristics that relate to that country's historical development. To carry out their functions, governments are organized in a variety of ways. Most large countries have several different levels of government. These usually include a national or central government, as well as the governments of smaller internal divisions such as provinces, states, counties, cities, towns, and villages.

Unitary System A **unitary system** of government gives all key powers to the national or central government. This structure does not mean that only one level of government exists. Rather, it means that the central government creates state, provincial, or other local governments and gives them limited sovereignty. The United Kingdom and France both developed unitary governments as they emerged from smaller territories during the late Middle Ages and early modern times.

Federal System A **federal system** of government divides the powers of government between the national government and state or provincial governments. Each level of government has sovereignty in some areas. The **United States** developed a federal system after the thirteen colonies became independent from Great Britain.

Another similar type of government structure is a confederation, a loose union of independent territories. The United States at first formed a confederation, but this type of political arrangement failed to provide an effective national government. As a result, the U.S. Constitution established a strong national government, while preserving some state government powers. Today, other countries with federal or confederal systems include Canada, Switzerland, Mexico, Brazil, Australia, and India.

Types of Governments

Governments can be classified by asking the question: Who governs the state? Under this classification system, all governments belong to one of the three major groups: (1) autocracy—rule by one person; (2) oligarchy—rule by a few people; or (3) democracy—rule by many people.

Autocracy Any system of government in which the power and **authority** to rule belong to a single individual is an **autocracy** (aw•TAH•kruh•see). Autocracies are the oldest and one of the most common forms of government. Most autocrats achieve and maintain their position of authority through inheritance or by the ruthless use of military or police power.

Supporters of the market system claim, however, that without free decision making and incentives, businesses will not innovate or produce products that people want. Customers will be limited in their choices, and economies will stagnate. As a result of these problems, command economies often decline. The Soviet Union, as described below by a Russian observer, provided an example of this situation:

> *In 1961 the [Communist] party predicted . . . that the Soviet Union would have the world's highest living standard by 1980. . . . But when that year came and went, the Soviet Union still limped along, burdened by . . . a stagnant economy.*
>
> Dusko Doder, "The Bolshevik Revolution," *National Geographic,* October 1992

By 2000, Russia and the other countries that were once part of the Soviet Union were developing market economies. **China** and **Vietnam** have allowed some free enterprise to promote economic growth, although their governments tightly control political affairs.

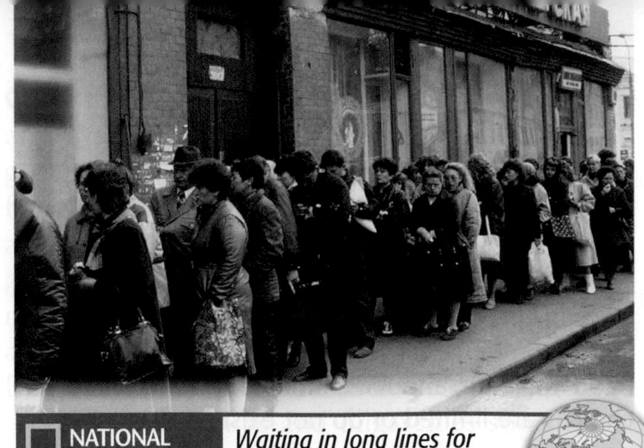

NATIONAL GEOGRAPHIC *Waiting in long lines for limited amounts of goods was common for people living in the Soviet command economy.*

Place How do communism and socialism differ?

Socialism Socialism allows a wider range of free enterprise. It has three main goals: (1) an equitable distribution of wealth and economic opportunity; (2) society's control, through its government, of decisions about public goods; and (3) public ownership of services and factories that are essential. Some socialist countries, like those in Western Europe, are democracies. Under democratic socialism, people have basic human rights and elect their political leaders.

READING Check **Place** On what idea is a market economy based?

SECTION 3 REVIEW

Vocabulary

1. Explain the significance of: unitary system, federal system, autocracy, monarchy, oligarchy, democracy, traditional economy, market economy, mixed economy, command economy.

Main Ideas

2. Define the three major economic systems. What are the three basic economic decisions that all economic systems must make?
3. Use a table like the one below to summarize the features of government–levels of government and types of government–that exist around the world.

Level of government	Example	Type of government	Example

Critical Thinking

4. **Answering the Essential Question** Explain the different ways in which an autocracy, an oligarchy, and a democracy exercise authority.
5. **Comparing and Contrasting** What different roles might local citizens have in government decision making under a unitary system, a federal system, and a confederation?
6. **Analyzing Information** How does a market economy affect other economies in a region?

Writing About Geography

7. **Expository Writing** Write a one-page essay explaining the human and physical geographic characteristics that can influence a country's foreign policy. Use Iraq, Israel, Japan, and the United Kingdom as examples.

Geography ONLINE

Study Central™ To review this section, go to glencoe.com and click on Study Central.

SECTION 4

 section audio

 spotlight video

Guide to Reading

Essential Question

How does the availability and use of natural resources affect economic activities and the environment?

Content Vocabulary
- natural resource (*p. 106*)
- industrialization (*p. 108*)
- developed country (*p. 108*)
- newly industrialized country (*p. 108*)
- developing country (*p. 108*)
- free trade (*p. 109*)
- pollution (*p. 109*)

Academic Vocabulary
- ensure (*p. 106*)
- conduct (*p. 109*)
- benefit (*p. 109*)

Places to Locate
- Malaysia (*p. 108*)
- European Union (*p. 109*)

Reading Strategy

Organizing As you read about natural resources, complete a web diagram similar to the one below by listing types of renewable energy resources.

Renewable Energy Resources

Resources, Trade, and the Environment

The growth of the global economy continues to make the world's peoples increasingly interdependent, or reliant on each other. Natural resources are extracted and traded around the world. And although important to modern life, they are misused and often cause pollution and damage to the environment.

NATIONAL GEOGRAPHIC Voices Around the World

"Below more than a mile of ocean and three more of mud and rock, the prize is waiting. At the surface a massive drilling vessel . . . shudders every few minutes as its thrusters put out a burst of power to fight the strong current. The PA system crackles, warning of small amounts of gas bubbling from the deep Earth. And in the shadow of the 23-story-tall derrick, engineers and managers gather in worried knots. 'We've got an unstable hole,' laments Bill Kirton, who's overseeing the project. . . . [I]f the wells live up to expectations, each will eventually gush tens of thousands of barrels a day. 'That's like a well in Saudi Arabia,' says [a drilling supervisor]. 'We hardly get those in the U.S. anymore.'"

—Tim Appenzeller, "The End of Cheap Oil," *National Geographic*, June 2004

A worker on an oil drillship

UNIT 2 The United States and Canada

Grand Canyon, Arizona, United States

NGS ONLINE To learn more about the United States and Canada visit www.nationalgeographic.com/education.

Why It Matters

The United States and Canada are peaceful neighbors, sharing the longest undefended border in the world. These two countries have many things in common, including similar ways of life and a democratic heritage. In recent years, free trade has brought their economies closer together. In each country, one finds an increasing number of products that were made in the other country.

The United States and Canada

PHYSICAL GEOGRAPHY The United States and Canada span North America, stretching from the Pacific Ocean to the Atlantic. These two huge countries share many physical features. Mountains frame their eastern and western edges, cradling a central region of vast plains.

When people first arrived on these plains, they found an immense sea of grass. Beneath the gently rolling landscape lay dark, fertile soil that settlers eventually transformed into some of the world's most productive farmland. To the east of the plains stand the ancient, rounded Appalachian Mountains. To the west are the much younger Rocky Mountains. Almost every imaginable type of climate—from tundra to desert to tropical wet—can be found within the borders of these two diverse countries.

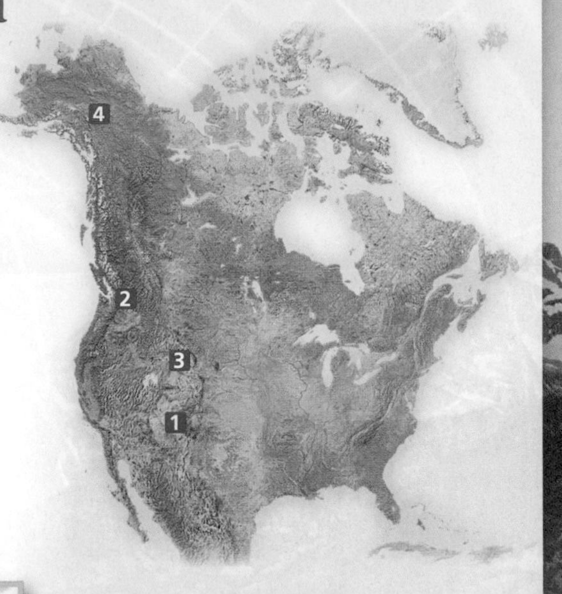

1 PLAINS AND PLATEAUS
Located along the Utah-Arizona border, the sandstone rock forms known as the Two Mittens attract tourists from around the world.

2 LAKES AND RIVERS Long rivers, such as the Fraser located in British Columbia, have played an important role in trade and industry in the United States and Canada.

3 MOUNTAINS The Rocky Mountains are the longest mountain range in North America, stretching from British Columbia in Canada to New Mexico in the United States.

4 NATURAL RESOURCES Oil is a vital resource. The Trans-Alaska Pipeline transports approximately 1 million barrels of crude oil a day across 800 miles (1,300 km) of rough Alaskan terrain to the port of Valdez in southern Alaska.

UNIT 2 WHAT MAKES THIS A REGION?

United States and Canada

CULTURAL GEOGRAPHY North America is a land of immigrants. It is believed the first peoples, ancestors of the Native Americans, came to the region from Asia. In the 1500s, immigrants began arriving from Europe. In the centuries that followed, others came from Africa, Asia, and Latin America. Many made this land their home by choice. Others were forced to come as exiles or slaves. Together these groups have shaped the culture of the region.

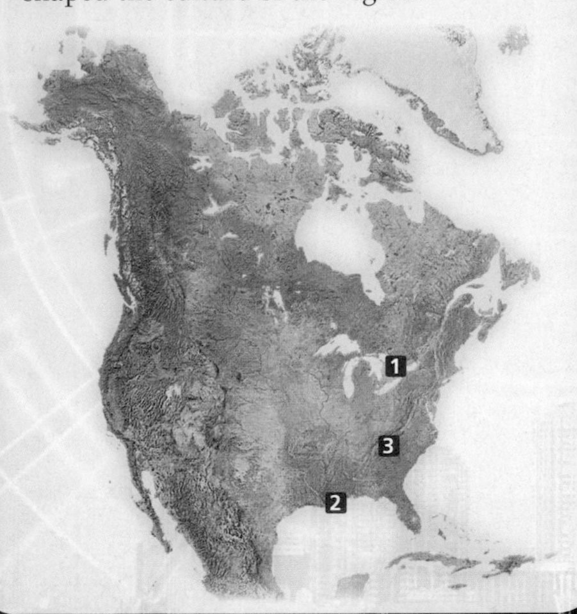

1

REGIONAL TIME LINE

10,000 B.C. Land bridge crossing/early settlement of indigenous peoples

A.D. 970–1020 Leif Eriksson first European in Americas

A.D. 1607 First English settlement at Jamestown

A.D. 1756–63 Seven Years' War; Britain awarded all French possessions in North America

A.D. 1775–1783 American Revolution

A.D. 1776 U.S. independence declared

George Washington

A.D. 1789 George Washington is elected president of the United States

10,000 B.C. | A.D. 1000 | A.D. 1600 | A.D. 1700

1 ECONOMY Today the service industry employs most of the workers in the United States and Canada. Many of these jobs are located in urban centers such as Toronto, Canada's largest city.

2 CULTURE Immigration to the United States and Canada from other areas of the world has had a dramatic effect on the cultures of the two countries. Parades celebrating these cultural roots are common throughout the region.

3 PEOPLE This Cherokee woman is creating beadwork. Despite being the first peoples to inhabit the region, Native Americans were pushed from their ancestral lands by European settlers.

A.D. 1800 — A.D. 1900 — A.D. 2000

A.D. 1861–1865 American Civil War

A.D. 1863 Abraham Lincoln presents the Emancipation Proclamation

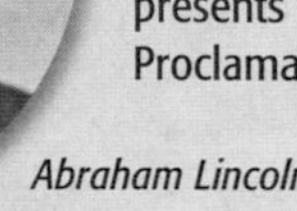

Abraham Lincoln

A.D. 1906 Earthquake in San Francisco and subsequent fires destroy 80 percent of city; more than 3,000 die

A.D. 1929 Great Depression begins

A.D. 1931–1939 "Dust Bowl" in U.S. Midwest and Southern Plains

A.D. 1939–1945 World War II: (U.S., 1941–1945; Canada, 1939–1945)

A.D. 1982 Pierre Trudeau is Prime Minister of Canada; United Kingdom transfers all powers relating to the constitution back to Canada

A.D. 2001 Terrorist attacks in the United States by al-Qaeda

A.D. 2008 Barack Obama is elected president of the United States

PHYSICAL United States and Canada

ARCTIC OCEAN
Greenland Sea
Greenland (Kalaallit Nunaat)
ARCTIC CIRCLE
Ellesmere Island
Queen Elizabeth Islands
Devon I.
Baffin Bay
Baffin Island
Davis Strait
Banks Island
Victoria Island
Beaufort Sea
Point Barrow
Bering Sea
Aleutian Is.
BROOKS RANGE
Yukon R.
Mt. McKinley 20,320 ft. (6,194 m)
ALASKA RANGE
Kodiak I.
Gulf of Alaska
Mt. Logan 19,551 ft. (5,959 m)
MACKENZIE MTS.
Mackenzie R.
Great Bear Lake
Great Slave Lake
Southampton I.
Hudson Strait
Ungava Peninsula
Labrador Sea
Labrador
Hudson Bay
COAST MOUNTAINS
Queen Charlotte Is.
Peace R.
Athabasca R.
Lake Athabasca
INTERIOR PLAINS
Nelson R.
Saskatchewan R.
CANADIAN SHIELD
Newfoundland
Vancouver I.
FRASER PLATEAU
ROCKY MOUNTAINS
Lake Winnipeg
Cape Breton I.
Sable I.
Columbia R.
CASCADE RANGE
COLUMBIA PLATEAU
GREAT PLAINS
Lake Superior
St. Lawrence R.
Lake Huron
Lake Ontario
PACIFIC OCEAN
COAST RANGES
SIERRA NEVADA
Snake R.
Black Hills
Missouri R.
Great Salt Lake
GREAT BASIN
Lake Michigan
Cape Cod
Platte R.
CENTRAL LOWLANDS
Lake Erie
APPALACHIAN MOUNTAINS
Mt. Whitney 14,494 ft. (4,418 m)
Colorado R.
COLORADO PLATEAU
Ohio R.
Chesapeake Bay
Death Valley −282 ft. (−86 m)
Arkansas R.
OZARK PLATEAU
Mississippi R.
PIEDMONT
Cape Hatteras
Red R.
Rio Grande
COASTAL PLAIN
ATLANTIC OCEAN
TROPIC OF CANCER
The Everglades
Gulf of Mexico
LATIN AMERICA
Caribbean Sea
180°
160°W
140°W
120°W
100°W
80°W
60°W
40°W
20°W
0°
80°N
60°N
40°N
20°N

Elevations

Feet	Meters
13,100	4,000
6,500	2,000
1,600	500
650	200
0	0

National boundary
State or provincial boundary
▲ Mountain peak
▼ Lowest point

160°W 155°W
Hawaiian Islands
Niihau
Kauai
Kaula
Oahu
Molokai
Lanai
Maui
Kahoolawe
20°N
Hawaii
PACIFIC OCEAN
0 200 kilometers
0 200 miles
Albers Equal-Area projection

N S E W

0 1,000 kilometers
0 1,000 miles
Lambert Azimuthal Equal-Area projection

Obstacles and Opportunities

The landscapes of the United States and Canada are marked by a variety of physical features that act both as obstacles and as opportunities for progress. As you study the maps and graphics on these pages, look for the geographical features that make the region unique. Then answer the questions below on a separate sheet of paper.

1. What physical features have acted as barriers to settlement in the United States and Canada?
2. What benefits has the Great Lakes–St. Lawrence Seaway System provided to the cities located along the Great Lakes?
3. What has contributed to the wearing away of the Appalachian Mountains? What predictions can be made about the future of other geographically younger mountain ranges such as the Rockies?

Very Different Mountain Ranges

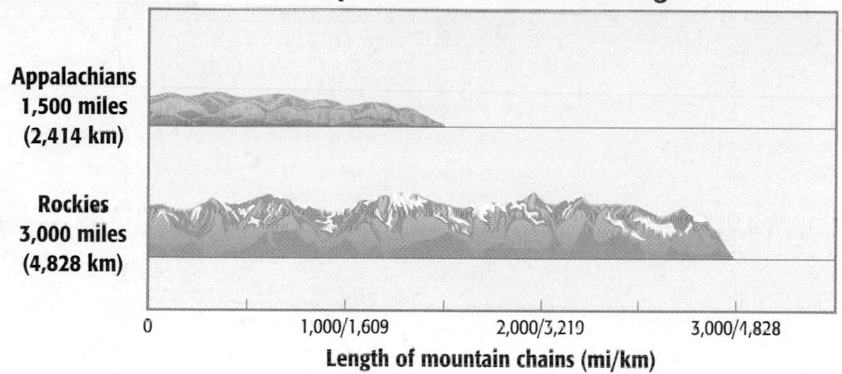

Great Lakes System Profile

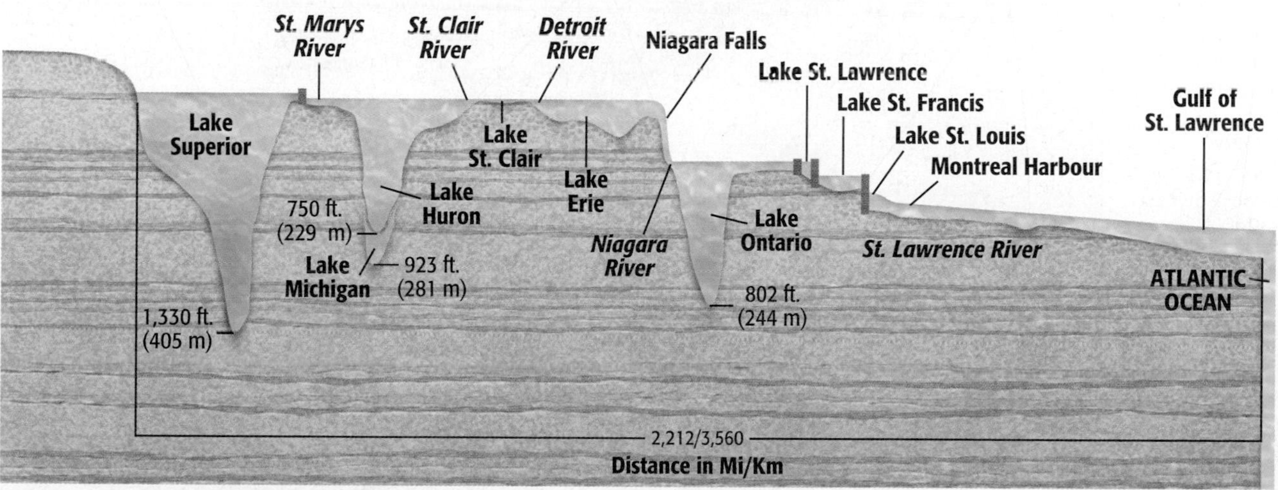

Source: U.S. Army Corps of Engineers, Detroit District.

POLITICAL United States and Canada

ARCTIC OCEAN
Greenland Sea
Bering Sea
Beaufort Sea
Ellesmere Island
GREENLAND (KALAALLIT NUNAAT) Den.
Baffin Bay
Banks Island
Victoria Island
Baffin Island
Davis Strait
Nuuk
ALASKA
Anchorage
Yukon R.
Gulf of Alaska
YUKON TERRITORY
Whitehorse
Mackenzie R.
NORTHWEST TERRITORIES
Yellowknife
NUNAVUT
Iqaluit
Southampton I.
Hudson Strait
Labrador Sea
Hudson Bay
NEWFOUNDLAND AND LABRADOR
St. John's
BRITISH COLUMBIA
ALB.
Peace R.
Athabasca R.
SASK.
CANADA
Nelson R.
Edmonton
MANITOBA
Saskatchewan R.
QUEBEC
Calgary
Vancouver
Victoria
Seattle
Regina
Winnipeg
ONTARIO
St. Lawrence R.
P.E.I.
Charlottetown
N.B.
NOVA SCOTIA
Quebec
Fredericton
Halifax
Montreal
Ottawa
MAINE
PACIFIC OCEAN
WASH.
Columbia R.
Missouri R.
OREGON
MONT.
N. DAK.
MINN.
MICHIGAN
VT.
N.H.
MASS.
R.I.
CONN.
IDAHO
WIS.
Toronto
N.Y.
S. DAK.
WYO.
PA.
New York City
N.J.
DEL.
MD.
Washington, D.C.
NEVADA
CALIFORNIA
IOWA
Chicago
Philadelphia
San Francisco
UTAH
NEBR.
ILL.
IND.
OHIO
Denver
COLO.
KANSAS
St. Louis
Ohio R.
W. VA.
VA.
KY.
Colorado R.
UNITED STATES
MO.
Los Angeles
ARIZ.
NEW MEXICO
TENN.
N.C.
OKLA.
ARK.
Arkansas R.
S.C.
Phoenix
Rio Grande
Mississippi R.
ALA.
Atlanta
GA.
ATLANTIC OCEAN
TEXAS
MISS.
LA.
FLA.
Houston
Miami
TROPIC OF CANCER
Gulf of Mexico
LATIN AMERICA
Caribbean Sea
ARCTIC CIRCLE
180°
160°W
140°W
120°W
100°W
80°W
60°W
40°W
20°W
0°
20°N
40°N
60°N
80°N

- National capital
- Provincial/territorial capital
- Major city
- National boundary
- State or provincial boundary

160°W
155°W
Niihau
Kauai
Oahu
Kaula
Molokai
Honolulu
Lanai
Maui
HAWAII
Kahoolawe
20°N
Hawaii
PACIFIC OCEAN
0 200 kilometers
0 200 miles
Albers Equal-Area projection

N
W
E
S

0 1,000 kilometers
0 1,000 miles
Lambert Azimuthal Equal-Area projection

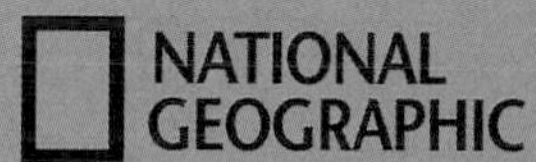

Comparing Past and Present

The cultures of the United States and Canada have been impacted by the cultures of those who have settled the region. As you compare the maps on these pages, look for patterns that may provide information about the cultures of the United States and Canada today. Then answer the questions below on a separate sheet of paper.

1. What conclusions can be drawn about the European settlement of the United States and Canada? What factors contributed to these settlement patterns?
2. Which country had a greater variety of Native American groups? What factors may have contributed to this situation?
3. How may location contribute to the differences between the Native American culture groups?

NATIVE AMERICAN SETTLEMENT PATTERNS, 1500s–1800s

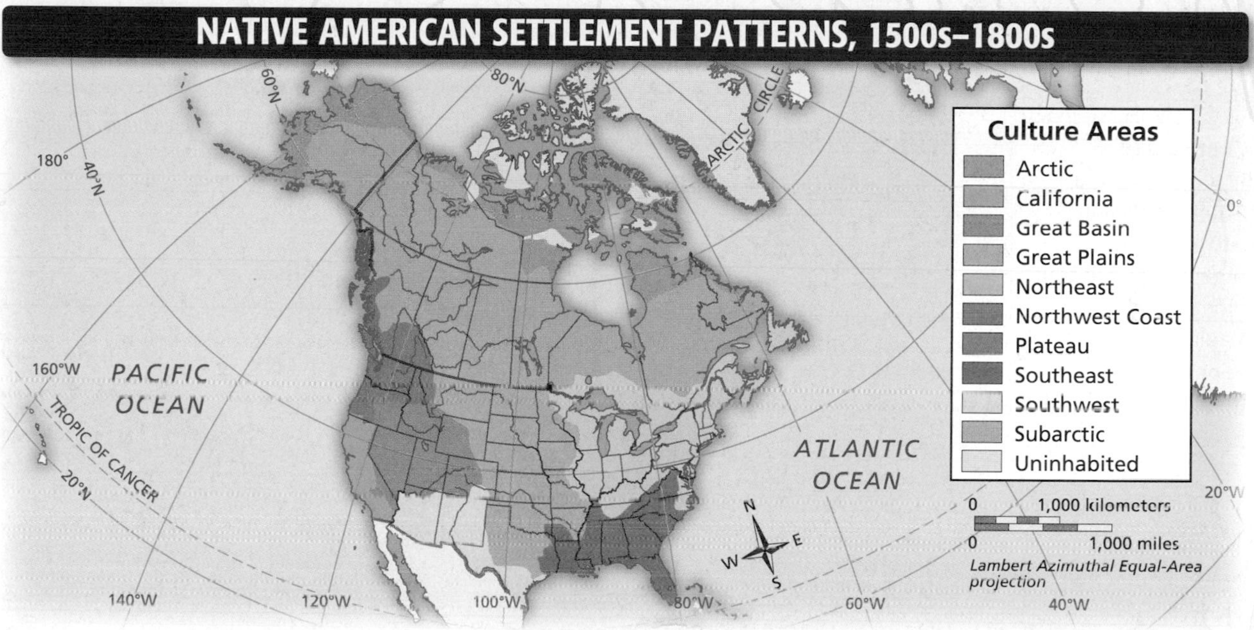

EUROPEAN SETTLEMENT PATTERNS

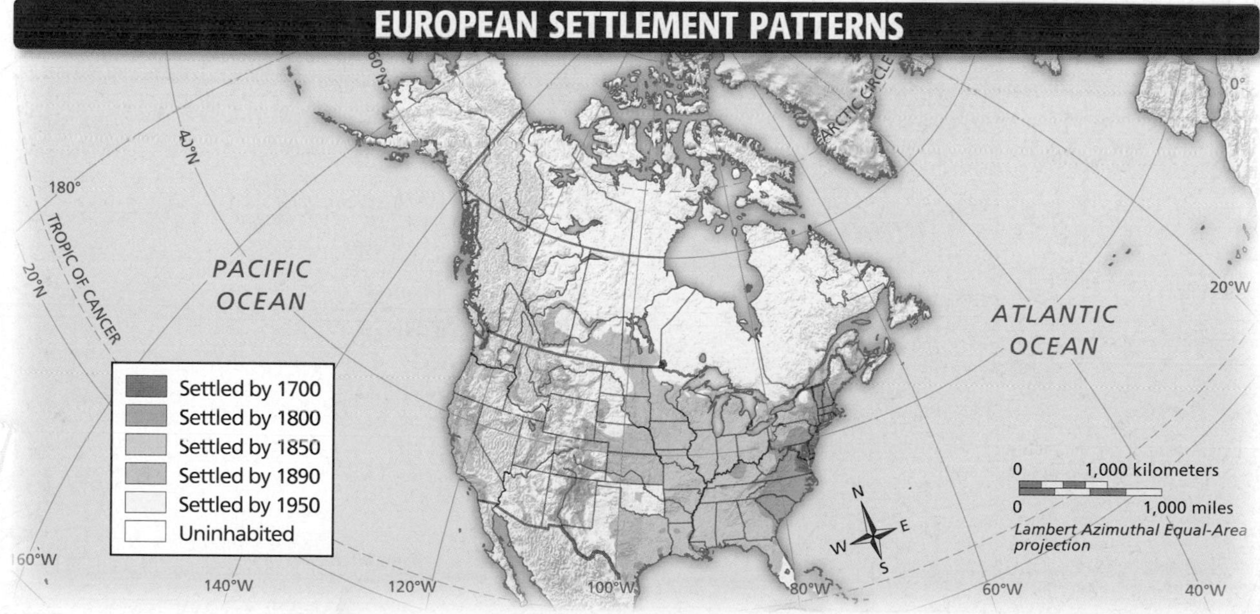

UNIT 2 REGIONAL ATLAS

ECONOMIC ACTIVITY United States and Canada

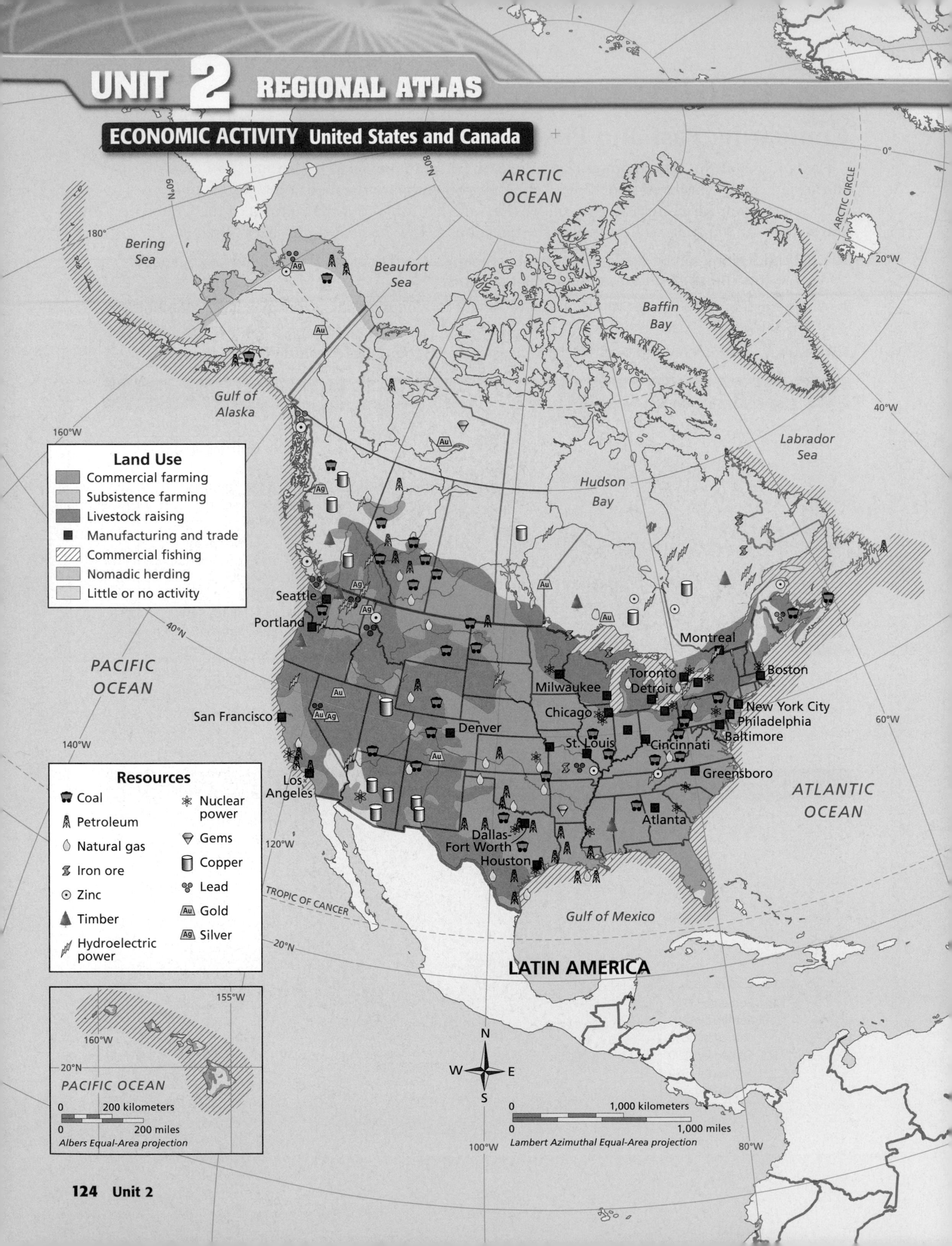

Land Use
- Commercial farming
- Subsistence farming
- Livestock raising
- Manufacturing and trade
- Commercial fishing
- Nomadic herding
- Little or no activity

Resources
- Coal
- Petroleum
- Natural gas
- Iron ore
- Zinc
- Timber
- Hydroelectric power
- Nuclear power
- Gems
- Copper
- Lead
- Gold
- Silver

ARCTIC OCEAN
Bering Sea
Beaufort Sea
Baffin Bay
Gulf of Alaska
Labrador Sea
Hudson Bay
PACIFIC OCEAN
ATLANTIC OCEAN
Gulf of Mexico
LATIN AMERICA
ARCTIC CIRCLE
TROPIC OF CANCER
180°
160°W
140°W
120°W
100°W
80°W
60°W
40°W
20°W
0°
60°N
80°N
40°N
20°N

Seattle
Portland
San Francisco
Los Angeles
Denver
Dallas-Fort Worth
Houston
Milwaukee
Chicago
St. Louis
Toronto
Detroit
Cincinnati
Montreal
Boston
New York City
Philadelphia
Baltimore
Greensboro
Atlanta

155°W
160°W
20°N
PACIFIC OCEAN
0 200 kilometers
0 200 miles
Albers Equal-Area projection

N
W E
S

0 1,000 kilometers
0 1,000 miles
Lambert Azimuthal Equal-Area projection

Industrialization and the Environment

The United States and Canada have used their vast energy resources to industrialize their countries. Industrialization has in turn had an impact upon the environment. As you study the maps and graphics on these pages, look for the effects of industrial pollution such as acid rain upon the region. Then answer the questions below on a separate sheet of paper.

1. Why might the eastern section of the United States and Canada experience higher levels of acid rain than the extreme northern and western sections of the region?
2. Where is Canada's greatest concentration of fossil fuel resources located?
3. Describe the process by which pollution becomes acid rain. What effect does acid rain have upon surrounding vegetation?

ACID RAIN

How Acid Rain Is Created

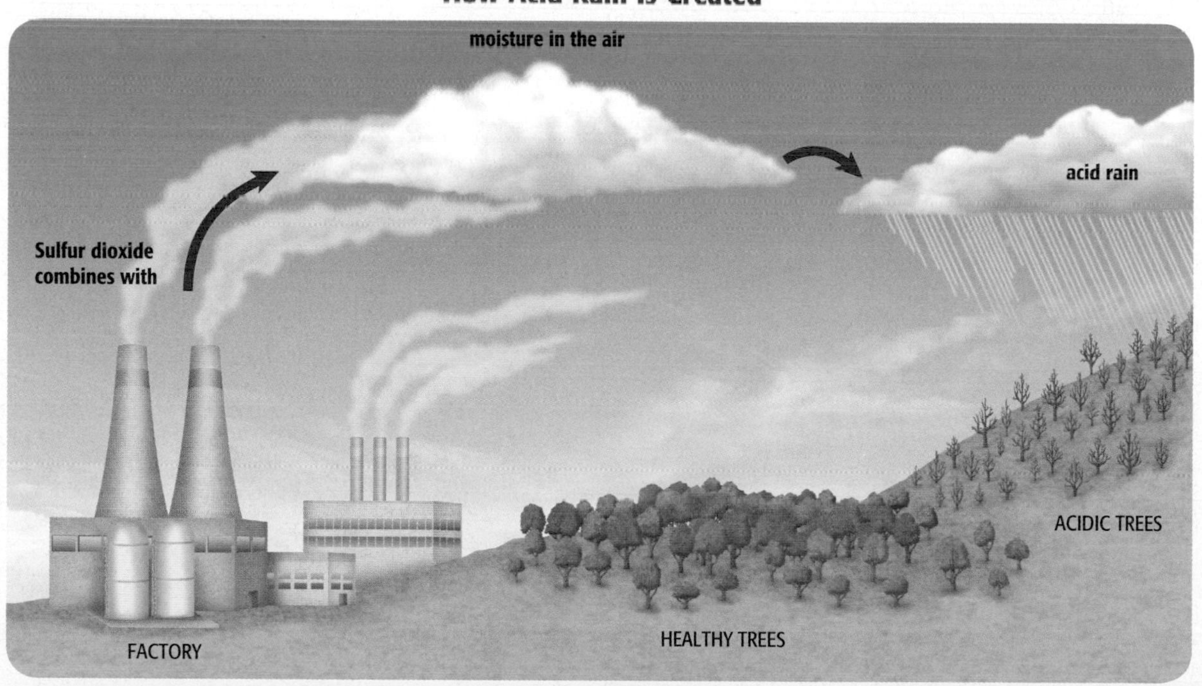

UNITED STATES AND CANADA

COUNTRY PROFILES

Country, Capital, & Area	Population & Density	Life Expectancy at Birth	GDP Per Capita*	% Urban	Literacy Rate (%)	Years of Compulsory Education	Phone Lines/ Cell Phones (per 1,000 people)	Internet Users (per 1,000 people)	Flag & Language
UNITED STATES Washington, D.C. 3,537,453 sq. mi. 9,161,966 sq. km	306,800,000 87 per sq. mi. 33 per sq. km	78 yrs.	$47,800	79	99.0	12	606/680	630.0	English
CANADA Ottawa 3,511,021 sq. mi. 9,093,507 sq. km	33,700,000 10 per sq. mi. 4 per sq. km	81 yrs.	$39,300	79	99.0	11	566/514	520.0	English/French

Sources: Central Intelligence Agency, *World Factbook,* 2009; Population Reference Bureau, *World Population Data Sheet,* 2009; UNESCO Institute for Statistics; United Nations, *Human Development Report,* 2007/2008.

U.S. State Names: Meaning and Origins

ALABAMA — Montgomery — "thicket clearers" (Choctaw)

ALASKA — Juneau — "the great land" (Aleut)

ARIZONA — Phoenix — "small spring" (O'odham/Pima)

ARKANSAS — Little Rock — "south wind" (Ohio Valley Native Americans' name for the Quapaws)

CALIFORNIA — Sacramento — named after Calafia, a place in a romantic Spanish story

COLORADO — Denver — "colored red" (Spanish)

CONNECTICUT — Hartford — "long river place" (Mohegan)

DELAWARE — Dover — named for Virginia's colonial governor, Thomas West, Baron De La Warr

FLORIDA — Tallahassee — "feast of flowers" (Spanish)

GEORGIA — Atlanta — named for England's King George II

HAWAII — Honolulu — unknown (Native Hawaiian)

IDAHO — Boise — unknown

ILLINOIS — Springfield — "tribe of superior men" (Algonquian)

INDIANA — Indianapolis — "land of Indians" (European American)

IOWA — Des Moines — name of a Native American group

KANSAS — Topeka — "people of the south wind" (Sioux)

KENTUCKY — Frankfort — "land of tomorrow," "cane and turkey lands," or "meadow lands" (Native American/ Iroquoian)

LOUISIANA — Baton Rouge — named for France's King Louis XIV

MAINE — Augusta — nautical term distinguishing the mainland from islands

MARYLAND — Annapolis — named in honor of the wife of England's King Charles I

MASSACHUSETTS — Boston — "at or about the great hill" (Native American)

MICHIGAN — Lansing — "great lake" (Ojibwa)

MINNESOTA — Saint Paul — "water that reflects the sky" (Dakota)

MISSISSIPPI — Jackson — "father of waters" (Chippewa)

*The CIA calculates per capita GDP in terms of purchasing power parity. This formula allows us to compare the figures among different countries.

Note: Land areas and flags are not drawn to scale.

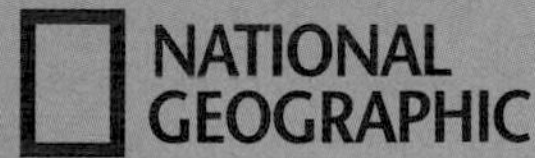

MISSOURI
Jefferson City
"town of the large canoes" (Sioux)

MONTANA
Helena
"mountainous" (Spanish)

NEBRASKA
Lincoln
"flat water" (Oto)

NEVADA
Carson City
"snowcapped" (Spanish)

NEW HAMPSHIRE
Concord
named for Hampshire, a county in England

NEW JERSEY
Trenton
named for Isle of Jersey, a British territory

NEW MEXICO
Santa Fe
named for the state's former colonial ruler, Mexico

NEW YORK
Albany
named in honor of the English Duke of York

NORTH CAROLINA
Raleigh
named in honor of England's King Charles I

NORTH DAKOTA
Bismarck
"friend" (Sioux); the Dakota were a Sioux people

OHIO
Columbus
"great river" (Iroquoian)

OKLAHOMA
Oklahoma City
"red people" (Choctaw)

OREGON
Salem
unknown meaning (Native American)

PENNSYLVANIA
Harrisburg
"Penn's woodland," named for the father of Pennsylvania's founder, William Penn

RHODE ISLAND
Providence
named for the Greek island of Rhodes

SOUTH CAROLINA
Columbia
named for England's King Charles I

SOUTH DAKOTA
Pierre
"friend" (Sioux); the Dakota were a Sioux people

TENNESSEE
Nashville
named for tana-see, "the meeting place" (Yuchi)

TEXAS
Austin
"friends" (Caddo)

UTAH
Salt Lake City
"people of the mountains" (Ute)

VERMONT
Montpelier
"green mountain" (French)

VIRGINIA
Richmond
named for the unmarried Queen Elizabeth I of England, known as "the Virgin Queen"

WASHINGTON
Olympia
named in honor of George Washington

WEST VIRGINIA
Charleston
began as the western part of Virginia before becoming a state in 1863

WISCONSIN
Madison
"river of red stone" (Algonquian)

WYOMING
Cheyenne
"upon the great plain" (Delaware)

Canadian Province and Territory Names: Meaning and Origins

ALBERTA
Edmonton
named for the daughter of England's Queen Victoria

BRITISH COLUMBIA
Victoria
named for the province's British heritage and the Columbia River

MANITOBA
Winnipeg
"the strait of the spirit" (Cree)

NEW BRUNSWICK
Fredericton
named for English royal family of Brunswick-Luneberg

NEWFOUNDLAND AND LABRADOR
St. John's
"new found land," named by explorer John Cabot in 1497; *lavrador*, "landholder" (Portuguese)

NORTHWEST TERRITORIES
Yellowknife
named for lands north and west of Lake Superior

NOVA SCOTIA
Halifax
Latin term for "New Scotland," based on province's Scottish heritage

NUNAVUT
Iqaluit
"our land" (Inuktitut)

ONTARIO
Toronto
"beautiful lake" or "sparkling beautiful water" (Native American)

PRINCE EDWARD ISLAND
Charlottetown
named for the son of England's King George III

QUEBEC
Quebec
"place where the river narrows" (Algonquian)

SASKATCHEWAN
Regina
"fast flowing river" (Cree)

YUKON TERRITORY
Whitehorse
"great river" (Native American)

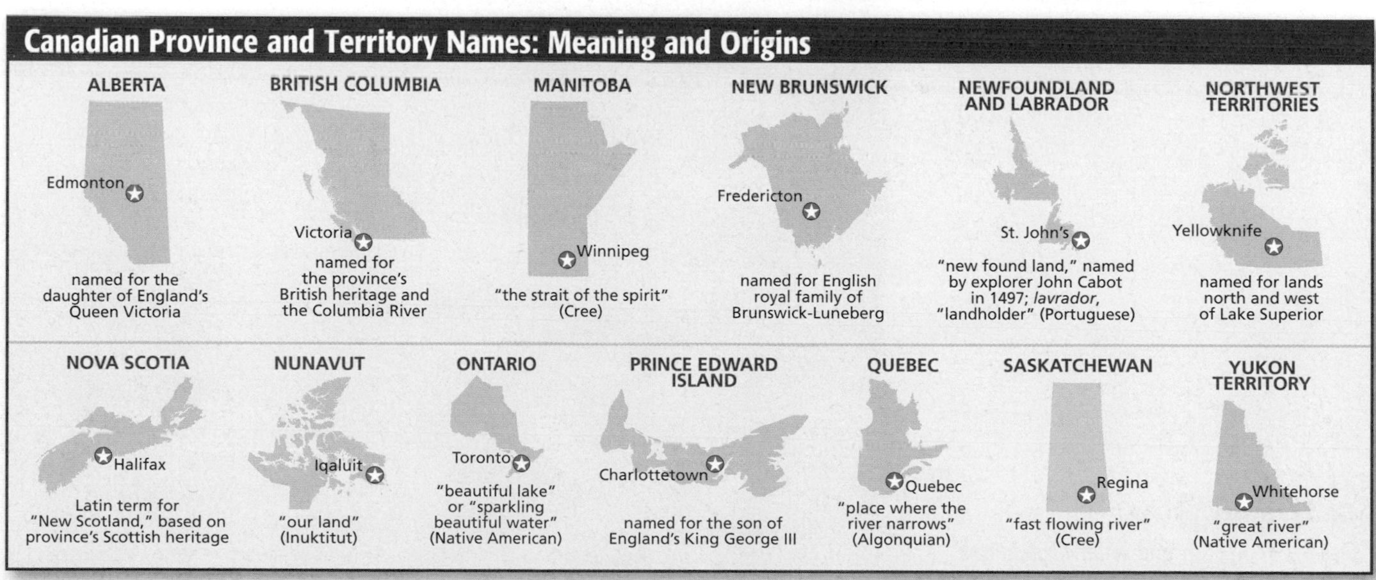

CHAPTER 5

PHYSICAL GEOGRAPHY OF The United States and Canada

BiGIdea

Geography and the environment play an important role in how a society is shaped over time. Stories in the news highlight the importance of the environment in the United States and Canada. Studying the physical geography will explain the significance of the region's natural features and how the environment affects the region's people.

Essential Questions

Section 1: The Land
How has physical geography affected the development of the United States and Canada?

Section 2: Climate and Vegetation
What factors cause variations in climate and vegetation in most of the United States and Canada?

Visit glencoe.com and enter ***QuickPass***™ code WGC9952C5 for Chapter 5 resources.

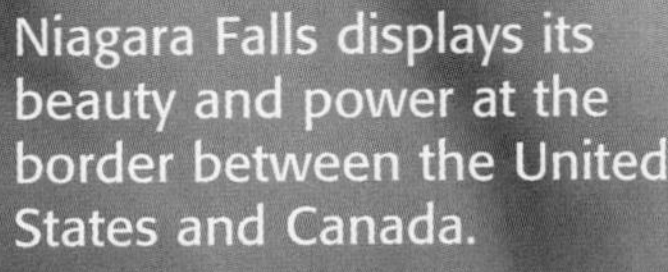

Niagara Falls displays its beauty and power at the border between the United States and Canada.

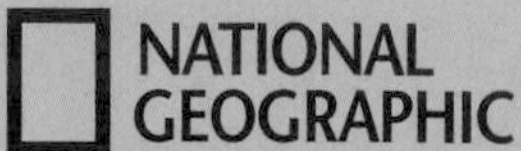

Compare and Contrast Make a Three-Tab Book with a Venn diagram to help you compare and contrast the physical geography of the United States and Canada.

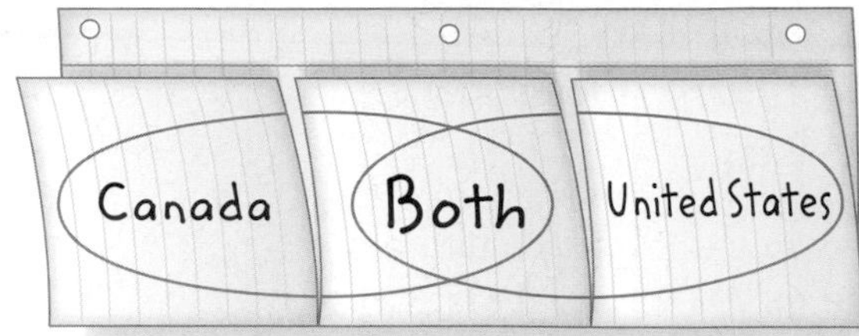

Reading and Writing As you read this chapter, write about physical features of the two countries that are similar under the center tab of the diagram. Note features and areas that differ under the outer tabs.

UNITED STATES AND CANADA

SECTION 1

The Land

 section audio  spotlight video

Guide to Reading

Essential Question

How has physical geography affected the development of the United States and Canada?

Content Vocabulary

- divide *(p. 132)*
- headwaters *(p. 132)*
- tributary *(p. 132)*
- fall line *(p. 133)*
- fossil fuel *(p. 134)*
- fishery *(p. 135)*
- aquaculture *(p. 135)*

Academic Vocabulary

- accumulated *(p. 132)*
- enormous *(p. 133)*
- crucial *(p. 133)*

Places to Locate

- Mount McKinley *(p. 131)*
- Rocky Mountains *(p. 131)*
- Canadian Shield *(p. 132)*
- Appalachian Mountains *(p. 132)*
- Mississippi River *(p. 132)*
- Great Lakes *(p. 133)*

Reading Strategy

Organizing Complete a web diagram similar to the one below by listing the major minerals found in the United States and Canada.

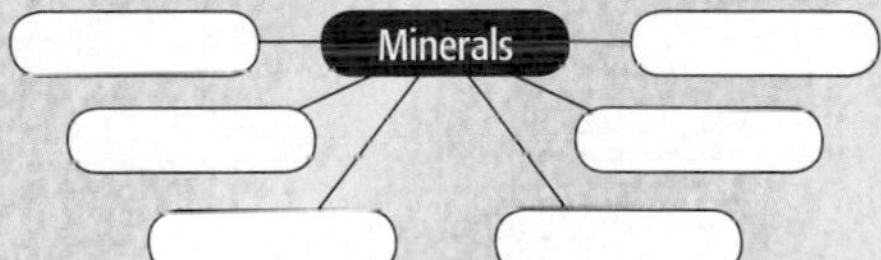

The United States and Canada form a geographic region of enormous physical variety and natural wealth. This wealth includes breathtaking landforms shaped by the forces of water, wind, and geology over millions of years. These landforms, such as the rugged, mountainous areas near Yellowstone National Park, have attracted adventurers and inspired writers for decades.

NATIONAL GEOGRAPHIC Voices Around the World

"From my cabin in Teton Valley, Idaho, . . . [t]here is a snow-covered meadow, and beyond that a stand of bare grey aspen trees, and beyond that a spill of sun-stunned white until the Earth rears back on itself and makes the Rocky Mountains. It is a landscape that has inspired . . . a great many acts of poetry, but I measure it by its ordinary day-to-day gifts. . . . Today . . . the slipping hold of winter is still evident. . . . And flies, giddy with the promise of longer days, seep out of the logs of my cabin and fall in exhausted layers on the windowsills. Life, in all its dangerous, complicated, annoying glory, has returned to this corner of the sun-tilted world."

—Alexandra Fuller, "Yellowstone & Grand Teton National Parks," *National Geographic*, November 2003

Hiking the Rocky Mountains

Landforms

MAIN Idea Water, wind, and geologic forces shape the landscapes of the United States and Canada.

GEOGRAPHY AND YOU What landforms are part of the area in which you live? Read to learn about the many different landforms that exist in the United States and in Canada.

As the physical map on page 120 shows, mountains rise at the eastern and western edges of the United States and Canada. In the west, young, sharp-edged mountain ranges tower above plateaus that descend to vast plains. These plains extend across the continent to meet the lower, more eroded mountains in the east.

Western Mountains, Plains, and Plateaus

Collisions between tectonic plates millions of years ago thrust up a series of sharp-peaked mountains called the Pacific Ranges. These ranges include the Sierra Nevada, the Cascade Range, the Coast Range, and the Alaska Range. The Alaska Range gives rise to the highest point on the continent, **Mount McKinley,** at 20,320 feet (6,194 m).

Like the Pacific Ranges, the **Rocky Mountains** grew as geologic forces heaved slabs of rock upward. The Rocky Mountains link the United States and Canada, stretching more than 3,000 miles (4,828 km) from New Mexico to Alaska. Some peaks of the Rockies soar to more than 14,000 feet (4,267 m).

Dry basins and plateaus fill the area between the Pacific Ranges and the Rockies. The Columbia Plateau in the north was formed by lava that seeped from cracks in the earth. The heavily eroded Colorado Plateau displays flat-topped mesas and the majestic Grand Canyon of the Colorado River. At its deepest, the canyon's steep walls plunge 6,000 feet (1,829 m). The Great Basin cradles Death Valley, the lowest place in the United States. Canada's plateaus are colder and narrower than those in the United States.

East of the Rockies, the land falls and flattens into the Great Plains, which extend 300 to 700 miles (483 to 1,126 km) across the center of the region. The Great Plains reach elevations up to 6,000 feet (1,829 m). Although the plains appear flat, the land slopes downward at about 10 feet per mile (about 2 m per km) to the Central Lowlands along the Mississippi River.

NATIONAL GEOGRAPHIC *The shape and location of the Pacific Ranges affect conditions on the Columbia Plateau.*

Location What type of vegetation could you expect to find in the flat, plateau areas between mountain ranges?

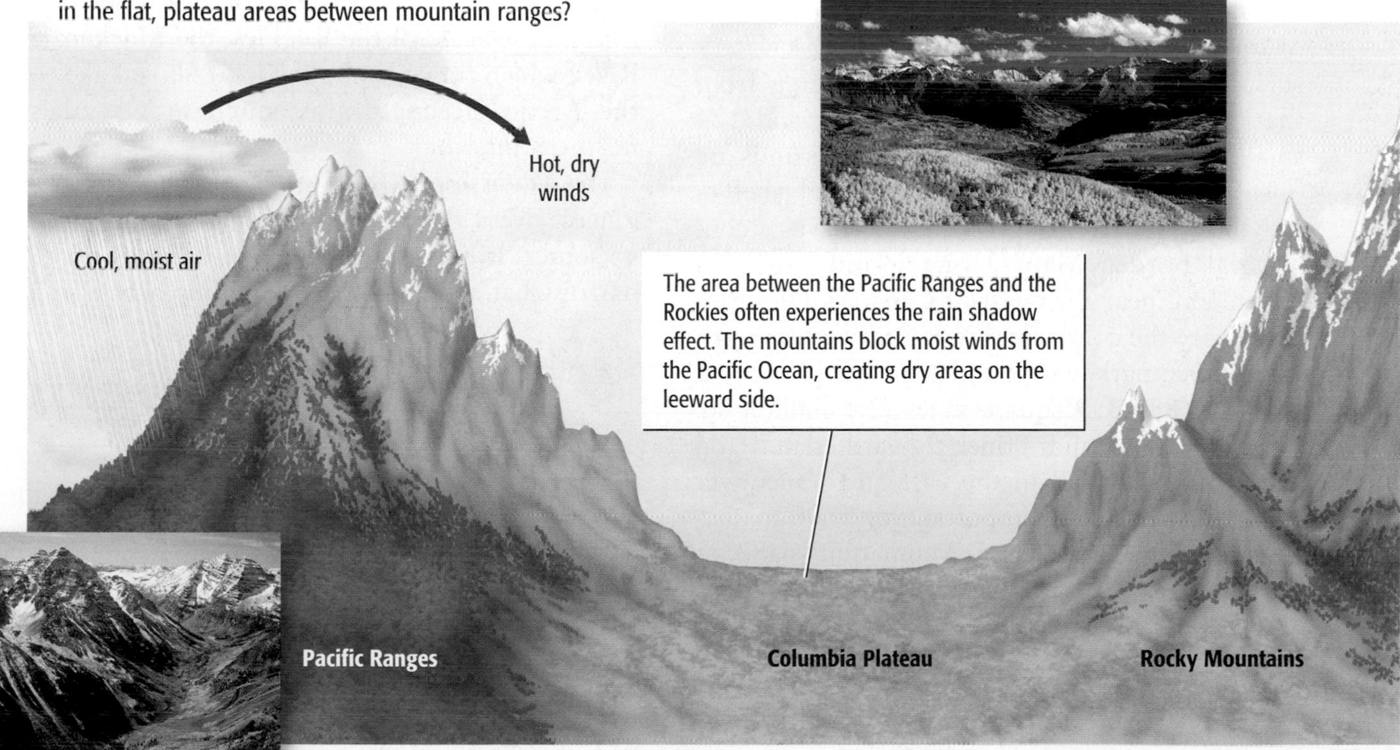

UNITED STATES AND CANADA

Eastern Mountains and Lowlands

East of the Mississippi, the land rises slowly into the foothills of the Appalachian Mountains. At the edge of the Canadian plains, the **Canadian Shield,** a giant core of rock centered on the Hudson and James Bays, anchors the continent. This stony land makes up the eastern half of Canada and the northeastern United States. In northern Quebec the Canadian Shield descends to the Hudson Bay.

The heavily eroded **Appalachian Mountains** are North America's oldest mountains. They are the continent's second-longest mountain range, extending about 1,500 miles (2,414 km) from Quebec to central Alabama. The Appalachians were formed by powerful upheavals within the Earth's crust and shaped over time by ice and running water. Coastal lowlands lie east and south of the Appalachians. Between the mountains and the coastal lowlands is the Piedmont, a wide area of rolling hills. Many rivers cut through the Piedmont, flowing east across the Atlantic Coastal Plain in the Carolinas. In the southeast, the Gulf Coastal Plain extends westward to Texas.

Islands

The islands of the United States and Canada were created in part by geologic forces. Oceanic islands, such as Hawaii, are volcanic. With each volcanic eruption, lava **accumulated** on the floor of the ocean until it pushed through the water's surface. Volcanic mountaintops emerging from the Pacific Ocean formed the 8 major and 124 smaller islands of Hawaii. Continental islands are unsubmerged parts of the continental shelf—a shallow, underwater platform that forms a continental border. Many larger islands, such as Greenland, near the coast of Canada's Ellesmere Island, are the continental type. An overseas territory of Denmark, Greenland is the world's largest island at 839,399 square miles (2.1 million sq. km). Newfoundland, Prince Edward Island, and Cape Breton Island in the east and Vancouver Island in the west play important roles in Canada's economy. New York City's Manhattan Island, at the mouth of the Hudson River, is a major U.S. and world economic center.

READING Check **Regions** What are some important similarities between the physical geography of the United States and the physical geography of Canada?

Water Systems

MAIN Idea Lakes and rivers in the United States and Canada are important to economic development in the region.

GEOGRAPHY AND YOU Can you name major U.S. cities and towns that are located close to waterways? Read to learn how waterways encourage the growth of cities and industrial centers.

Freshwater lakes and rivers have helped make the United States and Canada prosperous. Abundant water satisfies the needs of cities and rural areas, provides power for homes and industries, and moves resources across the continent.

Rivers

In North America the high ridge of the Rockies is called the Continental Divide, or the Great Divide. A **divide** is a high point or ridge that determines the direction in which rivers flow. East of the Continental Divide, waters flow toward the Arctic Ocean, Hudson Bay, the Atlantic Ocean, and the Mississippi River system, which empties into the Gulf of Mexico. To the west, waters flow into the Pacific Ocean. Rivers—such as the Colorado and the Rio Grande—have their **headwaters,** or source, in the Rockies. Many **tributaries,** or smaller rivers and streams, connect with one of these two large rivers. Northeast of the Rockies, the Mackenzie River, which flows from the Great Slave Lake to the Arctic Ocean, drains much of Canada's northern interior.

The **Mississippi River,** one of North America's longest rivers, flows 2,357 miles (3,792 km) from its source. It begins in Minnesota as a stream so narrow that a person can easily jump across it:

> *"When I was nine years old, I jumped across the Mississippi. . . . My parents let me know this modest stream I'd taken in stride was actually one of the Earth's great corridors, dominion of paddleboats and Huck Finn, prime mover of food, fertility, and commerce across our land."*
>
> —Barbara Kingsolver, "San Pedro River: The Patience of a Saint," *National Geographic,* April 2000

From a narrow stream at its source, the Mississippi River reaches a width of 1.5 miles (2.4 km) as it empties into the Gulf of Mexico. The river drains 1,200,000 square miles (3,108,000 sq. km) of land, including all or part of 31 U.S. states and 2 Canadian provinces. This **enormous** reach makes the Mississippi one of the world's busiest commercial waterways.

In the eastern United States, a boundary called the **fall line** marks the place where the higher land of the Piedmont drops to the lower Atlantic Coastal Plain. Along the fall line, eastern rivers break into rapids and waterfalls, blocking ships from traveling farther inland.

Canada's St. Lawrence River flows for 750 miles (1,207 km) from Lake Ontario to the Gulf of St. Lawrence in the Atlantic Ocean, forming part of the border between Canada and the United States. The Canadian cities of Quebec, Montreal, and Ottawa grew up along the St. Lawrence River and its tributaries and depend on these waters for trade.

Niagara Falls, on the Niagara River, forms another part of the border between Canada and the United States. Two separate drops form the falls—the Horseshoe Falls, adjoining the Canadian bank of the river, and the American Falls, adjoining the U.S. bank. The falls are a major source of hydroelectric power for both countries.

Lakes and Other Waterways

In northern Canada, glacial dams created Great Bear Lake and Great Slave Lake. Glaciers also gouged the Canadian Shield and tore at the central section of the continent, leaving glacial basins that filled and became the **Great Lakes.** Large deposits of coal, iron, and other minerals near the lakes favored the development of industries and urban growth in the area.

Providing a link between inland and coastal waterways has been **crucial** to the economic development of North America. The greatest of these connections is the Great Lakes–St. Lawrence Seaway System—a series of canals, the St. Lawrence River, and other inland waterways that link the Great Lakes and the Atlantic Ocean. The seaway helped make cities along the Great Lakes powerful trade and industrial centers.

READING Check Human-Environment Interaction What major waterways do people in the United States and Canada depend on for the purposes of trade?

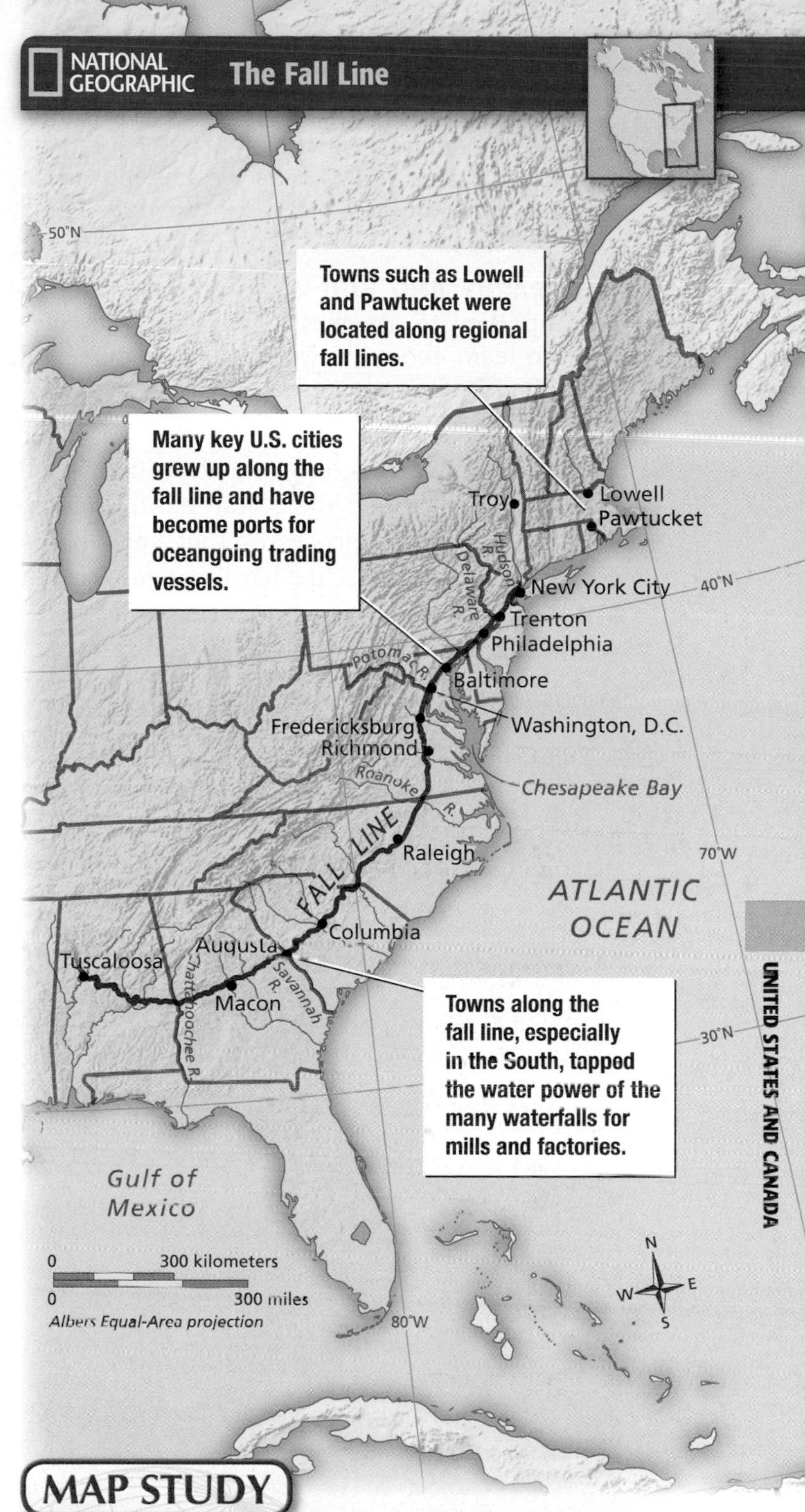

UNITED STATES AND CANADA

MAP STUDY

The fall line of the eastern United States indicates where the higher land of the Piedmont drops to meet the Atlantic Coastal Plain.

1. **Human-Environment Interaction** Why did cities such as Philadelphia, Baltimore, and Washington, D.C., spring up along the fall line?
2. **Regions** How did location along the fall line affect the economies of towns in the South?

Maps In MOtion Use **StudentWorks™ Plus** or glencoe.com.

Natural Resources

MAIN Idea Abundant natural resources have made the United States and Canada wealthy, but these resources and the areas in which they are found need protection.

GEOGRAPHY AND YOU What natural resources are important to activities in your everyday life? Read to learn about the vital natural resources of the United States and Canada.

Ample freshwater is only one of the many natural resources of the United States and Canada. The same geologic processes that shaped the North American landscape left the region rich in a wide variety of resources. Access to this natural wealth has helped speed industrialization.

Fossil Fuels and Minerals

The United States and Canada have important energy resources, such as petroleum and natural gas. Texas and Alaska rank first and second in petroleum reserves in the United States. Texas also has the greatest reserves of natural gas. Most of Canada's petroleum and natural gas reserves lie in or near Alberta. Coal in the Appalachians, Wyoming, and British Columbia has been mined for more than 100 years.

Coal, petroleum, and natural gas are forms of **fossil fuels.** Such fuels were formed in the Earth from the buried plant and animal remains of a previous geologic time hundreds of millions of years ago. Fossil fuels must be conserved because they are nonrenewable, which means they cannot be replaced naturally in a short period of time. Currently there is much interest in finding new sources of fossil fuels in North America, in Alaska for example, without disrupting the natural environments in which they are found.

Mineral resources are also plentiful. The Rocky Mountains yield gold, silver, and copper. Parts of the Canadian Shield are rich in iron and nickel. Iron ore exists in northern Minnesota and Michigan. Canada's minerals include 33 percent of the world's production of potash (a mineral salt used in fertilizers), 4 percent of its copper, 4 percent of its gold, and 5 percent of its silver. Conservation and land preservation are important issues for today's mining industry.

Like fossil fuels, mineral resources are nonrenewable and could become depleted. Because mining involves heavy equipment, uses large quantities of water, and moves a great deal of rock and other natural materials, it can damage land, water, and air systems. In the past, people did not pay a great deal of attention to preserving the environment while mining. Today, the challenge for mining companies in the United States and Canada is finding ways to remove and process minerals and metal resources with the least disruption to surrounding ecosystems. One aspect of these efforts involves restoring land used in mining when mining operations in a particular area have finished. This reclaimed land can then be used for activities such as wildlife parks, tree farms and orchards, public hunting and fishing areas, and grazing livestock.

NATIONAL GEOGRAPHIC *Coal mining in West Virginia is important to the state's economy and well-being.*

Human-Environment Interaction How are mining companies working to reduce the impact of mining on the environment?

Geography ONLINE

Student Web Activity Visit glencoe.com, select the *World Geography and Cultures* Web site, and click on Student Web Activities—Chapter 5 for an activity about the natural resources of North America.

Timber and Fishing

Timber is a vital resource for the United States and Canada. Forests and woodlands once covered large expanses of both countries. Today, however, forests cover about 34 percent of Canada and only about 33 percent of the United States. Commercial lumber operations face the challenge of harvesting the region's precious timber resources responsibly.

Trees are a renewable resource, but only if people take steps to protect forests and the ecosystems they sustain. Positive efforts to preserve forests include replanting trees to replace those cut for lumber, cooperating to protect the 1,000 species of native forest animals, and preserving old-growth forests.

The coastal waters of the Atlantic and Pacific Oceans and the Gulf of Mexico have been essential to the region's economy. Rich with fish and shellfish, these waters were important **fisheries,** or places for catching fish and other sea animals. The Grand Banks, once one of the world's richest fishing grounds, covers about 139,000 square miles (360,000 sq. km) off of Canada's southeast coast. In recent years, overfishing has caused fish stocks to decrease rapidly, leading the Canadian government to ban cod fishing. Both countries of the region now are working to protect species that have been or are in danger of becoming overfished. After a steep decline in fishing, **aquaculture,** or fish farming, has become a growing economic activity. The line graph below shows that thousands of tons of fish are produced each year in the region.

The Growth of Aquaculture

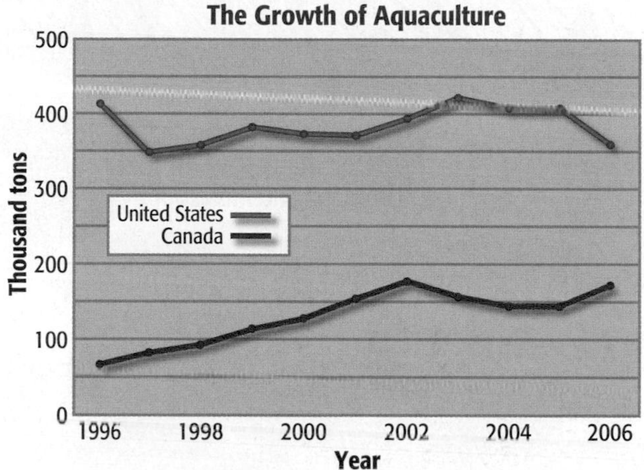

Source: OECD Factbook 2009: Economic, Environmental, and Social Statistics.

READING Check **Place** What industries are supported by the natural resources of the United States and Canada?

SECTION 1 REVIEW

Vocabulary

1. Explain the significance of: divide, headwaters, tributary, fall line, fossil fuel, fishery, aquaculture.

Main Ideas

2. Explain how each of the following factors—water, wind, and tectonic forces—has influenced landscapes in the United States and Canada.
3. What types of natural resources have made the United States and Canada wealthy? Why do such resources need protection? What efforts have been made to preserve forests in the United States and Canada?
4. How are lakes and rivers important to economic development in the United States and Canada? On a sheet of paper, fill in a chart like the one below that lists examples of each for both countries.

	Lakes	Rivers
United States		
Canada		

Critical Thinking

5. **Answering the Essential Question** Explain how the fall line in the eastern United States influenced economic development in the region.
6. **Making Inferences** Why is it in the best interest of industries to use natural resources responsibly?
7. **Analyzing Visuals** Study the line graph above. In which year did the United States produce the largest amount of fish? How much less did Canada produce during that same year?

Writing About Geography

8. **Expository Writing** Write a paragraph describing the effects of a physical process, such as weather or gravity, on the flow of rivers in the United States and Canada.

Study Central™ To review this section, go to glencoe.com and click on Study Central.

Story of a HURRICANE

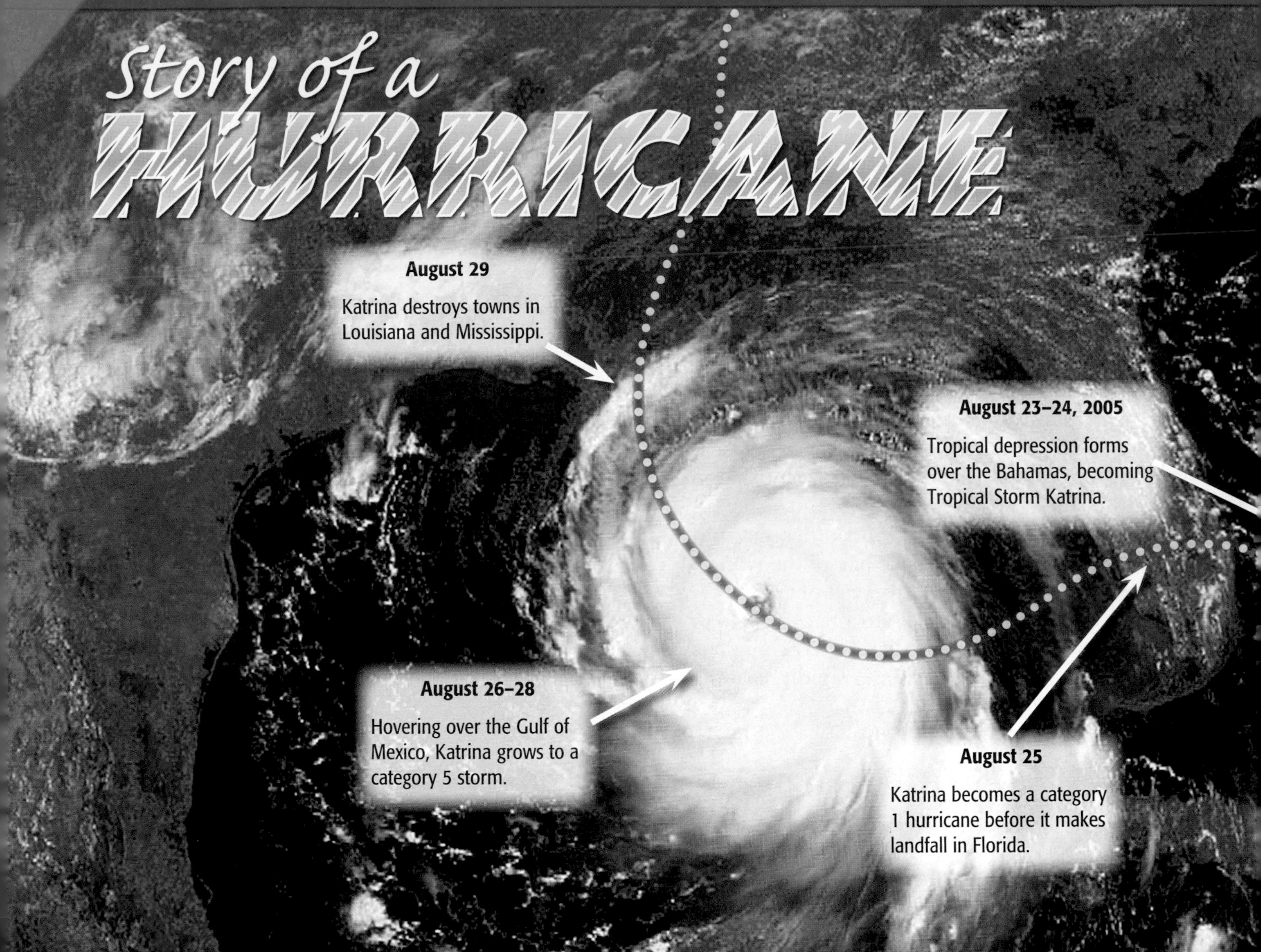

August 29

Katrina destroys towns in Louisiana and Mississippi.

August 23–24, 2005

Tropical depression forms over the Bahamas, becoming Tropical Storm Katrina.

August 26–28

Hovering over the Gulf of Mexico, Katrina grows to a category 5 storm.

August 25

Katrina becomes a category 1 hurricane before it makes landfall in Florida.

Awful Aftermath

Disaster A record-breaking 26 named tropical storms, including 13 hurricanes, formed during the 2005 Atlantic hurricane season. Katrina, Rita, and Wilma devastated the Gulf Coast with destructive winds, mountainous waves, torrential rains, storm surges, and tornadoes. After Katrina's storm surge breached the levees, 80 percent of New Orleans flooded. Along the coast, whole towns were wiped out. The storm's effects were felt as far north as Ontario, Canada.

Gulfport, Mississippi, woman outside her home

What was the cost of Katrina? Katrina was the most destructive and the costliest natural disaster in the history of the United States, with property damage estimates of $75 billion. To clean up the debris and toxins and to rebuild will cost billions more.

Hurricane Glossary	
eye	calm center of storm, surrounded by the strongest winds
hurricane	also known as cyclone, typhoon, tropical cyclone
hurricane season	June 1 to November 30
landfall	when the center of the storm crosses the coastline
storm surge	dramatic rise in sea level that causes water to crash into coast

Awesome Power

Katrina was born from a cluster of thunderstorms near the Bahamas. Like other hurricanes, Katrina formed from high humidity, light winds, and water temperatures of at least 80°F (27°C). Katrina became the largest hurricane of its strength ever to hit the United States. With 125 mph winds, gusting to 215 mph, and a 34-foot (10.4-m) storm surge, Katrina had the energy of 10,000 nuclear bombs.

Louisiana residents try to reach safety.

Louisiana officer rescues a terrified child.

What was the human toll? Thousands were left battered, displaced, and homeless. The official death toll was 1,383, but months later, more than 4,000 people were still unaccounted for. For more information, visit *Beyond the Textbook/Hurricane Katrina* at **glencoe.com**.

How Hurricanes Are Classified

Category	Wind Speed	Storm Surge (above normal tide)
1	74–95 mph	4–5 feet
2	96–110 mph	6–8 feet
3	111–130 mph	9–12 feet
4	131–155 mph	13–18 feet
5	above 155 mph	above 18 feet

Source: Saffir-Simpson Hurricane Scale

THINKING GEOGRAPHICALLY

1. **Environment and Society** Conduct research to learn how individuals and communities prepared for Katrina. Then create a multimedia presentation detailing your findings.
2. **Human Systems** Why might people not want to leave their homes before a hurricane hits? Why might people choose to rebuild homes in areas often affected by hurricanes?

SECTION 2

 section audio  spotlight video

Guide to Reading

Essential Question

What factors cause variations in climate and vegetation in most of the United States and Canada?

Content Vocabulary

- hurricane *(p. 139)*
- chaparral *(p. 139)*
- prairie *(p. 140)*
- supercell *(p. 140)*
- timberline *(p. 141)*
- chinook *(p. 141)*
- blizzard *(p. 142)*

Academic Vocabulary

- distinct *(p. 139)*
- methods *(p. 141)*
- visibility *(p. 142)*

Places to Locate

- Everglades *(p. 139)*
- Death Valley *(p. 139)*
- Great Plains *(p. 140)*
- Newfoundland *(p. 142)*
- Yukon Territory *(p. 142)*

Reading Strategy

Organizing Complete a graphic organizer similar to the one below by listing the factors that contribute to the varying climate and vegetation found in the northern areas of the United States and Canada.

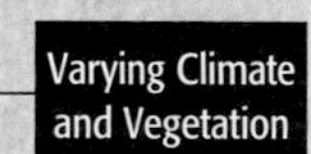

Varying Climate and Vegetation

Climate and Vegetation

Diversity of climate and vegetation characterizes the region of the United States and Canada. Conditions in this vast region include the wet and dry seasons of the southern United States, the bitter cold of high-latitude areas, the radically changing seasons of the interior regions, and the cool, wet climates of the Pacific Coast.

NATIONAL GEOGRAPHIC Voices Around the World

"Off the west coast of British Columbia's Vancouver Island, Bob Van Pelt tramped ahead across a smaller isle named Meares. We were in woods as old, quiet, green, and wet as a forest can be. Even the air felt soaked. It was hard to tell how much of the moisture came from the chilly rain, how much was fog, and how much was steam rising off the burly figure of a bearded Van Pelt, also known as Big Tree Bob. . . . When we reached a giant that the locals call Big Mother, Van Pelt . . . took precise measurements . . . and announced that this western red cedar would probably rank among the ten largest known on the continent."

—Douglas H. Chadwick, "Pacific Suite," *National Geographic*, February 2003

Old-growth cedar, Vancouver

Southern Climates

MAIN Idea Location near the coast, as well as prevailing wind patterns, results in the warm and wet climates and warm and dry climates of the southern United States.

GEOGRAPHY AND YOU Have you ever been at the beach and experienced warm winds blowing off the water? Read to learn how the Atlantic Ocean and the Pacific Ocean influence climate in some parts of the United States.

Subtropical, tropical, desert, and Mediterranean climates are found primarily in the southern United States. The climate map on page 140 shows that these climate zones are part of an area that extends from about 25° N to about 40° N.

Warm and Wet Climates

The humid subtropical climate of the Southeast is rainy with long, muggy summers and mild winters. Because the Southeast borders a major source of water—the Atlantic Ocean—there is no dry season. Deciduous forests extend as far south as Louisiana, but land has been cleared for farming along the Mississippi River. Wetlands and swamps like Florida's **Everglades** shelter a great variety of vegetation and wildlife. In late summer and early autumn, **hurricanes**—ocean storms hundreds of miles wide with winds of 74 miles per hour (119 km per hour) or more—can pound the region's coastlines.

Within the contiguous United States, the 48 states whose borders touch, only the extreme southern tip of Florida has a tropical climate that has a **distinct** dry season. Florida's dry season is in winter. Hawaii, about 2,400 miles (3,862 km) west of the mainland, and the Caribbean island of Puerto Rico have tropical wet climates. These tropical wet climates support lush rain forests.

Warm and Dry Climates

The rain shadow effect creates desert areas when dry air moves down the leeward side of a mountain. This effect keeps the plateaus and basins between the Pacific Ranges and the Rocky Mountains hot and dry. Such a climate often contributes to problems with water quantity. Much of the area has a steppe or desert climate. Deserts in this area, including Death Valley, bake under the relentless sun. **Death Valley** has the highest temperature ever recorded in the United States, 134°F (57°C).

A Mediterranean climate is characteristic of central and southern California. Such a climate is confined to coastal areas and is characterized by mild, wet winters and hot, dry summers. The vegetation in this area is a drought-resistant woodland of twisted, hard-leafed trees. In this region of the world, such Mediterranean scrub vegetation is known as **chaparral** (SHA•puh•RAL). Under natural conditions, chaparral's growth depends on regular burning. However, these fires create a major hazard in the hills around Los Angeles and Oakland. Scheduled burning often escalates to widespread brush fires when the hot, dry Santa Ana winds blow down the mountain slopes from the inland plateaus.

READING Check **Regions** What four climates are found in the southern United States?

NATIONAL GEOGRAPHIC *Although the Florida wetlands (right) and Death Valley (left) are located in the southern United States, their climates are quite different.*

Regions Compare the climate of the Florida wetlands to that of Death Valley. How are they similar? Different?

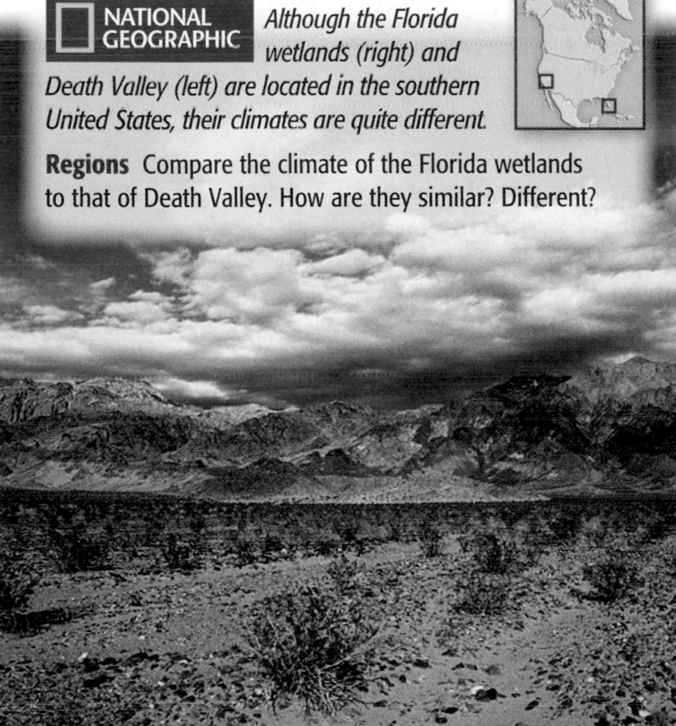

UNITED STATES AND CANADA

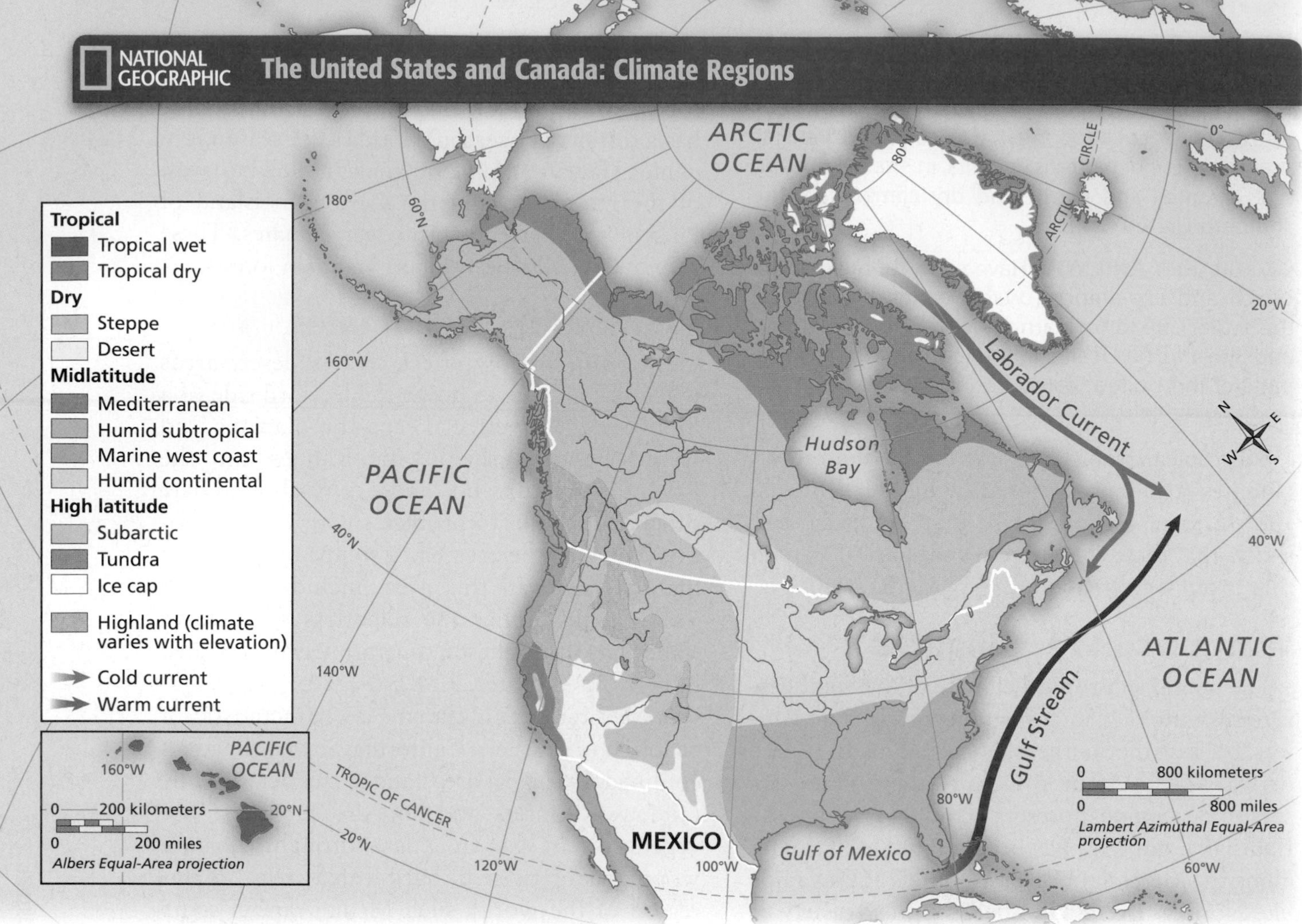

Northern Climates

MAIN Idea Variations in climate and vegetation in most of the United States and Canada are the result of the combined effects of latitude, elevation, ocean currents, and rainfall.

GEOGRAPHY AND YOU Have you ever experienced a tornado? Read to learn how the variation in climate in the region's interior can result in violent weather phenomena.

Most of the contiguous United States and the southern one-third of Canada—from about 40° N to 50° N—experiences variations in climate and vegetation. The area's climate ranges from hot and humid to cool and wet.

Interior Climates

Far from large bodies of water that tend to moderate climate, the **Great Plains**, in the center of the continent, has a humid continental climate with bitterly cold winters and hot summers. Although western mountains block moisture-bearing Pacific winds, the Great Plains benefits from warm, moist winds that blow north along the Rockies from the Gulf of Mexico and cold, moist winds that blow south from the Arctic. The climate map above shows that a humid continental climate extends into southern Canada. Such a climate also extends from the northeastern United States into southeastern Canada.

Prairies, naturally treeless expanses of grasses, spread across the Great Plains of the continent's midsection. Each year, rainfall ranging from 10 to 30 inches (26 to 76 cm) waters tall prairie grasses, such as switchgrass and bluestem. Towering 6 to 12 feet (1.8 to 3.7 m) high, these grasses can grow as much as half an inch (1.3 cm) a day. In the Great Plains and the eastern United States, violent spring and summer thunderstorms called **supercells** often spawn tornadoes, twisting funnels of air with winds that can reach 300 miles (483 km) per hour.

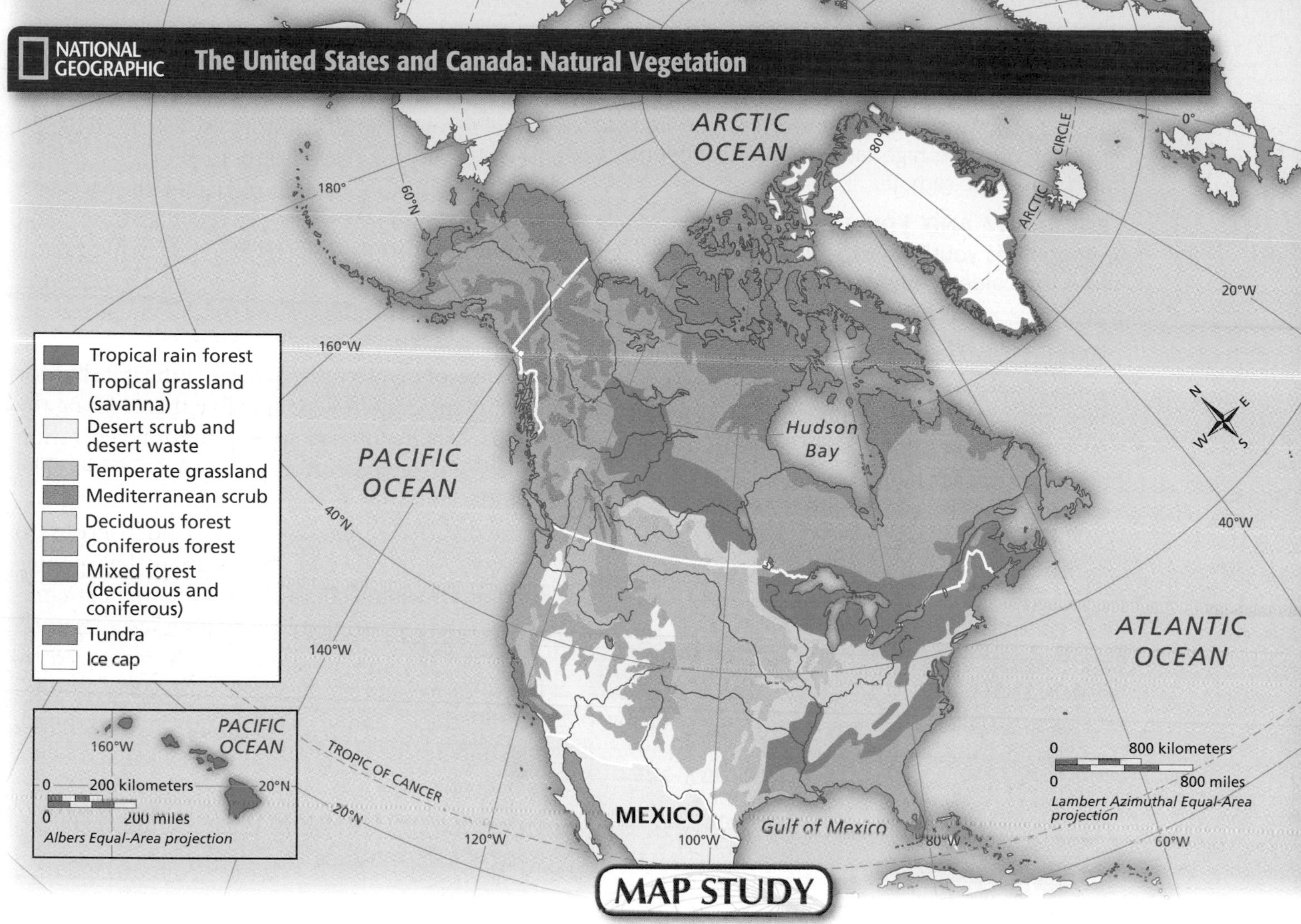

MAP STUDY

Natural vegetation reflects the region's climatic variety.

1. **Location** How does location affect the climate and vegetation of the western coast of Canada?
2. **Location** In what climate region are most of Canada's forests?

Maps In MOtion Use **StudentWorks™ Plus** or glencoe.com.

Settlers on the Great Plains broke up the densely packed sod to grow crops. When dry weather blanketed the plains in the 1930s, winds eroded the topsoil, reducing farmlands across several U.S. states to a barren wasteland called the Dust Bowl. The resulting economic hardships, made worse by the Great Depression, caused people to migrate out of the area. Since the 1930s, improved farming and conservation **methods** have restored the soil.

Some areas west of the Great Plains experience a steppe climate with a mixture of vegetation, depending on latitude or elevation. Steppe climates are transitional climates that occur between the dry desert climates and the humid interior climates.

Elevation gives the higher reaches of the Rockies and Pacific Ranges a highland climate. Coniferous forests cover the middle elevations of the ranges. Beyond the **timberline**, the elevation above which trees cannot grow, lichens and mosses grow. In early spring, a warm, dry wind called the **chinook** (shuh•NUK) blows down the eastern slopes of the Rockies, melting snow.

Coastal Climates

The interplay of ocean currents and westerly winds with the Pacific Ranges gives the Pacific coast from northern California to southern Alaska a marine west coast climate. The mountain barrier forces the warm, wet ocean air upward, where it cools and releases moisture. As a result, parts of this region receive more than 100 inches (254 cm) of rain each year. Winters are overcast and rainy. Summers are cloudless and cool. Ferns, mosses, and coniferous forests grow here.

READING Check **Location** How do the western Rocky Mountains affect climate and vegetation in the United States and Canada?

High-Latitude Climates

MAIN Idea Parts of the United States and Canada are located in the high latitudes and experience a harsh, subarctic climate.

GEOGRAPHY AND YOU Think about the coldest temperatures you have experienced. Read to learn about the coldest temperatures in North America.

Large parts of Canada and Alaska lie in the high latitudes and have a subarctic climate with frigid winters. Winter temperatures can fall to –70°F (–57°C) in some places. A high atmospheric pressure area that lingers over the Canadian subarctic spawns the cold winds that chill much of the United States during the winter.

Just How Cold Is It?

City	Average Number of Days Below 32°F/0°C	Average Winter Temperature
Chicago, Illinois	132	25°F/–3°C
Yellowknife, Northwest Territories	224	–10°F/–23°C

Source: www.weatherbase.com

Many parts of northern North America experience winter **blizzards** with winds of more than 35 miles per hour (56 km per hour), heavy or blowing snow, and **visibility** of less than 1,320 feet (402 m) for three hours or more.

The vegetation map on page 141 shows that a band of coniferous and mixed deciduous and coniferous forests sweeps from **Newfoundland** into the subarctic **Yukon Territory.** Lands along the Arctic coast fall into the tundra climate zone. Bitter winters and cool summers in this vast expanse of wilderness make it inhospitable for most plants, and few people live there. Along the coasts of Greenland, sparse tundra vegetation consists of sedge, cotton grass, and lichens. The island's small ice-free areas have few trees, but some dwarfed birch, willow, and alder scrubs do survive. As in other northern areas, few people inhabit Greenland because of its harsh climate.

The interior parts of Greenland have an ice cap climate. This type of climate is characterized by layers of ice and snow, often more than 2 miles (3 km) thick, that constantly cover the ground. The only form of vegetation that can survive here is lichens.

READING Check **Regions** What are the characteristics of a blizzard?

SECTION 2 REVIEW

Vocabulary

1. Explain the significance of: hurricane, chaparral, prairie, supercell, timberline, chinook, blizzard.

Main Ideas

2. How do location near the coast and prevailing wind patterns affect climate in the southern United States?
3. What causes variations in climate and vegetation in most of the United States and Canada?
4. Describe the climate and vegetation of the high-latitude regions in the United States and Canada.
5. Use a Venn diagram like the one below to compare and contrast the climate and vegetation of the United States and Canada.

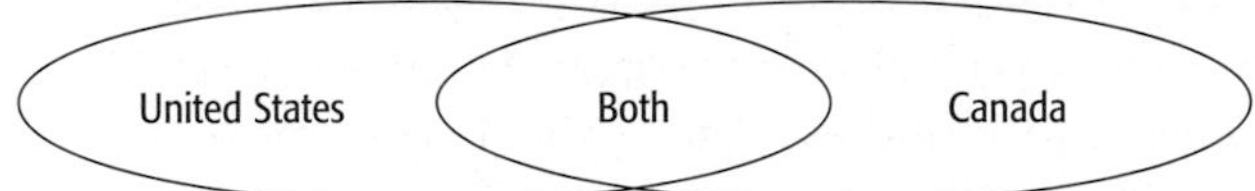

Critical Thinking

6. **Answering the Essential Question** How does location of a place near a large body of water influence its climate?
7. **Making Comparisons** How do the Pacific winds and the Arctic winds differ in their impact on climate?
8. **Problem Solving** How might the conditions that caused the 1930s Dust Bowl disaster have been avoided?
9. **Analyzing Visuals** Study the climate map on page 140. How does the climate pattern of the western United States reflect the occurrence of the rain shadow effect?

Writing About Geography

10. **Expository Writing** Write an essay describing how the climate and vegetation of the United States and Canada may have influenced human settlement in the region.

Geography ONLINE

Study Central™ To review this section, go to glencoe.com and click on Study Central.

STUDY TO GO — Study anywhere, anytime by downloading quizzes and flashcards to your PDA from glencoe.com.

A Fossil Fuels
- Petroleum and natural gas deposits in Texas, Alaska, and Alberta
- Coal mined in Appalachian Mountains, Wyoming, and British Columbia for more than 100 years
- Issues of using fossil fuels while protecting the environment

B Canadian Shield
- Located east of the Canadian plains
- Rocky core centered on the Hudson and James Bays
- Subarctic climate with coniferous forests

C Timber and Fishing
- Timber is important for the region, but lumber operations face the challenge of harvesting trees responsibly.
- Coastal waters are home to fisheries, but some areas have been overfished.

D Great Lakes
- Five lakes created by movement of glaciers
- Deposits of coal and iron fueled industrial development
- Linked to the Atlantic Ocean by the St. Lawrence Seaway

E Appalachian Mountains
- Extend from Quebec to central Alabama
- North America's oldest mountains shaped over time by ice, wind, and running water
- Midlatitude climates with coniferous and deciduous forests

A
B
D
C
F
G
E

F Rocky Mountains
- Stretch from New Mexico to Alaska, linking the United States and Canada
- Young mountains created through tectonic activity
- Highland climate varies with elevation

G Mississippi River
- Headwaters in Minnesota and mouth in Louisiana
- Drains all or part of 31 U.S. states and 2 Canadian provinces
- One of the world's busiest commercial waterways

CHAPTER 5

STANDARDIZED TEST PRACTICE

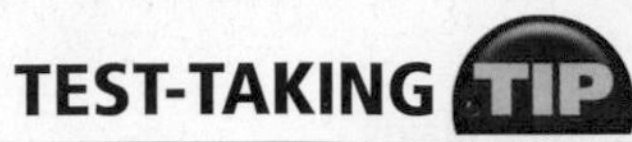

Tests are generally constructed to avoid a string of correct answers that are letters in alphabetical order. If you find that you have a lot of answers that are letters in alphabetical order, go back and check your work.

Reviewing Vocabulary

Directions: Choose the word or words that best complete the sentence.

1. At the ________ are rapids and waterfalls that blocked ships from traveling farther upstream.

- **A** Piedmont
- **B** coastal plain
- **C** Appalachian Mountains
- **D** fall line

2. ________ were formed in the Earth millions of years ago and can be burned for energy.

- **A** Renewable resources
- **B** Minerals
- **C** Fossil fuels
- **D** Mountains

3. Thunderstorms that can cause tornadoes are ________.

- **A** hurricanes
- **B** supercells
- **C** prairies
- **D** cold fronts

4. Trees cannot grow above ________.

- **A** the timberline
- **B** the fall line
- **C** latitude lines
- **D** boundaries

Reviewing Main Ideas

Directions: Choose the best answers to complete the sentences or to answer the following questions.

Section 1 *(pp. 130–135)*

5. Why is it especially important to conserve fossil fuels?

- **A** They are nonrenewable.
- **B** People do not understand how to find them.
- **C** People do not know how to use them.
- **D** The government owns all the sources.

6. As commercial fishing has declined, what activity has taken its place?

- **A** agriculture
- **B** lumbering
- **C** aquaculture
- **D** conservation

Section 2 *(pp. 138–142)*

7. The highest temperature ever recorded in the United States was in ________.

- **A** the Florida Everglades
- **B** Death Valley
- **C** Phoenix, Arizona
- **D** Las Vegas, Nevada

8. The interior parts of ________ have an ice cap climate.

- **A** Alaska
- **B** Yukon Territory
- **C** Northwest Territories
- **D** Greenland

GO ON

Critical Thinking

Directions: Choose the best answers to complete the sentences or to answer the following questions.

9. Aside from latitude, what other factor greatly influences climate in North America?

A trees

B grasslands

C large landforms

D rivers

Base your answer to question 10 on the map and on your knowledge of Chapter 5.

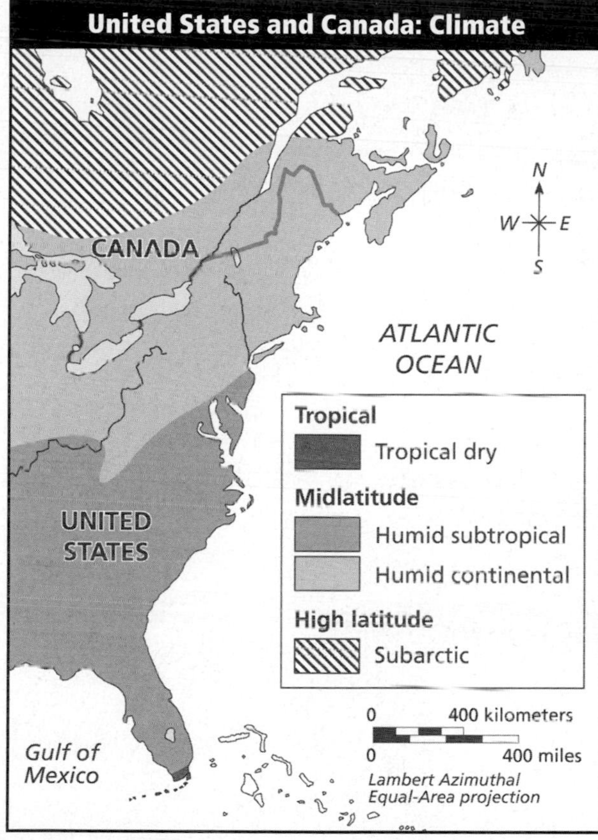

10. What type of climate dominates the extreme southeastern United States?

A tropical dry

B humid subtropical

C humid continental

D subarctic

Document-Based Questions

Directions: Analyze the document and answer the short-answer questions that follow the document.

This excerpt discusses mountaintop mining in the Appalachians.

> *Coal miner.*
>
> *Those words may conjure the image of a man with a light on his helmet and a pick in his hand. But more than two-thirds of this country's coal comes from surface mines—strip mines, or in their latest, largest incarnation, mountaintop removal mines.*
>
> *Instead of tunneling into a mountain and hauling out its coal, strip miners move chunks of mountain out of the way until the coal is at the surface.*
>
> *In mountaintop removal mining—just like it sounds—the mountaintop is pulverized to get at the coal. Begun in West Virginia and Kentucky in the late 1960s, the pace of mountaintop removal has picked up in the past decade as demand for coal has grown with the rise in the cost of other fuels. And with the increase in mountaintop removal has come greater outcry about the effects of the practice. . . .*
>
> *Opposition groups blame strip mining and the clear-cutting that precedes it for flooding. They say it damages wildlife habitat. They worry about sludge ponds, filled with the liquid waste created in the coal-cleaning process. . . .*
>
> —Tim Thornton, "Mountaintop Removal," *Roanoke Times*, July 2, 2006

11. How does strip mining differ from the traditional practice of tunneling into the mountain?

12. According to its opponents, what environmental problems are caused by the practice of mountaintop removal mining?

Extended Response

13. Exploring the BIG Idea
Describe the effect that lakes and rivers have had upon the economic development of the United States and Canada.

For additional test practice, use Self-Check Quizzes—Chapter 5 on glencoe.com.

Need Extra Help?

If you missed questions. . .	1	2	3	4	5	6	7	8	9	10	11	12	13
Go to page. . .	133	134	140	141	134	135	139	142	139	139	145	134	133

CHAPTER 6

CULTURAL GEOGRAPHY OF

The United States and Canada

BiG Idea

Certain processes, patterns, and functions help determine where people settle. Studying the cultural geography of the United States and Canada will introduce the peoples of this region—their history, how and where they live, and the reasons behind their choices.

Essential Questions

Section 1: The United States

What are three factors that have influenced the cultural geography of the United States?

Section 2: Canada

How have immigration and physical geography influenced the cultural geography of Canada?

Visit glencoe.com and enter *QuickPass*™ code WGC9952C6 for Chapter 6 resources.

A U.S. naturalization ceremony takes place in Freedom Park in Arlington, Virginia.

Organizing Information Make a Folded Chart to help you organize information about the cultural geography of the United States and Canada.

Chapter 6	Population Patterns	History and Government	Culture
United States			
Canada			

Reading and Writing As you read this chapter, make notes about the population patterns, history and government, and culture of the two nations. Write the information in the correct squares of the chart.

SECTION 1

section audio

spotlight video

Guide to Reading

Essential Question

What are three factors that have influenced the cultural geography of the United States?

Content Vocabulary

- immigration *(p. 149)*
- Sunbelt *(p. 149)*
- urbanization *(p. 150)*
- metropolitan area *(p. 150)*
- suburb *(p. 150)*
- urban sprawl *(p. 150)*
- megalopolis *(p. 150)*
- Underground Railroad *(p. 152)*
- dry farming *(p. 152)*
- bilingual *(p. 153)*
- literacy rate *(p. 153)*
- jazz *(p. 153)*

Academic Vocabulary

- discrimination *(p. 149)*
- expansion *(p. 150)*
- amendment *(p. 151)*

Places to Locate

- Los Angeles *(p. 153)*
- New York City *(p. 153)*

Reading Strategy

Organizing Complete a web diagram similar to the one below by listing the cities that comprise the Boswash megalopolis.

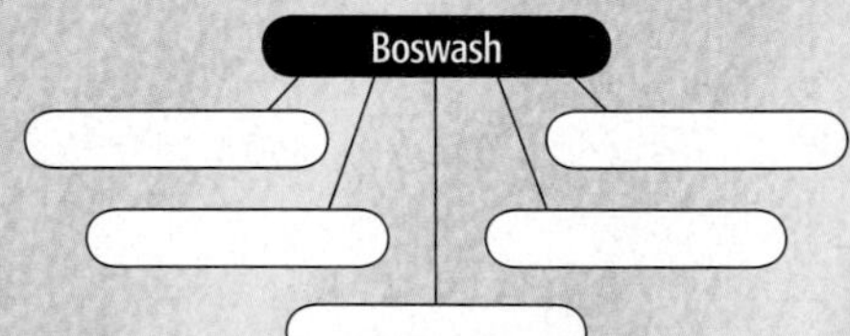

The United States

The United States may have a wider range of ethnic and cultural groups than most other countries in the world. The Mohawk people, a Native American group, are one of many groups that help define the population of the United States and add to the country's unique cultural landscape.

NATIONAL GEOGRAPHIC Voices Around the World

"I was on a train clattering south along the Hudson River, heading toward . . . New York City. . . . New York today is home to more than 85,000 Native Americans. About 85 percent of Indians in the United States now live off the reservation, and every large city in the U.S. has its own Indian community. . . . Brad Bonaparte is one of these urban Indians, a 42-year-old Mohawk artist and ironworker whose father and grandfather walked the high steel with wrenches and welding torches, making the city's skyline. Every workday he puts on a brown hard hat bearing the insignia of an eagle feather, a potent symbol of blessing and protection worn by many Mohawk ironworkers."

—Joseph Bruchac, "Indian: Scenes from a Renaissance," *National Geographic,* September 2004

A Mohawk construction worker

Population Patterns

MAIN Idea The United States is a country shaped by immigration, with a continually shifting population and densely populated urban areas.

GEOGRAPHY AND YOU How would you describe your community's population? Read to learn about population patterns in the United States.

The population of the United States is among the world's most diverse. This diversity reflects the influence of **immigration,** the movement of people into one country from another.

The People

There are more than 300 million people in the United States today, many of whom are immigrants or descendants of immigrants. Some arrived only recently. Others belong to families whose ancestors came to the region centuries ago. It is believed that the first peoples moved into the region from Asia about 20,000 years ago. Today their descendants, known as Native Americans, number over 2.5 million in the United States. Other peoples—Europeans, Asians, Africans, and Latin Americans—came later.

Some immigrants came to the United States to seek political and religious freedom and to find better economic opportunities. Others fled wars or natural disasters. Rich natural resources and the region's rapid industrial and economic development made the United States an attractive destination. Some immigrants faced **discrimination** at first, but they offered hard work, talent, enthusiasm, and diverse cultural practices.

Today many immigrants arrive seeking jobs and educational and career opportunities or refuge from difficult political situations. In 2008, 12.5 percent of the total U.S. population was foreign born. In recent years, more than half of the foreign-born population has been from Latin America.

Density and Distribution

The population density of the United States is about 87 people per square mile (33 people per sq. km). Outside of large urban areas, however, the population is widely distributed. The Northeast and Great Lakes regions are densely populated because they are the historic centers of American commerce and industry. The Pacific coast attracts people looking for a mild climate and economic opportunities, resulting in a population cluster there. The least densely populated areas of the country include the subarctic region of Alaska, the parched Great Basin, and parts of the arid and semi-arid Great Plains.

Since the 1970s the American South and Southwest, including California, Arizona, and New Mexico, have become some of the country's fastest-growing areas. Nicknamed the **Sunbelt** for its mild climate, this area draws people to its growing manufacturing, service, and tourism industries. The area's proximity to Mexico and the Caribbean also draws immigrants from these regions.

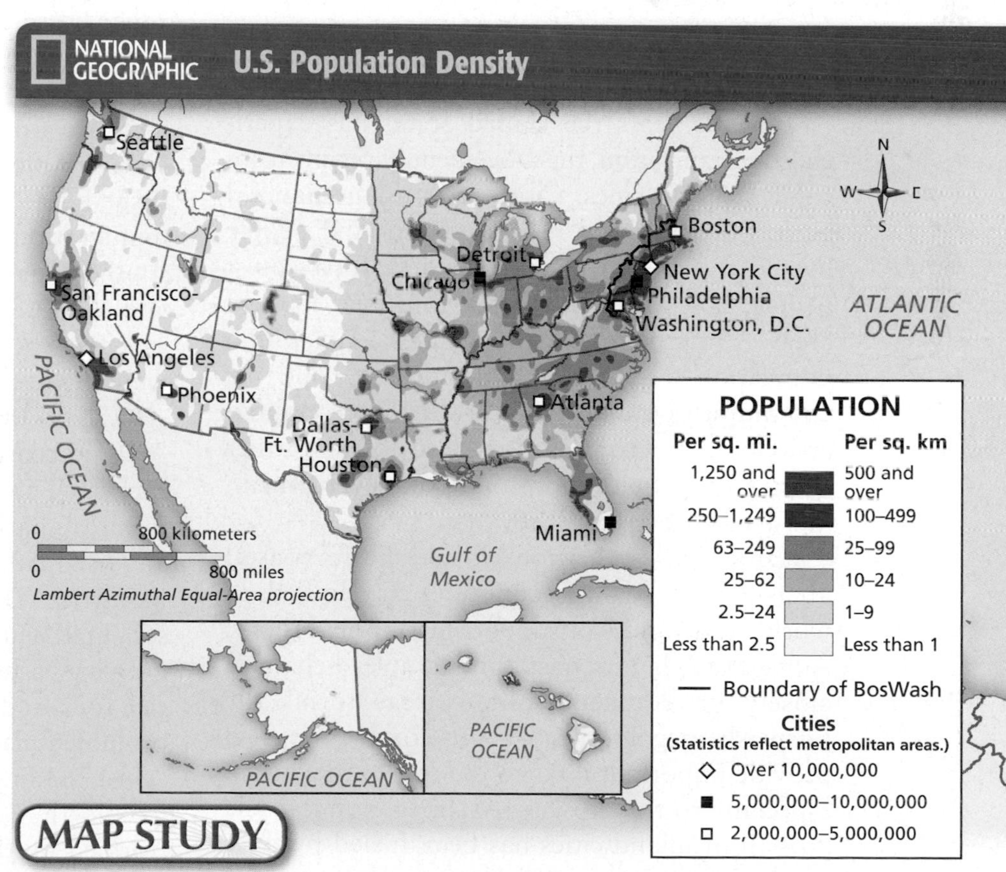

MAP STUDY

1. **Regions** What is the most densely populated area of the United States?
2. **Human-Environment Interaction** How has access to water affected the development of cities in the United States?

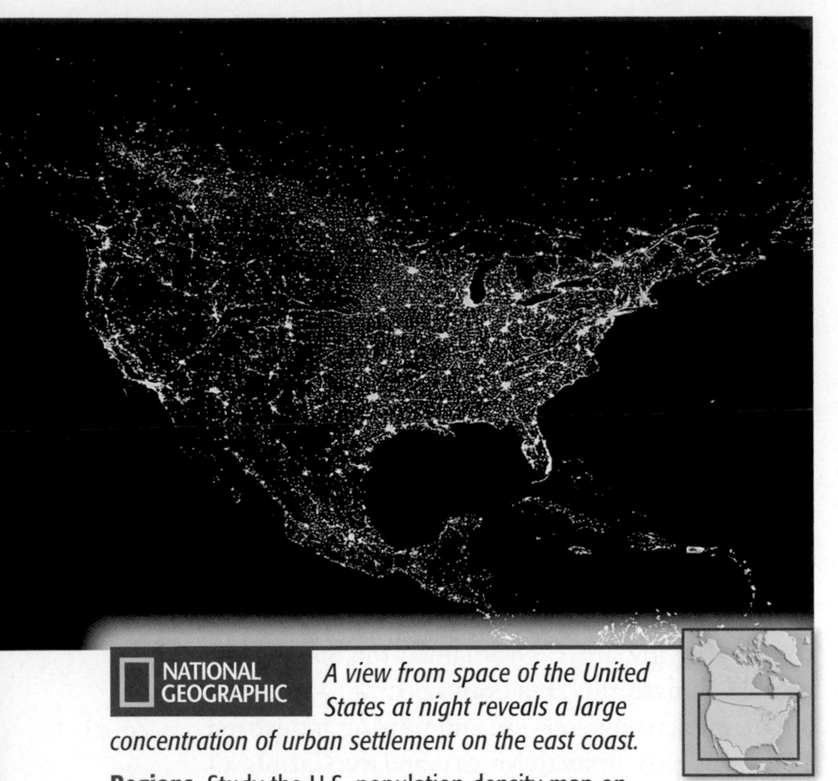

NATIONAL GEOGRAPHIC *A view from space of the United States at night reveals a large concentration of urban settlement on the east coast.*

Regions Study the U.S. population density map on page 149. How is this image similar to the map? How is it different?

Urban Areas

Over the years the United States has experienced **urbanization,** the movement of people from rural areas to cities. Cities grew as mechanized agriculture required fewer workers and people moved to cities in search of work. Today most people in the United States live in metropolitan areas. A **metropolitan area** includes a city with a population of at least 50,000 people and outlying communities called **suburbs.** As metropolitan areas become crowded, they lead to the further spread of people and suburban development, known as **urban sprawl.**

Many U.S. population clusters lie in coastal areas where strong economies are linked to world trade and can support large populations. Along the Atlantic coast, for example, a chain of closely linked metropolitan areas forms the Boswash **megalopolis.** Pacific coast cities also provide important links to the rest of the world, especially to the growing Asian economies. The growth of inland cities has been fueled by their proximity to rivers and lakes.

READING Check **Location** Why are many U.S. population clusters located near a coast?

History and Government

MAIN Idea Physical geography and a spirit of independence influenced U.S. development.

GEOGRAPHY AND YOU Why is your community located where it is? Read to learn what factors played a role in the settlement of the United States.

U.S. history has been influenced by Native Americans, European colonization, a war for independence, the creation of a new government, westward **expansion,** and industrial growth.

Early Nation Building

Archaeologists believe that nomads crossing a land bridge from Asia to what is now Alaska first settled in North America thousands of years ago. Recent evidence suggests, however, that nomads from Central and South America may have populated North America at about the same time.

The lives of Native Americans, the descendants of these early peoples, were shaped by location and climate. For example, people in the desert Southwest used irrigation to farm the dry land. Native Americans occupied the region undisturbed until the mid-1500s when European immigration began. The Spanish explored the southern parts of the region, setting up farms and cattle ranches, military posts, and missions. The French settled mostly in the northeast and were involved in the fur trade.

By the 1700s, Britain controlled land along the Atlantic coast. The New England Colonies had rocky soil and a short growing season. However, the area's harbors and abundant supply of timber and fish made shipbuilding and fishing important industries. The Middle Colonies had the fertile soil, mild winters, and warm summers needed for growing cash crops for export. The mild climate, rich soils, and open land of the Southern Colonies promoted plantation agriculture.

In 1763 France was forced to give up much of its North American empire to Great Britain (formed by the union of England and Scotland in 1707). Conflicts soon arose between Native Americans and colonial settlers. Many settlers pushed out Native American communities and nearly destroyed their cultures.

In the 1760s, the British government angered the colonists by imposing new taxes and limiting their freedoms. The thirteen colonies eventually fought for independence from Great Britain in the American Revolution (1775–1783). The outcome was an independent United States of America.

The U.S. government is a federal republic. The national government shares power with the states. The Constitution created a strong national government while preserving the rights of citizens. **Amendments,** or changes to the Constitution, have been made to meet the country's changing needs. The first 10 amendments—the Bill of Rights—guarantee the basic rights of citizens.

The national government has three separate but equal branches. The executive branch—president, vice president, and administrative departments—carries out law. Congress, the legislative branch, enacts law. The Supreme Court and federal courts are the judicial branch, which interprets law.

Growth, Division, and Unity

The map below shows that during the 1800s, the United States more than doubled its size and gained valuable land and natural resources. For Native Americans, however, westward expansion of settlers signaled the steady loss of lands and restrictions on traditional ways of life.

Industrialization transformed the United States in the 1800s. The first factories arose along the fall line in the Northeast, which had many waterfalls whose power could run machines. Later, large supplies of coal in the Midwest were used to fuel steam engines, making steam power cheap and manufacturing profitable. As a result, the Midwest became a leading center of industry.

MAP STUDY

1. **Movement** In which direction did U.S. expansion advance?
2. **Regions** What impact did the Louisiana Purchase have on the size of the United States?

Maps In MOtion Use **StudentWorks™ Plus** or glencoe.com.

NATIONAL GEOGRAPHIC **U.S. Expansion**

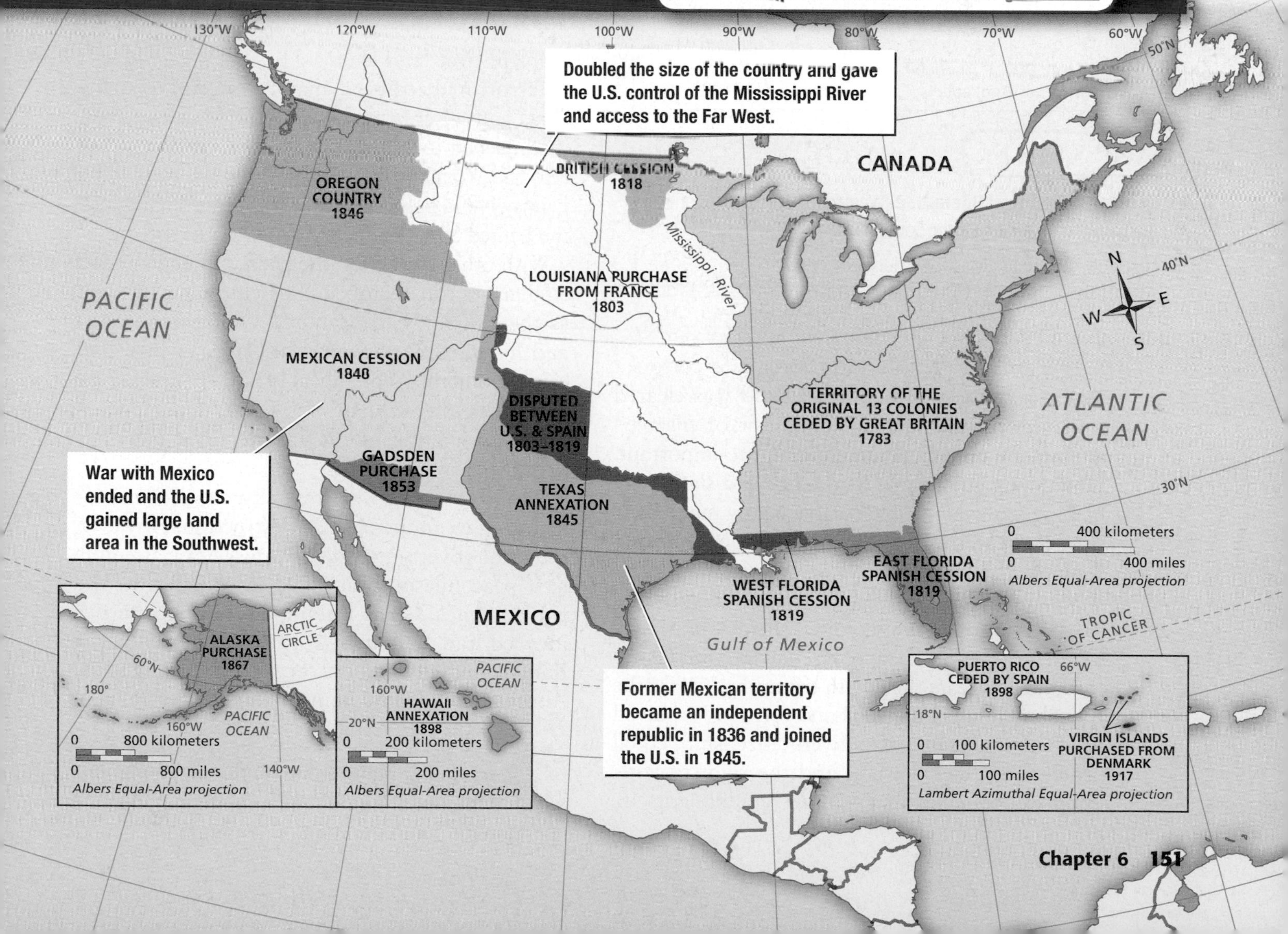

Doubled the size of the country and gave the U.S. control of the Mississippi River and access to the Far West.

War with Mexico ended and the U.S. gained large land area in the Southwest.

Former Mexican territory became an independent republic in 1836 and joined the U.S. in 1845.

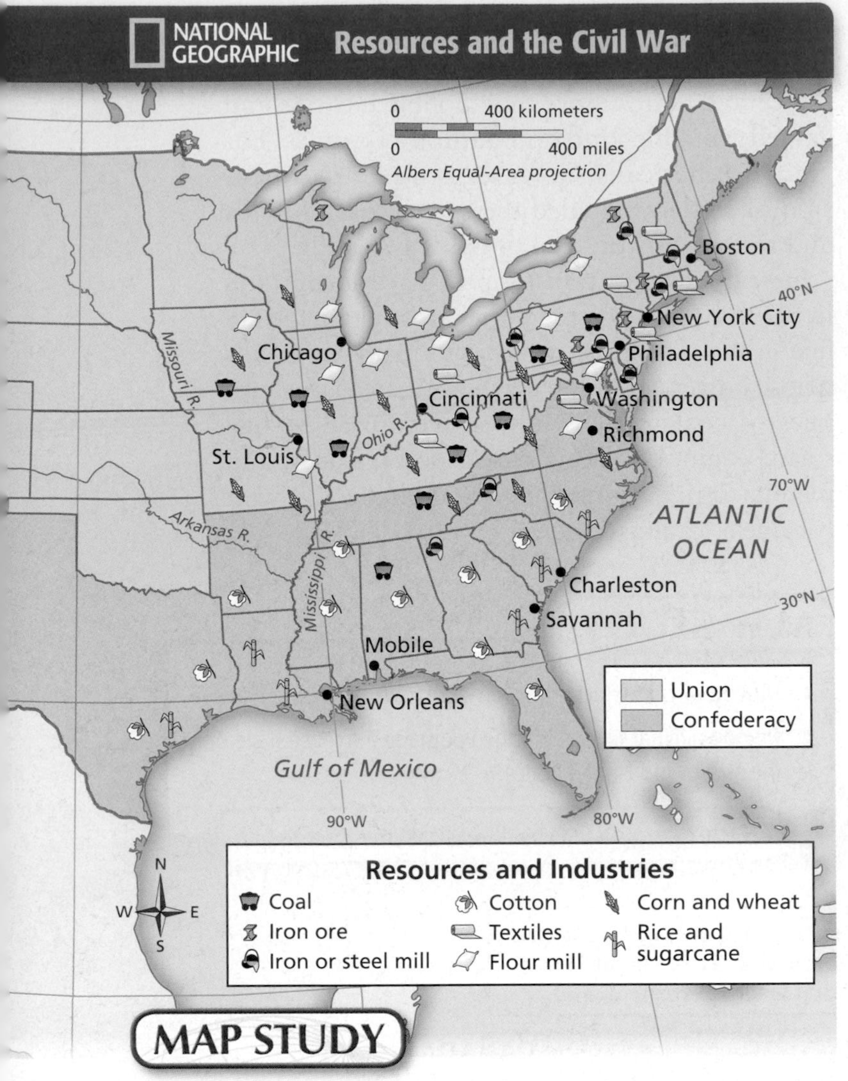

MAP STUDY

1. **Human-Environment Interaction** What kinds of resources enabled the North to triumph over the South? Explain.
2. **Location** Where are most cities located? Why?

The growing textile industry made cotton a major cash crop in the South. Land was cleared for more plantations, and the labor of enslaved African Americans became ever more important. However, some people worked to end slavery by helping enslaved people escape. Many made their way north to freedom through the **Underground Railroad,** an informal network of safe houses.

Tensions between the industrialized North and the agricultural South mounted steadily until they erupted in the American Civil War in 1861. After four bloody years the North triumphed. After the war, slavery was abolished and formerly enslaved African Americans were given citizenship, equal protection under the law, and the right to vote. Reunited, the country set about rebuilding itself.

Changes and Challenges

In the late 1800s, the U.S. government encouraged settlement of the Great Plains to ease crowding in Eastern cities. The government also needed people to farm the area to provide more food for cities. Due to dry conditions on the Great Plains, settlers developed **dry farming,** a method of cultivating land to catch and hold rainwater. Steel plows and steam tractors made planting and harvesting large areas easier.

Chinese, Irish, Mexican, and other immigrants helped build railroads in the United States, including the transcontinental lines. This network of railways transported manufactured goods from east to west and food products from west to east.

Two world wars spurred economic growth. Assembly lines increased efficiency and improved the standard of living. The population became more mobile and urbanized. By the 1990s, many manufacturing activities were less important than rising high-tech industries.

Social changes also took place. Immigration from Latin America and Asia increased. Minority groups began to participate in business and politics. Native Americans negotiated with the government over land claims.

Terrorism became a major concern of many Americans after September 11, 2001, when terrorists hijacked four passenger planes, crashing them into the World Trade Center, the Pentagon, and a Pennsylvania field. After such devastation and loss, the United States launched a war on terrorism.

With other countries, the United States invaded Afghanistan in October 2001. Its rulers, known as the Taliban, had harbored Osama bin Laden and his terrorist network, al-Qaeda, which had carried out the September 11, 2001 attacks. By December, the Taliban was forced from power. Renewed fighting by the Taliban since 2003 has kept U.S. forces there.

The next target was Iraq, suspected of manufacturing weapons of mass destruction. However, the Iraqi government denied this. After attempts at a peaceful solution failed, a coalition of the United States, Great Britain, and other countries invaded Iraq on March 20, 2003. By April 9, the Iraqi regime fell and U.S. forces occupied Baghdad. Although a new government was formed in May 2006, U.S. forces remain in Iraq.

READING Check **Human-Environment Interaction** Why did the Midwest become a center of industry?

Culture

MAIN Idea Immigration has influenced the diversity and culture of the United States.

GEOGRAPHY AND YOU What words and phrases come to mind when you think about culture in the United States? Read to learn about the country's unique cultural characteristics.

The immigrant roots of the United States give it a respect for diversity. Its varied cultures blend into new patterns and yet manage to maintain their individual qualities.

Language and Religion

English is the main language in the United States. Because of immigration from around the world, however, people also speak or use words and phrases from other languages. Some people are **bilingual** and learn to speak English in addition to their native language. The growing Latino population has made Spanish the second most commonly spoken language today.

Many religions flourish throughout the country. Most people who are members of an organized religion are Christians. Judaism, Islam, Hinduism, and Buddhism are among other religions practiced in the country.

Education and Health Care

Education is built on a network of public and private schools. Attending school is required until the age of 16. The **literacy rate,** the percentage of people who can read and write, is 99 percent.

A highly developed economy enables the United States to devote substantial resources to health care. Still, many people are unable to buy health insurance, and others cannot afford health care even with insurance. The role of government in providing health care for all citizens is under debate.

The Arts

The arts go back to the first Americans, who integrated art, music, and storytelling into daily life. After European settlement, the arts were dominated by European traditions. By the mid-1800s, however, people in the United States began to create art forms that reflected their own lives and cultures. For example, **jazz,** which blends African rhythms with European harmonies, developed in African American communities. Writers began writing about life and culture in different parts of the country.

U.S. cultural influence on the rest of the world is strongest in the area of popular entertainment. Hollywood, a **Los Angeles** district, is synonymous with the movie business. Broadway, a street in **New York City,** is identified with theater.

Culture

Architecture

Fallingwater, one of architect Frank Lloyd Wright's most acclaimed works, is integrated into the surrounding natural environment in western Pennsylvania.

A Anchored in the rocks next to the falls, each of the house's floors or "trays" mimics the natural pattern of the rock ledges.

B Wright used the color tan to imitate the autumn color of dying plants that surround the house, connecting it to the change of seasons.

CHAPTER 7

THE REGION TODAY

The United States and Canada

BiG Idea

Geography and the environment play an important role in how a society is shaped over time. The physical geography of the United States and Canada has shaped economic activities, which in turn have impacted the region's environment.

Essential Questions

Section 1: The Economy

How have the economies of the United States and Canada grown and changed according to where and how people live?

Section 2: People and Their Environment

How have human actions modified the environment in the United States and Canada?

Geography ONLINE
Visit glencoe.com and enter *QuickPass*™ code WGC9952C7 for Chapter 7 resources.

The busy port of Vancouver, British Columbia, handles much of Canada's trade with Asia.

FOLDABLES™ Study Organizer

Summarizing Information Create a Four-Door Book to summarize information about four important aspects of life in the United States and Canada today.

Reading and Writing As you read this chapter, write down key details about the region's economic activities, transportation and communications, trade and interdependence, and ways of managing resources under the appropriate door.

Agriculture

Farming in the United States and Canada is overwhelmingly commercial, with agricultural **commodities,** or goods, produced for sale. Large commercial corporations, however, account for only a small percentage of farm ownership. Most farms in the region are still owned by families, many of whom have formed cooperative operations.

Large areas of the region are involved in agricultural activities. About 920 million acres (372,311,120 ha) of land are used for agriculture in the United States. Although it has much less **arable** land, or land suited for farming, than the United States, Canada still devotes 167 million acres (67,586,739 ha) to agriculture.

While the average size of farms in the United States and Canada has increased since the 1950s, the number of farmers has decreased. Among the factors contributing to this **decline** are the high cost of farming, unpredictable consumer demand, the risk of natural disasters, and the time and hard work needed to run a farm.

Key Agricultural Products Cattle ranches operate mostly in the western, southern, and midwestern United States and in Canada's Prairie Provinces. Other important livestock-producing areas include the north-central United States, Quebec, and Ontario. Wheat is grown in the Prairie Provinces of Canada and on the Great Plains of the United States, a region often called the **Wheat Belt.** The **Corn Belt** stretches from Ohio to Nebraska. Corn is also grown in Quebec, Ontario, and Manitoba.

Agricultural Technology In the past, geographic factors often limited where certain types of agriculture could be carried out. Cattle ranching, for example, needed the wide-open spaces and natural grasses of the western prairies and plains. Most American dairy farms were concentrated in a belt of land stretching from upper New York State to Minnesota. This region, known as America's Dairyland, has cooler summers and native grasses ideal for dairy cattle.

The development of breeds of cattle that need less room to graze has opened up the southern United States to cattle ranching. Because of improved feed sources and automation, dairy farms can now be found in every American state and many Canadian provinces.

READING Check **Place** In what types of jobs are most of the region's workers employed?

Transportation and Communications

MAIN Idea People in the United States and Canada depend on reliable and continually improving transportation and communications systems.

GEOGRAPHY AND YOU What kind of transportation do you use to get to school? Read to learn how the United States and Canada depend on transportation and communications systems to sustain their economies.

The development of reliable transportation and communications systems in the United States and Canada was essential because of the large land area of both countries, their population distribution, and their need to move goods and services.

The Automobile

Since World War II, the most popular means of personal transportation in the region has been the automobile. This extensive reliance on cars has required heavy investment in highways, roads, and bridges, which are also important in the transport of goods. Mirroring population density, the network of U.S. roads is densest along the east and west coasts, as well as along the Mississippi and **Ohio River** valleys. Canada's smaller, more concentrated population relies on a smaller network of roads, most of which are located in the southern part of the country. The **Trans-Canada Highway** runs 4,860 miles (7,821 km) from Victoria, British Columbia, to St. John's, Newfoundland.

Reliance on the automobile, however, creates air pollution through the burning of gasoline that affects most urban areas. Automakers and government agencies are working together to reduce the use of autos in certain urban districts and to find cleaner, more efficient ways to use fuel.

Another challenge posed by auto use is traffic congestion in the region's cities. Mass public transportation reduces the number of vehicles. Cities such as Montreal and New York now have well-established subway systems. Los Angeles continues to expand a system that combines subways with elevated trains, and Seattle and Dallas both have monorail systems. Buses and commuter trains are also used to ease congestion.

Other Means of Transportation

For long-distance travel, many people use the region's busy network of airports. Atlanta's Hartsfield and Chicago's O'Hare International Airports vie for the title of the busiest U.S. airport. Toronto's Pearson International Airport is Canada's busiest. Passenger railroads and long-distance buses account for only a small portion of the region's passenger travel.

The transport systems of the region also move goods. Railroads haul about 36 percent of the region's freight, while about 13 percent is carried along inland waterways. The Great Lakes–St. Lawrence Seaway System is used by both the United States and Canada to transport goods. Barges and ships on the Mississippi River system transport U.S. grain and petroleum products. Long-haul trucks carry about 28 percent of the region's freight. Airplanes carry only a small portion of the region's heavy freight, but handle a growing amount of overnight delivery business. Pipelines carry about 20 percent of the region's freight in the form of gas and oil.

Communications

The success of the region's economy has been influenced by the development of communications networks. Cellular and digital services have made telephone communication more mobile. Business transactions and personal communications can be completed instantaneously using e-mail and the Internet. Thousands of television stations, radio stations, newspapers, and magazines provide local, national, and international news. While Canada's broadcasting and publishing services are publicly owned, private companies operate these services in the United States. Federal government regulations, however, ensure that there is no **monopoly,** the total control of an industry by one person or one company.

READING Check **Movement** What are the primary methods for moving goods throughout the region?

NATIONAL GEOGRAPHIC **The U.S. Interstate Highway System**

Seattle
Boston
Chicago
New York City
Denver
Washington, D.C.
Las Vegas
Los Angeles
Atlanta
1950
Houston
Miami

Seattle
Boston
Chicago
New York City
Denver
Washington, D.C.
Las Vegas
Los Angeles
Atlanta
2009
Houston
Miami

Source: www.fhwa.dot.gov/interstate

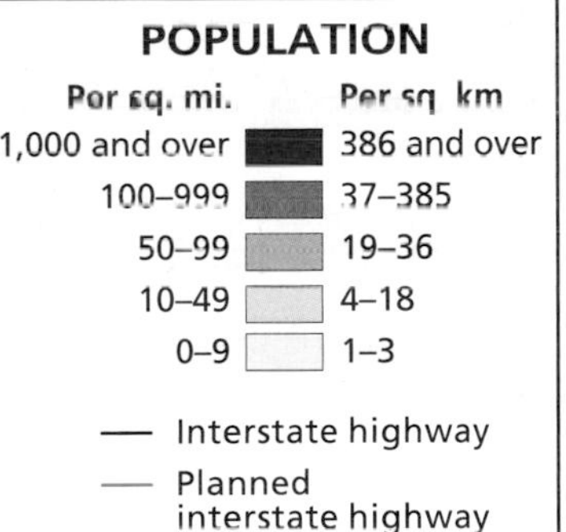

MAP STUDY

Interstate highways in the United States are used to move both people and goods.

1. **Place** Compare the maps above. How did the distribution of highways change from 1950 to 2009?
2. **Location** Which part of the country has the most highways? Why do you think this is so?

Maps In MOtion Use **StudentWorks™ Plus** or glencoe.com.

NATIONAL GEOGRAPHIC **Case STUDY**

THE GLOBAL ECONOMY: What happens when people overseas are used to fill jobs that Americans have been doing?

Factories close and operations move to another country. Office work is also being done by people in foreign countries. When you call tech support for help with your computer, there is a good chance you are talking to someone in India. When you board a plane and the gate agent scans your boarding pass, someone in another country compiles the information and e-mails it back to the airline. This transfer of work to people in other countries is called *outsourcing*.

Understanding the Issue

Outsourcing, and the involvement of the United States in this business practice, can be examined from various perspectives.

A Moral Issue Some people argue that U.S. companies have an obligation to keep their employees working and that laying them off, even for economic reasons, is morally wrong. Workers may have difficulty finding new jobs that offer the same pay and benefits as the jobs they lost. Older workers may not be able to find new jobs of any kind.

An Economic Dilemma U.S. companies today face competition from countries where wage rates are much lower and companies do not have to provide health insurance, which can be extremely expensive. Companies try to find the least expensive places to conduct business. Modern transportation and communications technology often make it cheaper to have goods made where wage rates are low and then transported to market than to make them in the United States. Information can be sent electronically in a split second at almost no cost, and many foreign countries are educating more of their people to higher levels than ever before. U.S. consumers benefit from lower prices, and shareholders benefit from larger corporate profits.

A Political Problem Laid-off workers in the United States face a lower standard of living. They often demand that the government provide such help as unemployment benefits and job training. They may become angry at immigrants who seem to be taking jobs at lower wages. People who believe that government aid simply makes people lazy and unwilling to work will oppose their demands.

Above right: Call center employees work in Bengaluru, India.
Above: Employees of General Motors strike in Flint, Michigan.

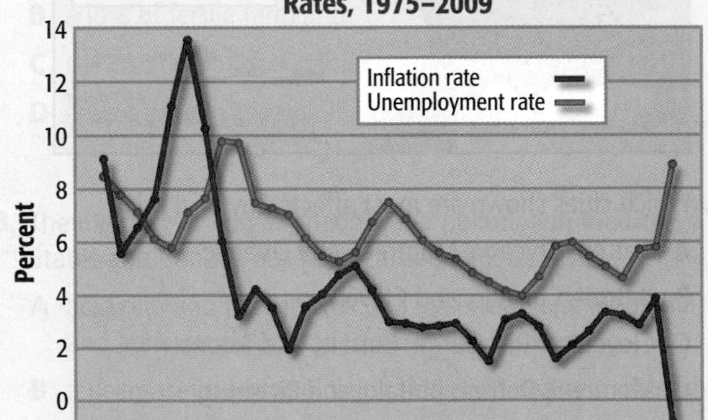

*Note: Data through September 2009.
Sources: www.inflationdata.com; Bureau of Labor Statistics, www.bls.gov

Possible Solutions

Outsourcing and its role in the global economy vary from one industry to another, as well as from region to region. So several solutions are possible.

Improved technology Technology such as computers and robots can make workers more productive. Companies that find moving overseas difficult have been active in developing such technology.

Educated workforce Advanced technology requires training to be used to its fullest potential. Better-educated workers can also solve problems more quickly, develop more efficient work routines, and produce higher quality goods and services.

Shifting to person-to-person services Some jobs are difficult to outsource. You cannot easily get a haircut in another country, for example. Concentrating on services that must be done where people are, while moving other jobs to a more cost-effective location, assures that workers have jobs.

A computer-operated robot is used to manufacture power tools in Louisiana.

Understanding the Case

The primary sources listed below give information about the effects of outsourcing. Use these resources, along with what you learned in Unit 2, to complete the activities listed on the next page.

The Situation

Primary Source 1

Excerpt from "Offshoring," www.economist.com, August 27, 2009.

There are differing opinions on the positive nature of outsourcing.

Offshoring—the wholesale shifting of corporate functions and jobs . . . to overseas territories—is what gave outsourcing a bad name. It is important, however, to note a crucial distinction between the two:

- *Outsourcing need not necessarily result in job losses in a particular territory or country. A job can simply be handed over to another organisation [organization] of the same nationality and geographical location where . . . it can be carried out more efficiently. Sometimes that other organisation may be in another country, but more often than not it is not.*
- *Offshoring, however, does involve shifting jobs to another country, but it may not involve transferring jobs to another organisation. For example, a company may simply decide to move its local customer services operation to one of its own subsidiaries abroad. That is offshoring, but it is not outsourcing.*

Possible Causes

Primary Source 2

Excerpt from "A Grand Goal for More U.S. Manufacturing Jobs," by Jessie Scanlon, *BusinessWeek*, August 31, 2009.

The problem with the loss of American jobs is the current U.S. trade policy of encouraging consumption of imports and incentives for outsourcing jobs.

The July employment report released on Aug. 7 by the Bureau of Labor Statistics, while better than expected, showed that the [manufacturing] sector had lost 52,000 jobs in July. That brings the total drop since December 2007 to 2 million jobs, or roughly 14.2% of that sector's employment. . . . The Alliance for American Manufacturing (which last month published a book, Manufacturing a Better Future for America, *laying most of the blame for the current state of affairs on U.S. trade policy) estimates that more than 40,000 factories across the nation have closed in the past decade.*

Employees in Manufacturing in the United States, 1968–2008

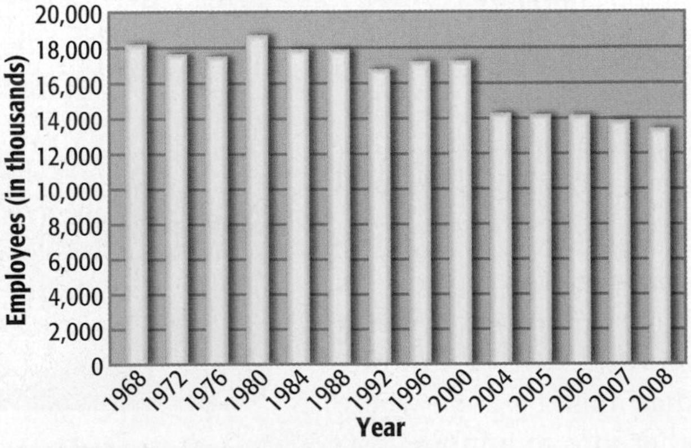

Source: Bureau of Labor Statistics, www.bls.gov/ces/home.htm#tables, Table B-1.

Possible Solutions

Primary Source 3

Excerpt from "Furniture makers mull moving back home," by Emily Kaiser, Reuters UK (www.uk.reuters.com), July 28, 2008.

Changes in the U.S. economy, as well as rising fuel and energy costs, have pushed some manufacturers to bring their operations back to the United States.

Furniture maker Carol Gregg got some puzzled looks when she went looking for a U.S. factory to make her Chinese antique reproductions.

. . . She used to ship American wood to a factory in China, which would make the furniture and then ship it back to the United States. Three years ago, even before oil hit $140 per barrel, she decided that was "really silly" and decided to move manufacturing back to the United States.

. . . In addition to cost, distance is also becoming a factor. Jobi Blachy, president of upscale furniture makers Edward Ferrell and Lewis Mittman, said one reason why his company manufactures in the United States is speed. Custom orders would take weeks longer if he had to ship them from overseas.

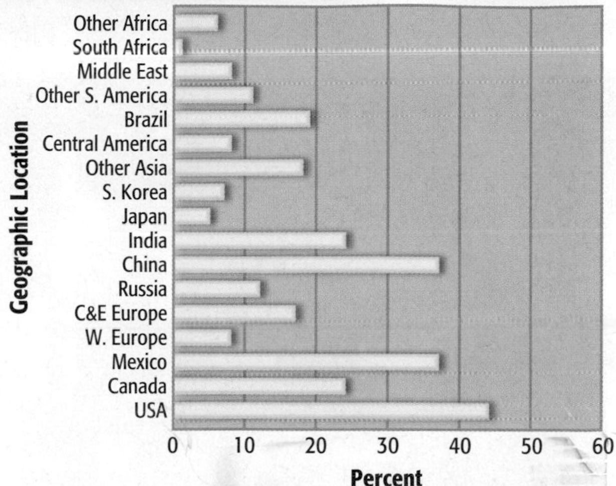

Source: "Made in North America" a Deloitte Research Manufacturing Study.

Blachy's customers spend as much as $26,000 for a dining room table, so paying a bit more for U.S. labor is no big deal. But even for lower-priced furniture retailers, which have scant pricing power when demand is weak, distance is starting to factor into the sourcing decision.

. . . "Retailers are trying to figure out how they can buy more products domestically and still not lose some of the economic value that imports bring to them," he said.

"That's quite a shift in some of the retail thinking because up until probably within the last year or two years, the retailers were continuing to import more and more directly from Asia and to be less dependent on domestic products."

Analyzing the Case

1. **Drawing Conclusions** Review the information in the primary sources above. How is outsourcing related to unemployment in the United States?
2. **Making Predictions** Is the trend of more manufacturing jobs moving back to the United States likely to continue? Explain your answer.
3. **Conducting a Debate** Use the following questions to conduct a class debate on outsourcing:
   - Why do companies want to close American facilities and open new facilities overseas?
   - What effects do these changes have on American families?
   - What role, if any, can the federal and state governments play in easing the problems that families face in these situations?
4. **Writing About the Case** Write an essay in which you argue for or against government action to halt the movement of jobs to other countries.

UNIT 3 Latin America

Ipanema beach, Rio de Janeiro, Brazil

Why It Matters

Latin America reflects a unique blend of world cultures, including Native American, European, and African. In turn, Latin America's diverse cultures have spread to other parts of the world. For example, the languages, music, foods, and arts of Latin America have profoundly influenced life in the United States. Today, many Americans are of Latin American descent and maintain close ties to their heritage. In addition, the United States and many countries in Latin America are close trading partners. They share democratic values based on human rights and revolt from European rule.

Latin America

PHYSICAL GEOGRAPHY Spanning more than 85 degrees of latitude, Latin America encompasses Mexico, Central America, the Caribbean islands, and South America. It is a region of startling contrasts.

High mountains run from northern Mexico through the heart of Central America. The higher peaks of the Andes course down South America's western side. Elsewhere, broad plateaus span huge areas. At still lower elevations, plains dominate the landscape. But when people think of Latin America, it's often rain forests that come to mind. Eternally wet, intensely green, and bursting with life, rain forests cover many parts of the region.

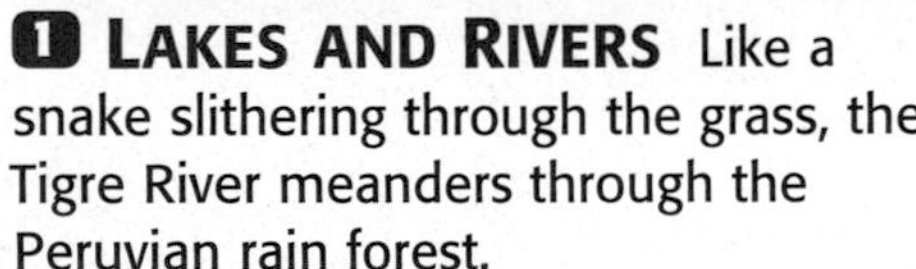

1 LAKES AND RIVERS Like a snake slithering through the grass, the Tigre River meanders through the Peruvian rain forest.

2 PLAINS AND PLATEAUS A wide plateau between two mountain ranges, the altiplano was created by intense volcanic activity.

3 MOUNTAINS The jagged peaks of Chile's Cuernos del Paine are part of the Andes, the world's longest mountain chain.

3

4

4 NATURAL RESOURCES Oil and natural gas are important natural resources in the region, with major deposits located along the Gulf of Mexico and in the southern Caribbean Sea.

UNIT 3 WHAT MAKES THIS A REGION?

Latin America

CULTURAL GEOGRAPHY Latin America is a region where cultures have collided. Native American civilizations flourished here long ago. Then Europeans arrived in the late 1400s and forced new laws, new languages, and a new religion onto the region's inhabitants. Yet native cultures survived by blending with those of the conquerors. Today the faces, costumes, and customs of many Latin Americans reveal their mixed heritage.

REGIONAL TIME LINE

200 B.C. — A.D. 1300 — A.D. 1400 — A.D. 1500

250 B.C. Teotihuacán becomes first major city in Americas

A.D. 1325 Aztec build Tenochtitlán on Lake Texcoco

A.D. 1492 Christopher Columbus lands on Hispaniola

Christopher Columbus

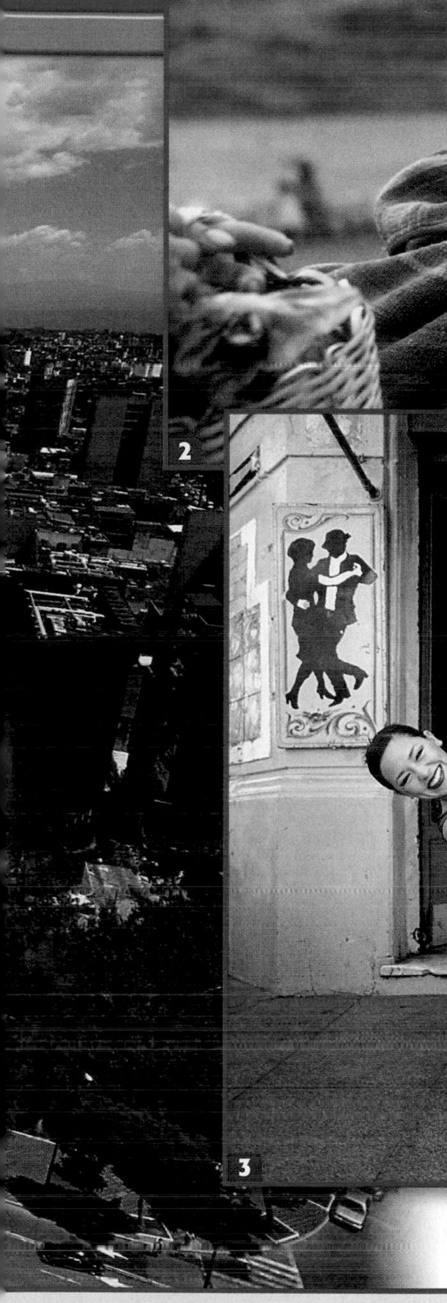

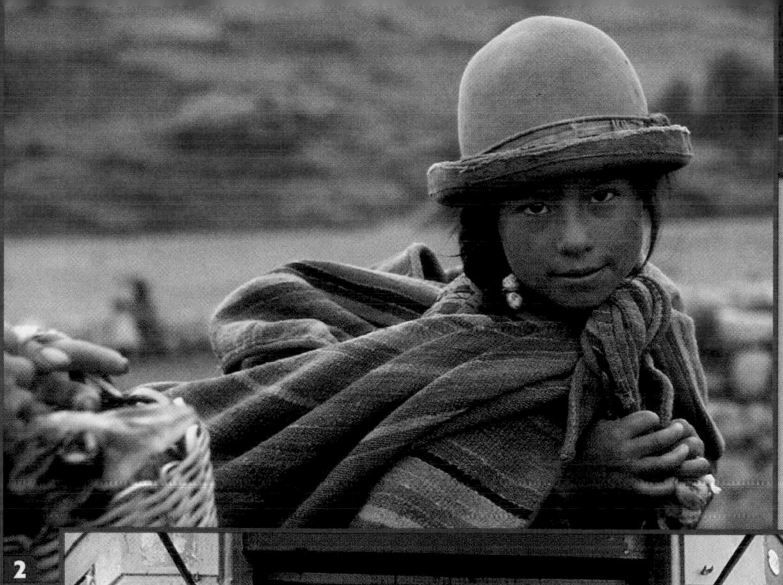

2

3

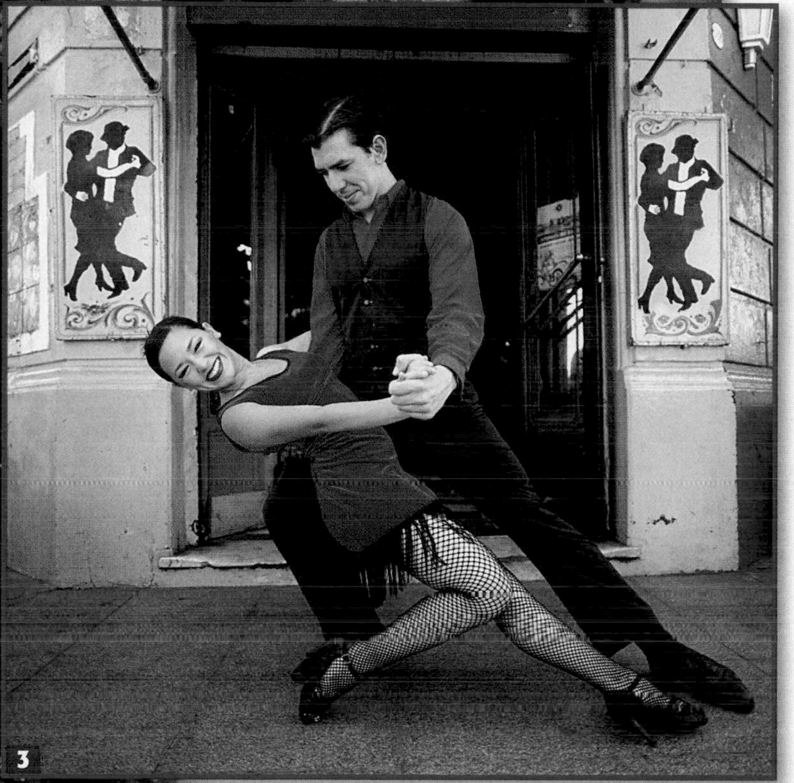

1 ECONOMY Many people migrate to urban centers such as Mexico City where work offers the possibility of economic advancement. This rapid growth forces cities to look for ways to provide their growing population with necessary resources.

2 PEOPLE A young Peruvian delivers produce to the marketplace in a nearby village. Many of the people of the Andes are Native Americans—descendants of the groups that flourished here before the arrival of the Europeans.

3 CULTURE The tango, which originated in Argentina, is one of many popular dances from Latin America. The tango became popular in the United States and Europe in the 1920s.

A.D. 1780 "Great Hurricane" hits Caribbean islands of Martinique, St. Eustatius, and Barbados

A.D. 1959 Fidel Castro comes to power in Cuba

A.D. 1973 Augusto Pinochet becomes dictator of Chile

A.D. 1999 Panama takes control of the Panama Canal

A.D. 2009 Military coup ousts Honduran president Manuel Zelaya

A.D. 1700 — A.D. 1800 — A.D. 1900 — A.D. 2000

A.D. 1813 Simón Bolívar establishes Venezuela's Second Republic

A.D. 1910 The Mexican Revolution, led by Emiliano Zapata and Francisco Villa, begins

Emiliano Zapata

A.D. 1985 Nevado del Ruiz erupts in Colombia; mudflow kills 25,000

A.D. 2004 Jean-Bertrand Aristide, Haiti's first democratically elected president, resigns

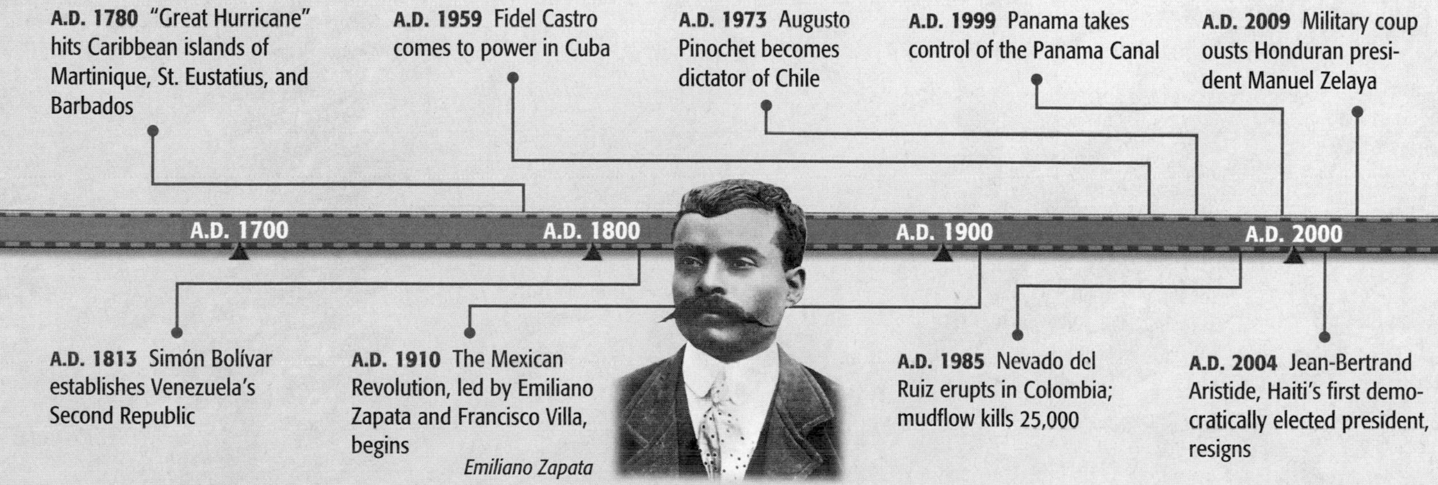

UNIT 3 REGIONAL ATLAS

PHYSICAL Latin America

120°W
80°W
60°W
40°W
20°N
0°
20°S
40°S
EQUATOR
TROPIC OF CANCER
TROPIC OF CAPRICORN

UNITED STATES
Río Grande
Baja California
SIERRA MADRE OCCIDENTAL
MEXICAN PLATEAU
SIERRA MADRE ORIENTAL
SIERRA MADRE DEL SUR
Gulf of Mexico
Yucatán Peninsula
Bermuda Islands
ATLANTIC OCEAN
WEST INDIES
Cuba
Hispaniola
Greater Antilles
Jamaica
Puerto Rico
Guadeloupe
Martinique
Lesser Antilles
Caribbean Sea
Mosquito Coast
Lake Maracaibo
Lake Nicaragua
Trinidad
Orinoco R.
Isthmus of Panama
LLANOS
GUIANA HIGHLANDS
ANDES
AMAZON BASIN
Putumayo R.
Río Negro
Amazon R.
Marajó Island
Galápagos Islands
Cape São Roque
PACIFIC OCEAN
La Montaña
SELVAS
Madeira R.
Catingas
Tocantins R.
São Francisco R.
MATO GROSSO PLATEAU
Lake Titicaca
Altiplano
Pantanal
BRAZILIAN HIGHLANDS
Atacama Desert
GRAN CHACO
Paraná R.
Campos
Cape São Tomé
Cape Frio
N
W
E
S
Paraguay R.
Aconcagua 22,834 ft. (6,960 m)
PAMPAS
Uruguay R.
Juan Fernández Islands
Río de la Plata
Chiloé Island
Valdés Peninsula
PATAGONIA
Falkland Islands (Islas Malvinas)
Tierra del Fuego
Cape Horn
South Georgia Island

Elevations

Feet	Meters
13,100	4,000
6,500	2,000
1,600	500
650	200
0	0

National boundary
▲ Mountain peak

0 — 1,000 kilometers
0 — 1,000 miles
Lambert Azimuthal Equal-area projection

Largest in the World

Latin America contains the longest mountain range and the largest river in the world. As you study the maps and graphics on these pages, look for these geographic features that make the region unique. Then answer the questions below on a separate sheet of paper.

1. What physical features could present barriers to the development of Latin America?
2. What challenges and benefits do you expect the Amazon River to provide people in this region?
3. What advantages can the Andes provide to people in this region?

The Stretch of the Andes

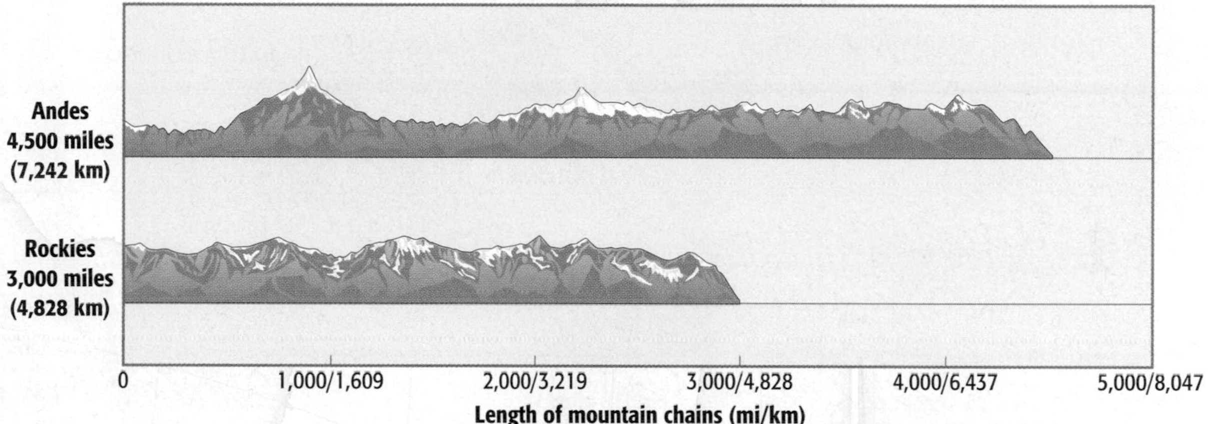

The Mighty Amazon

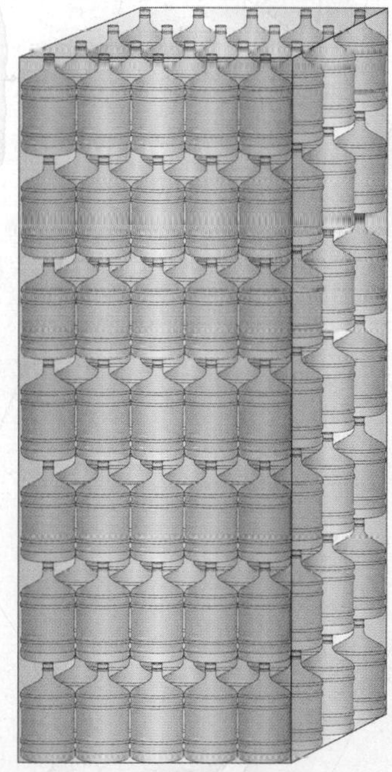

= 50,000 cubic feet per second

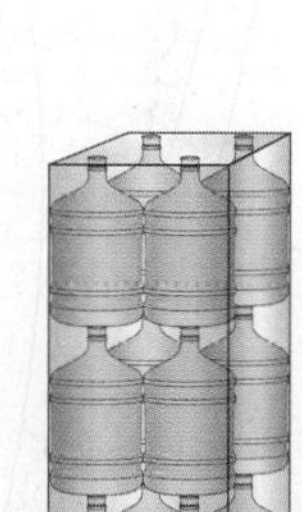

Amazon River
7,000,000 cubic feet per second
(198,217.9 cubic m per second)

Mississippi River
600,000 cubic feet per second
(16,990.1 cubic m per second)

UNIT 3 REGIONAL ATLAS

POLITICAL Latin America

120°W
80°W
60°W
40°W
UNITED STATES
ATLANTIC OCEAN
Bermuda U.K.
Puerto Rico U.S.
Virgin Islands U.S. & U.K.
Anguilla U.K.
ANTIGUA AND BARBUDA
ST. KITTS AND NEVIS
Montserrat U.K.
Guadeloupe Fr.
DOMINICA
Martinique Fr.
ST. LUCIA
BARBADOS
ST. VINCENT AND THE GRENADINES
GRENADA
Tijuana
Ciudad Juárez
Rio Grande
Chihuahua
Monterrey
Gulf of Mexico
BAHAMAS
Nassau
TROPIC OF CANCER
Havana
CUBA
MEXICO
Guadalajara
Mexico City
Puebla
Veracruz
Orizaba
20°N
Cayman Is. U.K.
DOMINICAN REPUBLIC
Port-au-Prince
JAMAICA
Kingston
HAITI
Santo Domingo
BELIZE
Belmopan
HONDURAS
Guatemala
Tegucigalpa
Caribbean Sea
Neth. Antilles Neth.
GUATEMALA
San Salvador
NICARAGUA
Aruba Neth.
Managua
EL SALVADOR
Port-of-Spain
TRINIDAD & TOBAGO
Panama
Caracas
Orinoco R.
San José
COSTA RICA
PANAMA
Medellín
VENEZUELA
Georgetown
Paramaribo
GUYANA
Cayenne
Bogotá
COLOMBIA
Cali
FRENCH GUIANA Fr.
SURINAME
EQUATOR
0°
Galápagos Islands Ecua.
Quito
ECUADOR
Rio Negro
Manaus
Belém
Amazon R.
Fortaleza
PACIFIC OCEAN
PERU
Madeira R.
BRAZIL
Tocantins R.
Recife
Lima
BOLIVIA
Lake Titicaca
Salvador
Arequipa
La Paz
Santa Cruz
Brasília
Sucre
20°S
PARAGUAY
Belo Horizonte
Paraná R.
TROPIC OF CAPRICORN
Rio de Janeiro
CHILE
Asunción
São Paulo
Curitiba
N
W
E
S
Paraguay R.
Uruguay R.
Pôrto Alegre
Valparaíso
Santiago
Rosario
URUGUAY
Juan Fernández Islands Chile
Buenos Aires
Montevideo
Río de la Plata
ARGENTINA

- National capital
- Territorial capital
- Major city

40°S

0 — 1,000 kilometers
0 — 1,000 miles
Lambert Azimuthal Equal-area projection

ATLANTIC OCEAN
Falkland Islands (Islas Malvinas) U.K.
South Georgia Island U.K.

Comparing Past and Present

The cultures in Latin America today show influences from the native cultures, Europe, and Africa. As you compare the maps on these pages, look for patterns that may provide information about the cultures in Latin America today. Then answer the questions below on a separate sheet of paper.

1. How do the borders of Latin America during European colonization compare with the borders of present-day Latin America?
2. What European countries still own territory in Latin America?
3. Based on the information on the maps, which ethnic groups would you expect to find in Mexico today? What issues do you predict may exist among these groups?

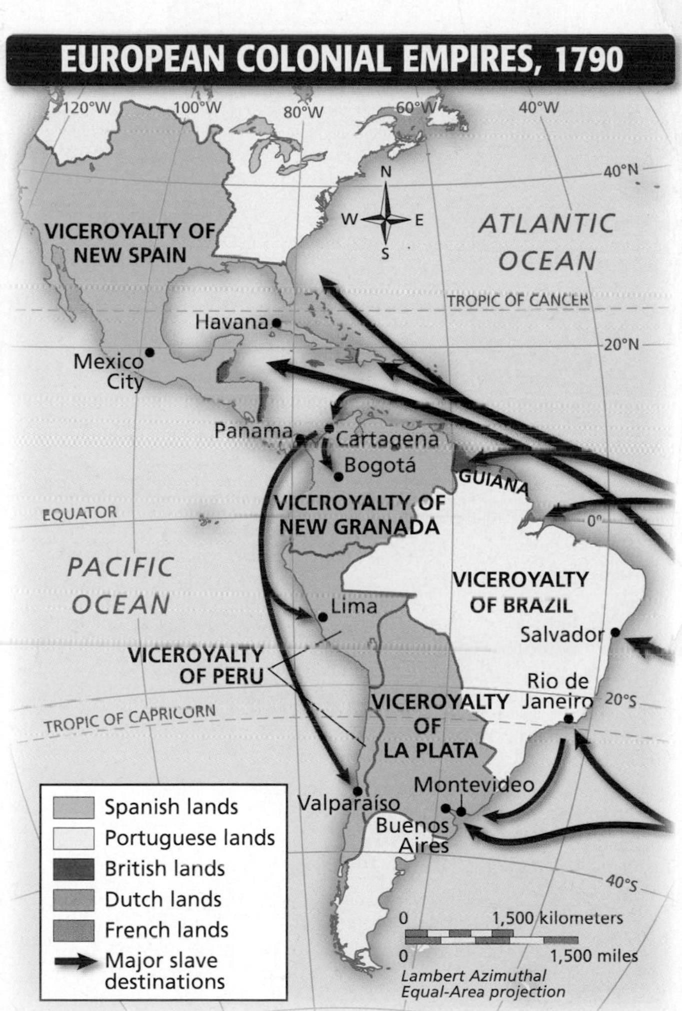

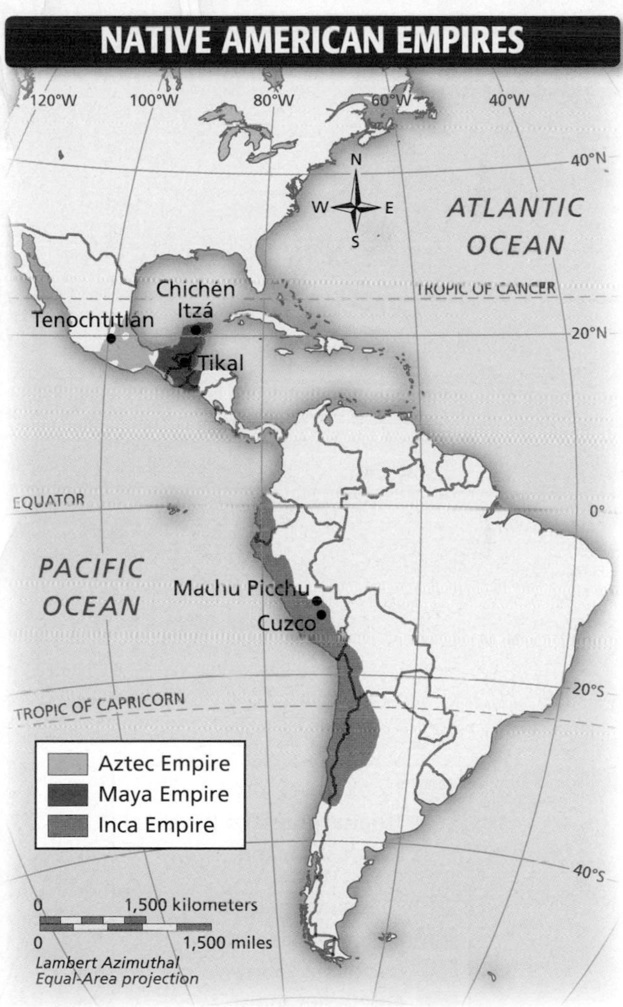

The Value of Resources

The Amazon rain forest is a valuable economic and environmental resource. As you study the maps and graphics on these pages, look for the importance of this resource to the region and to the world. Then answer the questions below on a separate sheet of paper.

1. What are the predominant types of natural vegetation between the Tropic of Cancer and the Tropic of Capricorn?
2. How has human activity threatened the rain forests of Latin America?
3. How do you think the region's culture and economy are affected as increasing areas of the rain forest are destroyed?

VEGETATION Latin America

NATIONAL GEOGRAPHIC
DEFORESTATION Latin America
Gulf of Mexico
80°W
60°W
40°W
TROPIC OF CANCER
20°N
N
W
E
S
GUATEMALA
BELIZE
HONDURAS
EL SALVADOR
NICARAGUA
PANAMA
COSTA RICA
Caribbean Sea
ATLANTIC OCEAN
PACIFIC OCEAN
Cash crops are grown as farmers clear large areas of the rain forests. Agriculture is a major source of employment in Latin America, as more than 25 percent of the labor force works in agriculture.
VENEZUELA
COLOMBIA
GUYANA
SURINAME
FRENCH GUIANA Fr.
The rate of deforestation is particularly severe in Brazil. International agricultural companies sponsor large-scale conversion of rain forests into large plantations.
EQUATOR
0°
ECUADOR
PERU
Although expanding pasture area contributes to deforestation, the region's economy depends heavily on livestock production. For example, Brazil has been the world's largest beef exporter since 2004.
BRAZIL
BOLIVIA
20°S
TROPIC OF CAPRICORN
PARAGUAY
40°S
Land Use
Forest
Pasture
Cropland
Shrubland
Hotspots of pasture expansion into forest
Hotspots of cropland expansion into forest
Hotspots are areas within a 6.2 mile (10 km) radius that are dominated by forests and 25 percent or more of which is in danger of being replaced by pasture or cropland.
0 800 kilometers
0 800 miles
Lambert Azimuthal Equal-Area projection
Source: Food and Agriculture Organization of the United Nations.
LATIN AMERICA

COUNTRY PROFILES Latin America

Country, Capital, & Area	Population & Density	Life Expectancy at Birth	GDP Per Capita*	% Urban	Literacy Rate (%)	Years of Compulsory Education	Phone Lines/ Cell Phones (per 1,000 people)	Internet Users (per 1,000 people)	Flag & Language
ANTIGUA AND BARBUDA St. John's 171 sq. mi. 443 sq. km	100,000 585 per sq. mi. 226 per sq. km	73 yrs.	$19,600	31	85.5	12	467/663	350	English
ARGENTINA Buenos Aires 1,056,641 sq. mi. 2,736,690 sq. km	40,300,000 38 per sq. mi. 15 per sq. km	75 yrs.	$14,200	91	97.2	10	227/570	177	Spanish
BAHAMAS Nassau 3,865 sq. mi. 10,010 sq. km	300,000 78 per sq. mi. 30 per sq. km	72 yrs.	$29,600	83	95.6	12	439/584	319	English
BARBADOS Bridgetown 166 sq. mi. 430 sq. km	300,000 1,807 per sq. mi. 698 per sq. km	77 yrs.	$19,100	38	99.7	11	500/765	594	English
BELIZE Belmopan 8,805 sq. mi. 22,806 sq. km	300,000 34 per sq. mi. 13 per sq. km	73 yrs.	$8,400	51	76.9	10	114/319	130	English
BOLIVIA La Paz Sucre 418,265 sq. mi. 1,083,301 sq. km	9,900,000 24 per sq. mi. 9 per sq. km	65 yrs.	$4,500	65	86.7	8	70/264	52	Spanish/ Quechua/ Aymara
BRAZIL Brasília 3,266,198 sq. mi. 8,459,417 sq. km	191,500,000 59 per sq. mi. 23 per sq. km	73 yrs.	$10,200	84	88.6	8	230/462	195	Portuguese
CHILE Santiago 287,191 sq. mi. 743,812 sq. km	17,000,000 59 per sq. mi. 23 per sq. km	78 yrs.	$14,900	87	95.7	8	211/649	172	Spanish
COLOMBIA Bogotá 428,227 sq. mi. 1,109,104 sq. kmz	45,100,000 105 per sq. mi. 41 per sq. km	72 yrs.	$8,800	75	90.4	10	168/479	104	Spanish
COSTA RICA San José 19,714 sq. mi. 51,060 sq. km	4,500,000 228 per sq. mi. 88 per sq. km	79 yrs.	$11,500	59	95.0	10	321/254	254	Spanish

*The CIA calculates per capita GDP in terms of purchasing power parity. This formula allows us to compare the figures among different countries.
Note: Countries and flags are not drawn to scale.

Country, Capital, & Area	Population & Density	Life Expectancy at Birth	GDP Per Capita*	% Urban	Literacy Rate (%)	Years of Compulsory Education	Phone Lines/ Cell Phones (per 1,000 people)	Internet Users (per 1,000 people)	Flag & Language
CUBA Havana 42,402 sq. mi. 109,820 sq. km	11,200,000 264 per sq. mi. 102 per sq. km	78 yrs.	$9,500	76	99.8	9	75/12	17	Spanish
DOMINICA Roseau 290 sq. mi. 751 sq. km	100,000 345 per sq. mi. 133 per sq. km	75 yrs.	$9,900	73	94.0	12	293/585	361	English
DOMINICAN REPUBLIC 18,656 sq. mi. 48,320 sq. km Santo Domingo	10,100,000 541 per sq. mi. 209 per sq. km	72 yrs.	$8,200	64	87.0	9	101/407	169	Spanish
ECUADOR Quito 106,889 sq. mi. 276,841 sq. km	13,600,000 127 per sq. mi. 49 per sq. km	75 yrs.	$7,500	63	91.0	10	129/472	47	Spanish
EL SALVADOR San Salvador 8,000 sq. mi. 20,721 sq. km	7,300,000 913 per sq. mi. 352 per sq. km	71 yrs.	$6,200	60	80.2	9	141/350	93	Spanish
FRENCH GUIANA (FRANCE) Cayenne 33,226 sq. mi. 86,504 sq. km	200,000 6 per sq. mi. 2 per sq. km	75 yrs.	$8,300	76	83.0	NA	NA	NA	French
GRENADA St. George's 133 sq. mi. 344 sq. km	100,000 752 per sq. mi. 291 per sq. km	74 yrs.	$12,900	31	96.0	12	309/410	182	English
GUATEMALA 41,374 sq. mi. 107,159 sq. km Guatemala	14,000,000 338 per sq. mi. 131 per sq. km	70 yrs.	$5,300	47	69.1	9	99/358	79	Spanish
Comparing Lands: ***Latin America is about three times the size of the contiguous United States.***									
	306,800,000 87 per sq. mi. 33 per sq. km	78 yrs.	$47,800	79	99.0	12	606/680	630	English

Sources: Central Intelligence Agency, *World Factbook,* 2009; Population Reference Bureau, *World Population Data Sheet,* 2009; UNESCO Institute for Statistics; United Nations, *Human Development Report,* 2007/2008.

COUNTRY PROFILES Latin America

Country, Capital, & Area	Population & Density	Life Expectancy at Birth	GDP Per Capita*	% Urban	Literacy Rate (%)	Years of Compulsory Education	Phone Lines/ Cell Phones (per 1,000 people)	Internet Users (per 1,000 people)	Flag & Language
GUYANA Georgetown 76,004 sq. mi. 196,849 sq. km	800,000 11 per sq. mi. 4 per sq. km	66 yrs.	$3,800	28	98.8	10	147/375	213	English
HAITI 10,641 sq. mi. 27,560 sq. km Port-au-Prince	9,200,000 865 per sq. mi. 334 per sq. km	58 yrs.	$1,300	43	52.9	6	17/48	70	French/Creole
HONDURAS Tegucigalpa 43,201 sq. mi. 111,890 sq. km	7,500,000 174 per sq. mi. 63 per sq. km	72 yrs.	$4,400	49	80.0	6	69/178	36	Spanish
JAMAICA 4,182 sq. mi. 10,831 sq. km Kingston	2,700,000 646 per sq. mi. 249 per sq. km	72 yrs.	$7,500	52	87.9	6	129/1,107	404	English
MEXICO 750,651 sq. mi. 1,943,945 sq. km Mexico City	109,600,000 146 per sq. mi. 56 per sq. km	75 yrs.	$14,200	77	91.0	10	189/460	181	Spanish
NICARAGUA 46,328 sq. mi. 119,990 sq. km Managua	5,700,000 123 per sq. mi. 48 per sq. km	71 yrs.	$2,900	58	67.5	6	43/217	27	Spanish
PANAMA Panama 28,703 sq. mi. 74,340 sq. km	3,500,000 122 per sq. mi. 47 per sq. km	75 yrs.	$11,700	64	91.9	9	136/418	64	Spanish
PARAGUAY 153,399 sq. mi. 397,302 sq. km Asunción	6,300,000 41 per sq. mi. 16 per sq. km	71 yrs.	$4,200	57	94.0	9	54/320	34	Spanish/ Guarani
PERU Lima 494,209 sq. mi. 1,279,996 sq. km	29,200,000 59 per sq. mi. 23 per sq. km	72 yrs.	$8,500	76	92.9	11	80/200	164	Spanish/ Quechua
PUERTO RICO (U.S.) San Juan 3,423 sq. mi. 8,870 sq. km	4,000,000 1,169 per sq. mi. 451 per sq. km	78 yrs.	$17,700	94	94.1	NA	NA	NA	Spanish

*The CIA calculates per capita GDP in terms of purchasing power parity. This formula allows us to compare the figures among different countries.
Note: Countries and flags are not drawn to scale.

Country, Capital, & Area	Population & Density	Life Expectancy at Birth	GDP Per Capita*	% Urban	Literacy Rate (%)	Years of Compulsory Education	Phone Lines/ Cell Phones (per 1,000 people)	Internet Users (per 1,000 people)	Flag & Language
ST. KITTS AND NEVIS Basseterre 101 sq. mi. 261 sq. km	50,000 495 per sq. mi. 192 per sq. km	70 yrs.	$19,500	32	97.8	12	532/213	212.8	English
ST. LUCIA Castries 234 sq. mi. 606 sq. km	200,000 855 per sq. mi. 330 per sq. km	73 yrs.	$11,100	28	90.1	11	NA/573	339	English
ST. VINCENT AND THE GRENADINES Kingstown 150 sq. mi. 389 sq. km	100,000 667 per sq. mi. 257 per sq. km	72 yrs.	$10,200	40	96.0	11	189/593	84	English
SURINAME Paramaribo 60,232 sq. mi. 156,000 sq. km	500,000 8 per sq. mi. 3 per sq. km	69 yrs.	$8,900	67	89.6	6	180/518	71	Dutch
TRINIDAD AND TOBAGO Port-of-Spain 1,980 sq. mi. 5,128 sq. km	1,300,000 657 per sq. mi. 254 per sq. km	69 yrs.	$23,600	12	98.6	7	248/613	123	English
URUGUAY Montevideo 67,574 sq. mi. 175,015 sq. km	3,400,000 50 per sq. mi. 19 per sq. km	76 yrs.	$12,400	94	98.0	10	290/333	193	Spanish
VENEZUELA Caracas 340,561 sq. mi. 882,050 sq. km	28,400,000 83 per sq. mi. 32 per sq. km	73 yrs.	$13,500	88	93.0	10	136/470	125	Spanish
VIRGIN ISLANDS (U.S.) Charlotte Amalie 134 sq. mi. 346 sq. km	109,800 819 per sq. mi. 317 per sq. km	79 yrs.	$17,200	95	95.0	NA	NA	NA	English
Comparing Lands: *Latin America is about three times the size of the contiguous United States.*									
	306,800,000 87 per sq. mi. 33 per sq. km	78 yrs.	$47,800	79	99.0	12	606/680	630	English

Sources: Central Intelligence Agency, *World Factbook,* 2009; Population Reference Bureau, *World Population Data Sheet,* 2009; UNESCO Institute for Statistics; United Nations, *Human Development Report,* 2007/2008.

CHAPTER 8

PHYSICAL GEOGRAPHY OF Latin America

BiG Idea

Certain processes, patterns, and functions help determine where people settle. A study of the physical geography of Latin America explains how landforms, climate, and natural forces influence human settlement.

Essential Questions

Section 1: The Land

How has physical geography affected where people have settled in Latin America?

Section 2: Climate and Vegetation

How do location and landforms affect climate in Latin America?

Geography ONLINE

Visit glencoe.com and enter *QuickPass*™ code WGC9952C8 for Chapter 8 resources.

Angel Falls descends from a high cliff in Venezuela's Canaima National Park.

Organizing Information Make a Three-Pocket Book to help you organize information about the physical geography of Latin America.

Reading and Writing As you read the chapter, write information about the physical features, climate, and vegetation of Latin America on note cards and place them in the correct pocket of your Foldable.

SECTION 1

 section audio

 spotlight video

Guide to Reading

Essential Question

How has physical geography affected where people have settled in Latin America?

Content Vocabulary
- cordillera *(p. 204)*
- altiplano *(p. 204)*
- escarpment *(p. 204)*
- llano *(p. 204)*
- pampa *(p. 204)*
- hydroelectric power *(p. 205)*

Academic Vocabulary
- isolate *(p. 203)*
- parallel *(p. 204)*
- volume *(p. 205)*

Places to Locate
- Andes *(p. 203)*
- Mexican Plateau *(p. 203)*
- Mato Grosso Plateau *(p. 204)*
- Brazilian Highlands *(p. 204)*
- Amazon River *(p. 205)*
- Río de la Plata *(p. 205)*
- Rio Grande *(p. 205)*

Reading Strategy

Organizing Complete a graphic organizer similar to the one below by listing the countries drained by the Amazon Basin.

Amazon Basin	

The Land

Latin America is a region of startling geographic contrasts. Low-lying plains and expansive river systems dominate the landscape in some parts of the region, influencing life in these areas. The beauty and magnificence of the high, rugged peaks of the Andes mountain ranges often contrasts sharply with the challenges faced by the people who have made the mountains their home.

NATIONAL GEOGRAPHIC Voices Around the World

"I was born in the Andes. . . . The mountains have been my lifelong companions, and I still make my home at their feet. To those of us who are their children, they are alive. We listen to them, learn to read their moods, and respect their power. Sometimes they welcome us with their solid embrace. Other times they shake with fury, and we know to stay away. Still sacred to some, they speak to the souls of all. . . ."

—Pablo Corral Vega, "In the Shadow of the Andes," *National Geographic*, February 2001

Quechua man

Landforms

MAIN Idea Latin America's rugged landscape has affected settlement of the region.

GEOGRAPHY AND YOU How have landforms affected where people live in your state? Read to learn how landforms in Latin America both attracted and isolated people in the region.

Located in the Western Hemisphere south of the United States, Latin America has an area of about 8 million square miles (20,720,000 sq. km), nearly 16 percent of Earth's land surface. Latin America is often divided into three subregions—Middle America (Mexico to Panama), the Caribbean, and South America.

Mountains and Plateaus

Latin America's most distinctive landforms are its mountains. Thrusting upward in countless folds and ridges, this mountain range begins in North America as the Rocky Mountains and extends to South America's southern tip. The mountains' names change; in Mexico they are the Sierra Madre; in Central America, the Central Highlands; and in South America, the **Andes.**

Latin America's rugged landscape is caused by its location along the Pacific Ring of Fire, where plates in the Earth's crust have collided for millions of years. These collisions have formed mountains and volcanoes and caused earthquakes. Plate movement continues to change the landscape. In 2010, for example, an earthquake struck Haiti's capital, Port-au-Prince. Thousands of buildings collapsed and large numbers of people were killed.

Despite this, humans have settled on Latin America's mountains and plateaus for thousands of years. The mountains' cooler climates and rich natural resources—water, volcanic soil, timber, and minerals—attracted settlers. Historically, the region's rugged terrain **isolated** communities. New technology—television, cell phones, and the Internet—is breaking down the physical barriers.

Mountains of Middle America and the Caribbean

The physical map on page 190 shows how Mexico's Sierra Madre mountain ranges surround the densely populated **Mexican Plateau.** The mild climate, fertile volcanic soil, and adequate rainfall in central Mexico have attracted humans for thousands of years.

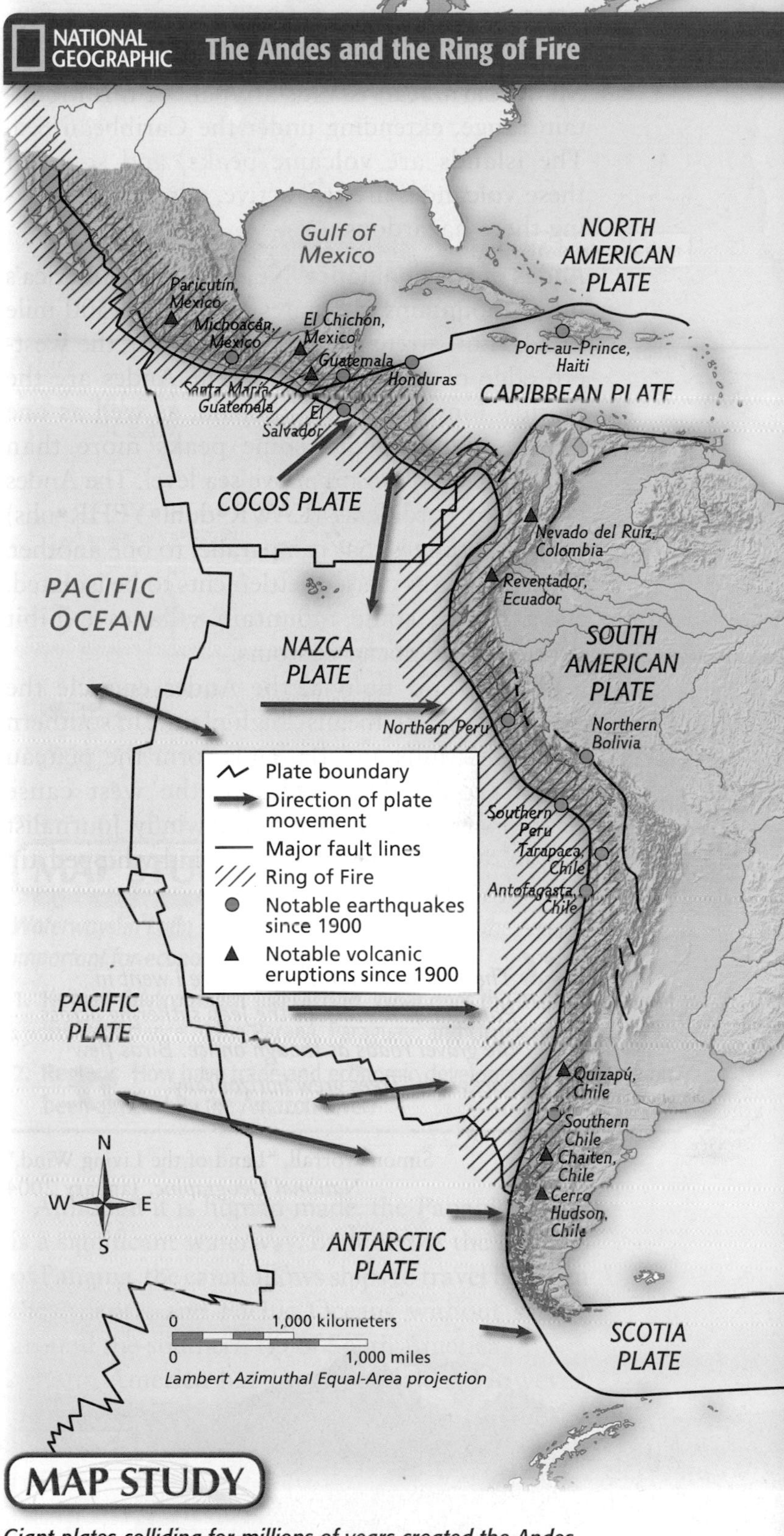

MAP STUDY

Giant plates colliding for millions of years created the Andes mountain ranges.

1. **Location** In which areas of Latin America have most of the volcanic eruptions occurred?
2. **Place** What factors influence volcanic eruptions in Latin America?

Maps In MOtion Use **StudentWorks™ Plus** or glencoe.com.

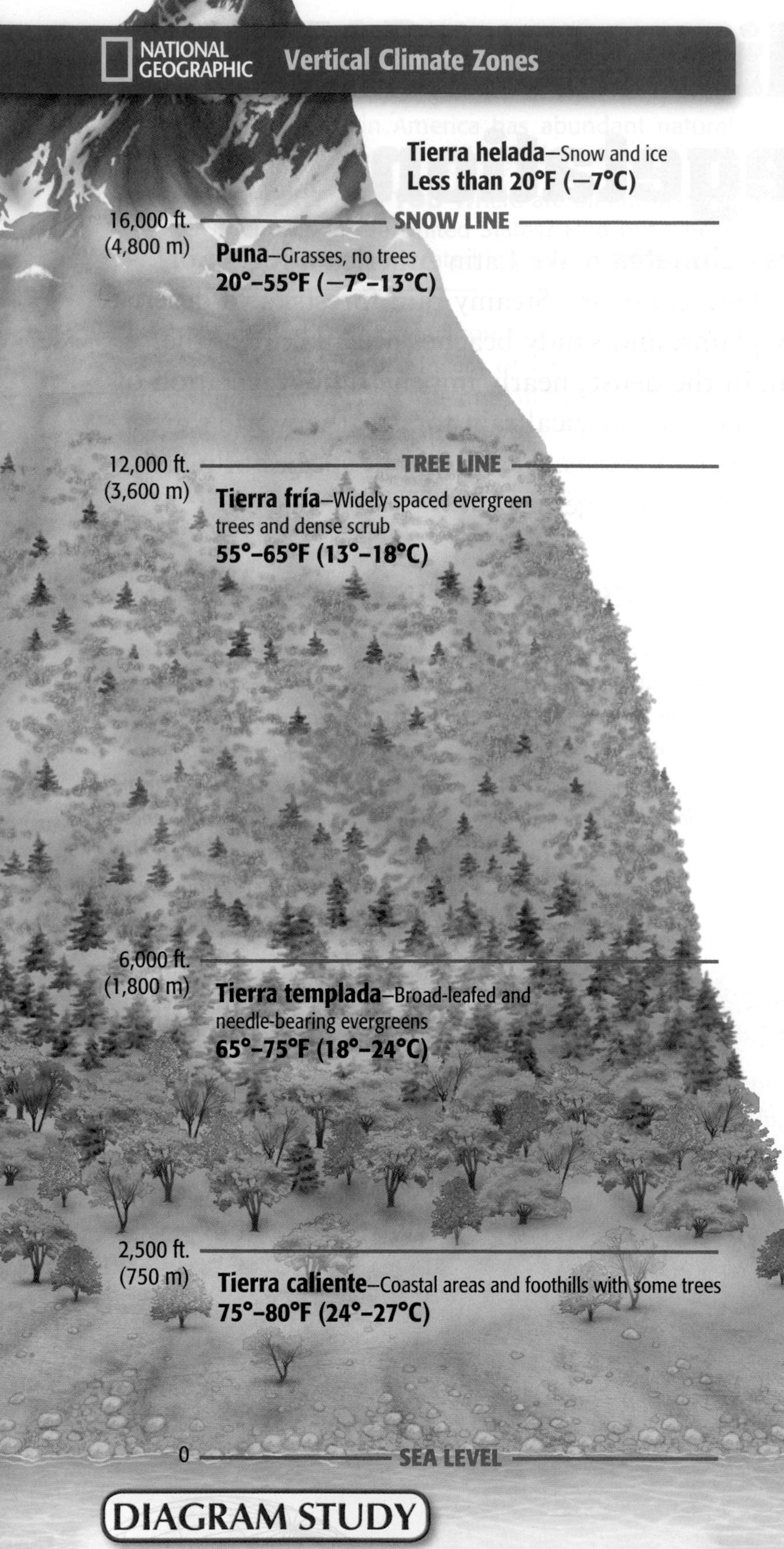

DIAGRAM STUDY

1. **Human-Environment Interaction** Why is the *tierra templada* the most densely populated of all the vertical climate zones?
2. **Location** How might increasing elevation affect the type of resources found in each vertical climate zone?

Elevation and Climate

MAIN Idea Differences in elevation create five diverse vertical climate zones in Latin America.

GEOGRAPHY AND YOU Have you traveled in the mountains and noticed how the temperature became cooler at higher elevations? Find out how altitude affects climate in Latin America.

Although located in the Tropics, some areas of Latin America are more affected by elevation than by distance from the Equator. Such areas have vertical climate zones in which soil, crops, livestock, and climate change as altitude increases. Spanish terms are used to describe the five vertical climate zones found in the highlands of Middle America and western South America.

Above the tree line, which ranges from 12,000 to 16,000 feet (3,600 to 4,800 m), are the ***tierra helada*** and the ***puna*** climate zones. The *tierra helada,* or "frozen land," is a zone of permanent snow and ice on the peaks of the Andes. The South American *puna* is a cold zone, which supports some grasses suitable for grazing sheep, llamas, and alpacas.

Land at 6,000 to 12,000 feet (1,800 to 3,600 m) is known as the ***tierra fría,*** or "cold land." At this elevation, winter frosts are common, but potatoes and barley grow well there. The *tierra fría* is Middle America's highest climate zone. Some of Latin America's largest cities—Bogotá, Colombia and Mexico City, Mexico—are located in this zone.

The ***tierra templada,*** or "temperate land," lies between 2,500 and 6,000 feet (750 and 1,800 m). Broad-leafed evergreens at lower altitudes give way to needle-leafed, cone-bearing evergreens at upper elevations. In the *tierra templada,* the most densely populated of the zones, coffee and corn are the main crops.

The ***tierra caliente,*** or "hot land," lies at elevations between sea level and 2,500 feet (750 m). In the rain forests of the *tierra caliente,* bananas, sugar, rice, and cacao are the main crops.

READING Check **Place** What type of agricultural activity takes place in the *puna* climate zone?

Climate and Vegetation Regions

MAIN Idea Although much of Latin America is located in the Tropics, the region has diverse climates and vegetation.

GEOGRAPHY AND YOU What kind of climate dominates the region where you live? Read to learn about the variety of climates in Latin America.

Much of Latin America has a tropical climate with lush vegetation. Compare the climate map with the physical map on page 190 to see how elevation affects both climate and natural vegetation.

Tropical Wet

A tropical wet climate with tropical rain forest vegetation **dominates** much of the region. High temperatures and abundant rainfall year-round result from the area's location on the Equator and the prevailing winds that carry warm, moist air from the Atlantic Ocean.

The world's largest expanse of tropical rain forest blankets the **Amazon Basin.** It is also the world's wettest tropical plain. Heavy rains drench much of the densely forested lowlands throughout the year. During the months of the rainy season, the sediment-laden Amazon River often floods.

In a rain forest, the trees grow close, forming a **canopy,** or dense layer of leaves. The canopy may soar to 150 feet (45 m) and keep sunlight from reaching the forest floor. The Amazon rain forest shelters more species of plants and animals per square mile than anywhere else on Earth.

Tropical Dry

A tropical dry climate is typical of the coast of southwestern Mexico, most Caribbean islands, and north-central South America. These areas have high temperatures and abundant rainfall but also experience an extended dry season. In many tropical dry areas, grasslands flourish. Some of these grasslands, such as the llanos of Colombia and Venezuela, are covered with scattered trees and are considered transition zones between grasslands and forests. In general, soils are not very fertile or suitable for large-scale agriculture. But flood control and water supply projects have turned some areas of the llanos into fertile farmland.

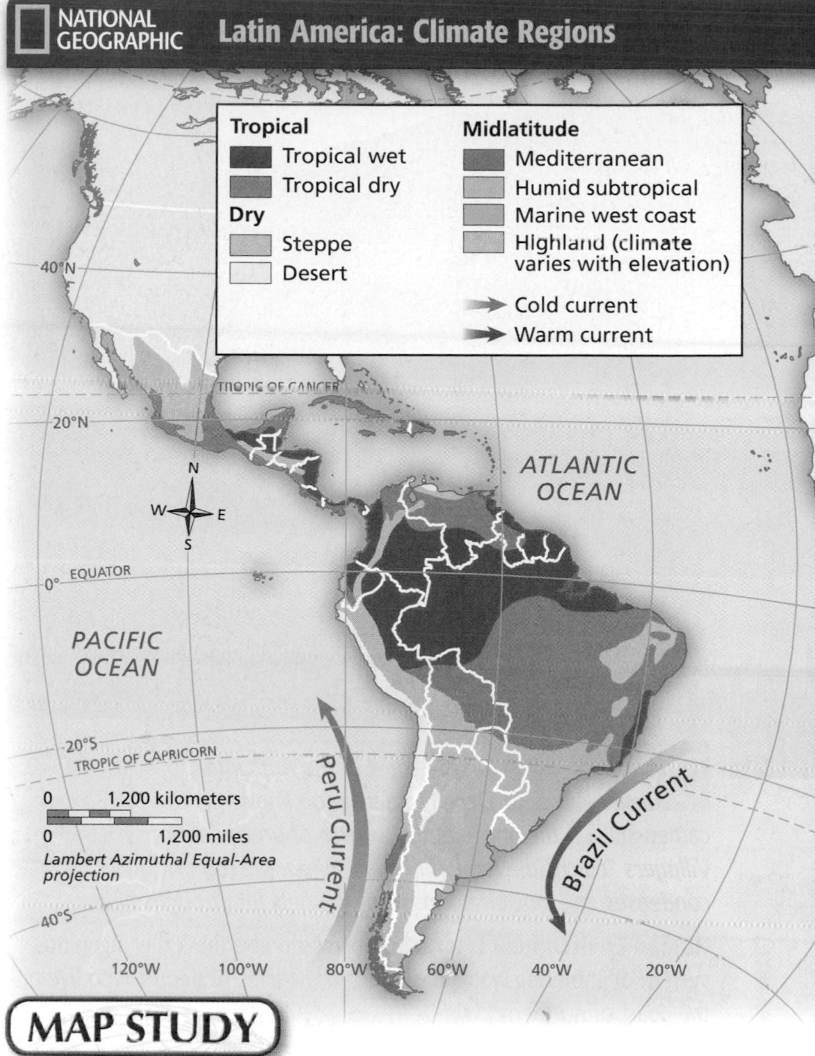

MAP STUDY

1. **Regions** Which areas are dominated by a tropical wet climate?
2. **Location** What are the predominant types of climates between the Tropic of Cancer and the Tropic of Capricorn?

Maps in MOtion Use **StudentWorks™ Plus** or glencoe.com.

Humid Subtropical

A humid subtropical climate prevails over much of southeastern South America. Winters here are short with cool to mild temperatures. Summers are long, hot, and humid. Rainfall is generally uniform throughout the year, but can be heavier during the summer.

Short grasses are the natural vegetation of these areas today. Although these plains were once home to large stands of trees, many were cleared by Spanish settlers for cattle ranching. Overgrazing eventually left only short clumps of grass to anchor the soil. Farmers now plant alfalfa, corn, and cotton for crops to hold the topsoil in place.

CHAPTER 

CULTURAL GEOGRAPHY OF Latin America

BIG Idea

Places reflect the relationship between humans and the physical environment. A study of the cultural geography of Latin America highlights how history, geography, and the blending of native and imported cultures have shaped the region.

Essential Questions

Section 1: Mexico

What groups have influenced the culture of Mexico?

Section 2: Central America and the Caribbean

How might colonization and slavery influence the culture of a region?

Section 3: South America

How have physical features influenced the culture of South America?

Visit glencoe.com and enter QuickPass™ code WGC9952C9 for Chapter 9 resources.

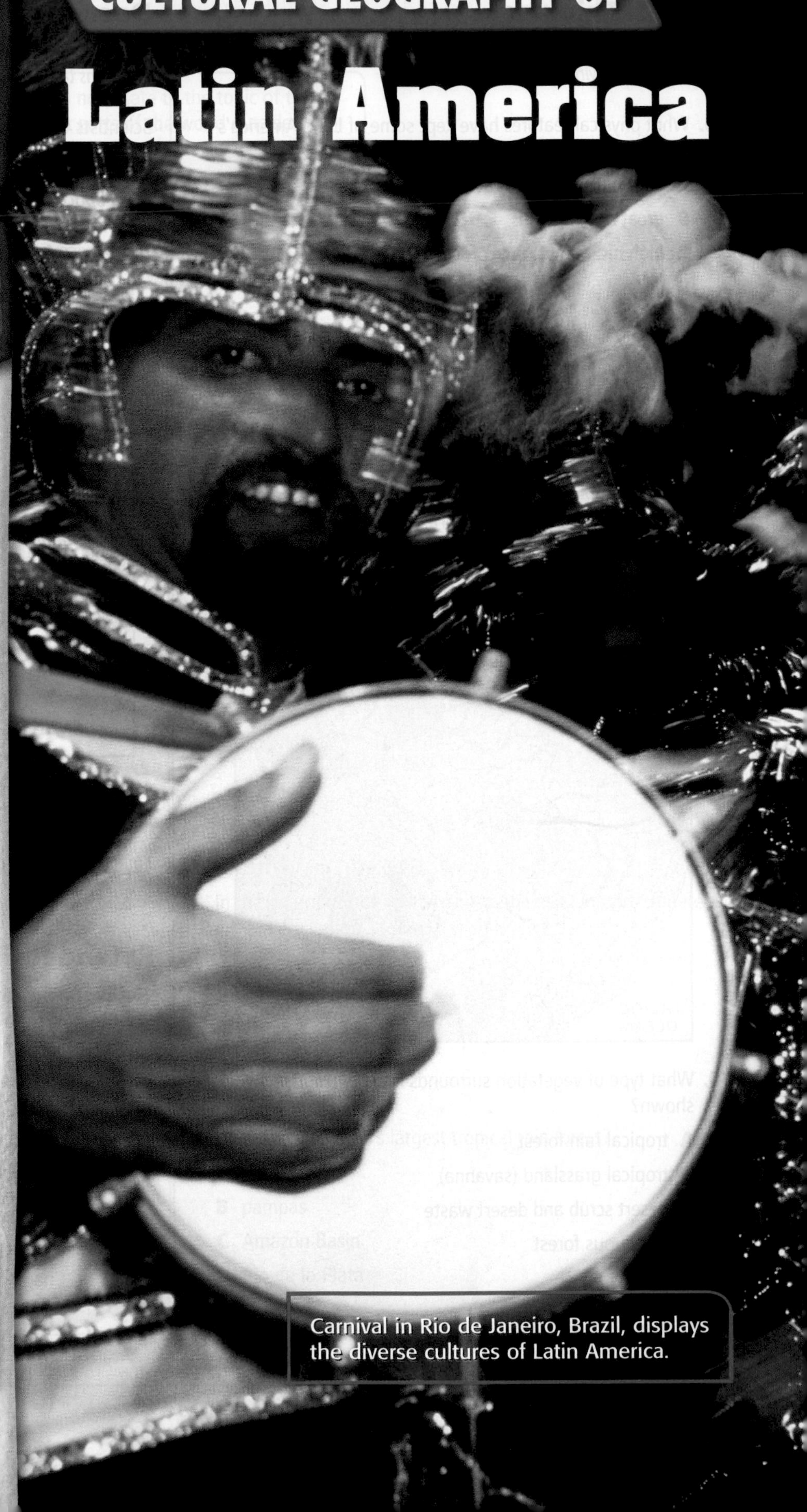

Carnival in Río de Janeiro, Brazil, displays the diverse cultures of Latin America.

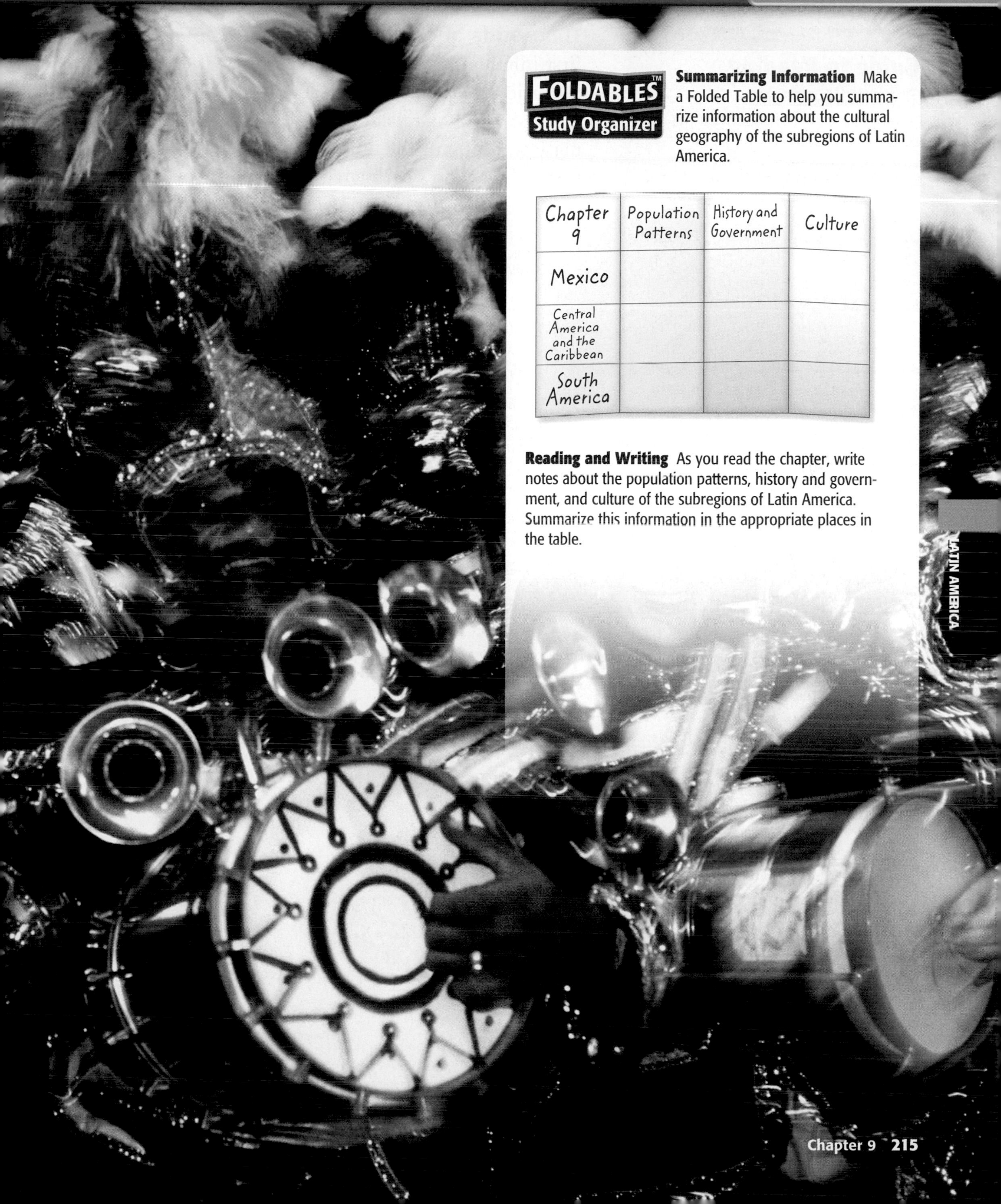

FOLDABLES™ Study Organizer

Summarizing Information Make a Folded Table to help you summarize information about the cultural geography of the subregions of Latin America.

Chapter 9	Population Patterns	History and Government	Culture
Mexico			
Central America and the Caribbean			
South America			

Reading and Writing As you read the chapter, write notes about the population patterns, history and government, and culture of the subregions of Latin America. Summarize this information in the appropriate places in the table.

SECTION 1

 section audio  spotlight video

Guide to Reading

Essential Question

What groups have influenced the culture of Mexico?

Content Vocabulary

- indigenous (p. 217)
- mestizo (p. 217)
- urbanization (p. 217)
- megacity (p. 217)
- primate city (p. 217)
- glyph (p. 218)
- *chinampas* (p. 218)
- conquistador (p. 219)
- viceroy (p. 219)
- caudillo (p. 219)
- syncretism (p. 219)
- malnutrition (p. 220)
- mural (p. 220)
- mosaic (p. 220)
- extended family (p. 220)

Academic Vocabulary

- external (p. 217)
- internal (p. 217)
- predict (p. 218)

Places to Locate

- Mexico (p. 217)
- Yucatán Peninsula (p. 217)
- Mexico City (p. 217)
- Tenochtitlán (p. 218)

Reading Strategy

Categorizing As you read, complete a graphic organizer similar to the one below by listing the reasons why many people migrate from rural areas to urban areas.

Reasons for Internal Migration

Mexico

Thousands of years ago, the descendants of Mexico's first inhabitants developed great civilizations. Over the centuries, there has been a blending of different ethnic groups, which is reflected in Mexico's present-day population. Many of the unique characteristics and traditions of Mexico's early civilizations, however, remain a vital part of the country's cultural geography today.

NATIONAL GEOGRAPHIC Voices Around the World

"Every March during the spring equinox at the ancient Maya ruins of Chichén Itzá, the plumed serpent god Kukulcán takes shape from light and shadow and descends the side of the great pyramid El Castillo. People come from all over the world to witness this event, which marks the renewal of the cycle of life on what many archaeologists believe is the Maya calendar constructed in stone. It also serves as a reminder of a great culture."

—Luis Albores, "Mexico: Five Cultural Bests," *National Geographic,* April 2004

Maya ruins of Chichén Itzá

Population Patterns

MAIN Idea Ethnic groups, migration, and urban growth have shaped population in Mexico.

GEOGRAPHY AND YOU What factors influence ethnic diversity in the United States? Read to learn how ethnic diversity has shaped Mexico's population.

Population Trends in Mexico

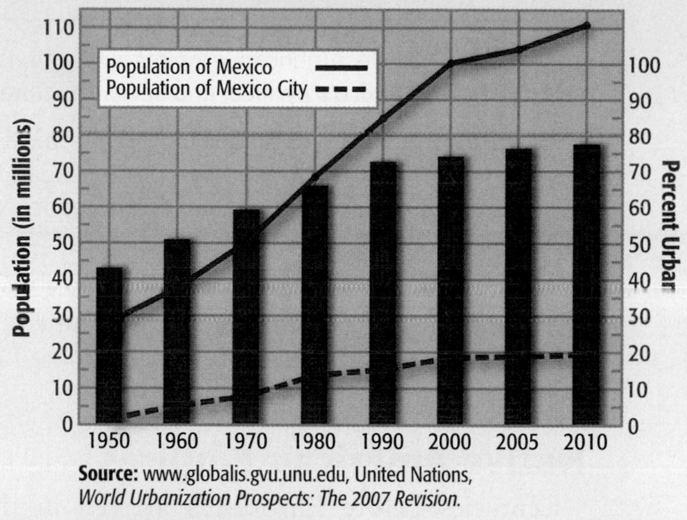

Source: www.globalis.gvu.unu.edu, United Nations, *World Urbanization Prospects: The 2007 Revision.*

GRAPH STUDY

1. **Place** Describe population growth in Mexico City from 1950 to 2010. How does it compare to the population growth of Mexico as a whole?
2. **Movement** Did Mexico's population grow at a faster rate between 1950 and 1970 or between 1990 and 2010?

Graphs In MOtion Use **StudentWorks™ Plus** or glencoe.com.

In **Mexico,** as in other Latin American countries, the ancestors of Native Americans were the first people to settle the region. For this reason, Native Americans today are known as **indigenous** (ihn•DIH•juh•nuhs) peoples, or peoples descended from an area's first inhabitants.

The People

Mexico's first inhabitants probably migrated from Asia 40,000 to 60,000 years ago. Because these first peoples were isolated from one another for centuries, they evolved into their own cultural groups and developed highly organized civilizations. These indigenous groups thrived in different areas, such as the Maya in the **Yucatán Peninsula.** The majority of these groups, however, lived in the southern part of the Mexican Plateau—the center of the Aztec Empire.

With the arrival of European settlers over the centuries, a blending of peoples of Native American and European descent created a new ethnic group called **mestizo.** In Mexico today, mestizos make up the largest part of the population.

Density and Distribution

With 109.6 million people, Mexico is the world's most populous Spanish-speaking country. A population density of 146 people per square mile (56 per sq. km) suggests a relatively uncrowded country. This figure is only an average, however. In **Mexico City,** about 19.5 million people live within an area of 597 square miles (1,547 sq. km). That makes its population density 32,663 people per square mile (12,605 per sq. km)!

Migration has shaped Mexico's population. The desire for job opportunities and improved economic conditions drives **external** migration to other areas of the world. **Internal** rural-to-urban migration has greatly changed the population distribution, with approximately 77 percent of the population now living in cities. People migrate to urban areas because of limited agricultural land and lack of access to social services. Many migrants move to cities along the U.S.-Mexico border. However, Mexico City remains the primary destination for most people who move within the country. This one-way migration from rural to urban areas results in ongoing, rapid **urbanization**—the migration of people from the countryside to cities and the natural population growth within urban areas.

Urban Areas

In some Latin American countries, growing cities have absorbed surrounding cities and suburbs to create **megacities,** cities with more than 10 million people. The region's largest megacity is Mexico City, with a population of about 19.5 million. By 2025, the city is expected to grow to more than 21 million people. Because of its size and influence, Mexico City is a **primate city,** an urban area that dominates its country's economy, culture, and political affairs.

READING Check **Movement** What two migration patterns have changed Mexico's population?

History and Government

MAIN Idea Native American empires and colonial rule influenced Mexico's political and social structures.

GEOGRAPHY AND YOU What U.S. people played key roles in the early history of the country? Read to learn about Mexico's history and the influence of the Maya and Aztec civilizations.

In Mexico today, as throughout Latin America, people struggle with unresolved issues rooted in the past.

Native American Empires

Centuries before Europeans arrived in the Americas, two Native American empires—the Maya and the Aztec—flourished in the area that is present-day Mexico. These civilizations left enduring marks on Mexican history and culture.

The Maya dominated southern Mexico and northern Central America from about A.D. 250 to 900. They established many cities and built terraces, courts, and temples. Priests and nobles ruled the cities and surrounding areas. The Maya based their economy on agriculture and trade.

Skilled in mathematics, the Maya developed accurate calendars and used astronomical observations to **predict** solar eclipses. They made **glyphs,** picture writings carved in stone, on temples to honor their gods and record history.

For reasons that are still a mystery, the Maya eventually abandoned their cities. Archaeologists have uncovered the ruins of more than 40 Maya cities. However, most of the glyphs remain untranslated. Descendants of the Maya still live in villages in southern Mexico and Central America, where they practice subsistence farming.

The Aztec civilization arose in central Mexico in the A.D. 1300s. Their capital, built on an island in a large lake, was named **Tenochtitlán** (tay•NAWCH•teet•LAHN). Today, it is the site of Mexico City. To feed the growing population, Aztec farmers cultivated all available land. They grew crops on ***chinampas***—floating "islands" made from large rafts covered with mud from the lake bottom.

The Aztec developed a highly structured class system headed by an emperor and military officials. High-ranking priests performed rituals to win the favor of the gods. At the bottom of Aztec society was the largest group—farmers, laborers, and soldiers.

Culture

The Arts

Diego Rivera is known for his wall paintings and frescoes, or large paintings done on wet plaster. He used his art to illustrate Mexico's history and culture. In the central arch detail from his fresco *From Conquest to 1930,* Rivera used images of peasants and workers to show foreign influence and Mexico's struggle for independence. This fresco is one in a series Rivera painted inside the National Palace in Mexico City depicting Mexico's history from Native American civilizations to just after Mexican independence.

An Independent Nation

The Aztec Empire was still expanding when Spanish explorers landed on the Yucatán Peninsula in 1519. Desiring wealth from Mexico's abundant resources of silver and gold, Spanish **conquistador,** or conqueror, Hernán Cortés defeated the Aztec in 1521 and claimed Mexico for Spain. For the next three centuries, Mexico was part of the viceroyalty of New Spain. It was governed under a highly structured political system by royally appointed officials known as **viceroys.**

In the late 1700s, resentment against European rule spread throughout Mexico and the rest of Latin America. The first Spanish-ruled country in Latin America to win independence was Mexico. A parish priest, Father Miguel Hidalgo, led Mexico's independence movement in 1810. After a long struggle, Mexico became independent in 1821.

In spite of its independence, Mexico's political and economic power remained in the hands of a small group of wealthy landowners, army officers, and clergy. Power struggles, public dissatisfaction, and revolts led to a chaotic situation in the new republic. It was during this time that a new kind of leader emerged—the **caudillo** (kow•DEE•yoh), or dictator. With the backing of military forces and wealthy landowners, caudillos became absolute rulers.

A new constitution in 1917 brought reforms and established Mexico as a federal republic with powers separated into three branches—executive, legislative, and judicial. A president is elected to a single six-year term. Beginning in 1929, however, one political party, the *Partido Revolucionario Institucional* (PRI), dominated the presidency and Mexican politics for nearly 70 years. The PRI's control ended in 2000 when Vicente Fox of the opposition party *Partido Acción Nacional* (PAN) won the election. The 2006 presidential election ensured PAN's continued control of the presidency when Felipe Calderon was declared the president.

Struggles for additional reforms and political power continue. Native Americans, farmers, and workers continue to pressure the government for greater inclusion in the political system.

READING Check **Regions** Why was the outcome of Mexico's 2000 presidential election so significant?

Culture

MAIN Idea The arts, traditions, and beliefs of indigenous peoples and of Europeans have influenced Mexican culture.

GEOGRAPHY AND YOU Do you know of any cultural traditions that are unique to Mexico? Read to learn about the culture of Mexico.

The culture of Mexico is an intricate blending of indigenous and Spanish influences.

Language and Religion

In Mexico, the official language is Spanish, which is spoken by more than 90 percent of the population. More than 50 indigenous languages, however, are still spoken by some. Nearly 90 percent of Mexico's population is Roman Catholic, although many indigenous peoples retain their traditional religious beliefs. Some indigenous peoples blend traditional Maya beliefs and Roman Catholic beliefs. This blending of beliefs and practices is called **syncretism.**

Education and Health Care

Education varies greatly in Mexico. Most public schools are in rural areas, but do not have the funding or qualified teachers of urban or private schools. Government promotion of adult literacy and funding for more schools has helped Mexico make gains in education.

NATIONAL GEOGRAPHIC *The Day of the Dead, a festival honoring loved ones who have died, has its origins in Aztec traditions.*

Place What other elements of Mexican culture have Native American influences?

LATIN AMERICA

As employment and education improve, health problems linked to poverty, lack of sanitation, and **malnutrition**—a condition caused by a lack of proper food—decrease. The federal government subsidizes health care, making it available to all citizens. However, the poor quality of public medicine causes many people to seek treatment in cities or other countries.

The Arts

Influences of indigenous cultures and the Spanish colonial period are evident in Mexico's arts. Early Native American architecture includes Maya pyramids and Aztec temples and palaces. Some of these buildings were decorated with **murals,** or wall paintings, and **mosaics,** pictures or designs made with colored stone or tile. The Spanish built churches and other buildings reflecting classic European architectural styles.

The twentieth century brought a renewed interest in precolonial history and culture. Diego Rivera painted murals of indigenous cultures and events in Mexico's history. Other noted Mexican artists include Frida Kahlo and José Clemente Orozco. The country's past and cultural identity have inspired writers Octavio Paz and Carlos Fuentes. Ballet Folklórico fascinates audiences by performing Native American and Spanish dances.

Family Life

Mexicans highly value the family. Each person is part of an **extended family** that includes great-grandparents, grandparents, aunts, uncles, and cousins as well as parents and children. Parents and children often share their home with members of the extended family. *Compadres,* or godparents, are chosen by the parents to sponsor their new baby and watch over his or her upbringing.

As in other Latin American societies, Mexican society still displays elements of machismo, a Spanish and Portuguese tradition of male supremacy. However, women have made rapid advances in recent decades.

Sports and Leisure

Spectators crowd into arenas to watch bullfighting, Mexico's national sport. People are also passionate about *fútbol,* or soccer. Baseball and jai alai (HY•ly), a game much like handball, have a large following as well.

One of the most popular leisure activities in Mexico may be celebrating. From friendly gatherings to special family dinners, religious feast days, and patriotic events, almost any social occasion is a party—a *fiesta,* or festival.

READING **Check** **Place** What role does the family have in Mexican society?

SECTION 1 REVIEW

Vocabulary

1. Explain the significance of: indigenous, mestizo, urbanization, megacity, primate city, glyph, *chinampas,* conquistador, viceroy, caudillo, syncretism, malnutrition, mural, mosaic, extended family.

Main Ideas

2. How have ethnic groups, migration, and urban growth shaped the population in Mexico?
3. Which ethnic group makes up the majority of Mexico's population?
4. Describe the influence Native American empires and European colonial rule have had on Mexico's political and social structures. Give examples.
5. Create a chart like the one below to help identify how the arts, traditions, and beliefs of indigenous peoples and of Europeans have influenced Mexican culture.

Indigenous Peoples	Europeans

Critical Thinking

6. **Answering the Essential Question** How have European settlers shaped Mexico's cultural geography? Give examples.
7. **Summarizing Information** Describe the factors that have influenced population patterns in Mexico.
8. **Analyzing Visuals** Study the graph on page 217. What generalization could you make about population growth in Mexico?

Writing About Geography

9. **Expository Writing** Write a paragraph suggesting suitable locations for constructing new cities to relieve population pressures that exist in Mexico City. What kinds of resources are required to sustain large populations?

Study Central™ To review this section, go to glencoe.com and click on Study Central.

SECTION 2

 section audio  spotlight video

Central America and the Caribbean

Guide to Reading

Essential Question

How might colonization and slavery influence the culture of a region?

Content Vocabulary
- dialect *(p. 225)*
- patois *(p. 225)*
- matriarchal *(p. 226)*

Academic Vocabulary
- diverse *(p. 222)*
- collapse *(p. 222)*
- transmission *(p. 223)*

Places to Locate
- Cuba *(p. 222)*
- Dominican Republic *(p. 222)*
- West Indies *(p. 223)*
- Hispaniola *(p. 223)*
- Panama *(p. 224)*
- Haiti *(p. 224)*
- Puerto Rico *(p. 224)*

Reading Strategy

Identifying As you read, complete a graphic organizer similar to the one below by identifying the many ethnic groups that make up the population of Central America and the Caribbean.

Island countries such as Cuba and Trinidad reflect in miniature the ethnic diversity that characterizes many parts of Central America and the Caribbean. In other locations, however, a majority ethnic group gives the population a strong cultural identity, with smaller groups adding their own unique flavor to the ethnic mix.

NATIONAL GEOGRAPHIC **Voices Around the World**

"Cuba is truly a melting pot, where Caribbean, African, and European stock mix and match in apparent harmony. Statistics bear that out: Over half of all Cubans are of mixed racial heritage, an astounding proportion by U.S. norms."

—Jon Bowermaster, "The Beat in Cuba," *National Geographic Traveler,* March 2004

The ethnic diversity of Cuban culture

NATIONAL GEOGRAPHIC *Expansion of the canal is taking place, with an expected completion date of 2014.*

Place Why did Panama provide the best place to create a shortcut to the Pacific Ocean?

Passage Through Panama Explorer Vasco Núñez de Balboa was the first to grasp the unique geographic features of the land known today as **Panama.** While exploring the isthmus, he climbed a peak and discovered a body of water as vast as the Atlantic Ocean. Centuries later, thoughts turned to building a waterway that would create a shortcut between the Atlantic and Pacific Oceans. In 1904, final construction on the Panama Canal began. Nearly 75,000 laborers from around the world built what is regarded as one of the engineering wonders of the world. The canal continues to be an important trade route.

Gaining Independence

In the late 1700s, Native Americans and Africans yearned for freedom from slavery and European rule. François Toussaint-Louverture (frahn•SWAH TOO•SAN•LOO•vuhr•TYUR), a soldier born to enslaved parents, led a revolt of enslaved Africans in **Haiti.** By 1804, Haiti had won its independence from France. Haiti went on to help independence movements in other Latin American countries.

Student Web Activity Visit glencoe.com, select the *World Geography and Cultures* Web site, and click on Student Web Activities–Chapter 9 for an activity about the Panama Canal.

Except for Haiti, Caribbean countries were the last territories in the region to achieve independence. Cuba, for example, won its independence from Spain in 1898, but remained under the protection of the United States until 1902. British-ruled islands, such as Jamaica and Barbados, did not gain independence until well into the 1900s. Even today some islands remain under foreign control. **Puerto Rico** and some of the Virgin Islands have political links to the United States. In Central America, Spain ruled until the nineteenth century.

However, struggles for independence ushered in a period of political and economic instability. During the 1800s, some leaders wanted to build political institutions and prosperous economies. In 1823, independent Central American provinces formed a federation called the United Provinces of Central America. The powerful elites opposed such a union, so the United Provinces separated into five separate countries—Guatemala, El Salvador, Honduras, Nicaragua, and Costa Rica.

Movements for Change

During the 1900s, many countries in Central America and the Caribbean experienced political, social, and economic changes. For example, after Panama became an independent country in 1903, the United States and Panama signed a treaty creating the Panama Canal Zone. The formation of industries, the building of railroads, and the expansion of trade brought new wealth to the upper classes. However, for the vast majority of people, progress was limited and demands for reform were ignored.

Reform did occur in Cuba, however, when a revolution in 1959 set up a communist state under Fidel Castro. Castro ruled Cuba for decades, until handing over power to his brother Raul in 2008. During the 1990s, military dictatorships gave way to democratically elected governments in several other countries. Today, many countries in Central America and the Caribbean are struggling to end corrupt politics and violence and bring economic benefits to all their citizens. In recent elections throughout Latin America people exercised their right to vote and demanded change.

READING Check **Place** Who led a revolt and helped Haiti gain its independence?

Culture

MAIN Idea The culture of Central America and the Caribbean has been influenced by the arts as well as the traditions and beliefs of indigenous peoples, Africans, and Europeans.

GEOGRAPHY AND YOU What kinds of music come to mind when you think of Central America and the Caribbean? Read to learn more about the culture of this subregion.

The people of Central America and the Caribbean express the diverse elements and unique mingling of their cultures through language, religion, and the arts.

Language and Religion

Spanish is the primary language of most countries in Central America. In the Caribbean, European languages spoken include English, Spanish, French, and Dutch. However, each country has its own **dialects,** or forms of a language unique to a particular place or group.

Millions of people speak Native American languages. Many people are also bilingual, while others speak one of many forms of **patois** (PA•TWAH), dialects that blend indigenous, European, African, and Asian languages. For example, Haitian Creole has a vocabulary based in French with other words of African and Spanish origin.

In Central America, four out of five people are Roman Catholic. In the Caribbean, most people living on the Spanish- and French-speaking islands are also Roman Catholic. Various forms of Protestant Christianity are found in English-speaking areas. Other faiths in the subregion include Hinduism and Islam. Scores of traditional Native American and African religions also thrive, often mixed with Christianity and other faiths. These mixed religions include Santería in Cuba and voodoo in Haiti and the Dominican Republic.

Education and Health Care

The quality of education varies greatly from country to country as well as within rural and urban areas of each country. Children generally are required to complete elementary school, but many do not because of long distances to school and lack of money for clothing and supplies.

Teen Life in Cuba

Communism has had a profound effect on the culture and development of Cuba. In a communist country, the government controls most aspects of daily life.

Did you know . . .

- English is a required course in Cuban secondary schools and is important to Cuba's tourist industry.
- Students attend school Monday through Saturday and are required to wear a uniform.
- School attendance is required between the ages of 5 and 12.
- Cubans have two surnames—one from their mother and one from their father. The mother's family name comes last, but people are commonly referred to by their father's name.
- Dining in restaurants is too expensive for most Cubans. With the exception of restaurants run out of private homes, all restaurants are owned by the government.
- The buying and selling of homes is strictly prohibited. Cubans can only purchase homes directly from the government or swap homes with other residents.

Communications in South America

Problem:

The countries of South America need an efficient, accessible, and cost-effective communications network to connect the residents of the region. A lack of such services has contributed to underdevelopment in parts of the region.

Inefficient and Costly It takes great time and effort to construct the poles and lines necessary to provide landline telephone service. There is added expense involved in keeping these networks in working order.

Lack of Services Before the explosion of cell phone technology, communications capabilities were limited in many South American countries. This man, living in Ecuador, sells the use of his working telephone to neighbors and passersby.

South America: Cell Phone Usage

NATIONAL GEOGRAPHIC

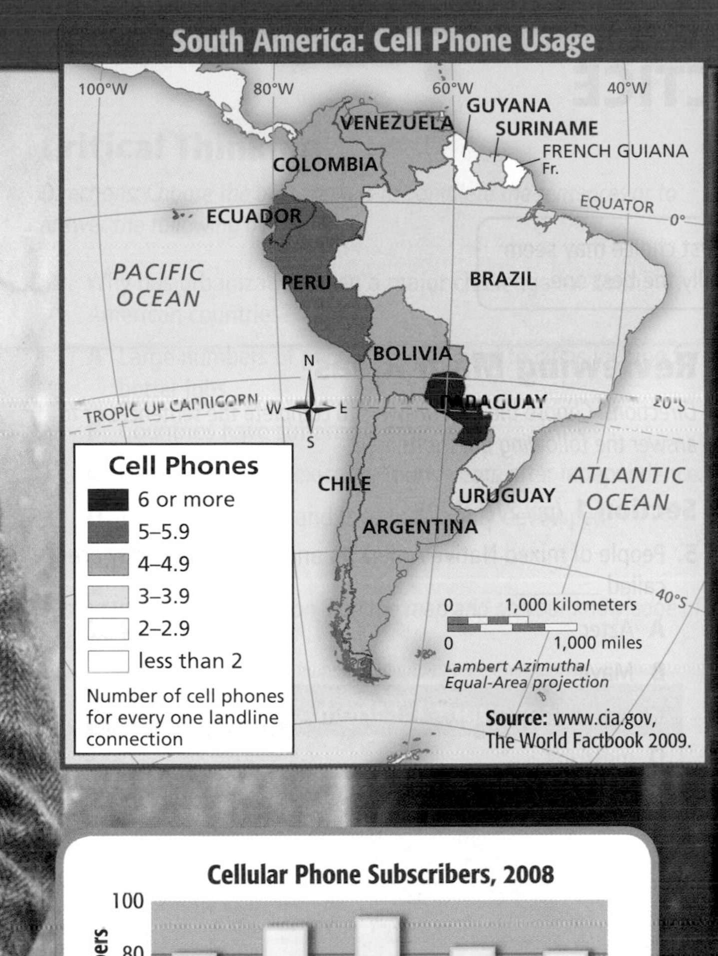

Source: www.cia.gov, The World Factbook 2009.

Cellular Phone Subscribers, 2008

Percent of all subscribers: 0, 20, 40, 60, 80, 100

Chile, Ecuador, Paraguay, Bolivia, Venezuela

Source: ITU World Telecommunications Indicators Database.

Connecting People

Establishing communications networks such as landline telephone services has long been a problem in South America. The rugged terrain, remote villages, and cost contribute to the region's limited access. Since the early 1990s, the number of Latin American households with cell phones has increased greatly, helping to connect the region's countries to each other and to the rest of the world.

Why has the construction been slow? The infrastructure needed to provide traditional landline service takes a lot of time, money, and labor to construct. The geography of the region makes this construction time-consuming and costly, particularly in remote areas. Residents in these areas often have little to no phone access at all.

What are the reasons for cell phone popularity? Cell towers can be constructed in remote areas more easily than landlines, allowing the residents in these areas to be better connected. The start-up costs are much less. Access to landline phone service, if available, costs between $50 to $200 in start-up fees. Economically this is out of the reach of many people. Cell phone providers offer service quickly and for a lower cost.

How do users control cost? Although the start-up fees associated with cell phones are less than the cost of regular phone service, the service does come with monthly fees. Increased usage and competition between cell phone providers have helped to bring these charges down, but they are still often too costly for many residents. To keep expenses manageable, most South American residents choose to prepay for service. Additional measures such as calling-party-pays (CPP) billing systems are also used.

Solution:

Much of South America, like many parts of Africa and Asia, has turned to cell phone technology to enhance the region's communications capabilities.

Cellular telephone towers, like this one, are now bringing phone service to those previously unable to receive these services because of remote location and cost.

THINKING GEOGRAPHICALLY

1. **Human Systems** How might the increased communications network offered by cell phones affect the economy of remote areas?
2. **The World in Spatial Terms** How might the physical geography of South America account for the greater number of cellular connections?

CHAPTER 10

THE REGION TODAY

Latin America

BiG Idea

Countries are affected by their relationships with each other. Latin America today faces the challenge of developing modern, industrial economies that promote regional and global trade, but do not destroy its natural resources.

Essential Questions

Section 1: The Economy
How might a region expand its economic growth?

Section 2: People and Their Environment
How do human activities place stress on a region's natural resources?

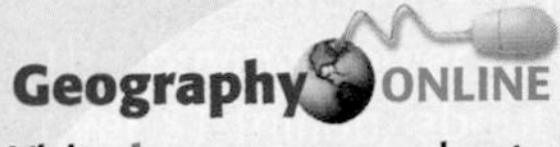

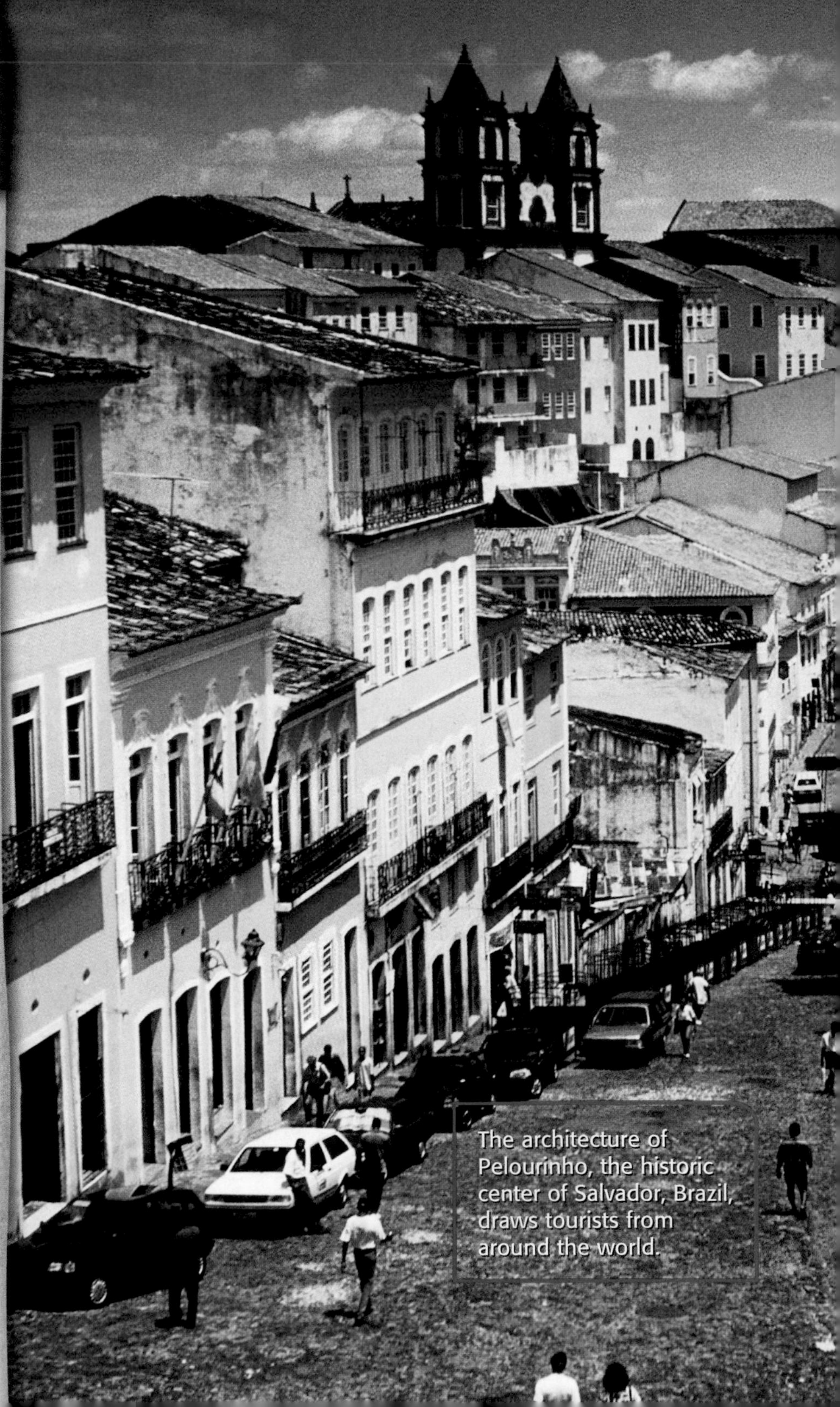

The architecture of Pelourinho, the historic center of Salvador, Brazil, draws tourists from around the world.

Geography ONLINE
Visit glencoe.com and enter *QuickPass*™ code WGC9952C10 for Chapter 10 resources.

Organizing Information Make a Four-Door Book to help you organize information about four key aspects of Latin America's economy today.

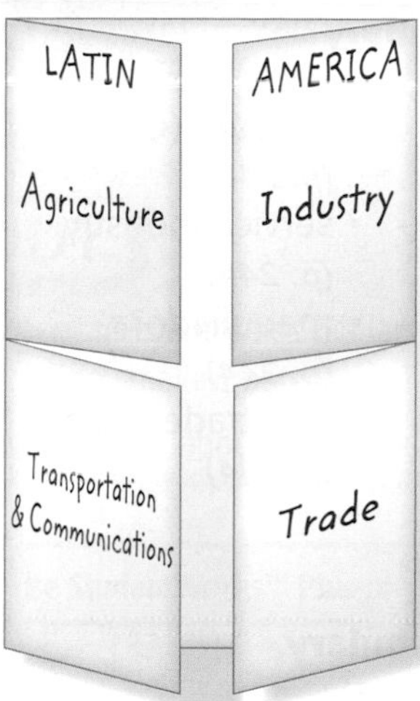

Reading and Writing As you read the chapter, write information about the four key aspects of Latin America's economy—agriculture, industry, transportation and communications, and trade. Include definitions for vocabulary terms related to these economic activities.

Tumucumaque Mountains National Park

One Problem:

Deforestation is a major issue for the world's rain forests. If this issue is not addressed, it is predicted that the rain forests will disappear within 40 years.

Northern bearded saki monkey in Tumucumaque

A Diverse Area Brazil's Tumucumaque Mountains region is so remote that its level of biological diversity is still not fully known. Estimates suggest that the park is home to at least 8 species of primates, 37 species of lizards, and 350 species of birds.

A protected jaguar

Safe Haven Jaguars, an endangered species, are now protected in Tumucumaque Mountains National Park.

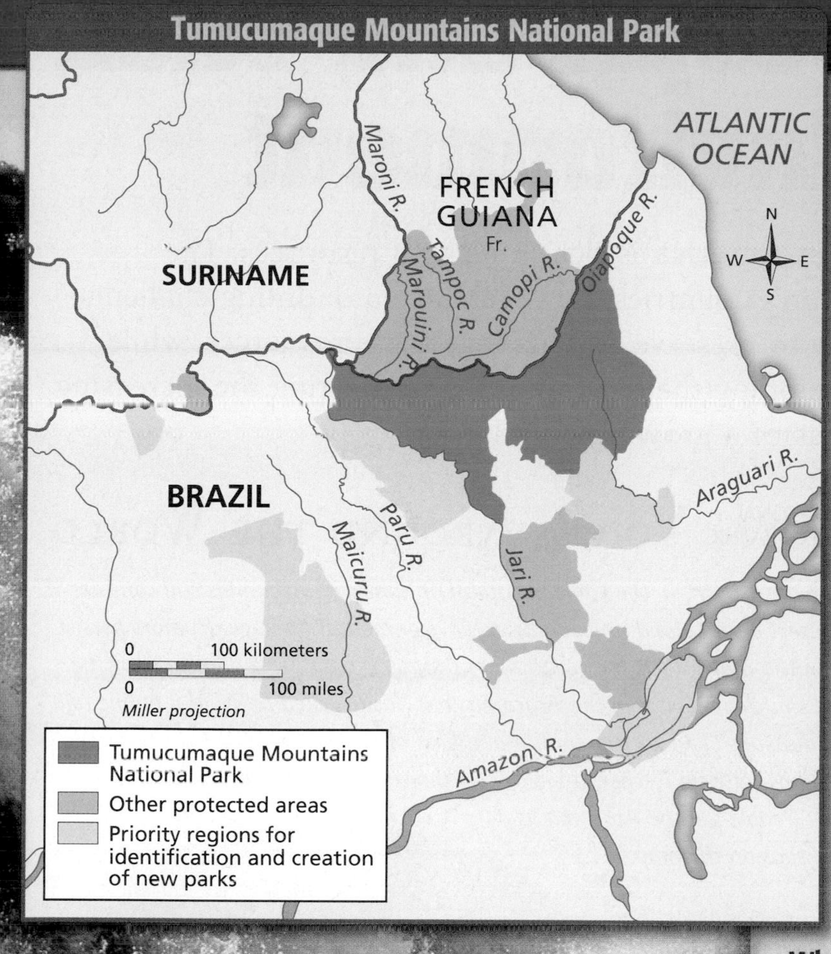

NATIONAL GEOGRAPHIC

Rain Forest Conservation

The Amazon rain forest is the largest and most diverse on Earth. Deforestation caused by logging, farming, and ranching threatens its depletion and the extinction of the plants and animals that call it home.

In August 2002 the Brazilian government and international partners took an important step toward preserving the Amazon by establishing the Tumucumaque (too•MOO•KOO•MAH•kay) Mountains National Park.

Why was this location chosen? Tumucumaque is one of only a few forest locations unaltered by humans. There are no indigenous peoples residing in the area. Also, the park is very difficult to access. There are no roads, and the swift currents and waterfalls in the area make the waterways nearly impossible to navigate. This remoteness helps to protect the park.

Why is the park important? At nearly 10 million acres (4 million ha)—the approximate size of Massachusetts and Connecticut combined—Tumucumaque is the largest tropical rain forest park in the world. It covers a full percent of the Brazilian Amazon. It is also home to approximately 30 percent of the world's animal, plant, and insect species.

What are the future plans for the Amazon? The Brazilian government and its partners continue to designate other areas of the Amazon for the establishment of national parks. These groups are working diligently to establish a trust fund to finance effective management of these new parks.

One Solution:

The Brazilian government established the world's largest tropical rain forest park in an effort to save the Amazon rain forest.

This waterfall is located in the Tumucumaque Mountains National Park. The park helps ensure that nature within its borders remains pristine and undisturbed.

THINKING GEOGRAPHICALLY

1. **Environment and Society** Look at the deforestation map on page 195. What types of human activity have led to deforestation? How might the creation of national parks such as Tumucumaque help counteract this activity?
2. **Places and Regions** What challenges does the Brazilian government face in the management and control of national parks?

SECTION 2

 section audio  spotlight video

Guide to Reading

Essential Question

How do human activities place stress on a region's natural resources?

Content Vocabulary
- deforestation *(p. 251)*
- sustainable development *(p. 251)*
- slash-and-burn farming *(p. 251)*
- reforestation *(p. 252)*
- shantytown *(p. 252)*

Academic Vocabulary
- apparent *(p. 252)*
- available *(p. 252)*
- conflict *(p. 253)*

Places to Locate
- Costa Rica *(p. 252)*
- Santiago *(p. 253)*
- Montserrat *(p. 254)*

Reading Strategy

Taking Notes As you read about concerns for the rain forest, use the major headings of the section to create an outline similar to the one below.

I. Managing Resources
   A.
   B.
II. Human Impact
   A.
   B.

People and Their Environment

Latin America is rich in natural resources. The region's countries, however, face a daunting challenge: how to preserve and manage these resources, while developing their economies and meeting the increasing needs of a growing population.

NATIONAL GEOGRAPHIC Voices Around the World

"To the far south, in the Chilean capital of Santiago, urbanites still consider the desert a wasteland, impervious to environmental damage. Rumors persist that in the mid-1980s the government proposed creating a dumpsite for the world's nuclear wastes in the Atacama, but backtracked to avoid a public relations disaster. 'There's a prejudice and lack of knowledge about the desert,' complains Patricio Fischer, a biology teacher in Iquique, one of the northern cities. 'People see the Atacama as a blank spot on the map.'"

—Priit J. Vesilind, "The Driest Place on Earth," *National Geographic,* August 2003

Chilean farmer in the Atacama Desert

Managing Resources

MAIN Idea Latin America is working to protect the environment while facing rapid urbanization and growing human needs.

GEOGRAPHY AND YOU Do you know of animal species that are threatened in the United States? Read to learn how Latin American countries are trying to protect animals in their rain forests.

Like rain forests in other regions of the world, those in Latin America are disappearing as a result of **deforestation**, the clearing or destruction of forests. Although the threats to the world's rain forests are well known, the proposed strategies for preserving them are hotly debated among various groups of people. One possibility is to work toward **sustainable development**—technological and economic growth that does not deplete the human and natural resources of a given area.

Farms Versus Forests

One of the most widespread activities in the Amazon Basin is the clearing of the rain forest to provide more land for farming and ranching. To prepare the land, farmers use an ancient technique known as **slash-and-burn farming**. All plants are cut down and any trees are stripped of bark. After the plants and trees have dried out, they are set on fire. The ash from the fire adds nutrients to the soil. Unfortunately, frequent rains leach away the benefits, and within one or two years, the soil loses its fertility. Crop yields decline, and farmers move on to clear new parts of the forest. The spent land supports little growth, and centuries-old rain forests have disappeared in just a few years. In recent years, Amazonian deforestation has accelerated under pressure from *latifundia* and large corporations to expand the land devoted to soybean cultivation because of growing global demand for Brazilian soybeans.

Farming and ranching are not the only activities that contribute to deforestation in the Amazon. Commercial logging operations harvest trees for timber and other products. Some estimates indicate that for every tree cut, two-thirds of the wood is not used or burned.

NATIONAL GEOGRAPHIC *Settlers in the Amazon Basin use the slash-and-burn technique to clear the land and add nutrients to the soil.*

Human-Environment Interaction What may happen to the rain forest if land clearing continues unchecked?

UNIT 4 Europe

The Louvre, Paris, France

NGS ONLINE To learn more about Europe visit www.nationalgeographic.com/education.

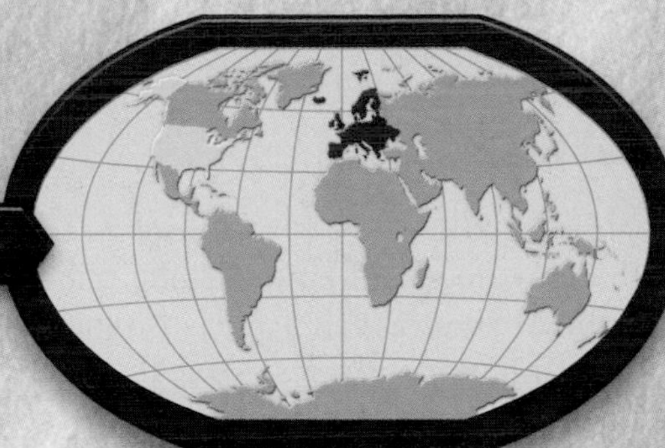

Why It Matters

In the 1990s, several nations of Europe formed the European Union, an alliance that works for the region's economic and political unity. Many European countries have replaced their national currencies with a common currency—the euro. As one of the world's leading economic powers, Europe has long had close political, cultural, and economic ties to the United States. Because of this important relationship, European ideas and practices have shaped your life and may continue to do so in the years ahead.

UNIT 4 WHAT MAKES THIS A REGION?

Europe

PHYSICAL GEOGRAPHY Europe is a peninsula of peninsulas, with many fingers of land sticking out into the Atlantic Ocean and the Mediterranean Sea. It also includes several large islands and island groups. Over the centuries, Europeans have taken advantage of their location, using the sea as a source of food and an avenue for trade and exploration.

Europe's landscape has three layers—highlands in the far north, a broad central lowland plain, and a more mountainous south. The climate tends to be mild because warm waters from the Gulf of Mexico reach the northwestern edge of Europe, and winds carry their warmth to the region.

1 PLAINS AND PLATEAUS The Northern European Plain, which stretches from southern France to Poland, is so large that it gives Europe the lowest average elevation of all continents.

2 LAKES AND RIVERS The Danube River flows from southern Germany to the Black Sea. Like many European rivers, it is an important commercial route as well as being scenic.

3 MOUNTAINS The Alps separate one part of Europe from another and provide the source for many of the region's major rivers.

3

EUROPE

4

4 NATURAL RESOURCES Quarries like this one in Greece produce marble, a fine stone used to make statues and buildings. Like Greece, Italy is famous for the quality of its marble.

Resources and Environmental Threats

Europeans have used the region's resources to develop strong economies, but they have also damaged their environment. As you study the maps and graphics on these pages, look for patterns connecting resources and environmental concerns. Then answer the questions below on a separate sheet of paper.

1. What sources of energy are prominent in Europe? What connection might there be between these energy sources and soil conditions?

2. What prospects do you see for European agriculture? Why?

3. Which country probably has the strongest wood products industry? Why?

ECONOMIC ACTIVITY Europe

Logging in Sweden

CHAPTER 

PHYSICAL GEOGRAPHY OF Europe

BiG Idea

Physical processes shape Earth's surface. A study of the physical geography of Europe will reveal the processes that have shaped the region's land and influence its climate.

Essential Questions

Section 1: The Land
Which European landforms do you think were shaped by the last Ice Age?

Section 2: Climate and Vegetation
What factors influence Europe's climate and vegetation?

Geography ONLINE
Visit glencoe.com and enter *QuickPass*™ code WGC9952C11 for Chapter 11 resources.

The town of Geiranger, Norway, is nestled in the crook of a winding fjord.

Categorizing Information Use a Top-Tab Book to review and categorize information about the physical geography of Europe.

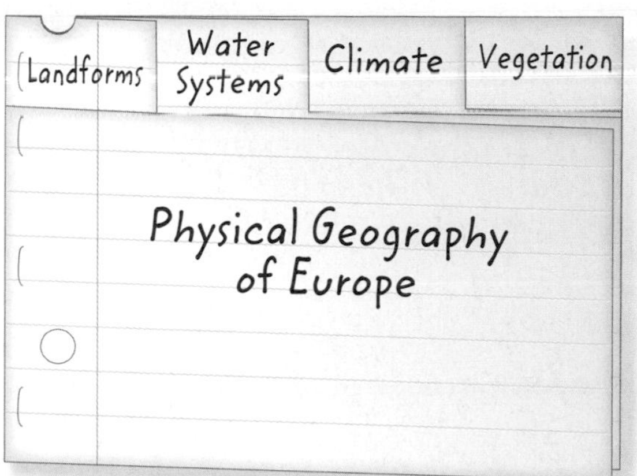

Reading and Writing As you read this chapter, write down key details about Europe's land, water systems, climate, and vegetation. Record details in the appropriate area of your Foldable.

SECTION 1

The Land

 section audio

 spotlight video

Guide to Reading

Essential Question

Which European landforms do you think were shaped by the last Ice Age?

Content Vocabulary

- glaciation *(p. 281)*
- loess *(p. 281)*
- dike *(p. 282)*
- polder *(p. 282)*
- fjord *(p. 282)*

Academic Vocabulary

- process *(p. 281)*
- located *(p. 281)*

Places to Locate

- Alps *(p. 281)*
- Rhine River *(p. 281)*
- Po River *(p. 281)*
- Northern European Plain *(p. 281)*
- Baltic Sea *(p. 282)*
- North Sea *(p. 282)*
- Mediterranean Sea *(p. 282)*
- British Isles *(p. 283)*
- Aegean Sea *(p. 283)*
- Danube River *(p. 283)*
- Black Sea *(p. 283)*

Reading Strategy

Organizing Complete a web diagram similar to the one below by filling in the natural resources found in Europe.

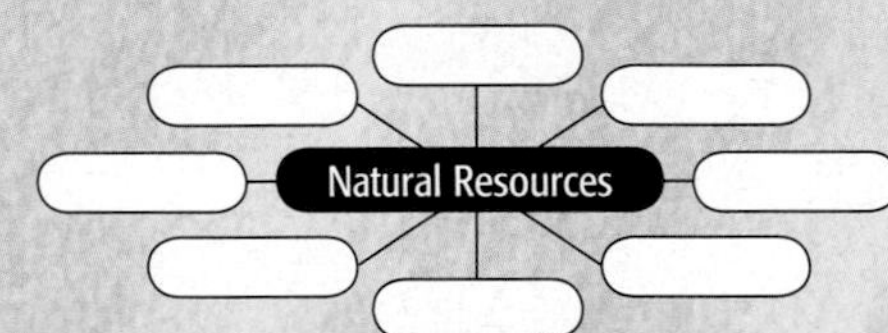

Corsica's varied landscape is like many other parts of Europe in the way it has influenced the history and lives of people who live there. Its craggy mountains have separated groups of people for centuries. In contrast, picturesque beaches and vineyards continue to attract visitors from around the world.

NATIONAL GEOGRAPHIC Voices Around the World

"In the chestnut forests of the Castagniccia region, tiny villages and grand buildings cling to mountainsides, eternally poised to repulse the waves of invaders that swept over Corsica for 2,000 years: Greeks, Carthaginians, Romans, Moors, Genoese, and, finally, the French. Beaches, deserts, and alpine forests are layered like the fromage de brebis *pastry that Corsicans serve to visitors. This diversity in a 115-mile-long outcropping of soil and rock once led an American diplomat, who had a house there, to say, "Corsica is not an island—it's a continent."*

—Peter Ross Range, "France's Paradox Island: Corsica," *National Geographic*, April 2003

A Corsican man near the village of Sainte Lucie de Tallano

Landforms

MAIN Idea Europe's landscape, created over time by physical processes, has shaped the lives and settlement patterns of Europeans.

GEOGRAPHY AND YOU What physical features come to mind when you think of Europe? Read to learn more about the landforms and other geographic features of Europe.

Wind, water, and ice are among the forces that have created Europe's landforms. These landforms have influenced in many ways the lives of people living in each region.

Mountains and Plains

The European landscape consists of plains interrupted by mountains running through its interior and along its northern and southern edge. Europe's northwestern mountains, rounded by millions of years of erosion and glaciation, feature relatively low peaks. **Glaciation** is the **process** in which glaciers formed and spread. Ben Nevis, the highest mountain in the British Isles at 4,406 feet (1,343 m), is part of these ranges. Extending from the Iberian Peninsula to eastern Europe, the central uplands consist of low, rounded mountains and high plateaus with scattered forests. This region includes the Meseta, Spain's central plateau, and the Massif Central, France's central highlands.

By contrast, southern Europe's geologically younger mountains are high and jagged. As the Earth's crust lifted and folded, the Pyrenees (PIHR•uh•NEEZ) were thrust upward to more than 11,000 feet (3,353 m). Created by glaciation and folding, the **Alps** mountain system forms a crescent from southern France to the Balkan Peninsula. Mont Blanc, the highest peak in the Alps, stands at 15,771 feet (4,807 m) in France on the border with Italy. Some of Europe's major rivers, such as the **Rhine** and the **Po,** have their source in the Alps. These mountains also separate the warm, dry climate of the Mediterranean region from the cooler climates of the north. The physical map on page 268 shows that another towering mountain chain, the Carpathians, runs through eastern Europe from Slovakia to Romania.

Europe's broad plains curve around the highlands. Scoured by Ice Age glaciers, the **Northern European Plain,** or Great European Plain, stretches

NATIONAL GEOGRAPHIC *The Alps are the region's most recognizable mountain system, running from southern France to the Balkan Peninsula.*

Location What is the highest peak in the Alps?

from southeastern England and western France eastward to Poland, Ukraine, and Russia. The plain's fertile soil and wealth of rivers originally drew farmers to the area. The southern edge is especially fertile because deposits of **loess,** a fine, rich, wind-borne soil left by glaciers, cover it.

Deposits of coal, iron ore, and other minerals found near the Northern European Plain led to western Europe's industrial development during the 1800s. Today many of Europe's largest cities, such as Paris and Berlin, are **located** on the plain.

Another fertile plains area, the Great Hungarian Plain, extends from Hungary to Croatia, Serbia, and Romania. Farmers cultivate grains, fruit, and vegetables and raise livestock in the lowlands along the Danube River.

Seas, Peninsulas, and Islands

Although Europe and Asia share a common landmass called Eurasia, Europe is a distinct region. Jutting westward from Asia, Europe has an unusually long, irregular coastline that touches many bodies of water, including the Atlantic Ocean and the Baltic, North, Mediterranean, and Black Seas.

Most of Europe lies within 300 miles (483 km) of a seacoast. This closeness to the sea has shaped the lives of its peoples over time. For example, in the Netherlands, water can be friend or foe.

winds, less precipitation falls in southern Europe than in northwestern Europe. Local winds in the region sometimes cause changes in the **normal** weather pattern. The **mistral,** a strong north wind from the Alps, can send gusts of bitterly cold air into southern France. By contrast, **siroccos** (suh•RAH•kohs), hot, dry winds from North Africa, may bring high temperatures to the region. The hot, dry summers in much of southern Europe encourage the growth of drought-resistant vegetation such as shrubs and small trees.

Much of eastern Europe has a humid continental climate with cold, snowy winters and hot summers. Warm ocean currents have less influence on climate in these areas farther from the Atlantic Ocean. As a result, summer and winter temperatures vary more widely. This humid continental climate supports vegetation that is generally a mix of deciduous and coniferous forests.

The Alps have a highland climate with colder temperatures and more precipitation than nearby lowland areas. Sudden changes can occur, however, when dry, winter winds called **foehns** (FUHNZ) blow down from the mountains into valleys and plains. Foehns can trigger **avalanches,** destructive masses of ice, snow, and rock sliding down mountainsides. Avalanches represent a serious natural hazard in the Alps. They threaten skiers, hikers, and villages, and destroy everything in their paths.

Dry Regions

Parts of southeastern and southwestern Europe have a dry steppe climate. The southeastern European steppe is a belt of dry, cold grassland that extends from **Hungary,** Serbia, Montenegro, and **Romania** through **Ukraine** and Central Asia. A steppe climate is characterized by hot summers and extremely cold winters. Levels of precipitation vary, as rainfall becomes scarce in the eastern areas of the steppe that are farther away from moist Atlantic Ocean winds. Farming is difficult in steppe regions because of the extreme temperatures, periods of drought, poor and easily eroded soils, and high winds.

Parts of the Meseta, an interior plateau located on the Iberian Peninsula, have a dry steppe climate. The Meseta extends over 81,000 square miles (210,000 square km). Madrid, Spain, is at its center.

READING Check **Place** What are three types of climates found in Europe's midlatitude regions?

SECTION 2 REVIEW

Vocabulary

1. Explain the significance of: permafrost, timberline, mistral, sirocco, foehn, avalanche.

Main Ideas

2. What factors influence climate in Europe? Give examples.
3. How do Europe's main climate regions—high latitude, midlatitude, and dry—differ?
4. Create an outline like the one below, showing the climate and vegetation found in three European countries.

Climate and Vegetation
I. Iceland A. Climates: subarctic, tundra, and permafrost B. Vegetation: conifers, lichens, moss

Critical Thinking

5. **Answering the Essential Question** How does latitude affect Europe's climate and vegetation in specific regions?
6. **Analyzing Information** What geographic factors contribute to vegetation differences between the highland area of the Alps and tundra climate regions?
7. **Analyzing Visuals** Study the vegetation and climate maps on pages 286–287. Where are most of Europe's coniferous forests located? What do these climate areas have in common that would make them favorable to this type of tree?

Writing About Geography

8. **Descriptive Writing** Describe the differences between two climate regions in Europe. Support your answer with evidence from the National Geographic excerpts on pages 280 and 285.

Study Central™ To review this section, go to glencoe.com and click on Study Central.

CHAPTER 11 VISUAL SUMMARY

Study anywhere, anytime by downloading quizzes and flashcards to your PDA from glencoe.com.

A ISLANDS

- Volcanic Iceland has hot springs and geysers; tundra and marine west coast climates
- British Isles consist of islands of Great Britain and Ireland; lie northwest of the European mainland
- Rugged mountains form islands of Sicily, Sardinia, Corsica, Crete, and Cyprus in the Mediterranean Sea

B RESOURCES

- Major petroleum and natural gas reserves in the North Sea
- Coal deposits in the United Kingdom, Germany, Ukraine, and Poland
- Heavy investment in nuclear power in France

C PENINSULAS

- Glaciation carved narrow, steep-sided fjords along the coasts of northern peninsulas.
- Warm ocean currents create marine west coast climate in coastal areas of northern peninsulas
- Southern peninsulas include the Iberian, Italian, and Balkan Peninsulas.
- Mediterranean climate characterizes the Iberian and Italian Peninsulas.

D RIVERS

- Rhine and Danube Rivers in the heartland of Europe are dominant rivers with large amounts of traffic.
- Seine and Rhone Rivers in France are important for transportation and urban development.
- Po River in Italy key in industrial development

E THE ALPS

- Located in southern Europe; young, high, jagged mountains
- Source of some of Europe's major rivers
- Form a barrier that separates warm, dry climate of the Mediterranean from cooler climates of the north

F NORTHERN EUROPEAN PLAIN

- Stretches from southeastern England and western France east to Poland, Ukraine, and Russia
- Mild climate, fertile soil, and access to rivers make it a highly productive agricultural area.

CHAPTER 12

CULTURAL GEOGRAPHY OF Europe

Big Idea

Cultures are held together by shared beliefs and common practices and values. Europe's peoples belong to many different cultural groups. A study of the cultural geography of Europe reveals how the region can be organized into subregions based on common experiences, beliefs, and practices.

Essential Questions

Section 1: Northern Europe
How did the Industrial Revolution affect life in northern Europe?

Section 2: Western Europe
What factor united much of western Europe during the Middle Ages?

Section 3: Southern Europe
What contributions from southern Europe formed the foundation for Western civilization?

Section 4: Eastern Europe
What major events in modern history have forced changes in the political and cultural makeup of eastern Europe?

Visit glencoe.com and enter QuickPass™ code WGC9952C12 for Chapter 12 resources.

Rome's Piazza Navona blends a bustling cosmopolitan atmosphere with art and history.

Summarizing Information Make a Three-Pocket Book to help you summarize information about population patterns, history and government, and culture of Europe's four subregions—northern Europe, western Europe, southern Europe, and eastern Europe.

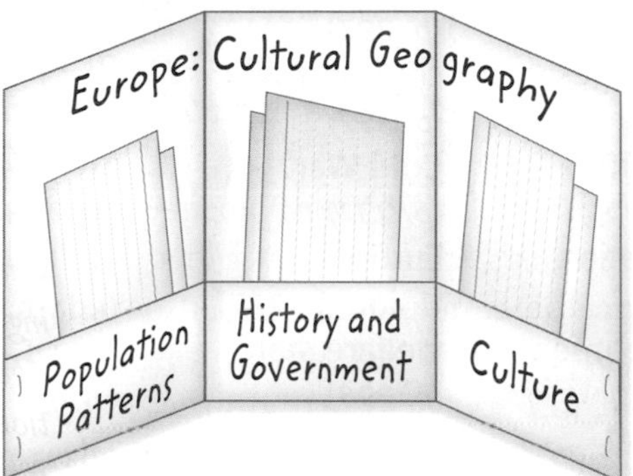

Reading and Writing As you read this chapter, write down key details about the cultural geography of Europe. Write information on labeled index cards and store the cards in the pockets of your Foldable.

EUROPE

NATIONAL GEOGRAPHIC *Although Norway is nearly 80 percent urban, traditional fish markets can be found in the middle of bustling cities.*

Place What is northern Europe's largest and oldest urban area?

Density and Distribution

The United Kingdom is the most densely populated country in the subregion. Its 61.8 million people create a population density of about 662 people per square mile (255 per square km). Denmark and Ireland have relatively high densities as well. These three countries have temperate climates and fertile soil that historically have supported larger populations. Scandinavia has lower population densities. Large areas of harsh terrain or climate have led its people to live along the coasts.

Internal and external migration have shaped the subregion. Internal migration has occurred from rural areas to urban areas, where people can find jobs. Often this urbanization is driven by poverty in rural areas.

Economic troubles may also drive external migration. For example, Ireland's economic depression and famine in the 1840s prompted 1.6 million people to leave the country.

Northern Europe's most important metropolitan areas are also its economic centers. The subregion's largest and oldest urban area is **London,** the capital of the United Kingdom. This multicultural city holds about 8.5 million people. Stockholm, Sweden, is home to about 1.3 million people. Copenhagen, Denmark, is the capital and a popular tourist destination.

READING Check **Human-Environment Interaction** Why do the countries of the British Isles have a larger population than the countries of the Scandinavian Peninsula?

History and Government

MAIN Idea The cultures of northern Europe have led its people to create powerful governments and thriving economies.

GEOGRAPHY AND YOU How does your community view the environment—to be used or protected? Read to learn how the land and the sea were used throughout the history of northern Europe.

Northern Europe was shaped by thousands of years of migrations and invasions. In recent times, **intensive** change has come more rapidly.

Early Peoples

Britain's earliest ancestors may have arrived from mainland Europe about 12,000 years ago, when a land bridge connected the two areas. Celts spread to the British Isles around 500 B.C. Their culture and language live on in the subregion today.

Scandinavia has also been home to people for thousands of years. The Sami are the native people of northern Norway, Sweden, and Finland. They are descendants of nomadic peoples who lived in northern Scandinavia for thousands of years.

The Rise of Northern Europe

Between 55 B.C. and A.D. 1066, the countries of northern Europe began to take shape. In A.D. 43, Romans invaded Britain, adding it to their empire. The Romans built towns, roads, and cities. They also brought Christianity to the subregion. The Roman Empire declined in the A.D. 300s and A.D. 400s, and its departure left Britain vulnerable to invading Germanic groups.

The Germanic Angles, Saxons, and Jutes invaded Britain when the Romans left. This period of rule is known as the early Middle Ages. The **Middle Ages,** lasting from about A.D. 500 to A.D. 1500, is known as the period between ancient and modern times.

Scandinavian powers raided the coasts of Europe during this period. These powerful Viking kingdoms led directly to the development of Denmark, Sweden, and Norway.

In 1066 Britain fell to the Normans. The new Norman king of Britain established **feudalism**—a system in which monarchs or lords gave land to nobles in return for pledges of loyalty.

Throughout the remainder of the Middle Ages, the region began to change as governments took shape. Britain laid the foundations for its strong government and built its foreign trade. Denmark, Norway, and Sweden were united throughout the 1400s but then divided and again were rivals fighting for control of Scandinavia.

The **Reformation** was a religious movement of the 1500s that lessened the power of the Roman Catholic Church and introduced Protestantism to Europe. The reform movement, led by the monk Martin Luther, began in Germany. It was popular with northern monarchies, where the power of the Church was often unwelcome.

Winds of Change

During the early 1700s, many educated Europeans embraced the **Enlightenment,** a movement to value reason and question tradition. At this time many northern Europeans fought for a voice in government. For example, in the late 1600s, the English Parliament, or lawmaking body, passed a Bill of Rights that limited the power of the monarchy. The English Parliament became a model for many governments around the world.

The **Industrial Revolution** transformed manufacturing in Europe by replacing human labor with machines. Widespread industrial and social changes created **industrial capitalism,** an economic system in which owners used profits to expand their companies. Although some people prospered, factory workers were poorly paid and lived in crowded, unhealthy conditions. These conditions led to the birth of **communism**—a philosophy that called for economic equality and ownership of resources by workers.

When a population undergoes sweeping transitions such as the Industrial Revolution, it changes in many ways. A demographic transition model describes these types of changes over time.

In the 1900s two world wars killed millions and left countries in ruins. After World War II, the division of Europe led to the **Cold War,** a power struggle between the Soviet-controlled Communist world and the non-Communist world.

A New Era

In the last few decades, northern European countries have built strong democracies and successful economies. Today, northern Europe belongs to the **European Union** (EU). The EU's goal is a united Europe in which goods, services, and workers can move freely among member countries.

READING Check **Movement** What group brought Scandinavian influence to the British Isles?

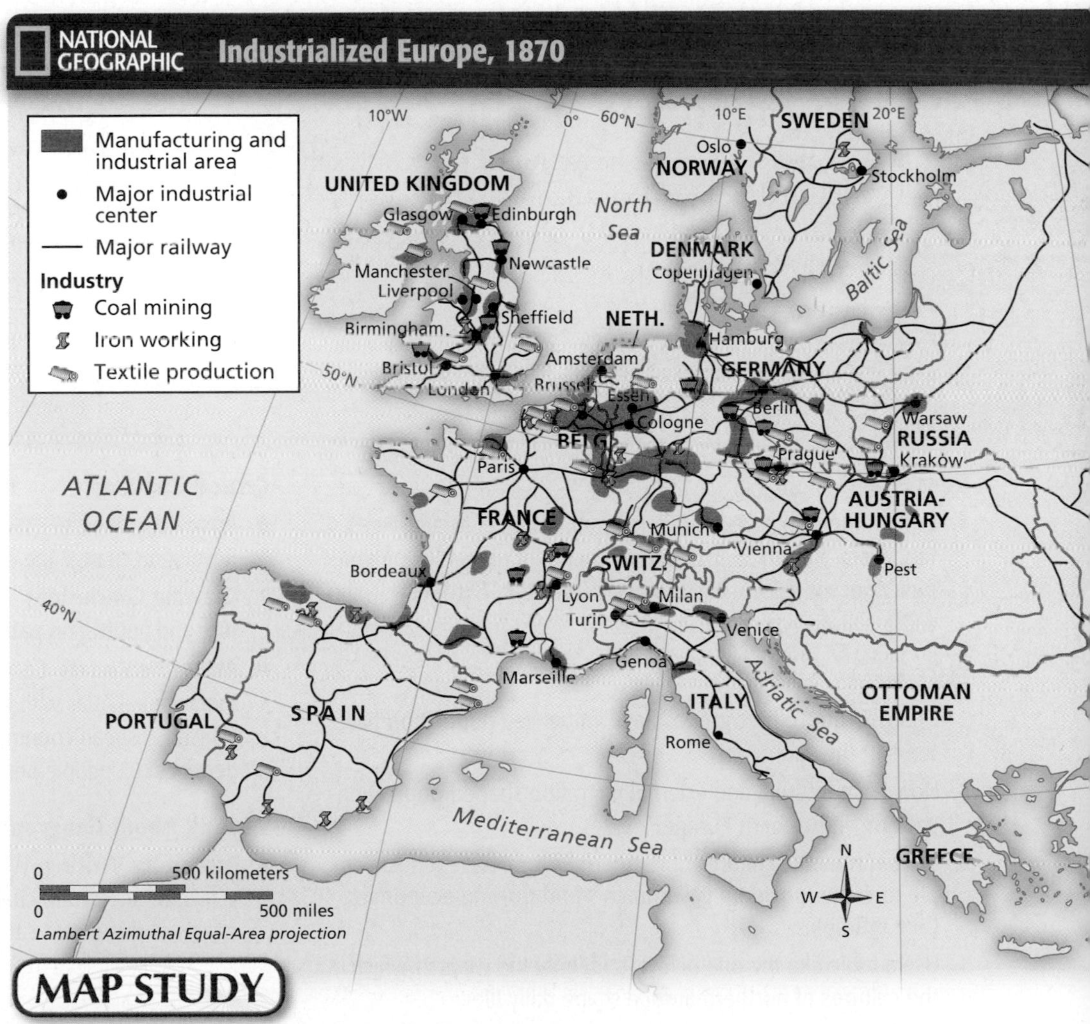

MAP STUDY

1. **Location** Where were most manufacturing and industrial areas located in 1870?
2. **Human-Environment Interaction** What relationship do you see between the location of coal mining and manufacturing and industrial areas in 1870?

Culture

MAIN Idea The centuries-old cultures of northern Europe shape daily life in the region today.

GEOGRAPHY AND YOU How many languages are spoken in your part of the country? Read to learn more about the languages of northern Europe.

The people of northern Europe value caring for their populations through education, quality health care, and other social programs.

Language and Religion

Northern Europe's languages are mostly Indo-European, such as Swedish and English. Many dialects, or local forms of languages, exist as well.

Although a single religion is dominant in every country of the subregion, these countries value religious freedom. Most countries are Protestant, but many minority religions exist.

Education and Health Care

Northern Europe has some of the world's most educated populations. School is mandatory for children for at least 10 years. Literacy rates are nearly 100 percent.

Northern Europeans have excellent health care. Countries such as Sweden offer complete social welfare programs to their citizens. These countries, known as **welfare states,** have health-care programs **funded** by governments.

The Arts

British writings have shaped the literary world. The works of William Shakespeare have become among the most widely read works of all time. In the late 1700s, the influential style of **romanticism** focused on emotions, stirring historical events, and the struggles of individuals. Denmark has a rich artistic tradition, as well. The fairy tales of Hans Christian Andersen (1805–1875) are known throughout the world.

READING Check **Regions** How are welfare states able to provide social services to all their citizens?

Student Web Activity Visit glencoe.com, select the *World Geography and Cultures* Web site, and click on Student Web Activities–Chapter 12 for an activity on religion in Europe.

SECTION 1 REVIEW

Vocabulary

1. Explain the significance of: immigrant, refugee, Middle Ages, feudalism, Reformation, Enlightenment, Industrial Revolution, industrial capitalism, communism, Cold War, European Union, welfare state, romanticism.

Main Ideas

2. Describe how ethnic groups have influenced population patterns in the subregion.
3. How have internal and external migration shaped population patterns in northern Europe?
4. Explain how the cultures of northern Europe have led its people to create powerful governments and thriving economies. Give examples.
5. Use a table like the one below to identify the ways in which the cultures of northern Europe shape daily life.

Element of Culture	Influence on Daily Life

Critical Thinking

6. **Answering the Essential Question** How did the Industrial Revolution change the economy of northern Europe?
7. **Drawing Conclusions** How has geography influenced the culture and population patterns of northern Europe? Give examples.
8. **Analyzing Visuals** Compare the political map on page 270 of the Regional Atlas with the population density map on page 295. Which European countries have large areas of low population densities (62 people per square mile [24 per sq. km] or less)?

Writing About Geography

9. **Expository Writing** Write a paragraph describing how the arrival of one group changed northern Europe. What cultural effects of this group can be seen today?

Study Central™ To review this section, go to glencoe.com and click on Study Central.

SECTION 2

 section audio

 spotlight video

Guide to Reading

Essential Question

What factor united much of western Europe during the Middle Ages?

Content Vocabulary
- guest worker *(p. 300)*
- Crusades *(p. 301)*
- reparations *(p. 302)*
- Holocaust *(p. 302)*
- realism *(p. 302)*
- impressionist *(p. 302)*

Academic Vocabulary
- contact *(p. 301)*
- subsidy *(p. 303)*

Places to Locate
- France *(p. 300)*
- the Netherlands *(p. 300)*
- Belgium *(p. 300)*
- Switzerland *(p. 300)*
- Germany *(p. 300)*
- Austria *(p. 300)*
- Paris *(p. 300)*
- Brussels *(p. 300)*
- Luxembourg *(p. 302)*

Reading Strategy

Taking Notes As you read about western Europe, use the major headings of the section to create an outline like the one below.

I. Population Patterns
  A.
  B.
II. History and Government
  A.
  B.

Western Europe

France, like other countries in western Europe, has long been a crossroads of cultures and a colorful mix of the old and the new. For example, the Parisian neighborhood known as the Marais is a collection of restored ancient buildings, museums, new and old restaurants and bakeries, street musicians, long-time residents, and fashionable newcomers.

NATIONAL GEOGRAPHIC Voices Around the World

"Cross a street in the Marais and you cross centuries and cultures. Such is the richness of the neighborhood that within 40 steps on the Rue des Ecouffes, you pass a synagogue, a kosher butcher . . . and an Internet café. . . . But at Jo Goldberg's, Korcarz, or any of the other Jewish bakeries or delicatessens, tradition rules.

'People come here to taste their roots,' says Florence Finkelsztajn, who runs a bakery at 19 Rue de Rosiers. But there's more to it than just cheesecake, she points out. It has to do with making connections. . . . It has to do with continuity. 'Our customers stay with us from generation to generation.' "

Cathy Newman, "Bohemian Rhapsody," *National Geographic*, August 2003

The Jewish quarter, Marais, Paris, France

Urban Growth and Transportation

One Problem:

Rapid population growth and an inadequate transportation network have caused problems for the commuters of metropolitan Toulouse.

Metropolitan Toulouse

Population boom The promise of jobs in the industries of Toulouse has drawn people to the area over the last several decades.

Traffic congestion

Commuting The rapid growth of Toulouse as an industrial center led to increased road traffic. This traffic made movement around the city difficult.

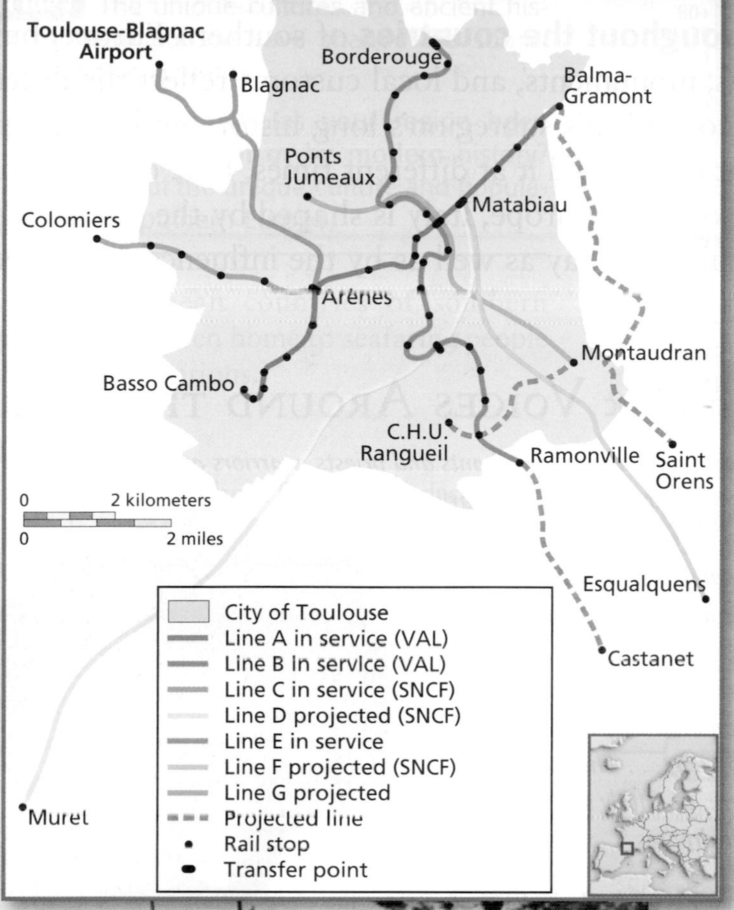

Moving About in Metropolitan Toulouse

Toulouse (tu•LOOZ), the center of Europe's aerospace industry, is also the fifth-largest metropolitan area in France. Because of its location in sunny southwestern France and its many high-tech industries, Toulouse is also one of Europe's fastest-growing metropolitan areas. Attracted by the climate and job opportunities, many people have made their home in Toulouse. Between 1960 and 2000 the city's metropolitan population doubled. With a bright future ahead for the aerospace and biotech industries, Toulouse continues to grow.

Economics and Industry This rapid growth, while contributing to a strong economy, also creates challenges for city planners. The high-tech industries that have made Toulouse so popular are sprinkled throughout the region, but poor roads inhibit travel to these job sites. In Toulouse, therefore, one of the most pressing issues involves upgrading public transportation to deal with commuter traffic.

Transportation Solution City officials have found an answer to Toulouse's transportation challenges—a network of rail and metro lines that complement the existing bus system. These lines include a high-frequency commuter rail, a steel-wheel surface tramway, and a modern metro system of rubber-tired, driverless (automatic) trains. Officials hope this reliable transportation option will prove popular with the ever-growing number of city residents commuting between home and work.

Possible Solution:

In 1993, the first driverless metro, Line A, opened in Toulouse. Additional lines have been opened since then, and many more are in the development and planning stages.

The metro lines help to connect the metropolitan area. The miles of line have been increasing yearly. There is approximately 42 miles (67 km) of line. More miles of track are planned for the future.

THINKING GEOGRAPHICALLY

1. **Environment and Society** Research public transportation in your own city, or in a nearby city, and compare it to that of Toulouse. Present your findings in a multimedia presentation comparing the similarities and differences.
2. **Places and Regions** If Toulouse continues to experience a high rate of growth, what other issues might it face?

WHY GEOGRAPHY MATTERS

GERMAN Reunification

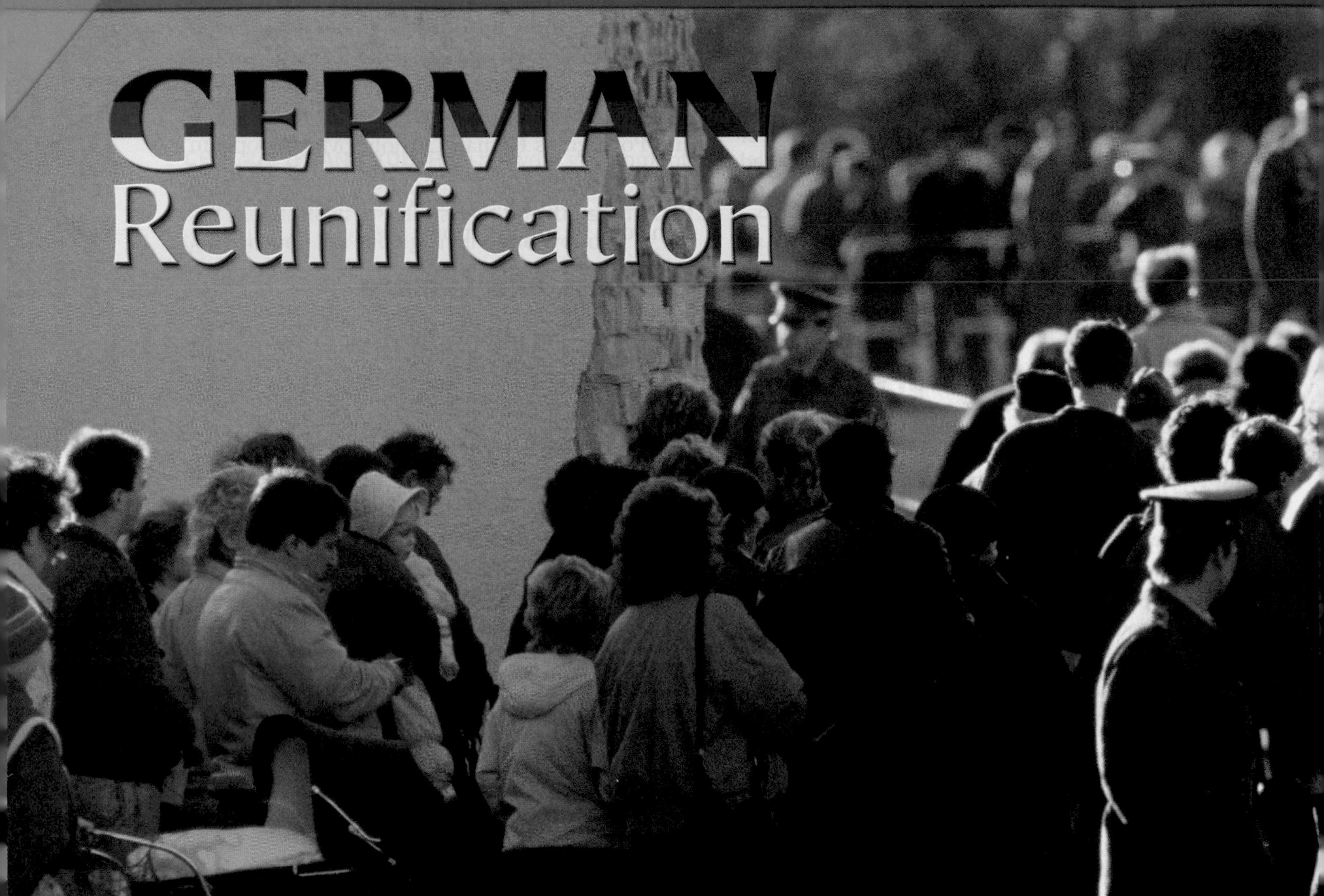

The Problem:

Since the reunification of Germany, eastern Germany has experienced deindustrialization, high unemployment, and the mass exodus of its youngest and most skilled citizens.

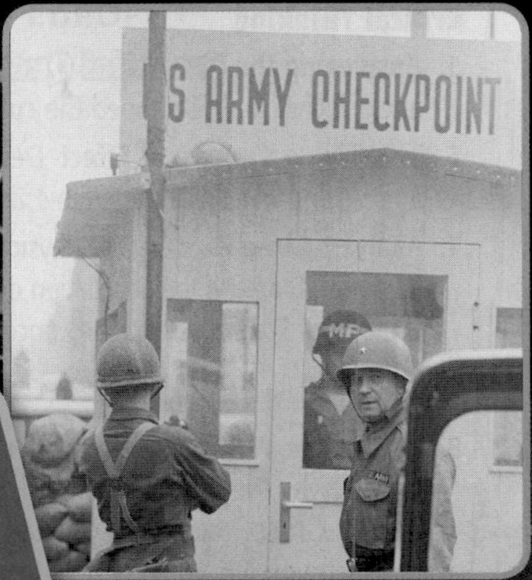
S ARMY CHECKPOINT
MP

Checkpoint Charlie

Separation Following World War II, Germany separated into two countries. Checkpoint Charlie was a border crossing between East and West Berlin during the Cold War.

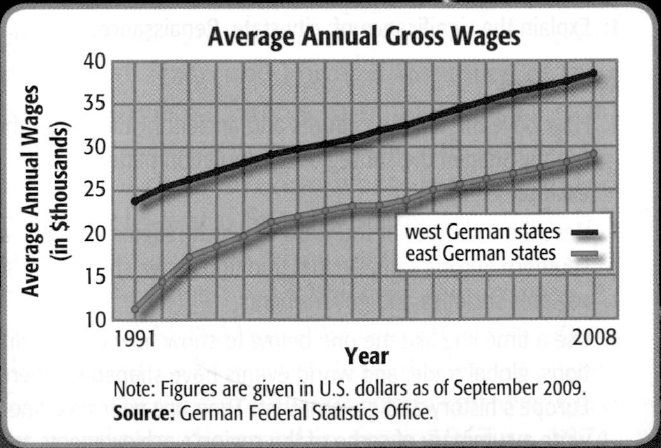
Average Annual Gross Wages
Average Annual Wages (in $thousands)
40
35
30
25
20
15
10
1991
2008
Year
west German states
east German states
Note: Figures are given in U.S. dollars as of September 2009.
Source: German Federal Statistics Office.

Wage Differences

Inequality After the reunification of Germany, the economic differences between the eastern and western regions became evident.

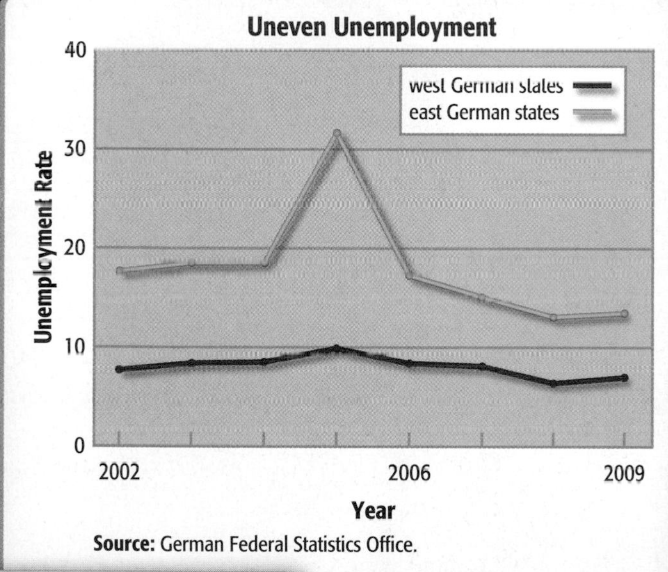

Source: German Federal Statistics Office.

The Fall of Communism

When communism fell in East Germany, residents expected their lives to improve. Unfortunately, years of communism had left the region far behind its western counterpart.

Why did citizens leave eastern Germany? As the border with the west opened, many of eastern Germany's youngest and most skilled citizens moved there to find work. After years of being stripped of their liberties, many longed for the prosperity that a successful capitalist environment offered. Overall, the east lost over 800,000 of its 17 million people, leaving the oldest and most unskilled segments of the population to rebuild the region.

What has happened to industry? Eastern factories were inefficient and unable to compete in the global capitalist market. As a result, many of them closed, causing widespread unemployment.

What has the government done to address unemployment and deindustrialization? Germany's government has invested the equivalent of almost one trillion U.S. dollars in eastern Germany in the hope of speeding its economic recovery. The money has been used to update and rebuild infrastructure. Additionally, the government has been promoting investment in the east and sponsors programs to retrain workers. All these efforts have begun to produce results as eastern industry has expanded, helping to close the economic gap between east and west.

Possible Solution:

Increased economic development in eastern Germany will help pull the region out of its current situation.

Although unemployment rates continue to be higher in eastern Germany, they have begun to decline.

THINKING GEOGRAPHICALLY

1. **Places and Regions** Years of Communist control have created a great social divide between citizens in eastern and western Germany. What methods other than economic development might be used to draw the two regions closer together?
2. **The World in Spatial Terms** Study the historical map of East and West Germany. What challenges might the government of East Germany have faced in trying to stop citizens from attempting to escape to the West?

SECTION 4

Eastern Europe

 section audio  spotlight video

Guide to Reading

Essential Question

What major events in modern history have forced changes in the political and cultural makeup of eastern Europe?

Content Vocabulary
- Balkanization *(p. 315)*
- ethnic cleansing *(p. 315)*

Academic Vocabulary
- trace *(p. 313)*
- circumstances *(p. 313)*
- unified *(p. 314)*

Places to Locate
- Serbia *(p. 313)*
- Montenegro *(p. 313)*
- Bosnia and Herzegovina *(p. 313)*
- Croatia *(p. 313)*
- Slovenia *(p. 313)*
- Macedonia *(p. 313)*
- Kosovo *(p. 313)*
- Czech Republic *(p. 313)*
- Poland *(p. 313)*
- Hungary *(p. 313)*
- Slovakia *(p. 313)*
- Ukraine *(p. 313)*
- Estonia *(p. 313)*
- Latvia *(p. 313)*
- Lithuania *(p. 313)*
- Bulgaria *(p. 313)*
- Belarus *(p. 314)*
- Romania *(p. 314)*

Reading Strategy

Organizing As you read, complete a web diagram like the one below by listing and describing the three main groups of Slavs.

Eastern Europe encompasses a diverse stretch of land and a vibrant mix of ethnicities. Over the centuries countries in this subregion have been faced with political, economic, and ethnic challenges. However, eastern Europe is also home to spirited communities rich with history and cultural traditions. Budapest, for example, is the capital of Hungary and is a celebrated cultural center.

NATIONAL GEOGRAPHIC Voices Around the World

"I realize that, although my return visit to Budapest has been filled with oversized personalities, hearty food, and more grand settings than I can count, what I'll remember most is the music. In this city by the Danube, music isn't confined to concert halls or dance clubs. Its spirit infuses every detail of life, from the passion of everyday conversations to the ebullience of the architecture. It swirls through . . . restaurants and darkening courtyards. I know it will always make me stop and listen."

—Amy Alipio, "Serenade in Budapest," *National Geographic Traveler*, April 2005

A mother and daughter in Budapest, Hungary

Population Patterns

MAIN Idea Eastern Europe's population patterns have been shaped by physical geography, migration, and political and ethnic struggles.

GEOGRAPHY AND YOU What ethnic groups can be found in your region of the world? Read to learn about eastern Europe's many ethnic groups.

Wars, migrations, and changing borders have played a role in shaping population patterns and characteristics of eastern Europe.

The People

Most eastern Europeans are ethnically Slavic. Slavs are descended from Indo-European peoples who migrated from Asia and settled in the region. Generally, they are divided into three groups—east Slavs, including Ukrainians, Russians, and Belarusians; west Slavs, including Poles, Czechs, and Slovaks; and south Slavs, including Serbs, Croats, Slovenes, and Macedonians.

The central portion of eastern Europe includes countries that once formed the country of Yugoslavia—**Serbia, Montenegro, Bosnia and Herzegovina** (BAHZ•nee•uh and HERT•seh•GAW•vee•nah), **Croatia, Slovenia, Macedonia,** and **Kosovo.** Ethnic and cultural groups living here include southern Slavic peoples such as Eastern Orthodox Serbs, Roman Catholic Croats, and Bosnian Muslims. Farther north are the countries of **Czech Republic, Poland, Hungary,** and **Slovakia.**

The easternmost portion of eastern Europe is home to people of Russian and Ukrainian origin. As the largest ethnic group in Europe, Russians can **trace** their roots to ancient Slavic goups near the Black Sea. Peoples of **Ukraine** came from Slavic groups as well. Also found here are the Baltic Sea countries of **Estonia, Latvia,** and **Lithuania.** One minority group that can be found in many eastern European countries, including **Bulgaria,** is the Roma people—once referred to as Gypsies.

Density and Distribution

Population density and distribution in eastern Europe are influenced by geographic factors. For example, Ukraine and Poland both have fertile soil and ample water resources that support large populations. Ukraine is the subregion's largest and most populous nation with 46 million residents and a population density of 206 per square mile (79 per sq. km).

In many countries of eastern Europe, difficult economic and political **circumstances** throughout history have prompted large-scale internal migration and emigration. After World War II, Poland's population decreased dramatically, partly from people fleeing Soviet control. Industrialization throughout the 1900s led to urbanization in eastern Europe.

Most of the population of eastern Europe today lives in and around large towns and cities. Since the end of Soviet control over the area, many cities are experiencing a rebirth. For example, Budapest, the capital of Hungary, is now a thriving center of business and culture.

READING Check **Movement** What factors have caused widespread emigration from eastern Europe?

MAP STUDY

1. **Place** Which country is the most ethnically diverse?
2. **Place** Which countries are home to Muslims?

History and Government

MAIN Idea Political, economic, and ethnic struggles have shaped the history and governments of eastern Europe.

GEOGRAPHY AND YOU Is your community connected to or isolated from neighboring communities? Read to learn more about eastern Europe and the process of Balkanization.

Today's Slavic people have their ethnic and cultural origins in central and eastern Europe's Celtic groups, Germanic groups, and Slavs.

Early Peoples and Empires

It is believed the earliest Slavs migrated from Asia thousands of years ago and settled in a region that now includes Ukraine and Poland. There they lived among migrating Celtic and Germanic groups. In the A.D. 400s and A.D. 500s, Germanic groups and Slavs began moving westward and southward.

Slavic Czechs settled in the historic region of Bohemia by the A.D. 500s. Moravia was settled by Slavic groups by the late A.D. 700s. They formed Great Moravia, a united empire that covered much of central Europe.

The Slavic groups living on the mountainous Balkan Peninsula established the independent states of Croatia, Serbia, and Slovenia. However, the Ottoman Empire ruled this area for nearly 500 years starting in 1400. They gave the peninsula its name—*Balkan,* or "mountains."

While other Slavic groups were moving westward as a result of Asian invasions from the east, the east Slavs settled in the forests and plains of northern-central Ukraine and southern **Belarus.** One such settlement was Kiev (now Kyiv) on the Dnieper River. The region around Kiev became known as Kievan Rus—the earliest of the east Slavic states.

The lands between the Carpathian Mountains and the Danube River were conquered by the Romans. Their customs and language led to the area's new name, **Romania.** Frequent invasions prevented Romania from being **unified** until the 1100s.

The eastern half of the Roman Empire became known as the Byzantine Empire. It lasted for a thousand years after the fall of Rome. The Byzantine Empire profoundly shaped the region. Byzantine

NATIONAL GEOGRAPHIC **Europe and the Cold War**

MAP STUDY

1. **Location** How might the location of Sweden and Finland have affected relations between NATO countries and Warsaw Pact countries?
2. **Place** Which southern European country remained neutral during the Cold War?

Maps In MOtion Use **StudentWorks™ Plus** or glencoe.com.

missionaries spread Eastern Orthodoxy—a form of Christianity—across eastern Europe. The empire also protected the region from invasions by Arabs and Turks. The empire fell in 1453 and came under the control of the Ottoman Empire.

Conflict, Union, and Division

The modern countries of eastern Europe have often suffered power struggles and ethnic divisions. The Balkan Peninsula, in particular, has long been a region of instability. Balkan Slavs overthrew the Ottoman Empire in the early 1900s but had a hard time uniting the region. Despite forming Yugoslavia, or "Land of the South Slavs," cooperation proved challenging for the Balkan nations. The term **Balkanization,** which has come to mean the division of a region into smaller regions that are often hostile with each other, first arose from the results of the Balkan wars.

Eastern Europe was the site of frequent battles and occupations during World War II. Throughout the war, an underground group in Yugoslavia fought against Germany. After the war, Yugoslavia emerged as a Communist country. Eastern Europe fell under Communist control of the Soviet Union at the end of the war. This division between Communist eastern Europe and democratic western Europe brought about the Cold War.

In the 1990s, Yugoslavia fractured along ethnic lines. Some republics declared independence, and ethnic hatreds sparked violence in Bosnia and Herzegovina and Kosovo. Following a policy called **ethnic cleansing,** Serb leaders expelled or killed rival ethnic groups in these areas. International peacekeeping efforts, however, have enabled many refugees to return to their homes in these areas.

A New Era

From the 1950s to the 1980s, revolts against Communist rule periodically swept eastern Europe. In 1989 public demonstrations led to the fall of the region's Communist governments. During the 1990s free elections installed democratic leaders, who encouraged the rise of market economies. Some eastern European countries are growing closer to the rest of Europe. Recently, they have joined the European Union.

READING **Check** **Regions** What area of eastern Europe experienced civil wars in the 1990s?

Teen Life in Hungary

Since the fall of communism in 1989, Hungarian teens have lived much like American teens. Hanging out with friends, watching television, and listening to music are some favorite pastimes.

Hungarian primary schools are similar to American schools—students study math, science, history, reading, writing, and geography. After primary school, students have the option of going to two different types of high schools. Some students attend an academic high school to prepare for university. Other students attend technical schools where they receive job training.

Did you know . . .

- Hungarians pull lightly on a person's earlobe when they wish them "Happy Birthday."
- Soccer, called *foci,* is Hungary's most popular sport.
- Students attend school between the ages of 6 and 16.
- Even though Hungary is landlocked, Hungarians are known for their skills at water sports.
- Guests are expected to take off their shoes when they enter someone's house.
- At Christmas, children put their winter boots on their windowsills for St. Nicholas to fill with candy.

Culture

MAIN Idea Religious and ethnic conflict have influenced culture in eastern Europe.

GEOGRAPHY AND YOU Why is tension over religion not a serious issue in the United States? Read to learn about the role of religion and ethnicity in eastern Europe.

The culture of eastern Europe today reflects its past as well as a hopeful future.

Language and Religion

Most people in eastern Europe speak Indo-European languages. Common Slavic languages include Polish and Czech. Baltic languages include Latvian and Lithuanian.

Roman Catholicism, Eastern Orthodoxy, and Islam are all common in eastern Europe. Religious and ethnic differences were at the heart of conflict in the Balkan Peninsula in the 1990s.

Education and Health Care

Literacy rates are high throughout the subregion. School is mandatory for children and free of charge. Some former Soviet-bloc countries, however, have faced funding challenges in their transition to democratic governments. The health-care system was challenged during the 1990s, as well. Despite this, most eastern Europeans have access to health care.

The Arts and Leisure

Many traditional forms of art and music exist throughout eastern Europe. Folk and classical music are particularly important among Czech, Hungarian, Slovak, and Slovene peoples. In larger cities, modern music from western Europe and the United States is also popular. Literature is also valued in the subregion. Eastern Europe has produced great fiction writers, such as Czech-born Franz Kafka (1883–1924).

As in other areas of Europe, families in many parts of eastern Europe tend to be smaller than they were in the past. However, some people still live in large extended families.

The former Soviet Union encouraged sports and physical education programs. Ukraine, for example, has many swimming pools and other facilities remaining from the Soviet era. Leisure time in eastern Europe also includes spending time with family and the preparing of traditional meals.

READING Check **Regions** What factors put a strain on health-care systems in eastern Europe?

SECTION 4 REVIEW

Vocabulary

1. Explain the significance of: Balkanization, ethnic cleansing.

Main Ideas

2. Describe the ways in which population patterns in eastern Europe have been shaped by physical geography.
3. How have migration and political conflict influenced population patterns in eastern Europe? Give examples.
4. What kinds of struggles have influenced the history and governments of eastern Europe?
5. What type of government dominated eastern Europe immediately following World War II?
6. Create a graphic organizer like the one below and use it to fill in key details for each aspect of eastern European culture.

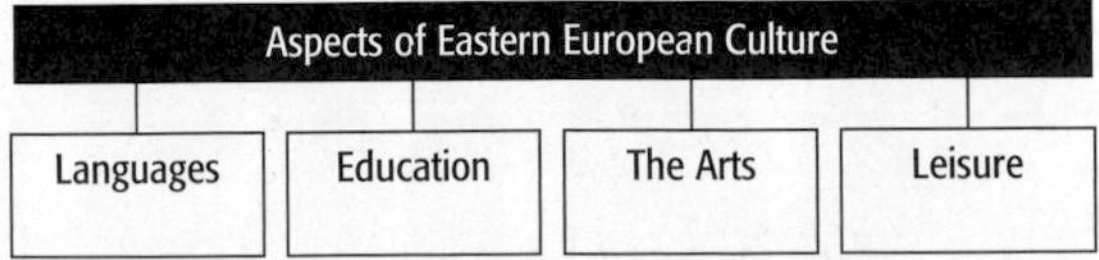

Critical Thinking

7. **Answering the Essential Question** What effects did Balkanization have on eastern Europe?
8. **Evaluating Information** What are some examples of the cooperation and the conflict that have shaped eastern Europe?
9. **Analyzing Visuals** Study the map of the Cold War on page 314. Which countries of northern Europe were neutral?

Writing About Geography

10. **Descriptive Writing** Write a paragraph describing the most densely populated areas of eastern Europe. Explain what would make a good location for a city. Think about location and natural features.

Study Central™ To review this section, go to glencoe.com and click on Study Central.

CHAPTER 12 VISUAL SUMMARY

STUDY TO GO

Study anywhere, anytime by downloading quizzes and flashcards to your PDA from .

The Former Yugoslavia

15°E
20°E
AUSTRIA
HUNGARY
SLOVENIA
Ljubljana
Zagreb
CROATIA
Danube River
ROMANIA
45°N
BOSNIA AND HERZEGOVINA
Belgrade
Sarajevo
SERBIA
Adriatic Sea
BULGARIA
MONTENEGRO
Pristina
KOSOVO
Podgorica
Skopje
MACEDONIA
ITALY
N
W
E
S
ALBANIA
Former boundary of Yugoslavia
40°N
0 200 kilometers
0 200 miles
Lambert Azimuthal Equal-Area projection
GREECE

A Diverse Region

- Europe contains a wide variety of ethnic groups and languages.
- Many ethnic groups, such as Germans, Italians, and Poles, have been in Europe for a long time.
- Other groups are more recent arrivals, a product of Europe's history of colonialism.
- Sometimes this diversity produces conflict. The former country of Yugoslavia broke up as a result of conflict between different ethnic and culture groups.

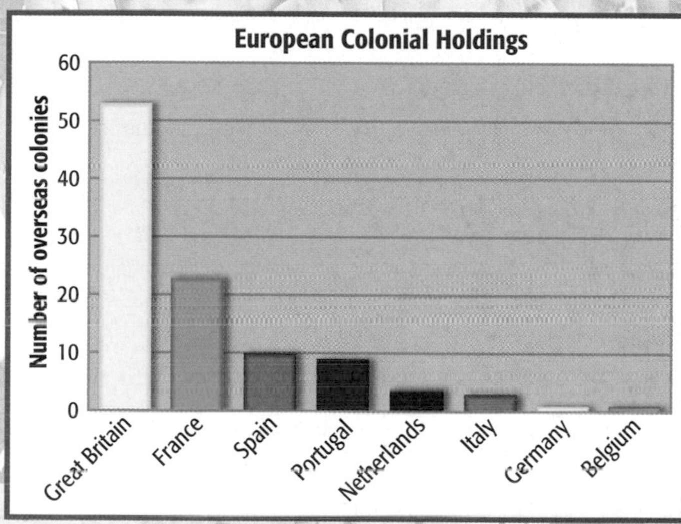

European History

- Greek and Roman civilizations influenced much of European history.
- Trade, colonization, and immigration spread European cultures to other regions.
- Some European countries colonized much of Africa, Latin America, and Asia.
- The Cold War divided Communist-controlled eastern Europe from non-communist western Europe.

A New Europe

- Although the countries of Europe are culturally diverse, they have been building closer ties through the European Union (EU).
- The EU allows goods, services, and people to move easily throughout Europe.
- Countries must meet certain economic and political standards to join the EU.
- Although many Europeans are excited about the newly united Europe, many fear they may lose their cultural heritage.

French posters supporting the EU Constitution

CHAPTER 12

STANDARDIZED TEST PRACTICE

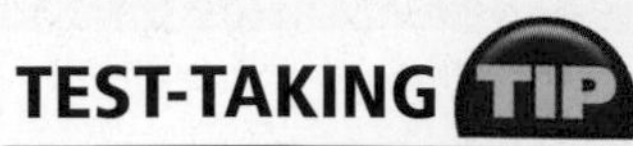

Before you begin answering a question, read all the choices. You can usually eliminate one answer by asking if it has to do with Europe. If it does not, it cannot be the right answer.

Reviewing Vocabulary

Directions: Choose the word or words that best complete the sentence.

1. ________ is an economic system in which business leaders use profits to expand their companies.

A Communism
B Democracy
C Industrial capitalism
D Agrarianism

2. People who work temporarily in a country are called ________.

A refugees
B guest workers
C emigrants
D unemployed

3. Ancient Greece was divided into separate communities called ________.

A boundaries
B metropolitan areas
C suburbs
D city-states

4. ________ is a policy of killing or expelling rival groups in an area.

A Intermarriage
B Acculturation
C Integration
D Ethnic cleansing

Reviewing Main Ideas

Directions: Choose the best answers to complete the sentences or to answer the following questions.

Section 1 *(pp. 294–298)*

5. The ________ transformed manufacturing by replacing human labor with machines.

A Enlightenment
B Reformation
C Industrial Revolution
D Middle Ages

Section 2 *(pp. 299–303)*

6. Which kingdom unified Germany for the first time?

A Prussia
B France
C Russia
D Great Britain

Section 3 *(pp. 306–309)*

7. Who runs the majority of businesses in Greece?

A the government
B international corporations
C foreigners
D families

Section 4 *(pp. 312–316)*

8. The ________ are a traditionally migratory minority people found in many eastern European countries.

A Hungarians
B Basques
C Roma
D Frisians

Critical Thinking

Directions: Choose the best answers to complete the sentences or to answer the following questions.

9. Beginning in the 1600s, Europe underwent a series of dramatic and lasting changes brought about by a variety of factors, including the ________.

A Holocaust

B fall of communism

C Enlightenment

D impressionists

Base your answer to question 10 on the map and on your knowledge of Chapter 12.

10. Study the locations of the Czech Republic, Slovakia, and Hungary. How might their locations affect their role in world trade?

A They are small.

B They are landlocked.

C They are Communist.

D They have many resources.

Document-Based Questions

Directions: Analyze the document and answer the short-answer question that follows the document.

The reunification of Germany forced many adjustments, especially in the former East Germany. Leipzig 15 years later shows this.

> *Investors whisper in baroque cafes, unfolding blueprints for this restless city where Faust sold his soul to the devil and Porsche and BMW factories rise like sleek music boxes on the outskirts.*
>
> *Leipzig is a dream unfinished. The Romantics spun poetry here, the communists came and went, but the town's penchant for commerce has endured, making it one of the few successes to emerge from the former East Germany. The Leipzig spirit is so bold that it challenged the capitals of the world to host the 2012 Olympics.*
>
> *London won, Leipzig shrugged. . . .*
>
> *From its ornate center to its coal-dusted neighborhoods, this is a metropolis of converging identities.*
>
> *One Leipzig is the numbing rectangles of communist architecture and unskilled factory workers who didn't flee west after the Berlin Wall fell. It's a place of 19% unemployment and 50,000 vacant apartments, where streets echo with the scrape of backhoes and bricks tumbling from buildings under demolition.*
>
> *The other Leipzig is a mix of martini-chic and bohemian aloofness, where young women buy the cheapest trinkets in designer stores just to walk out with the bag. Vegetable markets and old world facades sometimes sit incongruously next to the chrome angles of new offices housing Internet start-ups, law firms and venture entrepreneurs.*
>
> —Jeffrey Fleishman, "Leipzig: A City with Many Identities, and None"

11. Describe the differences between the two sections of Leipzig portrayed in this passage.

Extended Response

12. Exploring the BIG Idea
Compare the people of western and eastern Europe. How are they similar? How are they different?

For additional test practice, use Self-Check Quizzes—Chapter 12 on glencoe.com.

Need Extra Help?

If you missed questions. . .	1	2	3	4	5	6	7	8	9	10	11	12
Go to page. . .	297	300	308	314	297	302	309	313	302	283	314	302–303, 316

CONNECTING TO

THE UNITED STATES

A sign outside a shop in Polish Village, Chicago

St. Patrick's Day Parade in New York City

Just the Facts:

- More than 40 million Americans identify themselves as German Americans, the largest self-reported ethnic group in the United States.
- Cincinnati, Ohio, claims to hold the largest authentic Oktoberfest celebration outside of Munich, Germany, the festival's original location.
- City names throughout the United States, such as Detroit and Baton Rouge, reflect the French colonization of certain areas.
- Of the top 15 U.S. trading partners, six are located in Europe. Germany is the top European trading partner of the United States.

Oktoberfest celebration in Cincinnati, Ohio

Mediterranean Cuisine

Food	Country
Gyros	Greece
Spaghetti	Italy
Pizza	Italy
Lasagna	Italy
Baklava	Greece
Feta cheese	Greece
Mozzarella cheese	Italy

Making the Connection

The influence of Europe upon American culture has been dramatic. When Europeans came to America and settled, they brought with them the customs and traditions of their homelands.

Popular Foods Mediterranean cuisine—cuisine from Italy, Greece, and the surrounding Mediterranean area—has become a mainstay of the American diet. Pizza, spaghetti, and gyros have become as popular here as in the countries from which they originate.

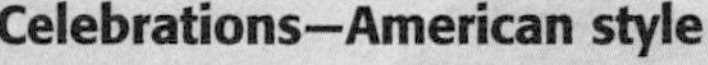

Originally from Italy, pizza is extremely popular in the United States.

Celebrations—American style St. Patrick's Day has, until recently, been a religious feast in Ireland. The idea of parades commemorating St. Patrick's Day originated in New York City with the first being held on March 17, 1762. These parades have continued throughout the United States as a celebration of Irish heritage. The New York City St. Patrick's Day parade is still the world's largest.

Other European festivals have also found their way to the United States. Oktoberfest and Mardi Gras celebrations take place throughout the country.

Ethnic Communities The United States has several ethnic communities. Some of these, including large Italian, Greek, and Polish communities, are located in New York City and Chicago. In Chicago's Polish Village, one can see signs and storefronts displaying the Polish language or even pick up one of several Polish language newspapers.

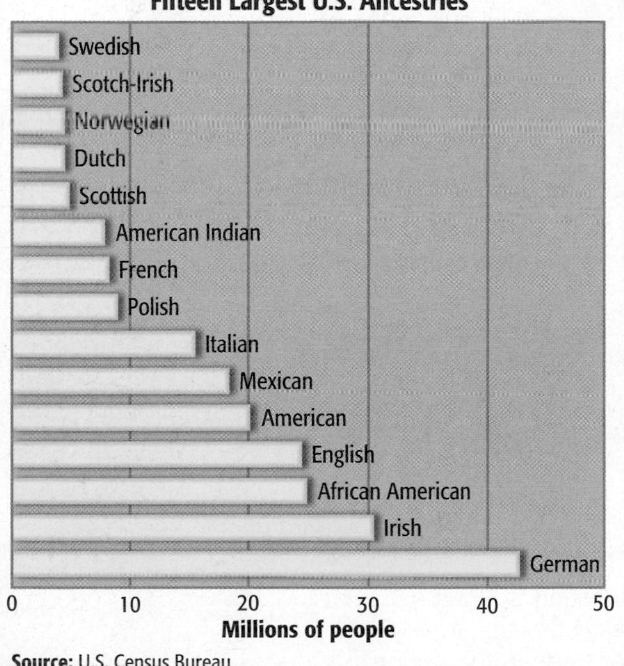

THINKING GEOGRAPHICALLY

1. **Human Systems** Research your state's history. What impact have European influences had on your state?
2. **The World in Spatial Terms** Study the chart of U.S. ancestry. About what percentage of U.S. citizens claim European ancestry?

EUROPE

CHAPTER 13

THE REGION TODAY

Europe

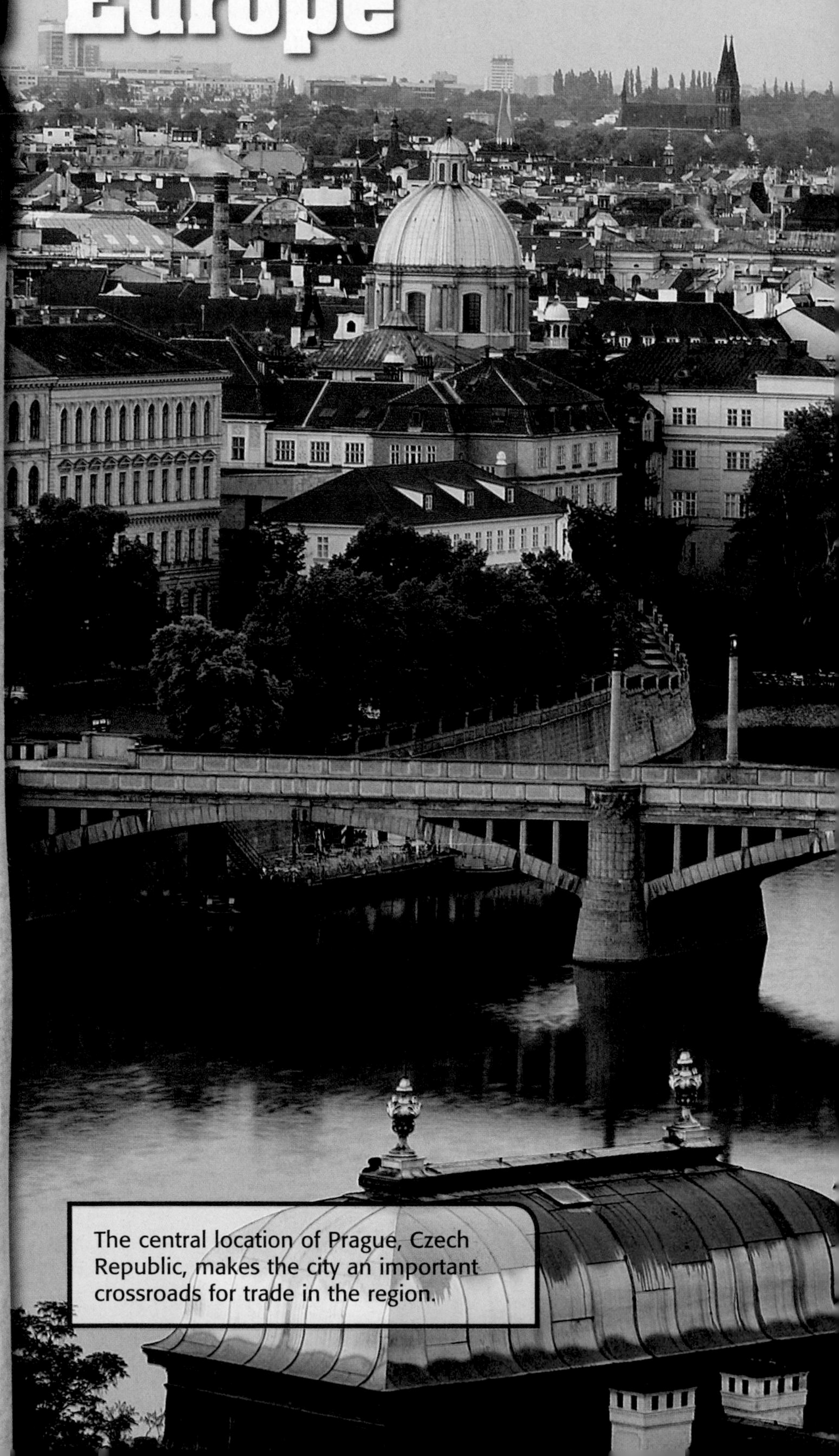

The central location of Prague, Czech Republic, makes the city an important crossroads for trade in the region.

BiG Idea

Economic systems shape relationships in society. The economies of Europe are changing and becoming more unified. At the same time, environmental issues are also related to the economic relationships in the region.

Essential Questions

Section 1: The Economy

How are European countries developing distinct identities while also uniting for greater economic strength?

Section 2: People and Their Environment

How are economics and environmental issues in the region related?

Geography ONLINE

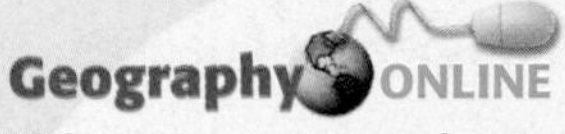

Visit glencoe.com and enter *QuickPass*™ code WGC9952C13 for Chapter 13 resources.

Summarizing Information Make a Layered-Look Book to help you summarize important information about Europe today.

Reading and Writing As you read about the economy and environment of Europe today, summarize information by taking notes about economic activities, transportation and communications networks, trade and interdependence, managing resources, and human impact on the environment.

SECTION 1

The Economy

 section audio  spotlight video

Guide to Reading

Essential Question

How are European countries developing distinct identities while also uniting for greater economic strength?

Content Vocabulary
- heavy industry *(p. 325)*
- light industry *(p. 325)*
- mixed farming *(p. 326)*
- farm cooperative *(p. 326)*
- genetically modified food *(p. 326)*
- organic farming *(p. 326)*
- Maastricht Treaty *(p. 328)*

Academic Vocabulary
- computer *(p. 325)*
- chemical *(p. 326)*
- media *(p. 327)*

Places to Locate
- Prague *(p. 325)*
- Budapest *(p. 325)*
- Paris *(p. 326)*
- Brussels *(p. 326)*
- London *(p. 326)*

Reading Strategy

Organizing As you read, use a web diagram like the one below to list the goals of the European Union.

Throughout Europe, people remain proud of their individual national identities, but they are also beginning to identify with the European Union and the region as a whole. At the same time, people want to maintain the distinct national identities that give Europe its unique and celebrated character.

NATIONAL GEOGRAPHIC VOICES AROUND THE WORLD

" . . . [O]ne of the more delicate issues facing the new Europe is how to create a common foundation without carpeting over the continent's rich tapestry of peoples, languages, cuisines, and cultures. Even Herr Eisenhauer [a European executive], despite . . . his commitment to unity, was worried about that. 'European culture is a bouquet de fleurs, *' [bouquet of flowers] he told me. . . . 'Together they are beautiful. But the rose is still a rose, and the tulip is a tulip. This must be preserved.'"*

—T. R. Reid, "The New Europe," *National Geographic,* January 2002

Celebrating Hungary's EU membership

Economic Activities

MAIN Idea European economies are based on different combinations of manufacturing, service and technology, and agriculture.

GEOGRAPHY AND YOU For what kinds of products and businesses is the U.S. economy known? Read to learn about the economic activities found throughout Europe.

Economic activities in Europe are related to the natural resources, people, and culture of the area. For example, some countries have been longtime leaders in manufacturing, while others are centers for international banking, insurance, and finance.

Industry

The Industrial Revolution made Europe the birthplace of modern industry. Today the region produces everything from **computers** to transportation equipment. However, service industries provide a large percentage of GDP.

Manufacturing The development of industry is often linked to the availability of raw materials. In the 1800s Europe's large deposits of coal and iron ore sparked the growth of **heavy industry**—the manufacture of machinery and industrial equipment. Today Europe's leading industrial centers include the Ruhr and the Middle Rhine districts in Germany, the Lorraine-Saar district in France, the Po basin in Italy, and the Upper Silesia-Moravia district in Poland and the Czech Republic. Countries lacking industrial raw materials specialize in **light industry**, such as high-end electronics and specialty tools.

Service and Technology Service industries employ a large percentage of the workforce in most European countries—in fact, more than 70 percent of workers in western Europe work in service industries. International banking and insurance rank among Europe's top service industries, with Switzerland and the United Kingdom the leaders in these fields. High-technology industries are a growing sector of western Europe's economy. Ireland, for example, has become a leading manufacturer of computer products and software.

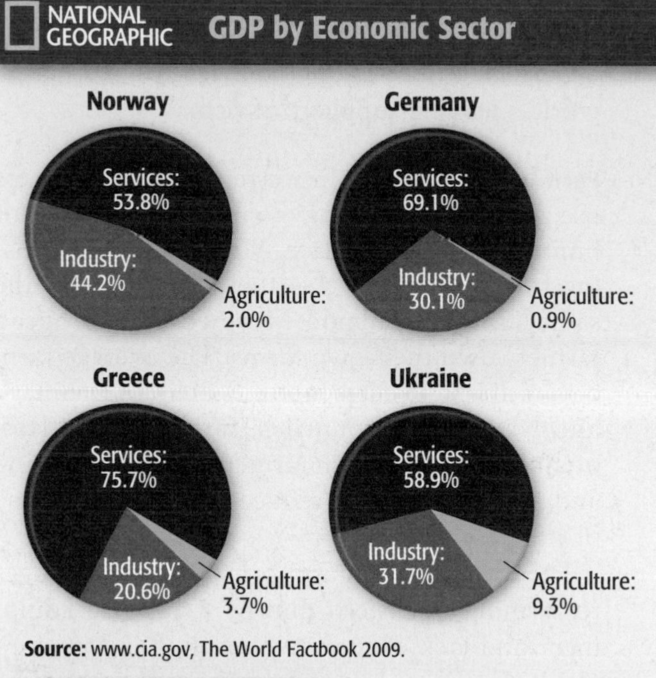

Source: www.cia.gov, The World Factbook 2009.

GRAPH STUDY

1. **Place** In which country do services account for the largest percentage of GDP? The smallest percentage?
2. **Place** How is Ukraine's GDP by sector different from that of Norway, Germany, and Greece? Why do you think this is so?

Graphs In MOtion Use **StudentWorks™ Plus** or glencoe.com.

Tourism is another large service industry in Europe, especially in Spain, Italy, and Greece. In recent years, changes in eastern Europe's political and economic landscape have also opened up tourism in urban areas. **Prague,** Czech Republic, and **Budapest,** Hungary, are two internationally popular eastern European cities.

Agriculture

Although largely industialized, Europe also has fertile farmland. As a result, many Europeans continue to earn a living farming. Yet the percentage of farmers in each country varies widely. For example, about 58 percent of Albania's workers are farmers, but in the United Kingdom, fewer than 2 percent engage in agriculture.

Agricultural crops vary from area to area. Olives, citrus fruits, dates, and grapes grow in warm Mediterranean areas. Farther north, in the cooler plains region, farmers grow wheat, rye, and other grains as well as livestock.

Northern countries, such as Denmark and the Netherlands, are major producers of dairy products. The Scandinavian countries are among the world's leading suppliers of fish.

Farming Techniques In western Europe, farmers use advanced technology to make the best use of limited agricultural space. **Mixed farming**—raising several kinds of crops and livestock on the same farm—is common. Most western European farmers own their own farms. The average farm covers about 46 acres (about 17 ha). In Denmark and some other countries, **farm cooperatives,** organizations in which farmers share in growing and selling products, reduce costs and increase profits.

The fall of communism brought many changes to farming in eastern Europe. Outdated equipment and lack of incentive resulted in low crop yields. Since the shift to democracy, private ownership of land and food production has risen. Yields and profits have increased through the use of modern equipment and fertilizers.

Agricultural Issues Throughout Europe new farming methods have not escaped criticism. Many Europeans, for example, oppose **genetically modified foods,** foods with genes altered to make them grow bigger or faster or be more resistant to pests. Opponents claim that little is known yet about the safety of these foods. Many consumers also avoid foods grown in fields treated with **chemicals** to control insects or weeds. So some farmers rely on **organic farming,** using natural substances instead of chemicals to increase crop yields.

Despite much success, agriculture in western Europe is still vulnerable to problems. For example, in 2001 an outbreak of foot-and-mouth disease in the United Kingdom required the killing of thousands of animals, severely crippling the country's livestock industry. Agricultural subsidies are also an issue. Billions of dollars are paid to farmers each year to supplement their income, support the agricultural industry, and protect prices. Many, however, argue that such subsidies create overproduction of crops and distort trade.

READING **Check** **Regions** What kinds of problems affect agriculture in western Europe?

Transportation and Communications

MAIN Idea The economies of Europe are supported by transportation and communications systems that are state-of-the-art in some areas and improving in other areas.

GEOGRAPHY AND YOU On what transportation and communications systems do people in your community rely? Read to learn about the role these systems play throughout Europe.

Europe's network of highways, railroads, waterways, and airline routes is among the best in the world. Modern communications systems also link most parts of Europe to one another and to the rest of the world.

Railways and Highways

Rail lines connect Europe's major cities and airports and link natural resources to major industrial centers. France pioneered the use of high-speed trains with its introduction in 1981 of *trains á grand vitesse* (TGVs), which means "very fast trains." TGVs cause less damage to the environment than most other forms of transportation. High-speed rail lines now also operate in Germany, Italy, and Spain. Bridges and tunnels carry traffic over or through barriers posed by water, mountains, or valleys. For example, in 2000 Denmark and Sweden opened a rail and road bridge linking Sweden to western Europe for the first time since the last Ice Age. A high-speed rail triangle links **Paris, Brussels,** and **London,** passing beneath the English Channel through the Chunnel, or Channel Tunnel.

A well-developed highway system also links Europe's major cities. Germany's super highways, called *autobahnen,* are among Europe's best roads. Europe has the highest number of automobile owners in the world except for the United States.

Comparing Transportation

Region	Land Area	Railroad	Paved Roads
Europe	3,998,086 sq. mi. 10,355,000 sq. km	175,546 mi. 282,514 km	3,611,279 mi. 5,811,790 km
United States	3,537,453 sq. mi. 9,161,966 sq. km	140,695 mi. 226,427 km	2,615,869 mi. 4,209,835 km

Seaports and Waterways

With its long coastline, Europe has a seafaring tradition. It handles a large amount of the world's international shipping. Rotterdam, the Netherlands, is the world's largest port in surface area, amount of freight handled, and numbers of ships that it can dock at one time.

Europe's many navigable rivers and canals allow for transporting goods at less cost than many other regions. The Rhine River and its tributaries carry more freight than any other river system in Europe. The economies of cities along the banks of the Danube River are dependent on it for trade. The Main-Danube Canal in Germany links inland ports between the North Sea and the Black Sea.

Communications

The International Telecommunications Satellite Organization uses a series of satellites (INTELSATs) to broadcast and receive television programs in Europe. Before the fall of communism, the Eurovision network linked most of western Europe, and the Intervision network operated in eastern Europe. The two networks merged in 1993 and are now both operated by the European Broadcasting Union.

Telephone service and print **media** vary throughout Europe. Western European telephone systems include cable and microwave radio relay, fiber optics, and satellite systems. Such high-quality service is not as available in eastern Europe. Most western Europeans use cell phones, electronic mail, and the Internet. Cell phone use is increasing in eastern Europe. Print media continue to shape public opinion.

READING Check **Movement** What role do rail lines play in supporting the economy of Europe?

Trade and Interdependence

MAIN Idea Trade within Europe and between Europe and the rest of the world is changing as a result of the European Union and changes in the political and economic landscape of eastern Europe.

GEOGRAPHY AND YOU What might be the challenges of many countries working together toward the same economic goals? Read to learn about the benefits and challenges of the European Union.

Europe's economies, like its peoples, are diverse and changing. The European Union (EU), which unites much of Europe into one trading community, enjoys a greater volume of trade than any single country in the world.

MAP STUDY

1. **Regions** What countries are the original members of the European Union?
2. **Regions** Why do you think Belarus and Ukraine are not members of the European Union?

Maps In MOtion Use **StudentWorks™ Plus** or glencoe.com.

The European Union

The movement for European unity arose from the ashes of World War II, as western European countries struggled to rebuild their ruined economies. In 1950 France proposed closer links among Europe's coal and steel industries, a move seen as the first step toward a united Europe. This was such a success that in 1957 Belgium, West Germany, Luxembourg, France, Italy, and the Netherlands created the European Economic Community to further integrate their economies.

Over the years more steps were taken toward that goal, but not until the 1990s did most Europeans agree that such a goal could ever be reached. In 1992 representatives from various European governments met in Maastricht, the Netherlands, to sign the **Maastricht Treaty,** which set up the European Union. This new body aimed to make Europe's economies competitive with those of the rest of the world by getting rid of restrictions on the movement of goods, services, and people across its members' borders. It also paved the way for a single European currency, a central bank, and a common foreign policy.

NATIONAL GEOGRAPHIC *Former prime minister of Portugal José Manuel Barroso is serving his second term as the president of the European Commission.*

Regions Why was the European Union created?

Since the EU was formed, member countries have worked to boost trade and to make their economies more efficient and productive. This process includes agreeing on a variety of issues directly and indirectly related to the economy.

In 2004 members attempted to establish an EU constitution. The original EU community needed to be restructured to accommodate the expanded membership. The acceptance of this new constitution came to a standstill when France and the Netherlands rejected it for various reasons in 2005.

Led by Germany, EU member countries began working on a treaty to replace the failed constitution. The Lisbon Treaty, signed in Lisbon, Portugal, in December 2007, retained parts of the draft constitution. Among the issues this treaty was intended to address were the establishment of the EU presidency and the consolidation of foreign policy representation for the EU. The treaty also tried to improve decision making within the European Union by requiring majority, rather than unanimous, approval of many decisions.

After a slow start, the treaty found success. Irish voters initially rejected the treaty, but approved it in a second referendum in October 2009. The treaty has been ratified by all 27 EU member countries, with the Czech Republic the last to do so in November 2009.

A New Eastern Europe

Since the fall of communism in 1989, eastern European countries have been moving from command economies to market economies. The graph on page 329 shows how GDP has changed in some countries. To compete in global markets, industries have had to overcome the obstacles of outdated equipment and inefficient production methods. Many laid-off workers were retrained, as industries tried to acquire new technology and adopt measures to reduce pollution. Eastern European governments are seeking to attract foreign investments and financial aid.

Change, however, is costly and difficult. Workers lost part of their social "safety net"—the free health care, lifetime jobs, and other social benefits—provided by the communist system. With reduced benefits, death rates have risen in some places, and life expectancy has declined. Despite these difficulties, however, people in eastern Europe are slowly adjusting to a new way of life.

In addition to benefits, membership in the European Union poses challenges for eastern European countries. To meet strict EU standards on trade, banking, business law, environmental issues, and human rights, countries seeking membership must examine and adjust their legal and financial systems and ways of doing business. As members, farmers and businesspeople must compete with existing EU members, who are in some cases years ahead of them in terms of productivity. In Poland, for example, farmers represent 17 percent of the workforce but generate less than 5 percent of the country's wealth.

READING Check **Regions** What economic change occurred in eastern Europe after the fall of communism in 1989?

Student Web Activity Visit glencoe.com, select the *World Geography and Cultures* Web site, and click on Student Web Activities–Chapter 13 for an activity about the European Union.

NATIONAL GEOGRAPHIC **Economies in Eastern Europe**

Note: Figures are given in U.S. dollars.
Source: International Monetary Fund.

GRAPH STUDY

1. **Place** Which country experienced the most dramatic increase in GDP after the fall of communism?
2. **Regions** What generalization can you make about the effect the fall of communism had on the economies in eastern Europe?

Graphs In MOtion Use **StudentWorks™ Plus** or glencoe.com.

SECTION 1 REVIEW

Vocabulary

1. Explain the significance of: heavy industry, light industry, mixed farming, farm cooperative, genetically modified food, organic farming, Maastricht Treaty.

Main Ideas

2. Describe examples of how the economies of European countries are built on different combinations of manufacturing, service and technology, and agriculture.
3. How is trade within Europe and between Europe and the rest of the world changing due to the influence of the European Union? How is it changing as a result of the political and economic landscape of eastern Europe?
4. Complete a chart like the one below to show how the economies of Europe are supported by transportation and communications systems.

Impact of Transportation and Communications	
Industry	
Service and technology	
Agriculture	

Critical Thinking

5. **Answering the Essential Question** In what ways could the merging of economies through the European Union give the entire region of Europe economic strength?
6. **Analyzing Information** What different challenges do eastern and western Europeans face as they move toward a more unified Europe?
7. **Analyzing Visuals** Study the table on page 326. How does Europe compare to the United States in terms of miles of railroad? Miles of paved roads?

Writing About Geography

8. **Narrative Writing** Imagine that you are a farmer in eastern Europe. Write a brief essay describing how your work has changed since the fall of communism. Discuss the pros and cons of command economies and market economies.

Study Central™ To review this section, go to glencoe.com and click on Study Central.

SECTION 2

 section audio  spotlight video

Guide to Reading

Essential Question

How are economics and environmental issues in the region related?

Content Vocabulary

- dry farming *(p. 331)*
- acid rain *(p. 332)*
- meltwater *(p. 332)*
- acid deposition *(p. 332)*
- environmentalist *(p. 333)*
- global warming *(p. 333)*

Academic Vocabulary

- cycle *(p. 331)*
- vehicle *(p. 333)*
- disposal *(p. 333)*

Places to Locate

- Strait of Gibraltar *(p. 333)*
- Białowieza Forest *(p. 334)*

Reading Strategy

Organizing Complete a graphic organizer like the one below by listing some of the reasons for eastern Europe's pollution problems.

People and Their Environment

For today's Europeans, the lasting effects of certain human activities on their environment are becoming all too clear. The Danube River, a vital waterway that passes through nine European countries, needs protection. Dams, dikes, and other types of development have damaged water quality and posed a threat to wetlands, fish, and bird populations.

NATIONAL GEOGRAPHIC Voices Around the World

"In the town of Tutrakan, Bulgaria, on a hill overlooking the Danube and Romania across it to the north, 65-year-old Dimo Kovachev sat idle on a tree stump. Below, on the waterfront, sat the town's crane, which he operated before retirement five years ago. . . . 'There's just no work here,' he said . . . Fishing has declined because of environmental degradation and habitat loss. 'Thirty years ago I caught four kinds of sturgeon in the Danube.' That was before wetlands on islands and along the riverbanks were diked to create cropland. 'In spring, when the water was up, the marshlands flooded,' providing a place for fish to spawn. 'If the marshes are restored, the fish will come back.'"

—Cliff Tarpy, "The Danube: Europe's River of Harmony and Discord," *National Geographic*, March 2002

A sturgeon fisher on the Danube River

Managing Resources

MAIN Idea Europeans realize their connection to the environment and the importance of managing resources wisely.

GEOGRAPHY AND YOU What measures do people in your community take to manage natural resources? Read to learn more about how natural resources are managed in Europe.

People in Europe face challenges posed by the physical environment. The frequent occurrence of earthquakes in countries such as Italy, Greece, and Macedonia indicates that tectonic changes continue to take place today. People in southern Europe also have to cope with low rainfall. For example, Spain's Meseta is so arid that streams dry up and drought is common. This arid climate makes dry farming necessary. **Dry farming** is a way of farming in dry areas that produces crops without any irrigation and relies on farming methods that conserve soil moisture.

Flooding

In recent years heavy rains have drenched much of Europe, causing widespread flooding and mud slides. This extreme weather has led to loss of life, property damage, and disruption of transportation networks. Some scientists claim that the natural climate **cycle** accounts for the rains. Others believe that global warming is responsible.

In northwestern Europe, violent Atlantic and North Sea storms strike countries that border the sea. In 1953 a severe Atlantic storm, combined with the North Sea's heavy spring tide, flooded the Netherlands, killing about 1,800 people. For nearly 30 years after this tragic event, engineers carried out the Delta Plan, a project aimed to prevent such severe flooding. Under the plan, a system of dams and dikes was built to seal off and protect the Netherlands' southwestern coast.

Soil Erosion

Humans' repeated misuse of the land has accelerated soil erosion in Europe. Activities that lead to rapid erosion include over-farming, bad farming practices, removing too much vegetation, and overgrazing of livestock.

NATIONAL GEOGRAPHIC *Heavy rains across eastern Europe in the summer of 2009 resulted in massive flooding.*

Human-Environment Interaction What do scientists believe to be the cause of heavy rains and flooding in Europe?

Soil erosion is a serious problem in the Mediterranean basin, and it affects the highly populated, sandy coastal areas of the Mediterranean countries. Since the 1980s, scientists have conducted studies to learn the extent of the problem and develop specific steps countries can take to reduce erosion. Forest destruction over time in the Mediterranean is one reason for the area's severe problems with erosion. As a result, reforestation is being promoted in some areas.

Deforestation

It is believed that about 80 percent of Europe was once covered by forest, and two-thirds of this forestland has been removed over time. Particularly in the Mediterranean region and in industrial Western Europe, people removed many trees in order to create cities and farms. In Europe today, the healthiest commercial forests are found in northern Europe.

To combat the effects of forest loss, many countries have taken steps toward reforestation, or the replanting of trees. Others have taken steps to manage the cutting of trees responsibly. Sweden, for example, has followed since the 1800s a strict system of cutting and replenishing trees. This is necessary because the average time to replace a full-grown spruce and pine is about 70 years in the south and about 140 years in the northern part of the country.

READING Check **Human-Environment Interaction** What activities have led to soil erosion in Europe?

Human Impact

MAIN Idea Population growth and industrialization have impacted Europe's environment in negative ways, and Europeans must find ways to reverse the damage.

GEOGRAPHY AND YOU What examples of worldwide pollution have you observed or read about? Read to learn how pollution and the other effects of human activities have impacted Europe.

Europe's high concentration of industry and population has had a devastating impact. For example, in the "black triangle"—a heavily industrialized area in Poland, eastern Germany, and the Czech Republic—soot covers the ground, and the air smells of sulfur from smokestacks.

Before 1989 eastern European countries had practically no laws to control pollution. With the Communist emphasis on rapid industrial growth—not environmental safety—pollution increased until it affected public health. Although efforts are now under way to clean up the environment, the "black triangle" still bears the scars of poorly considered development from the past. Countries in western Europe have also experienced environmental damage from the dumping of industrial wastes into the air and water. The European Union (EU) now requires environmental protection and cleanup from its members.

Acid Rain

In the 1970s and 1980s, European industries built tall smokestacks to carry pollution away from industrial sites. This worked locally, but pollution drifted across national borders. This pollution, containing acid-producing chemicals, combined with moisture in the air and fell as **acid rain.** Polluted clouds, drifting from the industrial belt of Europe, withered forests in other areas.

Many western European countries have made the switch from coal to natural gas, thus reducing the effects of acid rain. However, many eastern European countries continue to rely heavily on coal. As a result, the effects of acid rain are especially severe in eastern Europe.

Acid rain is not limited to forests; it also falls on lakes and rivers. In winter, snow carries the industrial pollution to the ground. In spring, **meltwater**—the result of melting snow and ice—carries the acid into lakes and rivers. As acid concentrations build, fish and other aquatic life die. Many lakes in Scandinavia have declining fish populations or even no fish at all. Some rivers in the Czech Republic and Slovakia cannot support aquatic life.

Automobile exhaust also adds acid-forming compounds to the atmosphere. **Acid deposition,** wet or dry acid pollution that falls to the ground, harms not only Europe's natural environment but also its historic buildings, statues, bridges, and stained glass windows.

1. **Regions** Which European countries are the most affected by acid rain?
2. **Human-Environment Interaction** Compare this map to the map of forests affected by defoliation on page 273 of the Regional Atlas. What relationship do you see between acid rain and defoliation?

The EU has set strict emissions regulations for industries and **vehicles.** This often involves equipping smokestacks and vehicle exhaust systems with devices that remove sulfur and nitrogen compounds. Many people believe that fossil fuels should be replaced with alternative energy sources. However, others believe that solar power and other energy sources are not yet realistic replacements for fossil fuels.

Air and Water Pollution

Air pollution is a problem throughout Europe. Traffic exhaust and industrial fumes cause eye irritations, asthma, and respiratory infections in people who live in industrial areas of western Europe. In eastern Europe, factories built in the Communist era belch soot, sulfur, and carbon dioxide into the air. Some former Communist countries are closing polluting factories. Yet they are also putting more cars on the road, increasing air pollution from traffic.

Water pollution is another issue facing Europe, particularly in the Mediterranean region. Countries bordering the Mediterranean Sea use the sea for waste **disposal,** dumping sewage, garbage, and industrial waste. In the past, bacteria in the Mediterranean Sea broke down most of the waste the sea received. In recent times, however, growing populations and tourism along the coast have increased the environmental problems. Small tides and weak currents tend to keep pollution where it is discharged. The Mediterranean Sea, open to the Atlantic only through the narrow **Strait of Gibraltar,** takes almost a century to renew itself completely.

Pollution contaminates marine and animal life and creates health hazards for people. The Mediterranean is overfished and cannot provide its former bounty. Native species of seaweed and shellfish compete with foreign species carried into the Mediterranean by ships.

Pollution also affects Europe's rivers and lakes. The Danube River, for example, is seriously affected by agricultural runoff. When fertilizers enter the river, they encourage algae growth. Algae, in turn, rob the river of so much oxygen that fish cannot survive. Another source of pollution is raw sewage, which is dumped into rivers in various places. Industries in western Europe deposit wastes into the Meuse and Rhine Rivers.

NATIONAL GEOGRAPHIC *In addition to industrial and agricultural pollution, trash also affects water and wildlife in Europe.*

Human-Environment Interaction What impact has pollution had on Europe's rivers and lakes?

Global Warming

The problems of air quality in Europe, like those in other industrialized regions, may have global consequences. Many **environmentalists**—people concerned with the quality of the environment—are studying the effects of increased carbon dioxide in the Earth's atmosphere. The burning of fossil fuels has raised the amounts of carbon dioxide in the atmosphere, increasing global temperatures. Scientists report an increase in the average global surface temperature of 1.1°F (0.6°C) during the twentieth century. Some forecasts show an increase of 3.2–7.2°F (1.8–4.0°C) by the end of the twenty-first century, a trend called **global warming.** A warmer climate will melt polar ice caps and mountain glaciers, causing ocean levels to rise and submerge coastal areas.

Facing the threat of global warming requires international cooperation. Many scientists, however, cannot agree on the exact causes of global warming, so the international community has done little to combat the problem. The EU, however, has consistently been a supporter of the Kyoto Protocol—an amendment to the international treaty on climate change designed to reduce the amount of greenhouse gases emitted by specific countries.

READING **Check** **Human-Environment Interaction** How has industrialization impacted Europe's environment?

Future Challenges

MAIN Idea European countries are uniting to develop ways of protecting and restoring the region's environment.

GEOGRAPHY AND YOU How might a group of many countries develop one plan to protect the environment? Read to learn what measures Europe is taking to address environmental concerns.

Few areas in Europe remain unchanged by the clearing of forests, the drainage of seas, or the building of canals. Although much of Europe has been altered by human activity, Europeans want to preserve what little wilderness area is left. One of the largest areas still in its natural state is the **Białowieza** (bee•ah•waw•vee•YAY•zhah) **Forest** in Belarus and Poland. It is home to animals such as the wolf, lynx, and European bison, all of which are now rarely seen elsewhere in Europe.

In recent decades Europeans have made more concerted efforts to clean up the environment. EU member countries can face legal action if they do not respect environmental protection laws. Individual countries are also addressing the consequences of pollution. For example, cities in western Europe now protect buildings and statues with acid-resistant coatings.

Pollution that crosses national borders presents a more complex situation. For example, pollution in the Danube River threatens wildlife in its outlet, the Black Sea. People recognize that improving water quality is a necessity, but directing and financing the cleanup is difficult when the process involves many countries.

The EU and others continue to develop ways to protect the environment. Many power plants now burn natural gas instead of lignite coal. By 2020 all EU member countries must lower emissions to 30 percent below 1990 levels to reduce greenhouse gases. Some countries are developing alternative fuels. For example, in 2005 Sweden introduced the first biogas-powered passenger train. Biogas, produced by decomposing organic material, is much less damaging to the atmosphere than fossil fuels.

To be admitted to the EU, eastern European countries are expected to meet EU environmental standards. Since cleanup will cost billions of dollars, eastern European countries are now seeking financial aid from EU countries in western Europe. Western Europe and U.S. companies are providing technology and investment to help modernize eastern Europe's industries.

READING Check **Human-Environment Interaction** How is western Europe encouraging eastern Europe to meet environmental standards?

SECTION 2 REVIEW

Vocabulary

1. Explain the significance of: dry farming, acid rain, meltwater, acid deposition, environmentalist, global warming.

Main Ideas

2. Describe three examples of how Europeans are managing their resources wisely.
3. How are European countries developing ways of protecting the region's environment? What are they doing to clean up the environment?
4. In a table like the one below, identify the ways in which growing populations and industrialization have impacted Europe's environment. Then describe steps to counteract pollution's effects.

Activities	Effects of Pollution	Steps Taken
Growing populations		
Industrialization		

Critical Thinking

5. **Answering the Essential Question** Why does eastern Europe have higher levels of pollution than western Europe?
6. **Drawing Conclusions** Why is cooperation among today's European nations necessary in order for cleanup and preservation of the environment to take place?
7. **Analyzing Maps** Draw a map of Europe from memory showing the western European countries most affected by acid rain.

Writing About Geography

8. **Narrative Writing** Imagine that you live in a polluted area of Europe. Write a letter to the editor of a newspaper there, suggesting steps to halt environmental damage.

Study Central™ To review this section, go to glencoe.com and click on Study Central.

STUDY TO GO — Study anywhere, anytime by downloading quizzes and flashcards to your PDA from .

The Region Today

Tuesday | Section B

The Birthplace of Modern Industry

- Europe's economy varies by country, particularly between eastern and western Europe.
- The Industrial Revolution made Europe the birthplace of modern industry.
- About 70 percent of western Europe's workforce is in service industries. Banking, insurance, and tourism rank among the region's top service industries.
- Although Europe is industrialized, many people still earn a living from farming.

Overcoming the Environment

Challenges

- Southern Europe has low rainfall, making farming difficult.
- Widespread flooding in Europe has destroyed property and transportation routes.
- Misuse of land over time has led to severe soil erosion in some places.
- Deforestation, especially in the Mediterranean region, has resulted in the loss of nearly two-thirds of Europe's forests.

Erosion in Europe

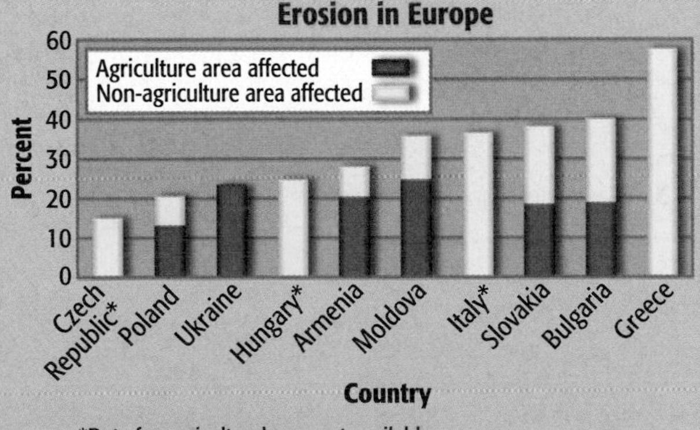

*Data for agricultural area not available.
Source: http://epaedia.eea.europa.eu, European Environment Agency.

Solutions

- Dry farming has made farming possible in arid locations.
- The Dutch have built a series of dikes and dams to protect the Netherlands from flooding.
- Studies have been done to identify and find solutions to combat soil erosion.
- Many European countries have taken steps toward reforestation to replenish the trees lost to deforestation.

EUROPE TODAY

CHAPTER 13

STANDARDIZED TEST PRACTICE

When you have finished, check to be sure you have answered all the questions.

Reviewing Vocabulary

Directions: Choose the word or words that best complete the sentence.

1. The manufacture of machinery and industrial equipment is ________.

A heavy industry
B light industry
C services
D economics

2. Organizations that share the cost of growing and selling crops are called ________.

A communes
B villages
C plantations
D farm cooperatives

3. When pollutants in the atmosphere combine with moisture, the result is ________.

A dumping
B acid rain
C soil erosion
D meltwater

4. When the Earth's average temperature rises, it is called ________.

A global cooling
B the greenhouse effect
C the acid rain effect
D global warming

Reviewing Main Ideas

Directions: Choose the best answers to complete the sentences or to answer the following questions.

Section 1 *(pp. 324–329)*

5. Today most Europeans work in ________.

A farming
B mining
C services
D manufacturing

6. The first step toward a more united Europe involved the ________ industry.

A textile
B agricultural
C coal and steel
D automobile

Section 2 *(pp. 330–334)*

7. Today the healthiest commercial forests are found in ________ Europe.

A northern
B southern
C eastern
D western

8. Waste disposal has caused severe pollution problems in the ________.

A North Sea
B Baltic Sea
C Black Sea
D Mediterranean Sea

GO ON

Critical Thinking

Directions: Choose the best answers to complete the sentences or to answer the following questions.

9. When eastern European countries close polluting factories, what is a negative result?

A There are more cars on the road.

B Workers go back to being farmers.

C People become unemployed and cannot get services they need.

D School enrollments go up.

Base your answer to question 10 on the map and on your knowledge of Chapter 13.

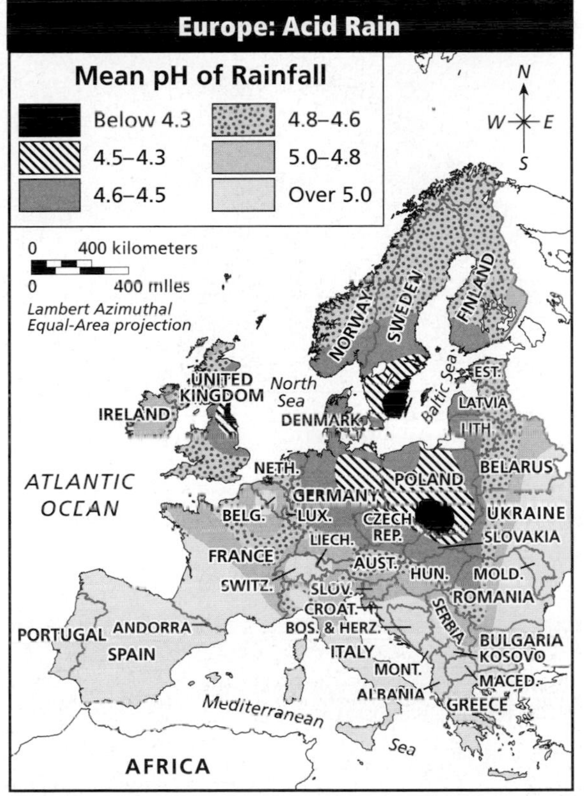

10. In which country is acid rain most concentrated?

A Czech Republic

B Germany

C Poland

D Slovakia

Document-Based Questions

Directions: Analyze the document and answer the short-answer question that follows the document.

The European Union proposed the Constitution Treaty, or Constitution of Europe, in 2004. It was not ratified, and so was replaced by the Lisbon Treaty. The excerpt below outlines some of the EU's powers as defined in the Treaty of Lisbon.

> *1. When the Treaties confer on the Union exclusive competence in a specific area, only the Union may legislate and adopt legally binding acts, the Member States being able to do so themselves only if so empowered by the Union or for the implementation of Union acts. . . .*
>
> *3. The Member States shall coordinate their economic and employment policies within arrangements as determined by this Treaty, which the Union shall have competence to provide.*
>
> *4. The Union shall have competence, in accordance with the provisions of the Treaty on European Union, to define and implement a common foreign and security policy, including the progressive framing of a common defence policy.*
>
> *5. In certain areas and under the conditions laid down in the Treaties, the Union shall have competence to carry out actions to support, coordinate or supplement the actions of the Member States, without thereby superseding their competence in these areas.*

11. How will foreign policy be addressed in the European Union?

Extended Response

12. Exploring the BIG Idea

Describe the current economy of eastern Europe. What effects has communism had upon the current economic situation?

For additional test practice, use Self-Check Quizzes—Chapter 13 on glencoe.com.

Need Extra Help?

If you missed questions. . .	1	2	3	4	5	6	7	8	9	10	11	12
Go to page. . .	325	326	332	333	325	328	331	333	332	337	337	328

NATIONAL GEOGRAPHIC **Case STUDY**

THE EUROPEAN UNION: What is next for this international organization that acts more and more like a single country?

At the end of World War II, Europe's cities lay in ruins, and its economy was destroyed. The rebuilding began with iron and steel and the raw materials that went into them: coal and iron ore. Germany had some of the best coal, while France had excellent deposits of iron ore. The European Coal and Steel Community, consisting of France, Germany, Belgium, the Netherlands, Luxembourg, and Italy, agreed to free trade in coal, iron ore, and steel. From this small start came today's European Union (EU), including almost all the countries on the continent. Today, as new members wait to join, the EU draws closer together, acting more and more like a single country.

Understanding the Situation

The creation and growth of the European Union is viewed from varying perspectives.

A Moral Issue Europeans needed a means to work together and end the series of wars that had destroyed their countries. As the EU developed, it established programs to help farmers and to end tariffs and other trade barriers for all products. These programs meant that people in richer countries were helping those in poorer ones, and some Europeans objected to these programs.

An Economic Dilemma Countries, farmers, and companies that had specialized in a certain crop or product because they were protected by tariffs and other regulations suddenly found themselves competing with lower-paid workers. Because people can move around in the EU and look for jobs in other countries, workers from low-wage countries began to move to high-wage countries. Workers in the high-wage countries felt threatened.

A Political Issue The EU has brought greater political unification among its members. For example, people today can get EU passports instead of passports from individual member countries. Some of the members have adopted a single currency called the euro, and there is a European Parliament that holds sessions in Brussels, Belgium, and Strasbourg, France. The European Parliament makes rules that the member countries must obey. The EU also has courts to enforce its laws.

Above right: The European Parliament Building, Strasbourg, France
Above: Euro coins and paper money

Average Monthly Minimum Wages in the European Union

(coin symbol) = 100 euros

EU Member Country	Average Monthly Minimum Wages (in euros), 2009
Bulgaria	122.7
Lithuania	231.7
Poland	281.2
Czech Republic	306.3
Greece	680.6
Spain	728.0
United Kingdom	1010.3
Luxembourg	1641.7

Source: www.epp.eurostat.ec.europa.eu

Challenges for the Future

As the EU has grown, its members have become more diverse. Some members that joined in 2004 and 2007 had been under Communist governments. Their transition to market economies and democracies caused economic problems. Unemployed workers wanted to move west to find jobs, and farmers had to compete with more technologically advanced farms.

Agricultural Policy Originally, the European Economic Community, the predecessor to the European Union, adopted policies to help farmers produce food as cheaply as possible so that citizens would be able to afford a healthy diet. Now, the policies aim to support farmers in producing high-quality foods using methods that protect the environment.

Workers In general, workers in EU member countries can move to any EU country to look for work. They must receive the same rights as other workers in that country. However, when ten new countries joined in 2004, some feared that many of their people would flood old member countries looking for work. So the new members agreed that for at least two years, their citizens would have to obtain work permits to move to old member states.

Tighter Organization The EU has become more involved in the lives of the people of its member countries. Most of its members use the euro, and its representatives take a more active role in world politics. Many EU leaders felt that it needed an updated document for its governance. A proposed constitution was rejected in 2005. By 2009, the Lisbon Treaty had been ratified by all 27 EU member countries.

European Union Membership

Year	Countries Joining
1951	European Coal and Steel Community: Germany, France, Italy, Belgium, Luxembourg, Netherlands
1973	Denmark, Ireland, United Kingdom
1981	Greece
1986	Spain, Portugal
1995	Austria, Finland, Sweden
2004	Cyprus, Czech Republic, Estonia, Hungary, Latvia, Lithuania, Malta, Poland, Slovakia, Slovenia
2007	Bulgaria, Romania
Negotiations beginning	Croatia, Macedonia, Turkey
Potential candidates	Albania, Bosnia and Herzegovina, Iceland, Kosovo, Montenegro, Serbia

UNIT 5 Russia

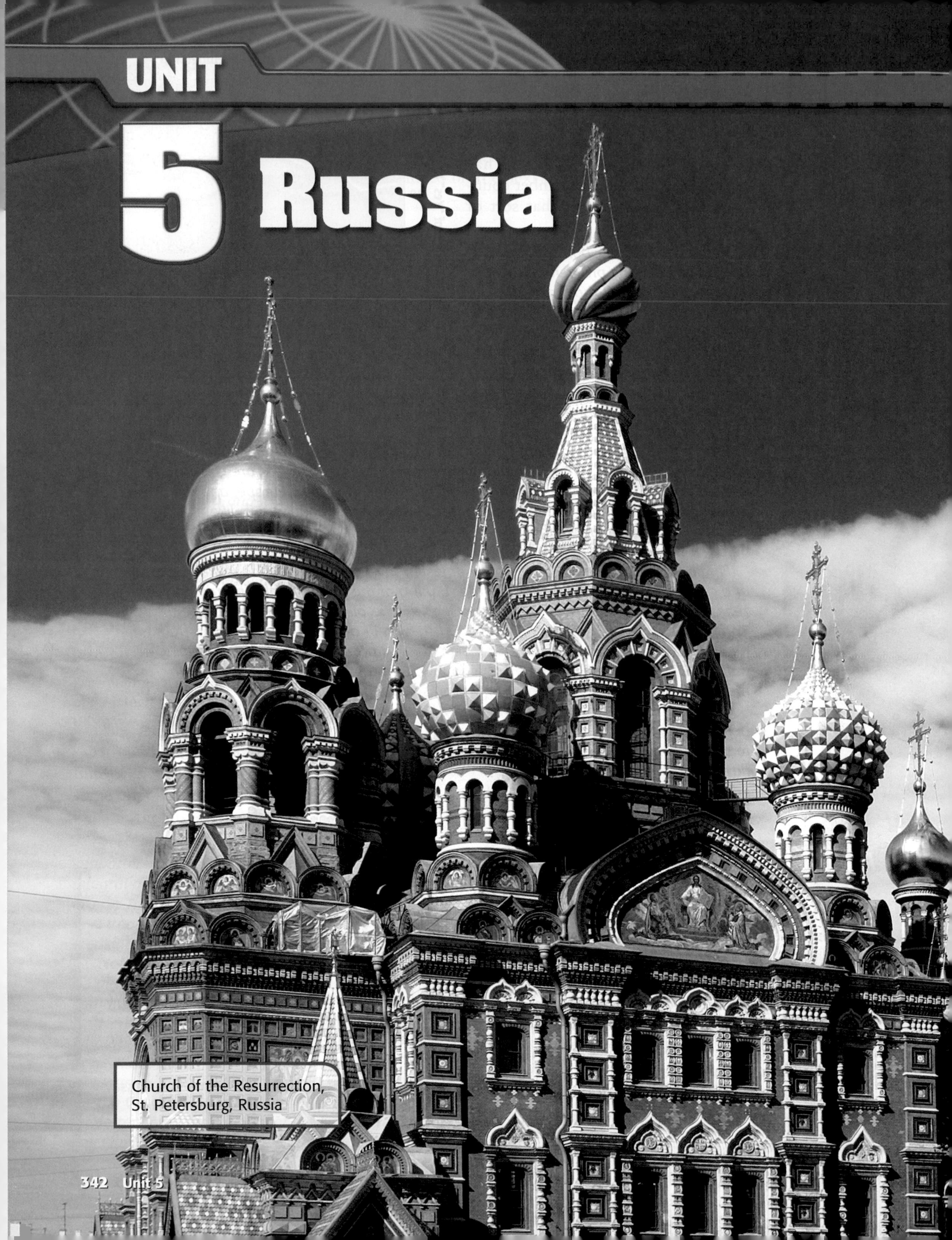

Church of the Resurrection, St. Petersburg, Russia

NATIONAL GEOGRAPHIC

NGS ONLINE To learn more about Russia visit www.nationalgeographic.com/education.

RUSSIA

Why It Matters

For most of the last century, Russia was part of the vast Soviet Union. Ruled by a Communist government, the Soviet Union challenged the United States and other democracies for global influence. Then the Soviet Union collapsed, and Russia emerged as an independent republic. Now Russia is struggling to build a stable democracy and free-enterprise economy. Because Russia is a key player in world affairs, its success—or failure—will affect your world in the years to come.

Russia

PHYSICAL GEOGRAPHY Russia is the largest country in the world, nearly twice as large as the United States. It is so large, it is a region unto itself. Two landforms dominate the landscape of this massive nation—lowland plains cover nearly half the country in the west, and plateaus rise over the rest.

Climate shapes settlement patterns. While winters in some parts of Russia are milder than others, winter in most of Russia is a challenge. The harsh Siberian winter has led most people to live in the western part of the region.

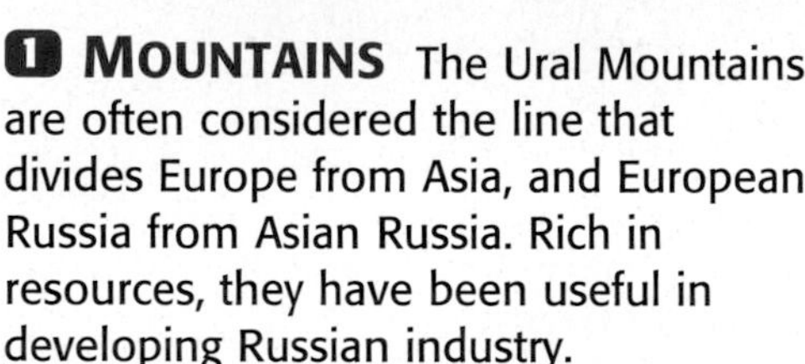

1 MOUNTAINS The Ural Mountains are often considered the line that divides Europe from Asia, and European Russia from Asian Russia. Rich in resources, they have been useful in developing Russian industry.

2 PLAINS AND PLATEAUS From the Ural Mountains to the Pacific, Russia is covered by the vast stretch of plains and plateaus called Siberia, a region of more than 2.5 million square miles (6.5 million sq. km).

3 LAKES AND RIVERS Lake Baikal, in southeastern Russia, is the world's oldest and deepest lake. With more water by volume than any other lake in the world, it holds about one-fifth of Earth's freshwater.

4 NATURAL RESOURCES Russia is one of the leading producers of oil and gas in the world. Much of this energy is found near the Ural Mountains and in Siberia.

UNIT 5 WHAT MAKES THIS A REGION?

Russia

CULTURAL GEOGRAPHY Russia is unified by its people. Though the region is home to dozens of ethnic groups, more than two-thirds of the people are ethnic Russians. They share a common language and a common history, including a long tradition of strong central government. For most of the 1900s, Russia was part of the Soviet Union, led by Communist dictators and having a government-controlled economy. In the early 1990s, the Soviet Union dissolved, and the Russians adopted democracy and a market economy. The transition to this new way of life has not been smooth, however.

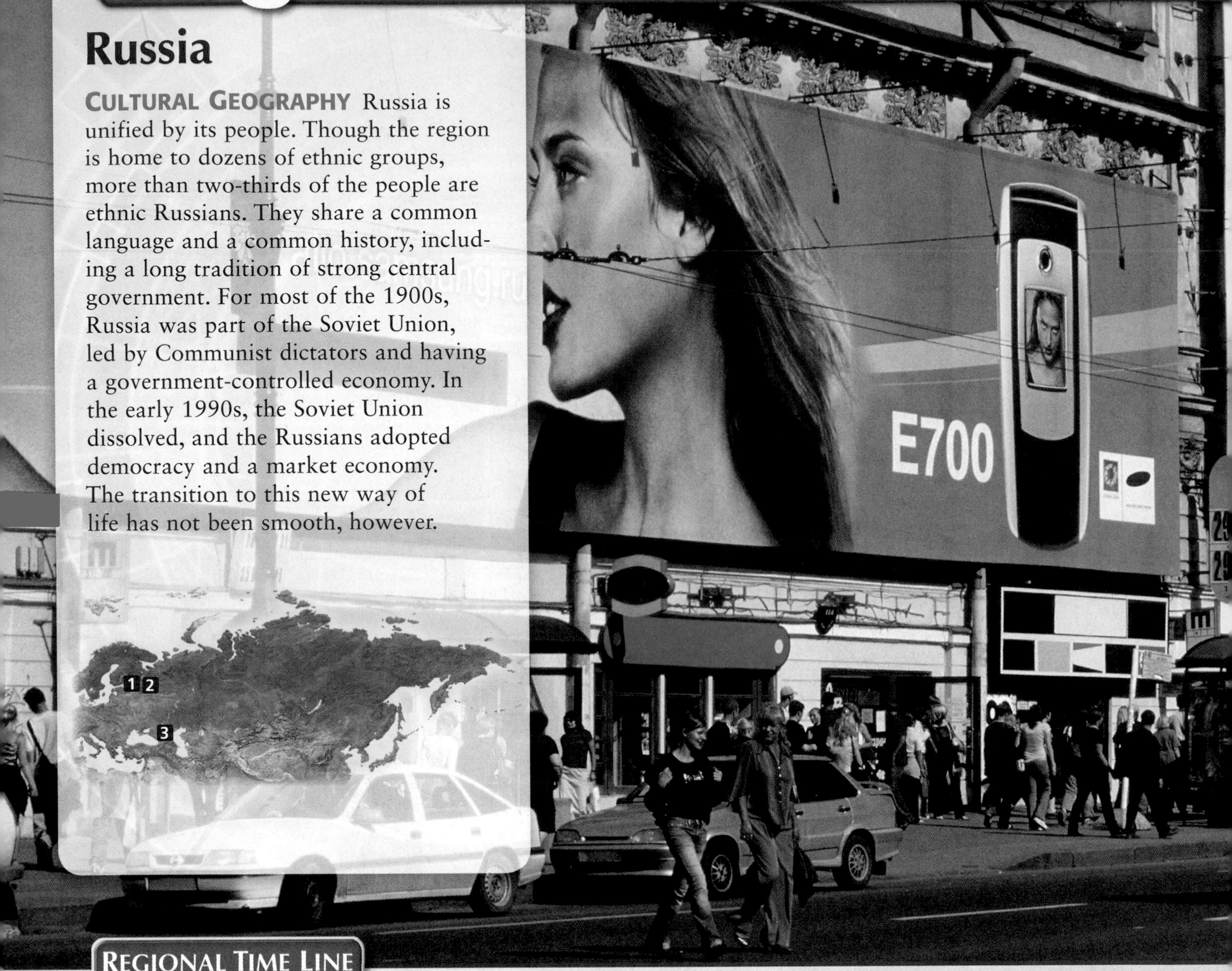

REGIONAL TIME LINE

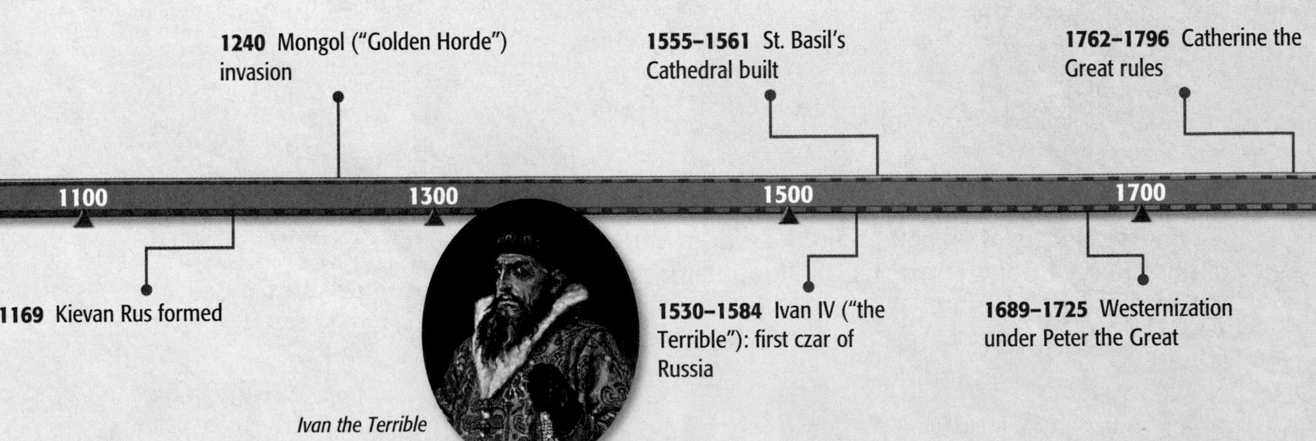

Ivan the Terrible

1 ECONOMY Western business practices like advertising became much more common after the shift to a market economy. Some Russians have gained great success in the new economic order, but others have suffered.

2 CULTURE A priest in the Russian Orthodox Church prays. The church has long been an important part of Russian identity, though membership was discouraged during Communist rule.

3 PEOPLE As in Europe, Russia faces a population challenge. Low birthrates are producing an aging population, and the growing number of older people will strain national resources.

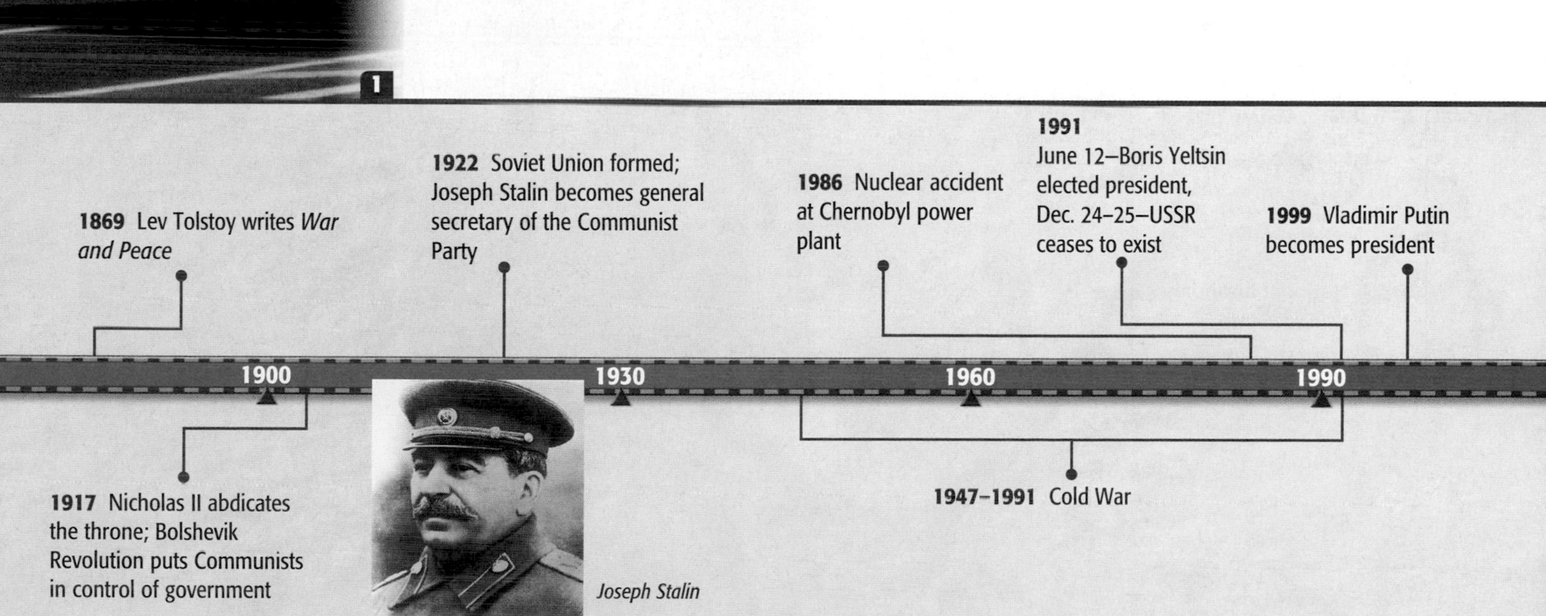

Joseph Stalin

UNIT 5 REGIONAL ATLAS

PHYSICAL Russia

GREENLAND
UNITED STATES
ATLANTIC OCEAN
20°W
ARCTIC CIRCLE
0°
NORTH POLE
160°W
180°
Chukchi Sea
Bering Strait
Wrangel I.
Chukchi Peninsula
60°N
ARCTIC OCEAN
80°N
160°E
East Siberian Sea
New Siberian Islands
140°E
Bering Sea
20°E
Franz Josef Land
40°E
60°E
80°E
Severnaya Zemlya
100°E
120°E
Laptev Sea
KOLYMA LOWLAND
KOLYMA MTS.
Kolyma R.
Klyuchevskaya Sopka 15,580 ft. (4,749 m)
Barents Sea
Novaya Zemlya
Kara Sea
Kola Peninsula
Taymyr Peninsula
Baltic Sea
Lake Ladoga
Gulf of Finland
Northern Dvina R.
Yamal Peninsula
VERKHOYANSKI MTS.
Lena R.
Kamchatka Peninsula
SIBERIA
Yenisey R.
EUROPE
Moskva R.
NORTHERN EUROPEAN PLAIN
Oka R.
URAL MOUNTAINS
WEST SIBERIAN PLAIN
Ob' R.
Lower Tunguska R.
Sea of Okhotsk
CENTRAL SIBERIAN PLATEAU
Aldan R.
Sakhalin Island
Kuril Islands
Don R.
STANOVOY RANGE
Sea of Azov
Black Earth Belt
Volga R.
Ural R.
Tobol R.
Irtysh R.
Angara R.
Lake Baikal
YABLONOVYY RANGE
Amur R.
Mt. Elbrus 18,510 ft. (5,642 m)
Caspian Depression
Black Sea
CAUCASUS MTS.
Caspian Sea shoreline -92 ft. (-28 m)
SAYAN MTS.
40°N
Caspian Sea
CENTRAL ASIA
ALTAY MTS.
EAST ASIA
Sea of Japan (East Sea)

Elevations

Feet	Meters
13,100	4,000
6,500	2,000
1,600	500
650	200
0	0

National boundary
▲ Mountain peak
▼ Lowest point

N
W E
S

SOUTH ASIA
TROPIC OF CANCER

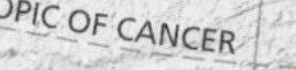

20°N

Fishermen on Lake Baikal

Arabian Gulf
Bay of Bengal
SOUTHEAST ASIA
South China Sea

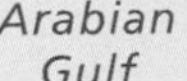

0 1,000 kilometers
0 1,000 miles
Two-Point Equidistant projection

A Vast, Cold Land

For a country as large as Russia is, it has little variety in landforms or climate. Living in a cold climate challenges Russians' creativity. Russians must adjust to the climate in all aspects of their lives—jobs, transportation, food and water supplies, heating, clothing, and plumbing. Businesses and industries also must adjust to the extreme cold by constructing buildings and machinery capable of withstanding these extreme temperatures. As you study the maps and graphics on these pages, look for the geographical features that make the region unique. Then answer the questions below on a separate sheet of paper.

1. Where is Russia located? How do you expect that to influence Russia's climate?

2. How does Yakutsk, Siberia, compare to Anchorage, Alaska, in latitude? How does it compare in climate? Which location do you think is nearer the ocean?

3. What challenges would Russians face in trying to gain economic benefits from their rivers?

The World's Deepest Freshwater Lake

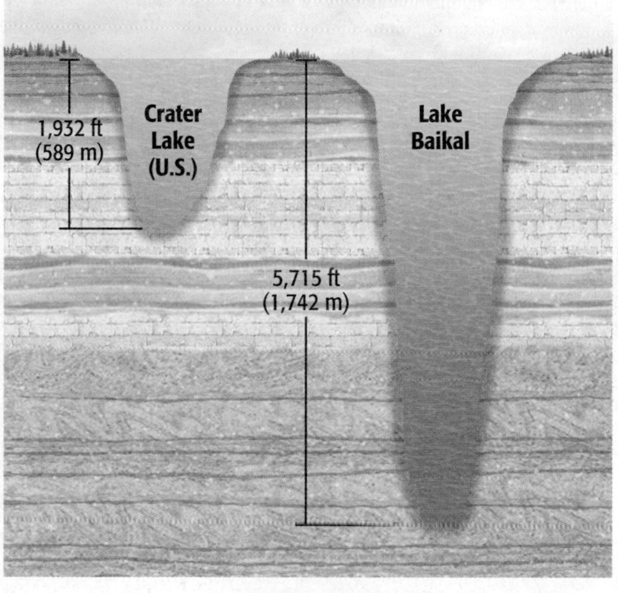

Comparing Climates

Yakutsk, Siberia

62° 13' N, 129° 49' E
338 ft. (103 m) above sea level

Average Monthly Temperature (°F/°C): 80/27, 60/16, 40/4, 20/–7, 0/–18, –20/–29, –40/–40, –60/–51

Average Monthly Precipitation (in./cm): 14/35.6, 12/30.5, 10/25.4, 8/20.3, 6/15.2, 4/10.2, 2/5.1, 0

J F M A M J J A S O N D

Month

Source: www.weatherbase.com

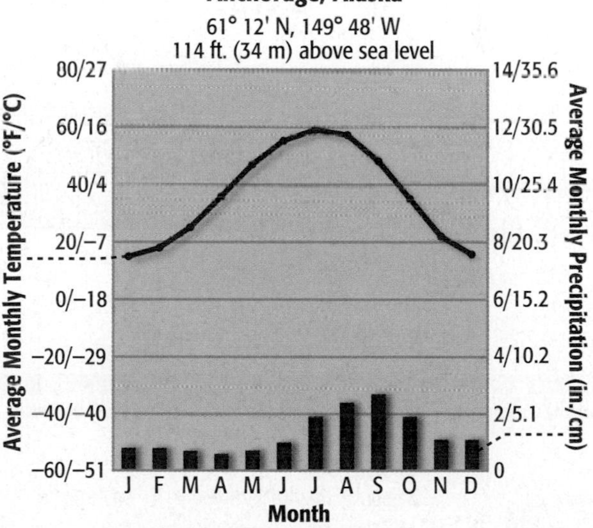

Source: www.weatherbase.com

UNIT 5 REGIONAL ATLAS

POLITICAL Russia

- ✪ National capital
- • Major city

Two-Point Equidistant projection

Country, Capital, & Area	Population & Density	Life Expectancy at Birth	GDP Per Capita*	% Urban	Literacy Rate (%)	Years of Compulsory Education	Phone Lines/ Cell Phones (per 1,000 people)	Internet Users (per 1,000 people)	Flag & Language
RUSSIA  6,323,479 sq. mi. 16,377,742 sq. km Moscow	141,800,000 22 per sq. mi. 9 per sq. km	68 yrs.	$16,100	73	99.4	10	280/838	152	Russian
Comparing Lands: ***Russia is about three times the size of the contiguous United States.***									
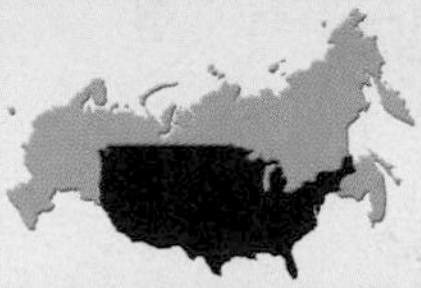	306,800,000 87 per sq. mi. 33 per sq. km	78 yrs.	$47,800	79	99.0	12	606/680	630	English

*The CIA calculates per capita GDP in terms of purchasing power parity. This formula allows us to compare the figures among different countries.

Note: Countries and flags are not drawn to scale.

Sources: Central Intelligence Agency, *World Factbook,* 2009; Population Reference Bureau, *World Population Data Sheet,* 2009; UNESCO Institute for Statistics; United Nations, *Human Development Report,* 2007/2008.

Changing Size

Russia's size and shape have changed over time. From 1921 to 1991, it was the main republic in the Soviet Union. As you study the maps on these pages, look for differences in the size of Russia over time. Then answer the questions below on a separate sheet of paper.

1. Where was Russia located before 1581, and in what directions did it expand in the following 100 years or so?
2. How much influence do you expect European culture to have on Russia? Why?
3. How does Russia today compare in size to the Soviet Union in 1945? What impact do you think that difference will have on the Russian people?

RUSSIA'S CHANGING BORDERS

THE SOVIET ERA

Resource-Rich but Environmentally Poor

Russia has a wealth of mineral resources, but exploiting them has often produced serious problems. As you study the maps on these pages, look for information about Russia's resources and its environmental problems. Then answer the questions below on a separate sheet of paper.

1. How would the location of Russia's resources pose challenges to making use of them?
2. Why are there manufacturing centers in the southern Urals and south central Siberia?
3. What environmental problems does Russia have? What areas are spared from these problems?

RUSSIA: ECONOMIC ACTIVITY

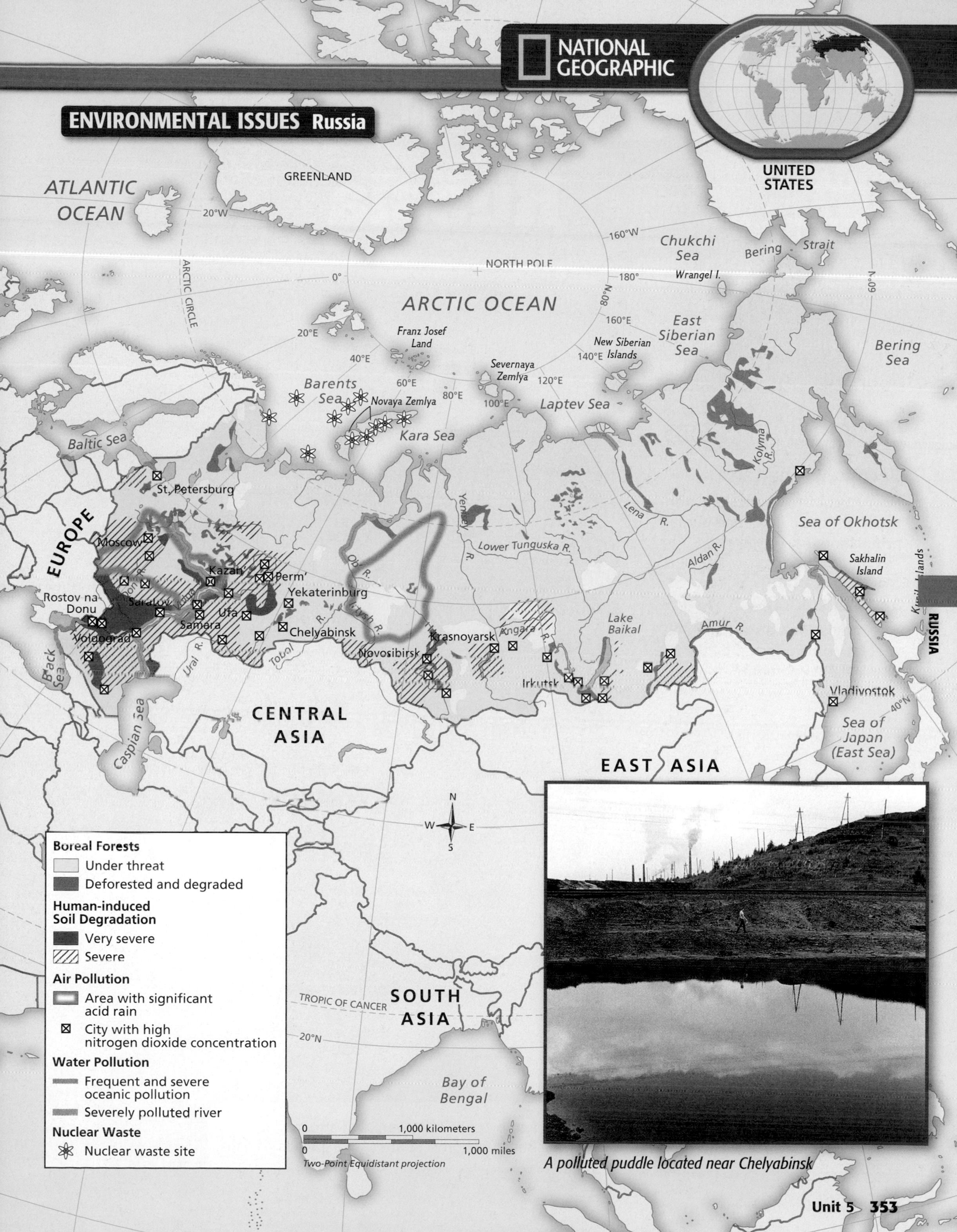

A polluted puddle located near Chelyabinsk

CHAPTER 14

PHYSICAL GEOGRAPHY OF Russia

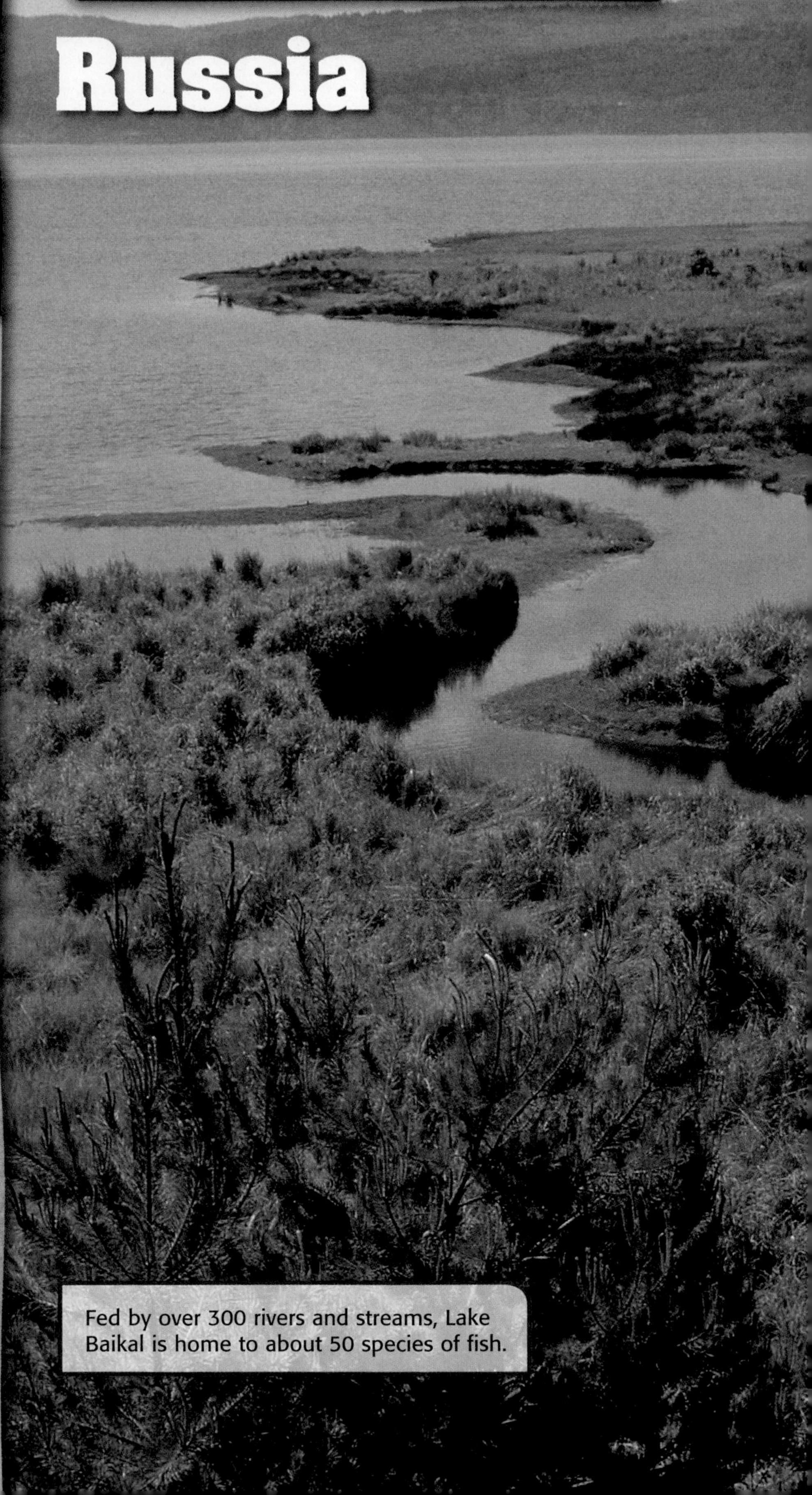

BIG Idea

Certain processes, patterns, and functions help determine where people settle. Russia's far northern location, interconnected plains and mountain ranges, large river systems, and extreme temperatures influence human settlement and activities.

Essential Questions

Section 1: The Land

How might the physical environment of Russia influence human populations?

Section 2: Climate and Vegetation

How might Russia's climate and vegetation be affected by the physical environment?

Geography ONLINE

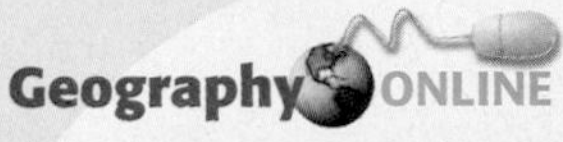

Visit glencoe.com and enter ***QuickPass***™ code WGC9952C14 for Chapter 14 resources.

Fed by over 300 rivers and streams, Lake Baikal is home to about 50 species of fish.

Identifying Information Make a Layered-Look Book to identify the key physical features found in Russia.

Reading and Writing As you read the chapter, identify specific features of Russia's physical geography and write them in the correct location in your Foldable.

SECTION 1

 section audio  spotlight video

The Land

Russia is a vast and varied land of plains divided and bordered by mountain ranges, tundra, subarctic forests, and wide rivers and seas. Within the borders of Russia's immense land area, there are places of such beauty and bounty that the wilderness seems boundless.

Guide to Reading

Essential Question

How might the physical environment of Russia influence human populations?

Content Vocabulary

- chernozem *(p. 357)*
- permafrost *(p. 359)*

Academic Vocabulary

- series *(p. 357)*
- estimate *(p. 358)*
- link *(p. 358)*

Places to Locate

- Ural Mountains *(p. 357)*
- Caucasus Mountains *(p. 357)*
- Central Siberian Plateau *(p. 357)*
- Siberia *(p. 357)*
- Northern European Plain *(p. 357)*
- West Siberian Plain *(p. 357)*
- Volga River *(p. 358)*

Reading Strategy

Taking Notes As you read about Russia's physical landscape, use the major headings of the section to create an outline similar to the one below.

I. Landforms
  A.
  B.
II. Water Systems
  A.
  B.

NATIONAL GEOGRAPHIC Voices Around the World

"I came to know the world's largest boreal forest through years of work in Russia, journeying from the taiga's heavily logged southern fringe on the Chinese border to beyond the Arctic Circle. . . . I had also been to the tropical forests of South America, where there is more of everything—more trees, more animals, more insects, more tumult. But I prefer the understated charms of the boreal, with its limitless expanse of lakes and ponds and its gentle gradations of green: the pale hues of the reindeer lichen, the black-green of the spruce, the lighter, almost chartreuse tints of aspen and birch. More than anything, perhaps, I am partial to the light of the north woods—slanting rays that in the warmer months cast long evening shadows and suffuse the landscape with a crystalline glow."

—Fen Montaigne, "The Great Northern Forest," *National Geographic*, June 2002

An indigenous Nenets woman in western Siberia

Landforms

MAIN Idea Russia's interconnected mountain ranges and plains shape human activities.

GEOGRAPHY AND YOU Why might mountains affect where people live? Read to learn how Russia's mountains have influenced life in this country.

In both total land area and geographic extent, Russia is the world's largest country. Covering about 6.3 million square miles (16.3 million sq. km), Russia stretches across parts of two continents—Europe and Asia.

Mountains and Plateaus

Mountains and plateaus punctuate the generally flat landscape of Russia. The **Ural Mountains** mark the traditional boundary between European Russia and Asian Russia. The Urals are an old, worn-down **series** of mountain ranges with an average height of about 2,000 feet (about 610 m). Though modest in height, the Urals are rich in iron ore and mineral fuels, such as oil and natural gas.

In southwestern Russia, the rugged **Caucasus** (KAW•kuh•suhs) **Mountains** lie between the Black and Caspian Seas. This area of moderate climate near the Black Sea has long attracted human settlement. The Caucasus Mountains reach their highest elevation at Mount Elbrus, an extinct volcano that reaches 18,510 feet (5,642 m), Russia's highest point.

Mountain ranges also form a rugged natural boundary between Russia and China. These mountains mark the southeastern edge of the **Central Siberian Plateau.** This rolling plateau has elevations ranging from 1,600 to 2,300 feet (490 to 700 m). Throughout the plateau's expanse, swiftly flowing rivers have carved out canyons.

Still farther east, mountains and basins extend through **Siberia** to the Pacific Ocean. In eastern Russia, the Kamchatka Peninsula has more than 100 volcanoes, including 29 that are active.

Plains Areas

Most of European Russia is part of the **Northern European Plain,** also known as the Russian Plain, that sweeps across western and central Europe into Russia. In Russia, the northern part of this plain is very flat and poorly drained, resulting in many swamps and lakes. By contrast, the southern part has navigable waterways and a rich black soil, known as **chernozem** (cher•nuh•ZYAWM), that supports the production of wheat, barley, rye, oats, and other crops. About 80 percent of the Russian population lives on the Northern European Plain. This region holds Russia's most populous cities, including Moscow and St. Petersburg.

Farther to the east, the Ural Mountains divide the Northern European Plain from another vast plains area—the **West Siberian Plain.** With almost 1 million square miles (2.6 million sq. km), the West Siberian Plain is one of the world's largest areas of flatland. At its widest, this plain stretches from the Arctic Ocean to the grasslands of Central Asia. Its lowland areas are poorly drained, with many swamps and marshes.

READING Check **Place** Which mountains form a natural boundary between European Russia and Asian Russia?

NATIONAL GEOGRAPHIC *The fertile soils of the Northern European Plain (top) provide quality farmland. Poor drainage makes many areas of the West Siberian Plain (bottom) swampy.*

Regions What causes the variation in vegetation in Russia's plains areas?

RUSSIA

Water Systems

MAIN Idea Russia's large river systems are vital for irrigation, transportation routes, electric power, and industries, such as fishing.

GEOGRAPHY AND YOU Can you name some major U.S. rivers? How are they important to the economy? Read to learn how Russia's waterways influence human activities throughout the country.

Russia's water systems include long coastlines, seas, and lakes that provide access to other parts of the world. Russian rivers are also important for economic activities in the country.

Coasts, Seas, and Lakes

Russia has the longest continuous coastline of any country in the world. Stretching 23,400 miles (37,650 km), Russia's coastline touches both the Arctic and Pacific Oceans. Other coasts lie along the Baltic Sea, Black Sea, and Caspian Sea.

The Black Sea provides Russia with a warm-water outlet to the Aegean and Mediterranean Seas through three Turkish-controlled waterways—the Bosporus, the Sea of Marmara, and the Dardanelles (DAHRD•uhn•EHLZ). A saltwater lake located in a deep depression, the Caspian Sea is the largest inland body of water in the world. Rivers flow into the Caspian, but there is no outlet to the ocean. Over time, water levels in the Caspian Sea vary due to evaporation and inflow of freshwater.

Lake Baikal (by•KAHL) lies in southern Siberia. At nearly 400 miles (644 km) long, 40 miles (64 km) wide, and over 1 mile (1.6 km) deep, Lake Baikal is the third-largest lake in Asia and the deepest freshwater lake in the world. It is **estimated** to contain about 20 percent of the Earth's total supply of freshwater.

Rivers

Some of the world's longest rivers flow through Russia, draining a large portion of the land and providing water for irrigation. They also serve as transportation routes or sources of electric power for many parts of the country. Most of Russia's longest rivers—which carry 84 percent of the country's water—are located in Siberia, where only 20 percent of the Russian population lives.

NATIONAL GEOGRAPHIC *Russia's far northern location causes most of its ports to be frozen for up to four months each year.*

Place Which sea provides Russia with a warm-water outlet to the Mediterranean?

The Volga River Western Russia's **Volga River** is the fourth-longest river in the country. Called *Matushka Volga,* or "Mother Volga," the river is vital to Russia. The Volga and its tributaries drain much of the eastern part of Russia's Northern European Plain. They connect Moscow to the Caspian Sea and, by way of the Volga-Don Canal, to the Sea of Azov and the Black Sea. Canals **link** the Volga to the Baltic Sea, providing a water route to northern Europe. The river provides hydroelectric power and water for drinking and irrigation.

Two-thirds of Russia's water traffic travels along the Volga. Heavy use of the river, however, has created challenges. Fed by melting snow, the Volga supplies about one-third of Russia's usable water, but half of it returns to the river carrying waste. Dams interrupt the river's flow, threatening wildlife and drinking-water supplies.

Siberian Rivers The Ob', Irtysh, Yenisey, and Lena are among the world's largest river systems. They flow north to the Arctic Ocean. Temperatures are warmer at the rivers' sources in the south than at their mouths in the north. Blocked by ice in the north, the meltwaters flood the land, creating swamps and marshes.

The Amur River, which drains eastward, forms the border between Russia and China for about 1,000 miles (1,610 km). Influenced by summer monsoon winds from the southeast, the Amur River valley is warmer than the rest of Siberia and is Siberia's main food-producing area.

READING Check **Regions** Which river provides western Russia with hydroelectric power?

Natural Resources

MAIN Idea Russia has an abundance of natural resources, but many are located in remote, inaccessible areas of the country.

GEOGRAPHY AND YOU What are some of the natural resources in the United States? Where are they located? Read to learn about natural resources in Russia.

Russia's physical geography is both a blessing and a challenge. The country holds an abundance of natural resources. Much of this wealth, however, lies in remote and climatically unfavorable areas and is difficult to tap or utilize.

Minerals and Energy

Russia has huge reserves of mineral resources. It is especially rich in fossil fuels. The country holds large petroleum deposits and 16 percent of the world's coal reserves. However, the country's biggest coal fields lie in remote areas of eastern Siberia. Russia is also a leading producer of natural gas, but much of this resource is located in northern Siberia. It also leads the world in nickel production and ranks among the top three producers of aluminum, gemstones, and platinum-group metals. Russia's rivers make it a leading producer of hydroelectric power.

Soil and Forest Land

Because of Russia's generally cold climate, only about 10 percent of its land can support agriculture. This is enough farmland to support the population with grains and vegetables. In the north and east, **permafrost,** a permanently frozen layer of soil, lies beneath the surface of the ground. A wide, fertile band called the Black Earth Belt covers about 250 million acres (100 million ha) and stretches from Ukraine to southwestern Siberia. Transporting crops from where they are grown in the chernozem soils of the south to cities in the north is a challenge since the distances are great.

MAP STUDY

1. **Location** Where is most of the threatened forested land in Russia located?
2. **Place** Compare the map below to the vegetation map on page 363. What type of vegetation is most common in Russia's boreal, or northern, forests?

Maps In MOtion Use **StudentWorks™ Plus** or glencoe.com.

RUSSIA

NATIONAL GEOGRAPHIC **Deforestation in Russia**

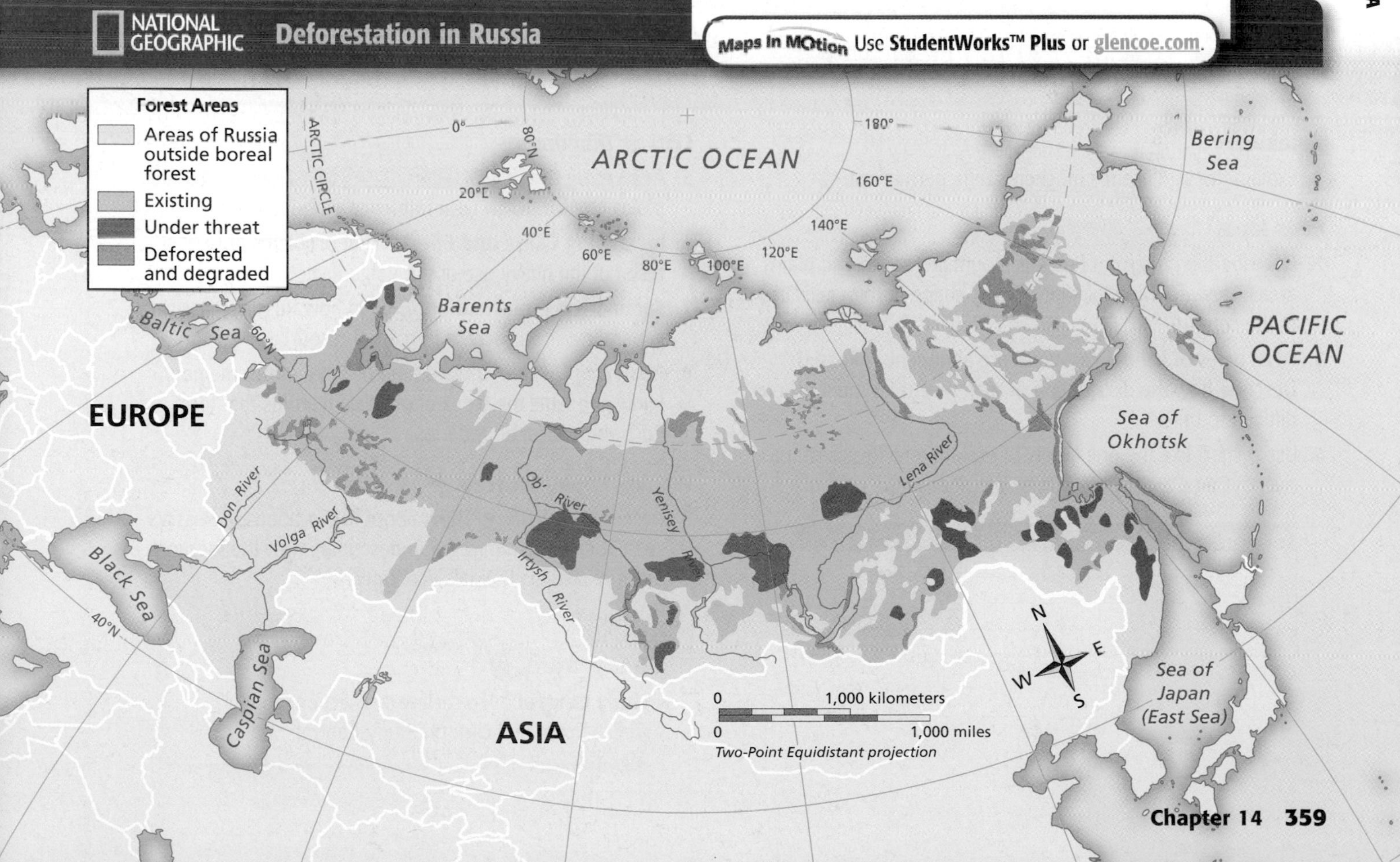

NATIONAL GEOGRAPHIC *Fishing is an important economic activity among the indigenous people of the Sakha Republic.*

Human-Environment Interaction Why has the supply of Russian caviar declined?

About one-fifth of the world's forest lands lie in Russia—75 percent of them in Siberia. Second only to the Amazon rain forest in the amount of oxygen returned to the atmosphere, Russian boreal, or northern, forests also supply much of the world's timber, mainly pine, fir, spruce, and cedar. As a result of commercial logging and wildfires, however, Russian forests shrink by almost 40 million acres (16 million ha) each year—a rate of loss higher than that of the Amazon Basin.

> *"If the tropical forests, which contain half the planet's woodlands, are one lung of the Earth, then the boreal forests are the other. Both play a vital role in regulating climate as they—along with the ocean, Earth's largest carbon repository—filter out billions of tons of carbon dioxide and other greenhouse gases during photosynthesis, storing the carbon in trees, roots, and soils."*
>
> —Fen Montaigne, "The Great Northern Forest," *National Geographic,* June 2002

The Fishing Industry

Fish are important to the Russian diet and economy. Salmon from the Pacific Ocean and herring, cod, and halibut from the Arctic Ocean support a flourishing fishing industry. However, the supply of world-famous Russian caviar, or salted fish eggs, has declined. Dams built on the Volga River have interrupted the migration of sturgeon, the fish that provide the eggs for caviar. Sturgeon is often fished illegally to meet the global demand for this delicacy.

READING Check **Regions** Why does some of Russia's wealth in coal remain untapped?

SECTION 1 REVIEW

Vocabulary

1. Explain the significance of: chernozem, permafrost.

Main Ideas

2. Describe the pattern of Russia's mountain ranges and plains areas. How do mountains and plains shape human activities in the country?
3. What types of natural resources are abundant in Russia? Describe how the distribution of such resources makes them difficult to utilize.
4. Use a chart like the one below to explain how Russia's large river systems are vital to human activities in the region.

River	How It Is Important

Critical Thinking

5. **Answering the Essential Question** How do Russia's plains areas differ in their influence on human settlement?
6. **Identifying Cause and Effect** What problems arise as a result of the heavy use of the Volga River?
7. **Summarizing Information** Explain how Russia's physical geography affects access to natural resources.
8. **Analyzing Visuals** Draw a map of Russia from memory. Show and label the types of physical features that form Russia's political boundaries.

Writing About Geography

9. **Descriptive Writing** Think about the locations of Russia's seas. Then write a paragraph describing how the locations of these seas affect Russia's economy.

Geography ONLINE

Study Central™ To review this section, go to glencoe.com and click on Study Central.

SECTION 2

 section audio  spotlight video

Guide to Reading

Essential Question

How might Russia's climate and vegetation be affected by the physical environment?

Content Vocabulary
- continentality *(p. 362)*
- tundra *(p. 362)*
- taiga *(p. 363)*
- steppe *(p. 364)*

Academic Vocabulary
- portion *(p. 362)*
- role *(p. 364)*
- enable *(p. 364)*

Places to Locate
- Arctic Circle *(p. 362)*

Reading Strategy

Categorizing As you read about Russia's physical geography, complete a graphic organizer similar to the one below by describing the climate and vegetation regions of Russia.

Region	Description
Tundra	
Humid continental	
Steppe	

Climate and Vegetation

Much of Russia experiences extreme cold and long winters because of its location in the high latitudes. In the country's remotest northern plateaus, the harsh terrain softens only during the very brief summer.

NATIONAL GEOGRAPHIC Voices Around the World

"The view was magnificent—mile after mile of flat-topped mountains receding to the Arctic horizon—though it was difficult to appreciate if you were on your hands and knees on the side of one of those plateaus, clinging to shards of scree. Vasily Sarana, chief of the Russian Geographic Society's Putorana expedition and a mountaineer who seemed capable of bounding up vertical walls, was not in such an undignified position, however. Standing tall, he turned around on the [dizzying] slope and scanned the horizon of his favorite corner of Russia."

—Fen Montaigne, "Remote Russia," *National Geographic*, November 2000

Hiker on the Kamchatka Peninsula

RUSSIA

CHAPTER 15 CULTURAL GEOGRAPHY OF Russia

BIG Idea

The movement of people, goods, and ideas causes societies to change over time. Europeans, Asians, and other peoples have shaped the cultural geography of Russia. The country's various ethnic groups have influenced its tumultuous history and changes in government over time.

Essential Questions

Section 1: Population and Culture

How might Russia's diverse population have influenced its cultural geography?

Section 2: History and Government

How have Russia's history and government been shaped by its many ethnic groups?

Geography ONLINE

Visit glencoe.com and enter QuickPass™ code WGC9952C15 for Chapter 15 resources.

Women in traditional clothes dance to celebrate the coming of spring in Moscow.

NATIONAL GEOGRAPHIC

Organizing Information Make a Three-Tab Book to help you organize information about the cultural geography of Russia.

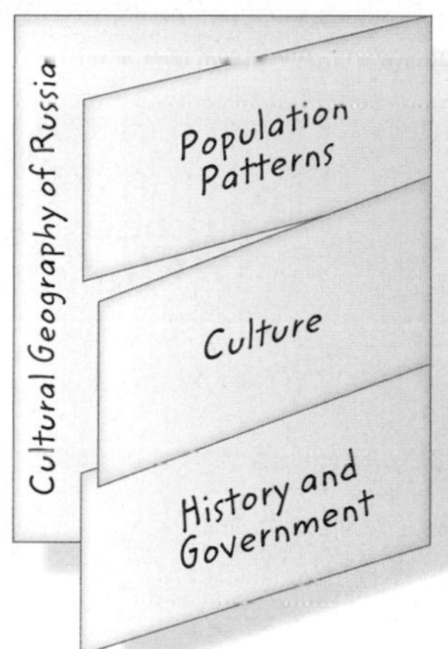

Reading and Writing As you read the chapter, use your Foldable to take notes about the population patterns, culture, and history and government of Russia.

RUSSIA

SECTION 1

 section audio  spotlight video

Guide to Reading

Essential Question

How might Russia's diverse population have influenced its cultural geography?

Content Vocabulary

- ethnic group *(p. 371)*
- Soviet era *(p. 371)*
- nationality *(p. 371)*
- sovereignty *(p. 372)*
- atheism *(p. 373)*
- pogrom *(p. 373)*
- intelligentsia *(p. 374)*
- socialist realism *(p. 375)*

Academic Vocabulary

- ethnic *(p. 371)*
- evident *(p. 375)*

Places to Locate

- Tatarstan *(p. 372)*
- Lena River *(p. 372)*
- Moscow *(p. 372)*
- Caspian Sea *(p. 373)*

Reading Strategy

Organizing As you read about the population and culture of Russia, complete a graphic organizer similar to the one below by filling in the different ethnic groups.

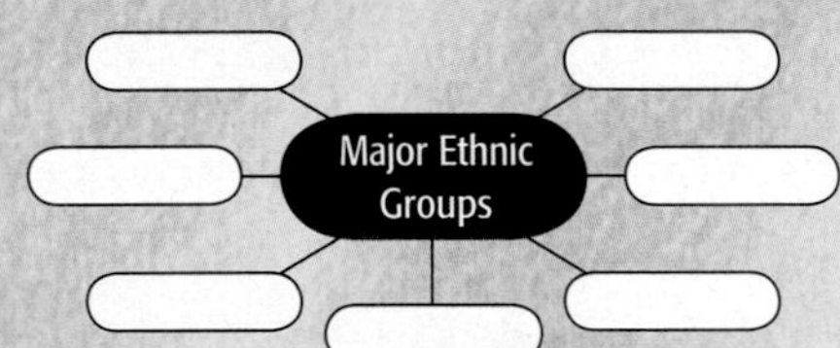

Population and Culture

Many of Russia's cultural traditions date back centuries. Maslenitsa began long ago as a folk holiday celebrating the coming of spring. It has since been absorbed into the Eastern Orthodox religion.

NATIONAL GEOGRAPHIC VOICES AROUND THE WORLD

"This . . . is the Russian version of Mardi Gras. All over the country people celebrate the last period of merrymaking before the Great Fast preceding Easter. The festival starts February 5, and for the next four days Russians—many in traditional costumes—build bonfires, enjoy sled rides, try to avoid injury during mock fistfights, . . . and eat pancakes, which symbolize the sun. The best places to take part in maslenitsa are . . . to the northwest in the Golden Ring, a region of old Russian cities offering fine examples of Russian culture, traditions, and architecture dating back to the 12th century."

Cassandra Franklin-Barbajosa,
"Russia: Five Cultural Bests,"
National Geographic, January 2004

Celebrating Maslenitsa in Red Square, Moscow

Population Patterns

MAIN Idea Ethnic groups, migrations, and invasions have shaped population patterns in Russia.

GEOGRAPHY AND YOU How has the ethnic diversity in your town or city changed over the years? Read to learn about the factors that have shaped Russia's population.

Russia today is home to one of the widest varieties of **ethnic** groups in the world—there are more than 120 different groups. An **ethnic group** shares a common ancestry, language, religion, customs, or a combination of these things.

The People

Over the centuries Russia grew from a territory to a multiethnic empire that stretched from Europe to the Pacific Ocean. In the process, many non-Russian ethnic groups came under its control. During the **Soviet era**—the period between 1922 and 1991 when Russia was part of the Union of Soviet Socialist Republics (USSR)—regional political boundaries often reflected the locations of major ethnic groups, or **nationalities.**

In 1991, after the fall of the Soviet Union, several of these larger republics, including Russia, became independent countries. Although Russia is ethnically very diverse—32 ethnic groups have their own republics or administrative territories within Russia—about 80 percent of the population is ethnic Russian.

The Slavs Ethnic Russians are part of a larger ethnic group known as Slavs, a linguistic and ethnic branch of Indo-European peoples that also includes Poles, Serbs, Ukrainians, and other eastern Europeans. The Russian Slavs have dominated the country's politics and culture.

Caucasian Peoples Another large group of diverse peoples is classified as Caucasian (kaw•KAY•zhuhn) because they live in the Caucasus region of southwestern Russia. Caucasian groups include the Chechens, Dagestanis, and Ingushetians.

MAP STUDY

1. **Place** Where in Russia do most of the Ukrainian peoples live? Turkic peoples?
2. **Movement** How have the settlement patterns of ethnic groups affected political boundaries?

Maps In MOtion Use StudentWorks™ Plus or glencoe.com.

NATIONAL GEOGRAPHIC **Ethnic Groups in Russia**

Source: C.I.A., *U.S.S.R. Summary Map*

Turkic Peoples Turkic-speaking peoples live in southwestern Russia in the Caucasus area and in the middle Volga area. The Turkic peoples of Russia include the Tatars, Chuvash, Bashkirs, and the Sakha. The most numerous of these groups are the Tatars, many of whom live in **Tatarstan** (TA•tuhr•STAN), a western republic. Russia has ruled Tatarstan since the mid-1550s. However, the republic, like other Russian republics, does have a limited amount of **sovereignty** (SAH•vuh•ruhn•tee), or self-rule.

The Sakha are a combination of local groups and Turkic peoples who originally settled along the middle **Lena River.** Formerly seminomadic, the Sakha in southern Siberia have expanded into northeastern Russia.

Density and Distribution

About 80 percent of all Russians live in western Russia. This is due in part to the rich soil, waterways, and a milder climate than that in eastern Russia. Densely settled western Russia includes the country's industrialized cities. The major industrial city is **Moscow,** Russia's capital. Since 1990, urban population growth in many industrialized centers has leveled off or decreased.

Population is more dispersed east of the Ural Mountains. Only 20 percent of Russia's population lives in Siberia, an area that accounts for about two-thirds of the country's land area. Frozen tundra, mountains, and forests make most of this part of Russia unsuitable for farming.

During the earlier years of the Soviet era, many ethnic Russians migrated to non-Russian republics of the Soviet Union. In the 1970s, however, this trend began to reverse. Since the breakup of the Soviet Union in 1991, more ethnic Russians have returned to their homeland. Most have settled in Moscow, St. Petersburg, and southwestern Russia. Because of this trend, the number of people moving into the country has been greater than the number of Russians leaving the country.

READING Check **Human-Environment Interaction** Why do most people live in western Russia?

MAP STUDY

1. **Human-Environment Interaction** What factors have contributed to the sparse population of eastern Russia?
2. **Location** Where are many of Russia's largest cities located? Why is this important?

Maps in Motion Use **StudentWorks™ Plus** or glencoe.com.

Culture

MAIN Idea The policies of the Soviet government have had lasting effects on Russia's culture.

GEOGRAPHY AND YOU How has religion influenced culture in the United States? Read to learn about the influence of religion in Russia.

Since the fall of the Soviet Union, millions of Russians are rediscovering their faiths and traditions and expressing themselves creatively.

Language and Religion

Although more than 100 languages are spoken in Russia today, Russian is the country's official language. Ethnic Russians generally speak only this language, while people belonging to other ethnic groups are bilingual and speak their own language and Russian. The Altaic family of languages is spoken by the Turkic peoples of western Russia. The primary languages spoken in eastern Russia are Russian and the Turkic languages spoken by the Sakha.

The Soviet government strictly discouraged religious practices and discriminated against certain groups. It actively promoted **atheism** (AY•thee•IH•zuhm), or the belief that there is no God or other supreme being. In the late 1980s, however, the government began to relax its restrictions on religion.

After the fall of the Soviet Union, the influx of many foreign missionaries from Western Christian denominations prompted lawmakers in 1997 to place restrictions on the activities of newly established religious groups. Only Russian Orthodoxy, Islam, Judaism, and Buddhism were allowed full liberty as traditional religions of Russia.

Christianity The Eastern Orthodox Church had been central to Russian culture for a thousand years before the Communist revolution in 1917. In 988 Prince Vladimir, leader of Kievan Rus, adopted Eastern Orthodox Christianity as Russia's official religion. By 1453 the Byzantine Empire, the center of the Eastern Orthodox Church, had fallen, and Russia asserted its claim as leader of the Orthodox Christian world.

Today, most Russians who claim a religious affiliation belong to the Russian branch of the Orthodox Church. Many of the churches that were looted or destroyed during the Soviet era have been repaired and rebuilt. Other Christian groups, including Roman Catholics and Protestants, have also reemerged.

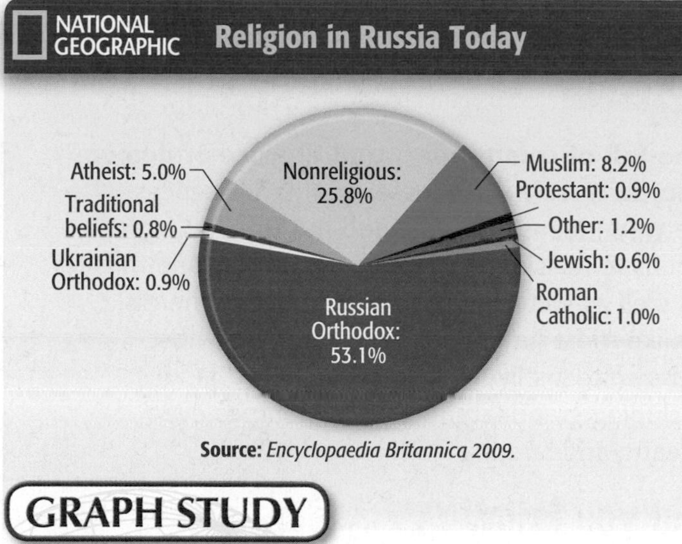

Source: *Encyclopaedia Britannica 2009.*

GRAPH STUDY

1. **Place** How does the number of Christians in Russia compare to the number of Muslims?
2. **Place** Why do you think so many Russians classify themselves as nonreligious?

Islam Islam is the second-largest religion in Russia and also enjoying a rebirth. The majority of people who practice Islam in Russia live in the Caucasus region and between the Volga River and the Urals. Most Russian Muslims follow the Sunni branch, which is practiced in most Arab countries as well as in Turkey and Afghanistan.

Judaism People practicing Judaism in Russia have long been persecuted. In czarist times, Jews could settle only in certain areas, could not own land, and were often the targets of organized persecution and massacres known as **pogroms.** Yet Jewish communities managed to thrive in many of Russia's cities.

Events in the twentieth century took a tragic toll on Russia's Jews. As a result, many Jews migrated to Israel or the United States. Despite lingering prejudice, Jewish communities in Russia are restoring their religious practices.

Buddhism The republics of Tuva, Buryatia, and Kalmykia (kal•MIH•kee•uh), near the **Caspian Sea,** have a large number of Buddhists, giving the religion traditional status in the country. A small number of Buddhists live in the larger cities such as St. Petersburg and Moscow, where they have access to urban Buddhist centers and facilities.

Teen Life in Russia

The fall of communism and the rise of democracy in Russia have lessened the differences between teens in Russia and the United States. Teens in Russia enjoy American movies and music, as well as watching TV with family and friends. Unlike most American teens, however, most Russian teens live with their extended family, including grandparents and sometimes even great-grandparents.

Did you know . . .

- Students attend school Monday through Saturday.
- Some students attend special schools where they can perfect a talent such as music or ballet.
- Soccer is a favorite sport. Ice hockey and ice skating are also very popular due to the cold climate in many parts of Russia.
- New Year's Day is the most popular holiday in Russia. Russians celebrate it in much the same way as many Americans celebrate Christmas, with a tree and presents.
- Most people use public transportation such as the bus, subway, trolley, or train.
- The voting age is 18.

Education and Health Care

During the Soviet era, education was free but mandatory. The emphasis was on math, science, and engineering rather than on language, history, and literature. This produced generations of technology-focused government officials. They, along with prominent educators, writers, and artists, made up the Soviet **intelligentsia** (in•TEL•luh•JEHN•see•uh), or intellectual elite.

The curriculum changed dramatically after the collapse of the Soviet Union. Schools began to emphasize a more balanced approach, including language, history, and literature. Today, students have a choice of different types of schools, but the country's unstable economy has limited school budgets. Teachers often abandon teaching because of the low pay and low morale. In addition, students focus on earning money rather than getting an education.

Russia is experiencing a demographic decline as the quality and availability of health care have worsened. Average life expectancy is 68 years, 10 years shorter than in the United States. The infant mortality rate, or the number of deaths per 1,000 births, is 9 compared to the U.S. rate of 6.6. Since 1992 the Russian death rate has exceeded the birthrate, resulting in negative population growth.

Disease, lifestyle choices such as smoking tobacco and drinking alcoholic beverages, and inefficient health-care systems all threaten the well-being of Russia's people. Additionally, an aging population, rising infertility, and increasing rates of infectious disease, including HIV and AIDS, have put a strain on an already struggling health-care system. The clinics and hospitals that are still run by the government are often inefficient and lose capable medical staff because of poor pay.

The Arts

The arts in Russia were often inspired by religion. For example, churches were crowned with onion-shaped domes that symbolized heaven in the Russian Orthodox tradition. Russian art changed its focus to nonreligious themes in the early 1700s when Peter the Great introduced European culture. By the early 1800s, Russia had entered an artistic golden age that lasted into the 1900s.

Russian arts are characterized by a list of well-known artists. Painters such as Viktor Vasnetsov and composers such as Pyotr (Peter) Tchaikovsky contributed to the richness of Russian culture. The works of poets Aleksandr Pushkin, Boris Pasternak, and Anna Akhmatova, and novelists Lev (Leo) Tolstoy and Fyodor Dostoyevsky have made Russian literature famous.

The Soviet government limited individual artistic expression and believed that it was the artists' duty to glorify the government's achievements in their works, an approach known as **socialist realism.** Artists who did not follow these guidelines were punished, as is **evident** in Aleksandr Solzhenitsyn's book *The Gulag Archipelago*.

Beginning in the mid-1980s, activity in the arts renewed as loosening government controls allowed the printing of previously unpublished works and new materials. During the height of Soviet repression, some works had been smuggled from Russia and printed in other countries.

Geography ONLINE

Student Web Activity Visit glencoe.com, select the *World Geography and Cultures* Web site, and click on Student Web Activities–Chapter 15 for an activity on Russian literature.

Family Life and Leisure

Living conditions in Russia affect family life. Due to a housing shortage, most families live in large apartment blocks. The apartments are usually very small—a bedroom, living room, kitchen, and bathroom—for a family of four. Extended family members often live together. For example, many newly married couples have to live with their parents until they can afford a home of their own. Grandmothers, or babushkas, may live with their adult children and grandchildren. Babushkas help with cooking, cleaning, and taking care of the children.

Attending concerts, the ballet, and the theater all provide popular entertainment. Sports, both amateur and professional, are popular with all age groups. Russia's tennis, track and field, and ice hockey athletes have had remarkable success in international events, as have figure skaters and gymnasts.

Today Russians observe May Day more as a spring festival than as a workers' holiday. Traditional religious holidays have also reemerged. In 1991, Christmas, celebrated by Eastern Orthodox Christians on January 7, became an official holiday in Russia for the first time since 1918.

READING Check **Regions** What contributed to a resurgence of the arts in the 1980s?

RUSSIA

SECTION 1 REVIEW

Vocabulary

1. Explain the significance of: ethnic group, Soviet era, nationality, sovereignty, atheism, pogrom, intelligentsia, socialist realism.

Main Ideas

2. How have the many ethnic groups in Russia shaped the country's population patterns?
3. Describe an example in which migration has shaped population patterns in Russia.
4. Create a graphic organizer like the one below, and use it to fill in the key details for each aspect of Russian culture during the Soviet era and today.

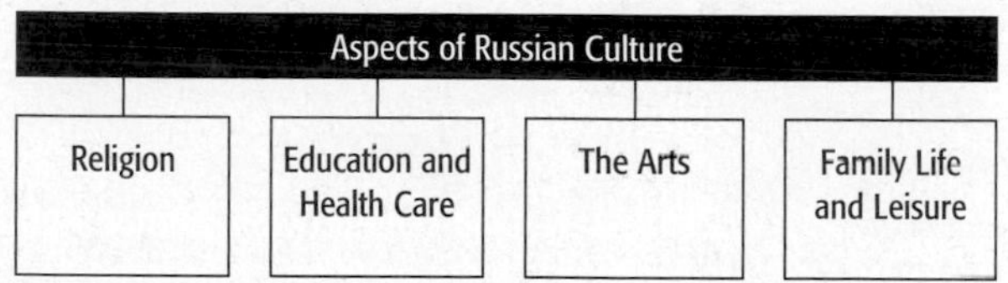

Critical Thinking

5. **Answering the Essential Question** How did the arrival of foreign missionaries affect religion in Russia?
6. **Comparing and Contrasting** Write a paragraph comparing the Russian education system during the Soviet era with the system today.
7. **Analyzing Visuals** Study the map of ethnic groups on page 371. Which ethnic groups are the most spatially concentrated?

Writing About Geography

8. **Expository Writing** Consider the ways in which physical geography influenced culture in Russia. In which part of the country do most followers of Islam live? Write a paragraph explaining why you think this is so.

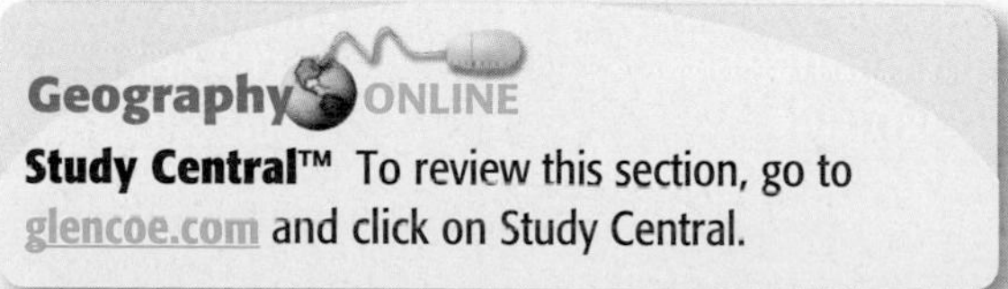

Geography ONLINE

Study Central™ To review this section, go to glencoe.com and click on Study Central.

Nationalism in Chechnya

Important Dates in Chechen History	
1991	USSR collapses; Chechnya declares independence
1994–1996	First Chechen War
1999	Russians invade; Second Chechen War begins
2003	Referendum approves new constitution giving Chechnya more autonomy
2007–2009	Ramzan Kadyrov becomes Chechen president; maintains support of Russia; claims insurgency has been crushed

Problem:

Since the beginning of Russian occupation, the Chechens have sought independence from foreign rule.

Occupation Russian occupation remains part of everyday life in Chechnya. These Russian soldiers talk as Chechen women travel through the rubble of Grozny, Chechnya's capital.

A memorial for Natalia Estemirova

Protest Natalia Estemirova, a human rights activist, was kidnapped and murdered in Chechnya in 2009. She was a critic of the Kremlin's policies in Chechnya.

Russia and the Republics

History of Tensions

Chechnya has historically been occupied by other powers. First controlled by the Turks and then by the Russians, Chechens have long strived for independence.

What is the history of the conflict? Although Russia has long controlled Chechnya, the republic has maintained its cultural heritage. Chechens have their own language and are predominately Sunni Muslim, and these differences have fueled the Chechens' desire for independence. Whenever internal or external conflict has weakened the Russian government, Chechen rebels have attempted to fight for autonomy, but their efforts have been unsuccessful. The Russian government has several reasons for maintaining Chechnya's status as a republic of the Russian Federation. Several oil and gas pipelines vital to the Russian economy run through Chechen territory. Also, if Russia were to grant Chechnya independence, it would likely face uprisings by other ethnic groups and republics within the Russian Federation.

What is the current political climate? In May 2000, Russian president Vladimir Putin established direct rule of Chechnya to try to stop the rebels. In 2003 a new constitution was passed in a referendum. Under this new constitution, Chechnya is still a republic within the Russian Federation, but it has been given a significant amount of autonomy. The new Chechen government is now struggling to recover from the violence waged by the Chechen rebels and to bring stability and peace to Chechnya and its people.

Solution:

Integrating Chechen representation into the Russian government may be the only solution to the ongoing animosity between the two groups.

Chechen president Ramzan Kadyrov (right) talks with Russian president Dmitry Medvedev about Chechen-Russian relations.

THINKING GEOGRAPHICALLY

1. **Human Systems** Research the conflict in Chechnya in more depth, and then write an essay detailing the diplomatic measures taken to try to bring stability to the republic.
2. **The World in Spatial Terms** How might Chechnya's location contribute to Russia's unwillingness to grant the republic its independence?

SECTION 2

section audio

spotlight video

Guide to Reading

Essential Question

How have Russia's history and government been shaped by its many ethnic groups?

Content Vocabulary

- czar *(p. 379)*
- serf *(p. 379)*
- Russification *(p. 380)*
- socialism *(p. 380)*
- Bolshevik *(p. 381)*
- communism *(p. 381)*
- satellite *(p. 381)*
- Cold War *(p. 381)*
- perestroika *(p. 382)*
- glasnost *(p. 382)*

Academic Vocabulary

- acquire *(p. 380)*
- policy *(p. 380)*
- aid *(p. 381)*

Places to Locate

- Moskva River *(p. 379)*
- St. Petersburg *(p. 380)*
- Vladivostok *(p. 380)*

Reading Strategy

Sequencing As you read about Russia's history, complete a time line similar to the one below by recording major events in the country's history.

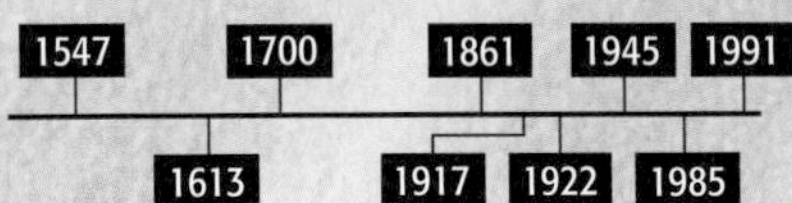

History and Government

Since the fall of the Soviet Union, Russians have struggled to move beyond this dark period of their past. Although difficult, the country is making steps toward democracy and a market economy.

NATIONAL GEOGRAPHIC VOICES AROUND THE WORLD

"A decade has passed since the U.S.S.R. ceased to exist, and during that time the Russian people have been subjected to nothing less than an economic and social revolution. Three-quarters of state enterprises have been fully or partly transferred to individual owners in a corrupt privatization drive. . . . [A]s a seven-week trip around Russia earlier this year showed, shoots of new life are springing up throughout the country."

—Fen Montaigne, "Russia Rising," *National Geographic,* November 2001

Signs of modern technology at an Internet café

Early History

MAIN Idea Russia's historical roots go back thousands of years and include many ethnic groups.

GEOGRAPHY AND YOU What do you know about the early history of the United States? Read to learn about the people involved in the rise of Russia.

Russia's historical roots go back to the A.D. 600s, when Slav farmers, hunters, and fishers settled near the waterways of the Northern European Plain.

Kievan Rus

During the 800s, Scandinavian warriors called the Varangians settled among the Slavs living near the Dnieper and Volga Rivers. Within a century the Varangians had adopted the Slav language and many Slav customs. The Slav communities were eventually organized into a loose union of city-states known as Kievan Rus. Ruled by princes, the leading city-state, Kiev, controlled a prosperous trading route, using Russia's western rivers to link the Baltic and Black Seas.

Eventually, fighting among the city-states weakened Kievan Rus. In the early 1200s, Mongols invaded Kiev and many of the Slav territories from their foothold in Central Asia. Although the Mongols allowed the Slavs self-rule, they continued to control the area for more than 200 years.

The Rise of Russia

Many Slavs fled the Mongol invasions and later settled along the **Moskva River.** One of their settlements grew into the city of Moscow, the center of a territory called Muscovy (muh•SKOH•vee), which was linked by rivers to major trade routes and surrounded by lands good for farming and fur trapping.

Muscovy's princes kept peace with the Mongols for about two centuries, but the peace did not last. Muscovy's Prince Ivan III then brought many Slav territories under his control, thus earning the nickname "the Great." Ivan's expanded realm eventually became known as Russia. In Moscow, Ivan built a huge fortress called the Kremlin and filled the city with churches and palaces. Russia's government today, the Russian Federation, uses the Kremlin as its executive headquarters.

In 1547 Ivan the Great's grandson, Ivan IV, became Russia's first crowned **czar** (ZAHR), or supreme ruler. He crushed all opposition to his rule and expanded his realm's borders into non-Slav territories, earning the name Ivan the Terrible.

After Ivan's reign, the country faced foreign invasion, economic decline, and social upheaval. When the Romanov dynasty came to power in 1613, the government gradually tightened its grip on the people. By 1650 many peasants had become **serfs,** a virtually enslaved workforce bound to the land and under the control of nobility.

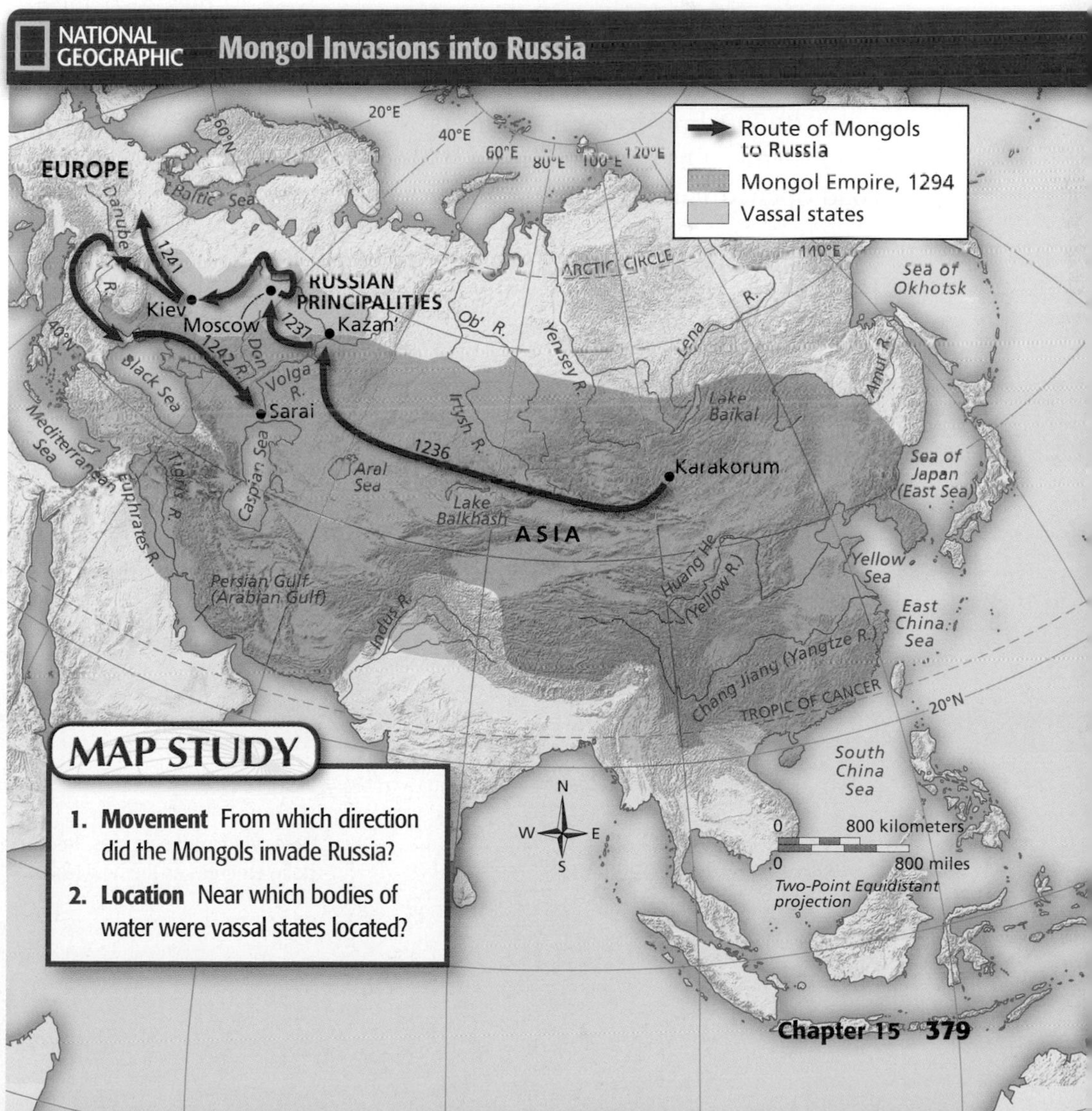

Romanov Czars and the Empire

While Russia struggled, western Europe moved forward and left Russia behind. Then in the late 1600s, Czar Peter I—known as Peter the Great—came to power, determined to modernize Russia. Under Peter I, Russia enlarged its territory, built a strong military, and developed trade with western Europe. To **acquire** seaports, Peter I gained land along the Baltic Sea from Sweden.

A new capital—**St. Petersburg**—was carved out of the wilderness along the Gulf of Finland, providing access to the Baltic Sea and giving Russia "a window on the West." Since most of Russia's other ports were icebound for almost half the year, St. Petersburg became a major port.

During the late 1700s, Empress Catherine the Great continued to expand Russia's empire and gained a long-sought-after warm-water port on the Black Sea. The Romanov expansion also brought many non-Russians under its rule. As the Russian nobility was adopting western European ways, a cultural gap developed between the nobility and the serfs. The serfs followed traditional Russian ways and faced poverty.

Beginning in 1891, under Czar Alexander III, Russia expanded into Siberia with the construction of the Trans-Siberian Railroad. Nearly 6,000 miles (9,700 km) long, it connects Moscow to **Vladivostok.** Once completed in 1916, the railroad opened Russia's interior to settlement.

READING Check **Location** How was St. Petersburg important to the expansion of the Russian empire?

Revolution and Change

MAIN Idea Discontent with inequality in Russian society led to revolution and freedom from generations of czarist rule.

GEOGRAPHY AND YOU What led to the American Revolution? Read to learn about the factors that led to the Russian Revolution and the rise of the Soviet Union.

A long cycle of halfhearted reforms, government repression, and the American and French Revolutions encouraged the desire among educated Russians to open up Russian society.

The Russian Revolution

Czar Alexander II's limited reforms, such as freeing the serfs in 1861 without providing for their education, caused many former serfs to move to cities. There they faced the poor conditions and meager wages of factory work. Non-Russian peoples also faced prejudice when the government introduced the **policy** of **Russification,** which encouraged people to speak Russian and follow Eastern Orthodox Christianity. Those who refused were harshly persecuted, especially the Jews, who were often blamed for Russia's problems.

This frustration and discontent led many Russian workers and thinkers to turn to **socialism,** a belief that calls for greater economic equality in society.

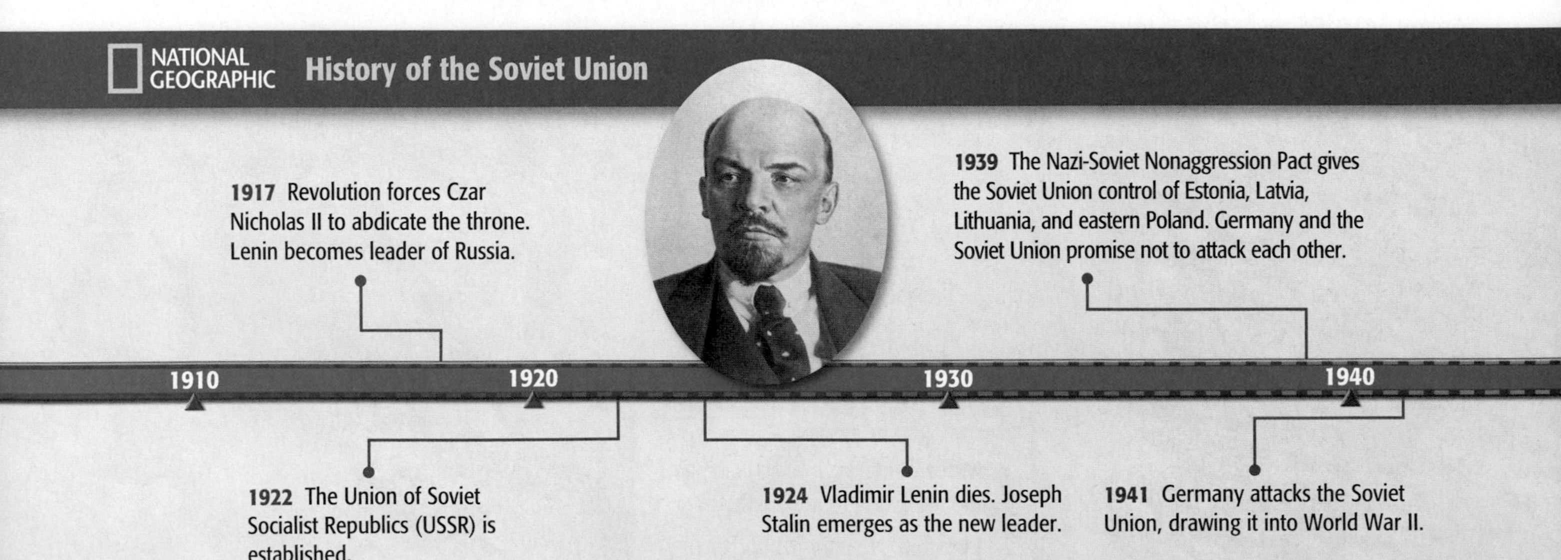

One of its biggest proponents, the German philosopher Karl Marx, advocated public ownership of all land and a classless society with an equal sharing of wealth. This would occur after a revolution, led by the working classes, against the wealthy.

Increasing discontent caused strikes and demonstrations to break out in the early 1900s. Then in 1917, the hardships of World War I brought even larger numbers of workers and now soldiers into the streets, demanding "bread and freedom." These actions forced Nicholas II to abdicate his throne, bringing czarist rule to an end. The following year, Czar Nicholas and his family were killed. Their murders signaled the demise of Europe's last absolute rulers and the emergence of communism in Russia.

The Soviet Era

The weak representative government established in 1917 made it easy for the **Bolsheviks,** a revolutionary group led by Vladimir Ilyich Lenin, to seize control. The Bolsheviks believed in **communism,** a philosophy based on Karl Marx's ideas that called for the violent overthrow of government and the creation of a society led by workers.

Promising the people "Peace, Land, and Bread!", the Bolsheviks withdrew Russia from World War I, surrendering much territory to Germany. They used their complete political control to take over industry, direct food distribution, and establish an eight-hour workday. Civil war eventually erupted between the Bolshevik Red Army and the anti-Bolshevik White Army.

The Bolsheviks won the civil war in 1922, and established the Union of Soviet Socialist Republics (USSR), or the Soviet Union. The Soviets gradually regained Ukraine, Belarus, and large parts of the Caucasus region and Central Asia.

After Lenin's death in 1924, Joseph Stalin became the leader of the Communist Party. Stalin took control of farms and factories as he made the USSR an industrial giant. He eliminated those who disagreed with him, and millions were either killed or died from hunger, physical hardships, or brutal conditions in labor camps.

A Superpower

The Soviet Union attained superpower status after World War II. The USSR controlled most of Eastern Europe at the war's end. By 1949 most of the region's countries had become **satellites,** countries controlled by the Soviet Union.

The next four decades saw the Soviet Union and the United States engaged in the **Cold War,** the struggle between two competing systems—communist and capitalist—for world influence and power. Although both countries built destructive nuclear weapons, the "weapons" used for this war were propaganda, the threat of force, and **aid** to developing countries.

READING Check **Regions** How did the size of Russia change after the Soviets gained control?

RUSSIA

GRAPH STUDY

1. **Regions** What did the Soviet Union gain from the Nazi-Soviet Nonaggression Pact?
2. **Place** What countries might the Soviet Union have had influence over after World War II?

1945 Germany is defeated in World War II. Postwar agreements give the USSR influence over eastern and southeastern Europe.

1961 Soviet astronaut Yury Gagarin becomes the first human to orbit the Earth.

1991 The USSR is dissolved.

1950 — 1960 — 1990

1953 Joseph Stalin dies. Nikita Khrushchev becomes the Communist Party leader.

1985 Mikhail Gorbachev comes to power and proposes *perestroika*, a restructuring of the economy.

Movements for Change

MAIN Idea New ideas about political and economic systems led to changes in Russia.

GEOGRAPHY AND YOU Do you believe that there is enough political openness in the United States? Read to learn how political openness helped Russia transition from its Communist past.

The breakup of the Soviet Union required leaders to change their ideas about governing a superpower in an increasingly interconnected world.

The Fall of the Soviet Union

The Soviet Union's weakening economy, along with great discrepancies between workers' wages and the privileges their leaders enjoyed, led to the breakup of the Soviet Union. In 1985 Mikhail Gorbachev, a reform-minded official, assumed power in the Soviet Union. He instituted a policy of economic restructuring called **perestroika** (PEHR•uh•STROY•kuh) and a policy of greater political openness called **glasnost** (GLAZ•nohst).

Many of the satellites overthrew their Communist rulers in 1989. A failed coup in 1991 to overthrow Gorbachev collapsed, but all the republics declared independence by year's end. Twelve of the 15 new countries became members of the Commonwealth of Independent States (CIS). The three Baltic countries did not. Boris Yeltsin was elected the first president of the Russian republic.

A New Russia

The new Russia began moving from a command economy to a market economy, closing outdated factories and restructuring agriculture. Separatist movements and ethnic conflict also threatened the stability of Russia. Beginning in the 1990s, Tatarstan, Dagestan, and other ethnic territories demanded greater self-rule. Some demands have erupted into war. Boris Yeltsin's successor, Vladimir Putin, inherited those conflicts when he became president in 1999.

Putin helped stabilize the economy by instituting reforms in labor, banking, and private property. He also helped Russia get involved with NATO through the NATO-Russia Council. After winning reelection in 2004, Putin instituted changes that were seen by many as taking a step away from democracy. Barred from a third term as president, Putin endorsed Dmitry Medvedev as his successor. After winning the election, Medvedev nominated Putin for prime minister. This allowed Putin to remain heavily involved in Russian government.

READING Check **Regions** Which Soviet republics formed the Commonwealth of Independent States?

SECTION 2 REVIEW

Vocabulary

1. Explain the significance of: czar, serf, Russification, socialism, Bolshevik, communism, satellite, Cold War, perestroika, glasnost.

Main Ideas

2. How did the Bolsheviks use their complete power to create the Soviet Union?
3. How did Mikhail Gorbachev help the transition away from communism?
4. Using a graphic organizer like the one below, list the key events in Russia or in the Soviet Union during each of the following time periods: Kievan Rus, Russian Empire, Soviet Union, and Russia.

Era	Dates and Key Events

Critical Thinking

5. **Answering the Essential Question** How did the migration of the Slavs and their interactions with other groups influence the history of Russia?
6. **Determining Cause and Effect** What led to the breakup of the Soviet Union? Write a paragraph explaining your answer.
7. **Analyzing Visuals** Study the map of Mongol invasions on page 379. How many miles separate Karakorum and Moscow?

Writing About Geography

8. **Expository Writing** Write a paragraph explaining why you agree or disagree with the following statement: "The Soviet Union was a 74-year-long experiment that failed."

Study Central™ To review this section, go to glencoe.com and click on Study Central.

VISUAL SUMMARY

Study anywhere, anytime by downloading quizzes and flashcards to your PDA from glencoe.com.

People and Ethnicity

- Russia is ethnically diverse.
- Most Russians are Slavic in origin.
- The diversity of people in Russia has led to many ethnic groups demanding greater self-rule or independence.
- In some places, like Chechnya, the groups have resorted to violent methods, such as terrorism.

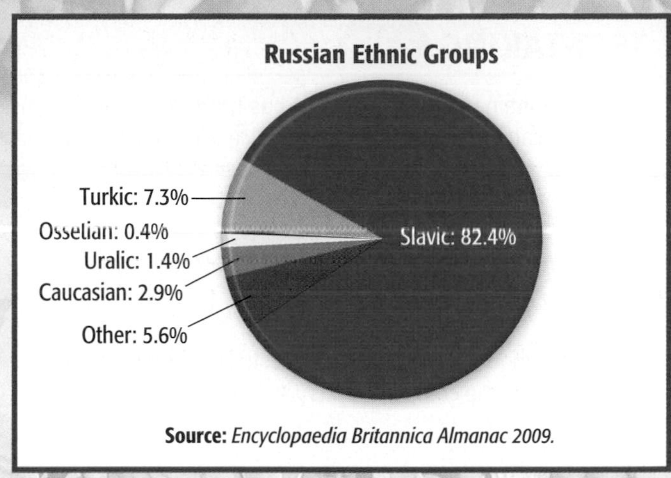

Nenets women cast their votes in the presidential election.

History and Government

- In Russia's early years it was ruled by czars, who were selected by birth.
- During the Soviet era, leaders were selected by a small group of Communist Party insiders.
- Democratic reforms were established in Russia following the collapse of the Soviet Union in 1991.
- Today Russia has free elections, but many fear that government is moving away from democracy.

Changing Economic Systems

- Under the czars, the economy was run by serfs bound to nobles.
- During the Soviet era, a command economic system was used.
- Capitalism was introduced to Russia after the fall of the Soviet Union in 1991.
- When Russia switched economic models, it caused instability in the marketplace, with high inflation and unemployment.
- Reforms put into place by Vladimir Putin have helped to stabilize the economy.

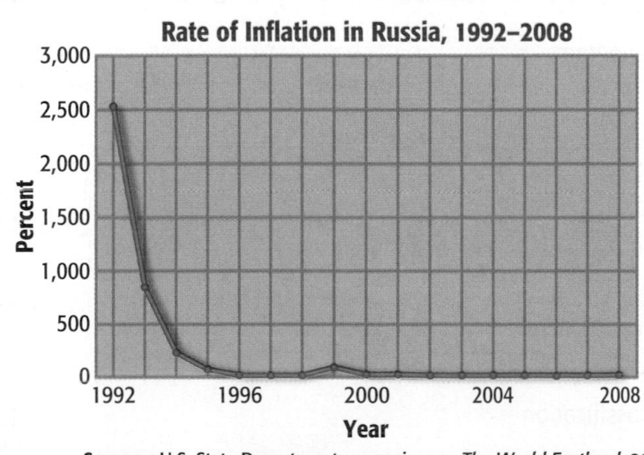

RUSSIA

CHAPTER 15

STANDARDIZED TEST PRACTICE

After you eliminate the choice that has nothing to do with Russia or the topic of the question, see if you can eliminate one other choice and thus narrow the possibilities down to two.

Reviewing Vocabulary

Directions: Choose the word or words that best complete the sentence.

1. A people who share a common ancestry, language, religion, customs, or some combination is ________.

A a soviet
B a nation
C an ethnic group
D a race

2. ________ is a belief that there is no God or other supreme being.

A Soviet
B Atheism
C Ethnicity
D Turkic

3. The supreme rulers of Russia were known as ________.

A czars
B Muscovites
C kings
D emperors

4. Gorbachev's economic restructuring plan was known as ________.

A glasnost
B communism
C perestroika
D Russification

Reviewing Main Ideas

Directions: Choose the best answers to complete the sentences or to answer the following questions.

Section 1 *(pp. 370–375)*

5. Where do most Russians live?

A western Russia
B eastern Russia
C the Ural Mountains
D the Caucasus Mountains

6. After the fall of the Soviet Union, what happened to religious life in Russia?

A The government continued to insist that everyone follow atheism.
B The Roman Catholic Church became the predominant religion.
C Most people became Protestant.
D Most people who claim a religious affiliation have returned to the Russian Orthodox Church.

Section 2 *(pp. 378–382)*

7. What development opened Russia's interior to settlement?

A the Trans-Siberian Railroad
B global warming
C the conquest of a warm-water port
D the defeat of the Mongols

8. Which group eventually gained power in Russia after the Russian Revolution?

A White Russians
B Socialists
C Communists
D Republicans

GO ON

Critical Thinking

Directions: Choose the best answers to complete the sentences or to answer the following questions.

9. Years of frustration led up to the Russian Revolution, but the event that finally brought it about was ________.

A a major famine in the countryside

B loss of freedom of the Russian Orthodox Church to practice its religion

C the killing of the czar and his family

D the suffering during World War I

Base your answer to question 10 on the map and on your knowledge of Chapter 15.

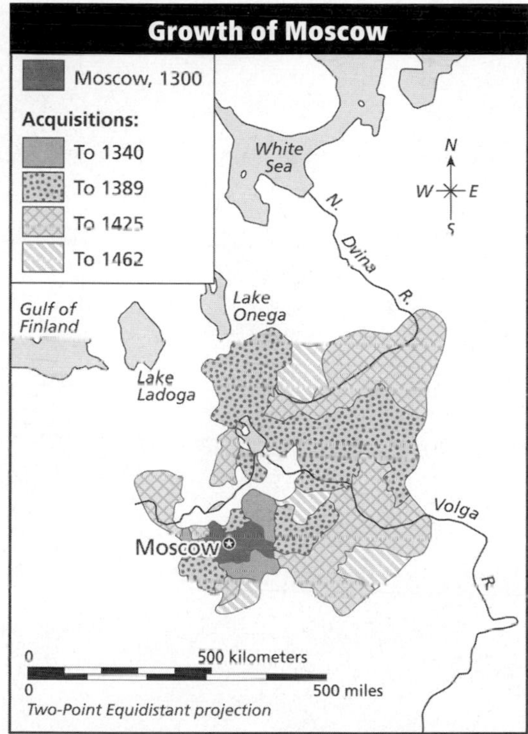

10. In what direction from Moscow were most of the acquired lands to 1389?

A north

B south

C east

D west

Document-Based Questions

Directions: Analyze the document and answer the short-answer questions that follow the document.

In the 1980s, Mikhail Gorbachev proposed *perestroika,* or restructuring, of the Soviet economic system. Here is his description of the reasons for the reforms.

> *In effect, we have here a new investment and structural policy. The emphasis has been shifted from new construction to the technical retooling of enterprises, to saving the resources, and sharply raising the quality of output. We will still pay much attention to the development of the mining industries, but in providing the economy with raw materials, fuel and power, the emphasis will now be on the adoption of resource-saving technologies, on the rational utilization of resources. . . .*
>
> *The economy has, of course, been and remains our main concern. But at the same time we have set about changing the moral and psychological situation in society. . . .*
>
> *We have come to the conclusion that unless we activate the human factor, that is, unless we take into consideration the diverse interests of people, work collectives, public bodies, and various social groups, unless we rely on them, and draw them into active, constructive endeavor, it will be impossible for us to accomplish any of the tasks set, or to change the situation in the country.*
>
> —Mikhail Gorbachev,
> *Perestroika: New Thinking for Our Country and the World*

11. What does Gorbachev name as the major concern of the Soviet Union?

12. According to Gorbachev, what is essential to making perestroika successful?

Extended Response

13. Exploring the BIG Idea
Describe the changes in Russia since the fall of communism. How has the economy changed?

For additional test practice, use Self-Check Quizzes—Chapter 15 on glencoe.com.

Need Extra Help?

If you missed questions. . .	1	2	3	4	5	6	7	8	9	10	11	12	13
Go to page. . .	371	373	379	363	372	373	380	381	381	385	385	385	382

CONNECTING TO

THE UNITED STATES

A production of The Nutcracker

A Russian Orthodox church in San Francisco, California

Just the Facts:

- The fall of communism in the Soviet Union led to Russian involvement in the International Space Station beginning in 1998.
- In 1794, the first Russian Orthodox Church in America was established in Alaska. Today the Orthodox Church in America has approximately 1 million members and 400 parishes countrywide.
- According to the 2000 U.S. Census, 706,242 Americans listed Russian as their spoken language.
- The Russian American population is estimated to be approximately 3 million people.

U.S. astronauts and a Russian cosmonaut aboard the International Space Station

Russian-born tennis player Maria Sharapova

Nobel prize-winning Russian author Aleksandr Solzhenitsyn

Making the Connection

Russian culture has influenced American culture through the arts, sports, religion, and space exploration. Russian influences, while not always obvious, have been significant.

Cold War Influences Russian cultural influence in the United States has in large part been based on the competition between the two nations as the result of the Cold War. Although never directly engaged in armed conflict, the United States and the Soviet Union were continually at odds and in competition with each other. This competition was exhibited through sporting events and the space race, each country striving to top the other.

Influencing the Arts Although ballet itself did not originate in Russia, the Russian ballet community made lasting changes to the art form and helped to popularize it throughout the world. Tchaikovsky (chy•KAWF•skee), a Russian composer who lived during the late 1800s, wrote several ballets that remain popular today. Most notable is *The Nutcracker,* a Christmas holiday favorite in the United States and around the world.

Sports Heroes—Russian Heritage Since the fall of communism, many Russian athletes have become popular American sports heroes. Russian influence is most notable in ice hockey and tennis. Some of the most recognizable of these Russian-born athletes are hockey players Sergei Fedorov and Alexander Ovechkin, and tennis player Maria Sharapova.

Russian Literature

Title	Author
War and Peace	Lev Tolstoy (tawl•STOY)
Anna Karenina	Lev Tolstoy
Crime and Punishment	Fyodor Dostoyevsky (DAHS•tuh•YEHF•skee)
Doctor Zhivago	Boris Pasternak (PAS•tuhr•NAK)
A Day in the Life of Ivan Denisovich	Aleksandr Solzhenitsyn (SOL•zhuh•NEET•suhn)

THINKING GEOGRAPHICALLY

1. **Human Systems** Research a Russian-born athlete mentioned above or one of your choosing. Describe the contributions that person has made to his or her sport.
2. **Environment and Society** Investigate Russian and American cooperative involvement with the International Space Station. What have both countries contributed to this project?

CHAPTER 16

THE REGION TODAY

Russia

BiG Idea

Disputes over ideas, resources, values, and politics can lead to change. Russia continues to adjust to difficult political and economic challenges, which require the country to find a way to balance its need for economic growth with preservation of the environment.

Essential Questions

Section 1: The Economy

How has Russia's transition to a market economy affected its relationships with other countries?

Section 2: People and Their Environment

How might human actions that have modified Russia's environment affect its growth?

Visit glencoe.com and enter QuickPass™ code WGC9952C16 for Chapter 16 resources.

Overlooking the Moskva River, the Moscow Kremlin is the official residence of the president of the Russian Federation. On the right is St. Basil's Cathedral.

Identifying Information Make a Four-Door Book to help you identify information about the Chernobyl disaster, which took place in the Soviet Union in 1986.

Reading and Writing As you read the chapter, write down details about the Chernobyl disaster. Imagine yourself as a news reporter on the scene in 1986, gathering details to present a complete report about the causes of the disaster.

Economic Activities

MAIN Idea Russia's transition to a market economy has transformed agriculture and industry.

GEOGRAPHY AND YOU Where are major manufacturing industries located in the United States? Read to learn about Russia's major industries.

Russia's transition to a market economy has included restructuring agriculture, privatizing industry, opening markets to foreign investors, and managing natural resources more efficiently.

Agriculture

Soviet-era farms were organized into state-controlled kolkhozes (kahl•KAW•zehz) and sovkhozes (sahf•KAW•zehz). **Kolkhozes** were farms worked by farmers who shared, to a degree, in the farm's production and profits. **Sovkhozes** were farms run more like factories, with the farmworkers receiving wages. However, the government controlled the prices and production in both the agricultural and industrial sectors. These sectors suffered because the system did not motivate workers.

In 1991 President Yeltsin began restructuring state-run farms so they could function better in a market economy. However, Russian farmers—accustomed to the **stability** of Soviet controls—continued to operate many of Russia's farms as kolkhozes or sovkhozes. Most farmers could not afford to buy land, and they worried that wealthy Russians or foreign investors might use the land for nonagricultural development. Because of these concerns, progress toward a market economy for agriculture has been slow. In 2002, however, a new land code made it easier for Russians to buy and sell farmland, which helped speed up the restructuring process. Over the last several years, Russian agricultural exports have increased, and the country has also become a large grain exporter.

NATIONAL GEOGRAPHIC **Contribution of Industry to Russia's GDP**

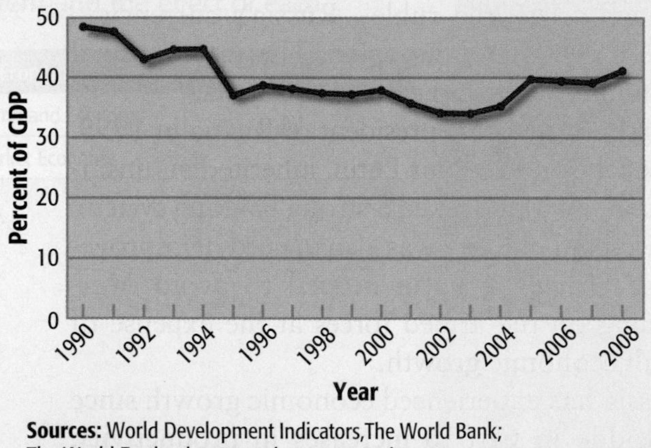

Sources: World Development Indicators, The World Bank; The World Factbook 2008 and 2009, www.cia.gov.

GRAPH STUDY

1. **Regions** Describe the trends in Russia's GDP from 1990 to 2008.
2. **Regions** In what year did the industrial sector's contribution to GDP reach its lowest point?

Industry

Russian industry has been transformed since the early 1990s. The graph at the left shows that the share of GDP from the industrial sector fell after the collapse of the Soviet Union. It has taken years to approach the Soviet levels. For many years, Russia's state-owned aerospace industry and its military-industrial system were its economic and technical focus. Many of these components have become privately owned and provide export income. Russia has also encouraged foreign investment by selling shares of ownership in some Russian companies and by opening Russia's markets to Western companies.

In general, Russia's economy is expanding, particularly in the areas of industry and services. Russia's most important industry is petroleum extraction and processing, and the country is one of the world's largest producers of crude oil. Russia's domestic oil provides its other industries with vital energy at a reasonable cost. While energy resources dominate Russian exports, minerals also provide important export income. The country is also a major producer of iron ore, manganese, and nickel. Huge forests in Russia produce one-fifth of the world's softwood, and Russian supertrawlers, or fish-factory ships, process catches from both the Atlantic and Pacific Oceans.

READING Check **Regions** What is Russia's most important industry?

Transportation and Communications

MAIN Idea Russia is developing and improving transportation and communications systems to help with the transition to a market economy.

GEOGRAPHY AND YOU How might Russia's physical geography and climate present obstacles to the country's transportation and communications systems? Read to learn how Russia is trying to overcome these obstacles.

In an age of speedy transportation and the Internet, Russia struggles to find funds for new highways and high-tech communications.

NATIONAL GEOGRAPHIC *The Trans-Siberian Railroad is the greater part of the rail route from Moscow to the Pacific port city of Vladivostok.*

Regions How does Russia's climate influence transportation in the country?

Transporting Goods

Russia's transportation systems must move resources great distances to reach consumers. A major highway system links Moscow with other major cities, but many roads are in poor repair. Melting snow and the spring thaw in places like **Siberia** often make roads impassable.

Because of its great size and climate extremes, Russia depends on railroads and waterways for most of its transportation needs. Russia boasts the world's longest continuous railroad line, the Trans-Siberian Railroad. Major cities are found where the Trans-Siberian Railroad crosses large rivers. Millions of tons of goods travel along thousands of miles of navigable inland waterways, which connect seaports and inland cities.

Pipelines are effective in transporting petroleum products, although constructing and maintaining them can be difficult in areas of harsh climate. Pipelines crisscross Russia, providing Russian cities and parts of western Europe with fuel. Minor pipelines run through **Chechnya** and **Dagestan** (DAH•guh•STAHN), republics in southwestern Russia. People in these republics are fighting for more self-rule from Russia, so control of the area's oil reserves and pipelines is a major concern.

Transporting People

Most Russians live in cities, and many do not own cars. Therefore, public transportation—such as trains, buses, and subways—is common. Private car ownership is a symbol of middle-class status, but public transportation remains a practical option for many. Some of the systems and equipment, however, need repair and improvements.

The Soviet Union used jet airplanes for passenger traffic, and the government financially supported air travel. The passenger airline Aeroflot was once the only one in the Soviet Union. Today, Aeroflot is the largest airline in Russia. It competes for business with other airlines, both domestic and foreign. Most major cities have national and international carriers.

Mass Communications

During the Soviet era, the state owned and controlled all mass communications systems. State agencies reviewed all print and broadcast materials to make sure they contained no criticism of the government. Since the fall of the Soviet Union, Russians have heard and read new voices and fresh views. Under Putin, however, the Kremlin exerted greater control over national television networks.

Russia has made progress in building the telecommunications infrastructure needed for a market economy. This has allowed the number of cell phone subscribers to jump from 1 million in 1998 to 188 million in 2007. And Russia ranks twelfth in the world in number of Internet users.

READING Check **Human-Environment Interaction** Why does Russia depend on railroads and waterways for most of its transportation needs?

Trade and Interdependence

MAIN Idea Russia is expanding international trade and is working to build political and financial relationships within the global community.

GEOGRAPHY AND YOU What role do energy resources have in U.S. involvement in the global economy? Read to learn about the importance of energy in Russia's international trade relationships.

Russia has focused on becoming a full partner in the global community by expanding trade and building international relationships.

Trade

Russia is a major source of energy and fuels, which account for over 50 percent of its exports. The European Union, other former Soviet republics, China, and Japan are among Russia's major trading partners.

Energy exports are expected to remain Russia's main item of international trade until its manufactured goods improve in quality and become more competitively priced. Working to strengthen its industries, Russia became a member of the Asia-Pacific Economic Cooperation (APEC) in 1998. In 2006, Russia and the United States signed a bilateral agreement for Russia's entry into the World Trade Organization (WTO).

International Relations

Russia works to maintain its role in world affairs. It benefits from occupying the former Soviet Union's seat in the United Nations Security Council. It has also joined European organizations that support security and cooperation. Russia has helped settle conflicts and has supported peace efforts in several countries. Even as Russia asserts itself internationally, however, economic problems have drained money from its military.

Adequate financial resources are vital to Russia's stability and progress in the global community. Other countries and world organizations have provided loans, and foreign investors have made funds available to Russian industry. With such help, Russia is trying to create secure and workable systems for banking, farming, manufacturing, transportation, and communications.

READING Check **Place** What natural resource accounts for the majority of Russia's total exports?

SECTION 1 REVIEW

Vocabulary

1. Explain the significance of: command economy, consumer good, black market, market economy, privatization, kolkhoz, sovkhoz.

Main Ideas

2. What challenges has Russia faced in its transition to a market economy?
3. How is Russia working to improve its transportation and communications systems?
4. What steps has Russia taken to expand international trade and build relationships in the global community?
5. Using a chart like the one below, fill in details about agriculture and industry in the Soviet command economy and in the Russian market economy. Then explain how these two sectors of the economy have been transformed.

	Soviet Command Economy	Russian Market Economy
Agriculture		
Industry		

Critical Thinking

6. **Answering the Essential Question** What actions has Russia taken to become part of the global economy?
7. **Identifying Cause and Effect** How did the transition from a command economy to a market economy affect the Russian people?
8. **Analyzing Visuals** Study the physical map on page 348 and the economic activity map on page 352 of the Regional Atlas, and the vegetation map on page 363. In what area is the raising of livestock concentrated? How is this related to the physical geography of the region?

Writing About Geography

9. **Expository Writing** Write a paragraph explaining how Russia's vast size affects the availability of natural resources and the country's ability to develop them.

Study Central™ To review this section, go to glencoe.com and click on Study Central.

SECTION 2

 section audio  spotlight video

People and Their Environment

Guide to Reading

Essential Question

How might human actions that have modified Russia's environment affect its growth?

Content Vocabulary

- nuclear waste *(p. 396)*
- radioactive material *(p. 396)*
- pesticide *(p. 397)*

Academic Vocabulary

- contribute *(p. 397)*
- resident *(p. 397)*
- cite *(p. 398)*

Places to Locate

- Barents Sea *(p. 396)*
- Baltic Sea *(p. 396)*
- Bering Sea *(p. 396)*
- Chernobyl *(p. 396)*
- Lake Baikal *(p. 397)*
- Noril'sk *(p. 397)*
- Kamchatka *(p. 397)*

Reading Strategy

Organizing As you read about Russia's environment, complete a graphic organizer similar to the one below by describing the environmental issues and concerns for each location.

Location	Description	Concerns
Chernobyl		
Lake Baikal		
Kamchatka		

Global demand for natural resources such as oil and timber has created a difficult challenge for Russia—how to manage its natural resources and balance economic growth with environmental conservation.

NATIONAL GEOGRAPHIC Voices Around the World

"I visited the Khabarovsk region . . . in mid-September. With Russia's economy and its wood products industry on the rebound after the 1990s—when timber harvesting dropped by 60 percent—there was abundant evidence of logging. Train cars stacked high with lumber rumbled down the Baikal-Amur Railway, much of it bound for China, where a growing economy and a near moratorium on logging have created high demand for Siberian wood. Khabarovsk's easily accessible forests have already been logged, forcing timber companies to forge ever deeper into the taiga. From the city of Komsomolsk on Amur I drove five hours to reach one of the closest logging areas. Along the way I passed numerous areas of burned forest—gray swaths of scorched tree trunks sprawling over the undulating landscape."

—Fen Montaigne, "Boreal: The Great Northern Forest," *National Geographic*, June 2002

A Russian timber worker

RUSSIA

Human Impact

MAIN Idea The environmental damage caused by Soviet-era industrialization continues to pose risks to natural resources and human health.

GEOGRAPHY AND YOU How would you describe the quality of the water and air in your community? Read to learn why pollution is such a major problem for Russia's water, soil, and air.

The Soviets' disregard for the environmental effects of industrialization damaged Russia's water, air, and soil.

Nuclear Wastes Between 1949 and 1987, the Soviet Union set off more than 600 nuclear explosions. Soviets developed and then stockpiled nuclear weapons throughout the Cold War. Today, the condition and fate of those weapons concern Russia and the rest of the world.

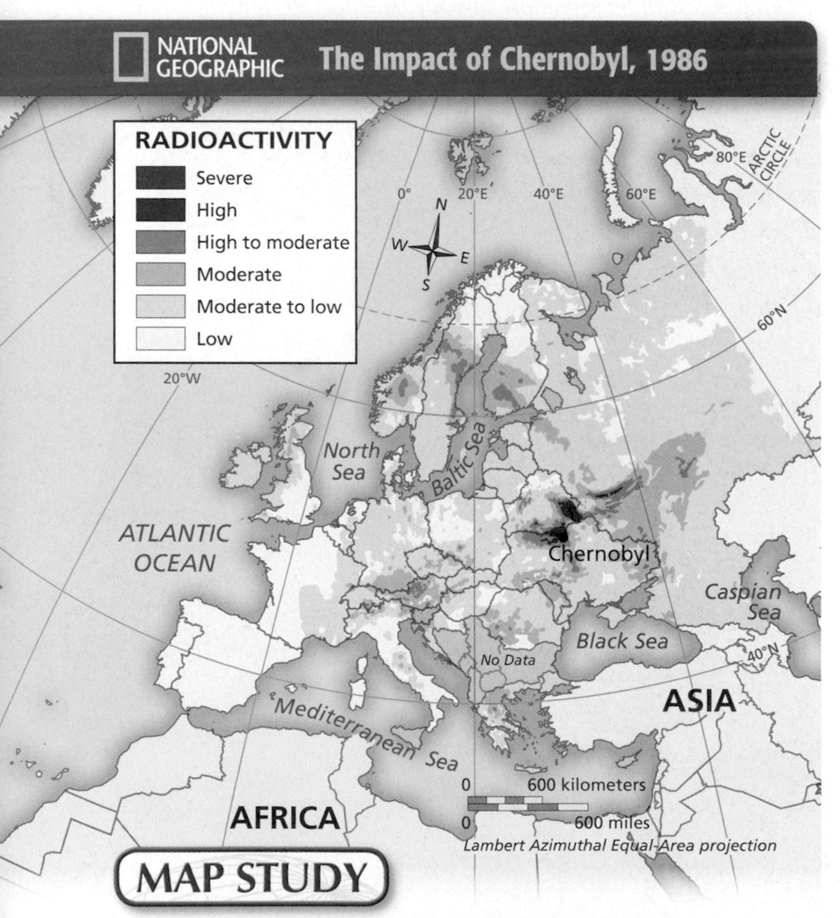

MAP STUDY

1. **Location** Which European countries had areas with high or severe levels of radioactivity?
2. **Regions** Which subregion of Europe–eastern or western–was more affected by radioactivity?

Nuclear wastes are the by-products of producing nuclear power and weapons. Some nuclear wastes can remain radioactive for thousands of years, posing dangers to people and the environment. The Soviets placed most nuclear wastes in storage facilities, but some **radioactive materials** —material contaminated by residue from the generation of nuclear energy and weapons—were dumped directly into the **Barents Sea,** the **Baltic Sea,** and the **Bering Sea.**

Chernobyl Past and Present In 1986 a fire in a nuclear reactor in the town of **Chernobyl** (chuhr•NOH•buhl), 60 miles (97 km) north of Kiev, Ukraine, released tons of radioactive particles into the local environment—400 times more radioactivity than was released at Hiroshima. Radiation covered tens of thousands of square miles of farmland and forests in the Soviet republics of Belarus, Ukraine, and Russia. Because of prevailing winds, other countries suffered as well.

Millions of people were exposed to deadly levels of radiation because Soviet officials were slow to alert the public to the crisis and did not evacuate people soon enough. Thousands of people died as a direct result of radiation poisoning. Millions more continue to suffer from cancer, stomach diseases, cataracts, and immune system disorders. Approximately 350,000 people were displaced from their homes. Today there is a tightly controlled exclusion zone surrounding the Chernobyl Nuclear Power Plant.

After the accident, international pressure prompted Soviet leaders to improve nuclear safety standards and to shut down dangerous plants. Despite concerns, 29 nuclear reactors continue to provide some of the country's electricity. Experts think that many remaining Soviet-era reactors are poorly designed and unsafe. Russia plans to expand its nuclear power industry by building more reactors and new power plants.

In late 2000, the remaining reactor at Chernobyl was shut down. In 2006, twenty years after the explosion at Chernobyl, hazardous work began to build a new structure to replace the fragile sarcophagus that entombs the deadly reactor. The new structure will not be a permanent solution. However, it could help prevent another Chernobyl disaster if the old structure were to collapse and release radioactive dust, or if water leaking into the old structure were to set off a nuclear chain reaction.

Water Quality Industrialization has polluted most of Russia's lakes and rivers. Fertilizer runoff, sewage, and radioactive material all **contribute** to poor water quality. The waters of the Moskva and Volga Rivers pose health risks. Dams along the Volga River trap contaminated water. Pollution also threatens the Caspian Sea.

Lake Baikal (by•KAWL) is the world's oldest and deepest lake. It contains one-fifth of the world's freshwater and 1,500 native species of aquatic plants and animals. Calling it "the Pearl of Siberia," Russians consider the lake a natural wonder. In 1957 the Soviet Union announced a plan to build a paper-pulp factory along its shores. Although this plan was opposed by people in the area, their protests were ignored, and the factory was built. This factory and others that followed dumped industrial waste into the lake.

However, in response to ongoing protests, the most serious polluters have been closed. Others are working to reduce pollution. Pollution levels in the lake are now relatively low compared with many lakes in Europe.

Soil and Air Quality For decades, toxic waste dumps and airborne pollution poisoned Russia's soil. Aging storage containers cracked, and toxic wastes leaked into the soil. Petroleum pipelines often broke, allowing petroleum to ruin the land. Overuse of fertilizers and **pesticides**—chemicals used to kill crop-damaging insects, rodents, and other pests—polluted farmland and water.

Russian experts believe that during the 1990s only 15 percent of Russia's urban population lived with acceptable air quality. Air pollution as far north as the remote Siberian town of **Noril'sk** has been a factor since the time of Stalin.

Industries, vehicle emissions, and burning coal are all sources of air pollution. In addition to releasing soot, sulfur, and carbon dioxide into the air, burning coal leads to acid rain. Acid rain and chemical pollution have reduced Russian forests dramatically.

READING Check **Human-Environment Interaction** What factors contribute to poor air quality in Russia?

Geography ONLINE

Student Web Activity Visit glencoe.com, select the *World Geography and Cultures* Web site, and click on Student Web Activities–Chapter 16 for an activity about Lake Baikal.

Managing Resources

MAIN Idea Russia is trying to repair past damage to the environment as well as manage natural resources without causing further harm.

GEOGRAPHY AND YOU What environmental issues are important in your community? Read to learn about the environmental challenges Russia faces today.

The World Bank's Sustainable Forestry Pilot Project is helping Russia manage its forests. Using land more wisely, protecting forests, planting new trees, and increasing private investment all help Russia's environment and economy. Increased employment opportunities in the forest industry and more stable local economies will be possible only if steps to conserve the forests are taken.

People have come together to oppose a mining operation in remote **Kamchatka** (kuhm•CHAHT•kuh) in eastern Russia. Environmental groups have demanded that the mining company meet strict environmental standards. The possible threat to the area's salmon spawning grounds prompted the local fishing industry to support the effort. The mine also caused concern among local **residents** because it was close to a protected wildlife area. Even with growing environmental awareness, economic pressure continues to open other regions to development.

READING Check **Human-Environment Interaction** Why is forest management important to Russia?

NATIONAL GEOGRAPHIC *Threats to salmon in Kamchatka also impact the area's brown bears, whose diet depends on the availability of salmon.*

Human-Environment Interaction How do economic development and environmental protection cause conflict in Russia?

RUSSIA

Challenges for the Future

MAIN Idea Russia faces many challenges as the country's growing economy and demand for natural resources impacts the environment.

GEOGRAPHY AND YOU What impact does the world's demand for oil have on the environment? Read to learn about some of the environmental challenges in Russia's future.

Fish populations are shrinking worldwide, and Russia's supertrawlers are largely to blame. Towing huge trawl nets—some large enough to scoop up a whale—supertrawlers are floating fish factories. These ships can catch and process more than 400 tons (360 t) of fish a day. Because supertrawlers want only certain kinds of fish, everything else hauled up in the nets gets discarded. Millions of fish and other marine animals die unnecessarily every year. Since smaller boats cannot compete with supertrawlers, the big ships threaten traditional fishing cultures. Supporters of supertrawlers **cite** the growing global demand for fish and fish products. Russian officials must balance the risk of destroying fish stocks with the need for a profitable fishing industry.

Pipelines built to transport oil and gas pass through wilderness areas and threaten the surrounding environment and wildlife. In 2006 Russia began constructing a highly controversial pipeline to carry oil from eastern Siberia to the Pacific Ocean. It will bring Russia billions of dollars from countries in the Asia-Pacific region. The pipeline will pass through a protected wilderness area near Lake Baikal. President Putin ordered that the proposed route be diverted farther away from the lake, but environmentalists still fear the irreversible damages that could be caused by an oil spill.

The widespread trend toward global warming is happening at a dramatic pace in western Siberia. An unprecedented thawing of the world's largest peat bog could release into the atmosphere billions of metric tons of methane, a powerful greenhouse gas. Where permafrost once covered the sub-Arctic region of western Siberia, shallow lakes now stand.

READING Check **Human-Environment Interaction** Why is a new pipeline linking eastern Siberia to the Asia-Pacific region controversial?

SECTION 2 REVIEW

Vocabulary

1. Explain the significance of: nuclear waste, radioactive material, pesticide.

Main Ideas

2. How is Russia trying to reverse past damage to its natural resources as well as manage them responsibly today?
3. Describe the challenges Russia faces as its growing economy and the demand for natural resources impact the environment.
4. Create a graphic organizer like the one below to explain how the environmental damage caused by Soviet-era industrialization continues to pose risks to natural resources and human health.

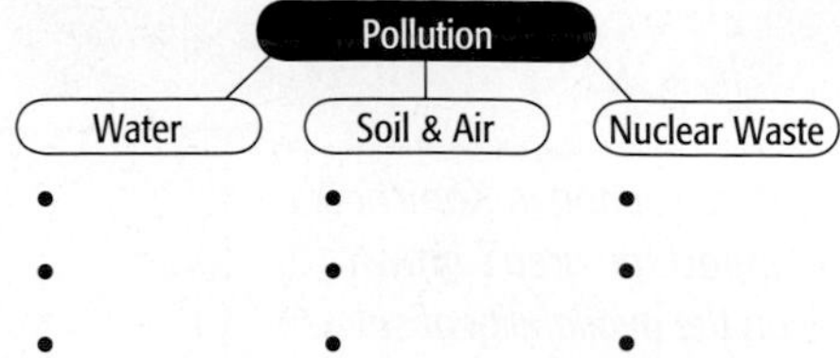

Critical Thinking

5. **Answering the Essential Question** How did the Soviet government's development of heavy industry affect the quality of Russia's environment?
6. **Making Generalizations** What generalizations can you make about the relationship between economic development and the environment in Russia?
7. **Analyzing Visuals** Study the economic activity map on page 352. Think about the regions of Russia in which pollution is a problem. Describe the relationship between the location of manufacturing centers and pollution.

Writing About Geography

8. **Expository Writing** Think about the challenges Russia faces concerning water quality. Write a paragraph explaining why Russians do not use more water from Lake Baikal to supply their freshwater needs.

Study Central™ To review this section, go to glencoe.com and click on Study Central.

The Region Today

Wednesday | Section C

An Economic Facelift

Command Economy

- Under Communist leadership, the Soviet Union operated as a command economy.
- The government emphasized heavy industry, making it an industrial giant and a world power.
- The country struggled with low wages, poverty, and scarcity.
- While other countries invested in technology, the Soviet Union continued to push heavy industry.

Russian GDP Per Capita

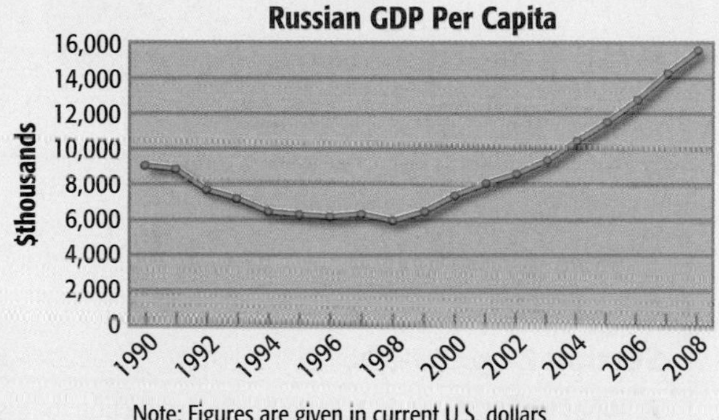

Note: Figures are given in current U.S. dollars.
Source: The World Bank Group, World Development Indicators.

Market Economy

- Russia shifted to a market economy in the 1990s.
- The government encouraged small businesses and foreign investment.
- Mass privatization resulted in a growing middle class.
- Today energy resources, particularly petroleum, dominate Russian exports.

Preserving Russia's Natural Resources

- Russia is a land of abundant natural resources, such as oil and timber.
- The World Bank's Sustainable Forestry Pilot Project is helping Russia protect against deforestation.
- The pollution of Russia's water supply has been reduced thanks to new industrial and pollution regulations.
- Nuclear wastes, left over from the Soviet era, are still a problem in some parts of Russia.

RUSSIA

CHAPTER 16

STANDARDIZED TEST PRACTICE

After you eliminate the choice that has nothing to do with Russia or the topic of the question, see if you can eliminate one other choice and thus narrow the possibilities down to two.

Reviewing Vocabulary

Directions: Choose the word or words that best complete the sentence.

1. In a ________ economy, the government makes key economic decisions.

A command
B market
C free
D declining

2. A ________ sells goods illegally, usually at higher prices than regulations allow.

A kolkhoz
B flea market
C black market
D hypermarket

3. Generation of nuclear energy produces polluting ________.

A heavy metals
B acid rain
C radioactive material
D pesticides

4. ________ are chemicals used to kill crop-damaging insects and other organisms.

A Nuclear wastes
B Acids
C Fertilizers
D Pesticides

Reviewing Main Ideas

Directions: Choose the best answers to complete the sentences or to answer the following questions.

Section 1 *(pp. 390–394)*

5. In the Soviet era, Russians endured shortages of ________.

A heavy industrial goods
B military hardware
C space exploration
D consumer goods

6. What is Russia's most important industry today?

A steel making
B automobiles
C petroleum extraction and processing
D textiles and clothing

Section 2 *(pp. 395–398)*

7. What factors have contributed to the reduction in Russian forests?

A increased large-scale farming and pipelines in the northernmost parts of Russia
B urban sprawl and industrialization
C overuse of fertilizers and pesticides
D acid rain and chemical pollution

8. After the accident at Chernobyl in 1986, what happened to energy production in the Soviet Union?

A All nuclear power plants were shut down.
B Nuclear power plants were inspected and renovated to a higher safety standard.
C Much of the country's electricity continues to come from the 29 remaining plants.
D The ruins of Chernobyl were encased in a totally secure structure to prevent radioactive leaks.

Critical Thinking

Directions: Choose the best answers to complete the sentences or to answer the following questions.

9. The Soviet era left a legacy of environmental pollution because ______.

A Soviet scientists were ignorant of environmental protection principles

B foreign countries were uninterested in protecting the environment

C industrialization was the main goal, and environmental protection was unimportant to the government

D the Soviets believed theirs was the world's best country

Base your answer to question 10 on the map and on your knowledge of Chapter 16.

10. What part of Russia has the most complete transportation network?

A western Russia

B eastern Russia

C northern Siberia

D the Pacific coast region

Document-Based Questions

Directions: Analyze the document and answer the short-answer question that follows the document.

Autocratic governments tend to be very secretive, not even distributing full and accurate information to their own people. Official Soviet government maps were incomplete and contained incorrect information. In 1989 the government cartographic office issued the first accurate street map of Moscow.

> *V. Yashchenko [chief of the Main Administration of Geodesy and Cartography] said: "Beginning last year, we have finally begun to get rid of the chronic complexes of spy mania and to get freed from the heavy burden which we had to carry since the Higher Directorate of Cartography came . . . to report to the NKVD (People's Commissariat of Internal Affairs). This is where it all began. . . ."*
>
> *However, times are changing. Fifty years were needed in order for this truth to become apparent: Why conceal from ourselves something that is no secret for others? So, work on declassification began. It appears that the world has not been profoundly surprised. . . .*
>
> *In March or April, a map of Moscow will be released. As V. Yashchenko says, all streets and lanes of the capital city have never been shown on a map available to the general public so diligently. . . .*
>
> *"Broad circles of the public have never seen such maps before. Believe me, they can rely on them completely," noted V. Yashchenko.*
>
> —G. Alimov, "Accurate Maps Reveal Former 'Secrets'"

11. Why did the Soviet Union publish incomplete maps?

Extended Response

12. Exploring the BIG Idea
Describe the steps that are being taken in Russia to repair the environmental damage of the Soviet era.

For additional test practice, use Self-Check Quizzes—Chapter 16 on glencoe.com.

Need Extra Help?

If you missed questions. . .	1	2	3	4	5	6	7	8	9	10	11	12
Go to page. . .	391	391	397	397	391	392	396	397	396	401	401	396

NATIONAL GEOGRAPHIC Case STUDY

PROTECTING PEOPLE AND THE ENVIRONMENT:

How can the environment and indigenous peoples be protected as the oil and natural gas industry expands in Russia?

The world is hungry for oil and natural gas, and Russia has about 5 percent of the world's known oil reserves and more than 20 percent of the known natural gas reserves. Even though Russia uses some of these resources, there is much more in the ground and discoveries of new oil fields are made regularly. As Russia develops its market economy, it looks to oil and natural gas exports as a way to earn income.

Understanding the Issue

The balance between using natural resources and preserving the environment and the ways of life of indigenous peoples can be viewed from several perspectives.

A Moral Issue Forests are often cut down to tap into the oil fields and transport the oil and natural gas. There is also a high risk for fires as oil may leak and can potentially catch fire, burning even more forests. Animal habitats may be destroyed, and the cultures of indigenous peoples disrupted. At the same time, money from selling oil and natural gas pays for infrastructure and other human needs. It also provides jobs.

An Economic Dilemma While oil and natural gas can bring great wealth to the country, Russia's oil fields are located far from Russia's population centers and other countries where the products can be sold. These locations also have extremely cold climates, and some are swampy or mountainous, making access difficult. Russia needs foreign investment to explore, extract, and transport oil and natural gas resources.

A Political Problem The United States, Western Europe, China, and Japan look to Russia to supply their growing energy needs. Russia wants to be sure to receive a fair price for their oil and natural gas and that their fragile northern environments are protected. Russian oil and natural gas may encounter further political problems. For example, in December 2005 the Russian government threatened to stop the flow of natural gas to Ukraine because of a political dispute. In January 2006, an agreement was reached, but the threat showed the vulnerability of Russian customers.

Top World Oil Producers, 2008

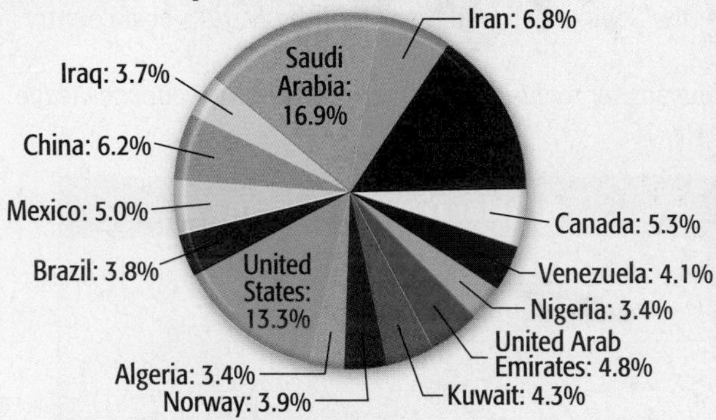

Source: Energy Information Administration, U.S. Department of Energy.

Proven Oil Reserves, 2009

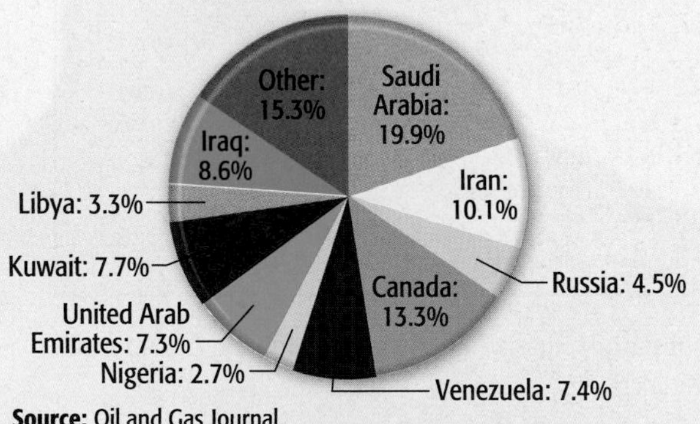

Source: Oil and Gas Journal.

World Natural Gas Production, 2008

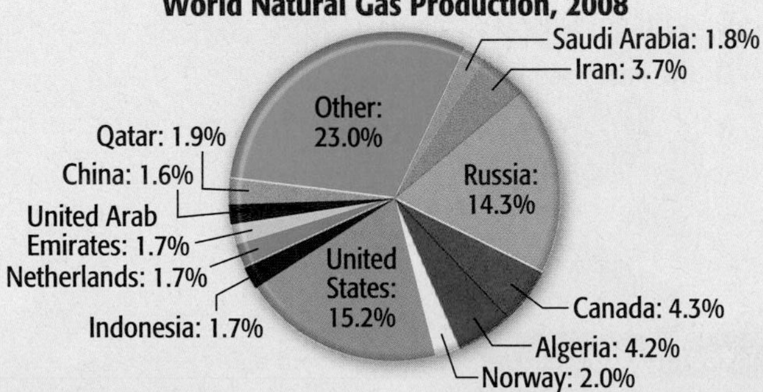

Source: Energy Information Administration, U.S. Department of Energy.

Workers construct a natural gas pipeline near Komsomol'sk-na-Amure in eastern Russia.

Possible Solutions

Solutions to the problem of using Russia's oil and natural gas must be approached in several ways.

Environmental Protection Monitoring, pollution control devices, and high-quality transportation equipment can help prevent spills, fires, and pollution from oil and natural gas operations. However, these measures are expensive to implement and require trained workers.

Conservation Russia must keep environmental and conservation measures in mind when establishing new oil fields. Other countries can reduce their dependence on oil and natural gas through more efficient vehicles, better insulation of buildings, and greater use of public transportation.

International Relations Russia wants to sell its oil and natural gas to other countries, and doing so requires that it develop good relations with those countries. Better relations between consuming countries and other producers would also take some of the pressure off of Russia. This would encourage it to preserve the environment and the ways of life of indigenous peoples.

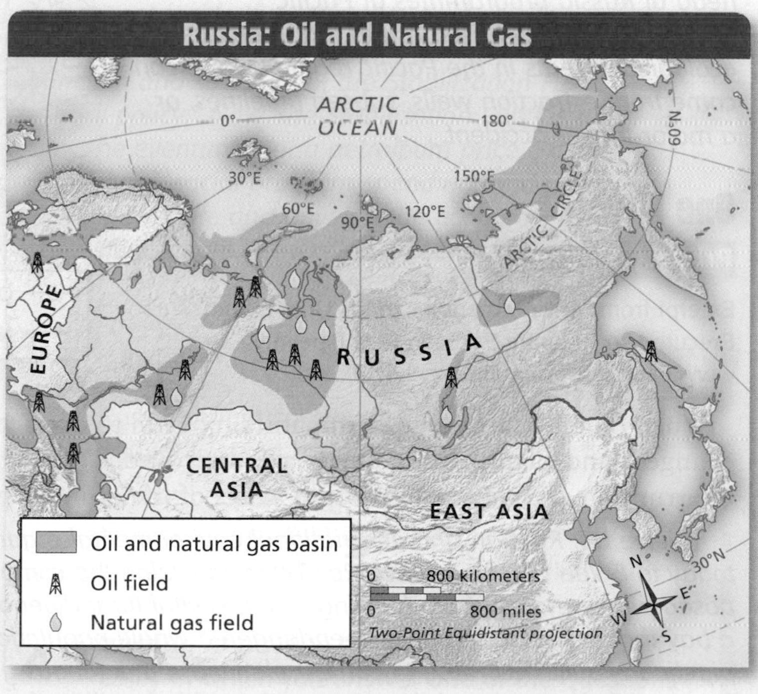

UNIT

6 North Africa, Southwest Asia, and Central Asia

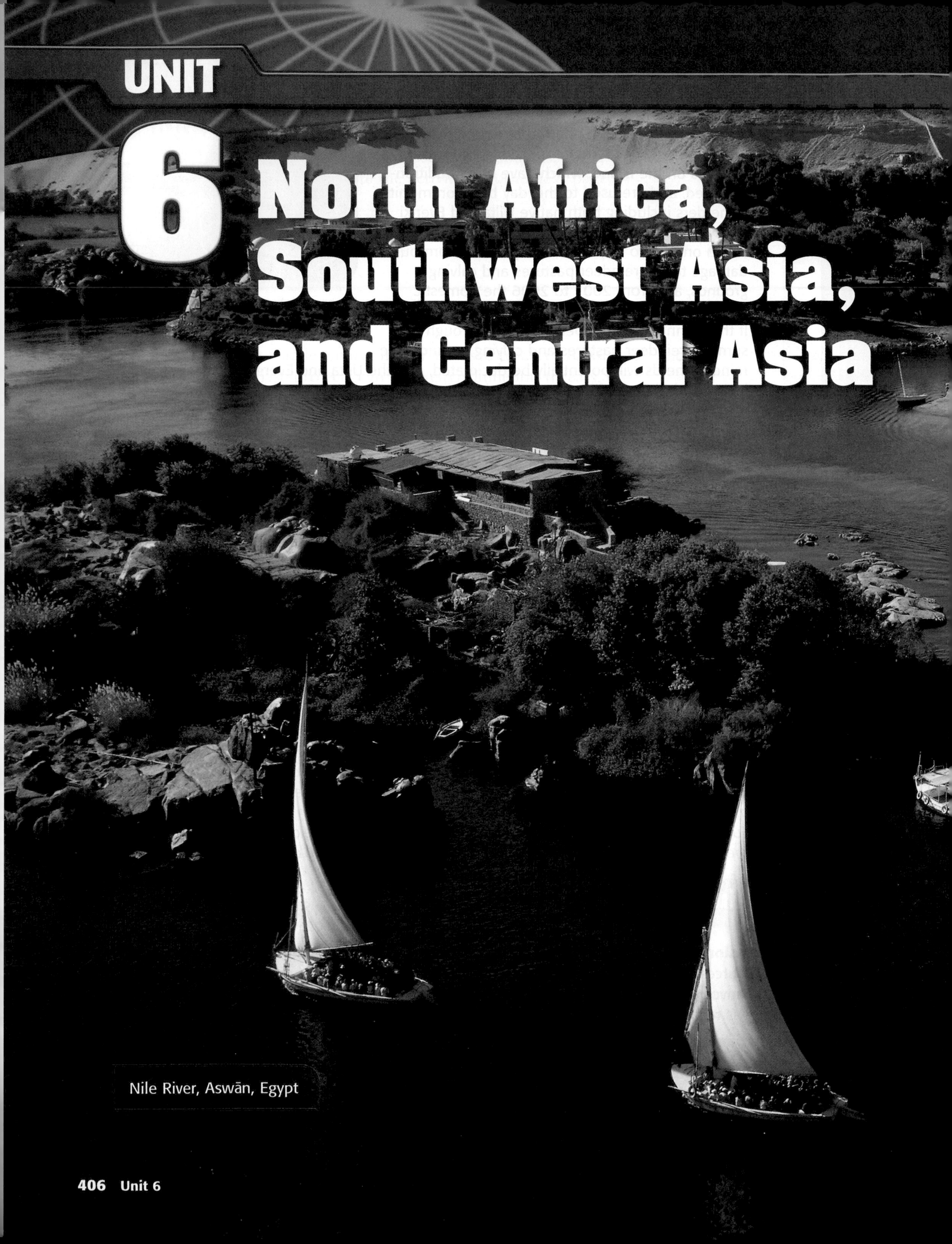

Nile River, Aswān, Egypt

NGS ONLINE To learn more about North Africa, Southwest Asia, and Central Asia visit www.nationalgeographic.com/education.

Why It Matters

Most Americans' modern lifestyle depends on oil. Without vehicles powered by gasoline, how would people get from one place to another, and how would goods be sent from warehouses to stores? Today, much of the world's oil comes from the region of North Africa, Southwest Asia, and Central Asia. Many American companies do business in the region. As a result, political, social, and economic changes there have a major impact on your daily life.

North Africa, Southwest Asia, and Central Asia

PHYSICAL GEOGRAPHY North Africa, Southwest Asia, and Central Asia spread across two continents, stretching from the shores of the Atlantic Ocean in the west to the border of China in the east. It is a region of mountains, plateaus, and plains and a largely dry climate that creates large deserts. The area is also home to two river valleys with rich soil that forms the basis for productive agriculture. In other parts of the region, though, sparse rain makes farming difficult. The water that people, plants, and animals need to survive and the oil that is abundant in the region are the area's most important resources.

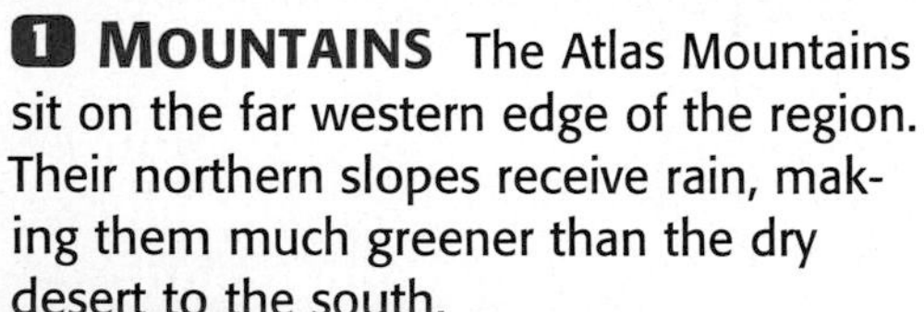

1 MOUNTAINS The Atlas Mountains sit on the far western edge of the region. Their northern slopes receive rain, making them much greener than the dry desert to the south.

2 LAKES AND RIVERS The Nile River flows north through Egypt into the Mediterranean Sea. The waters and soil of the Nile, the world's longest river, bring life to the Egyptian desert.

3 PLAINS AND PLATEAUS The Rub' al-Khali, or Empty Quarter, fills more than a quarter of Saudi Arabia. Virtually lifeless, the vast desert also holds the world's largest oil field.

4 NATURAL RESOURCES The region has more than half of the world's oil reserves, making it vitally important to the rest of the world and giving some countries in the region the opportunity to earn great wealth.

UNIT 6 WHAT MAKES THIS A REGION?

North Africa, Southwest Asia, and Central Asia

CULTURAL GEOGRAPHY The region of North Africa, Southwest Asia, and Central Asia is the birthplace to three world religions: Judaism, Christianity, and Islam. Muslims—followers of Islam—form the majority of the population in most of the region's countries. The teachings and traditions of Islam are an important force in many of these countries. While many people across the region follow traditional ways of life, modern ways are also increasingly prominent.

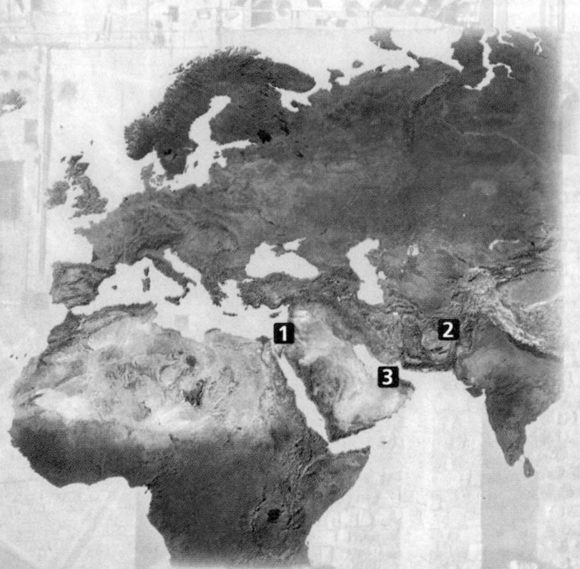

REGIONAL TIME LINE

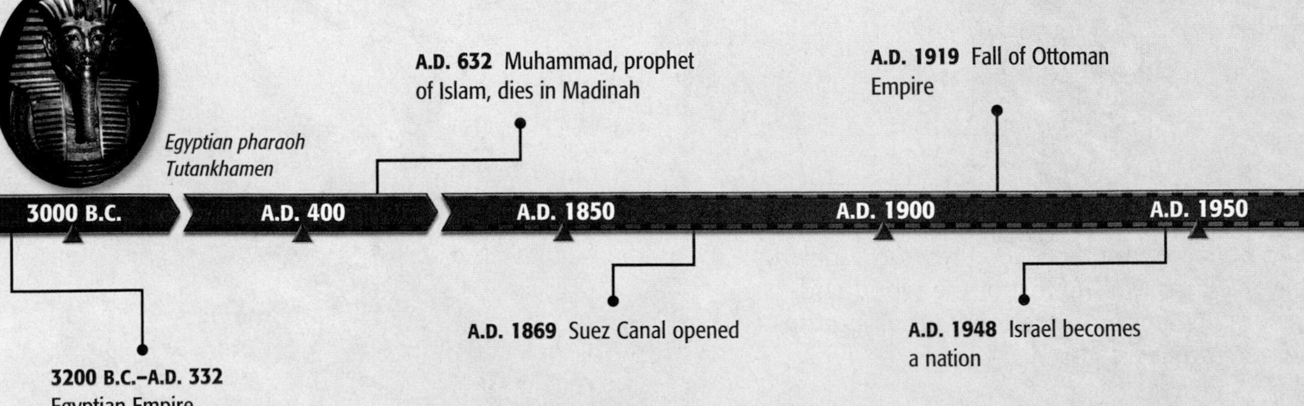

Egyptian pharaoh Tutankhamen

3200 B.C.–A.D. 332 Egyptian Empire

A.D. 632 Muhammad, prophet of Islam, dies in Madinah

A.D. 1869 Suez Canal opened

A.D. 1919 Fall of Ottoman Empire

A.D. 1948 Israel becomes a nation

1 CULTURE Jerusalem is a holy city to the followers of the three world religions that began in the region. Many Jews visit the Western Wall, the remains of the structure surrounding the Second Jerusalem Temple, while Muslims go to the nearby Dome of the Rock, a shrine with a golden dome.

2 PEOPLE Women wearing traditional robes called burkas vote in Afghanistan. In recent years, several nations in the region have moved to more democratic governments.

3 ECONOMY The leaders of Dubai, a small state on the Persian Gulf, have used the wealth earned from selling oil to build a modern port and hotels.

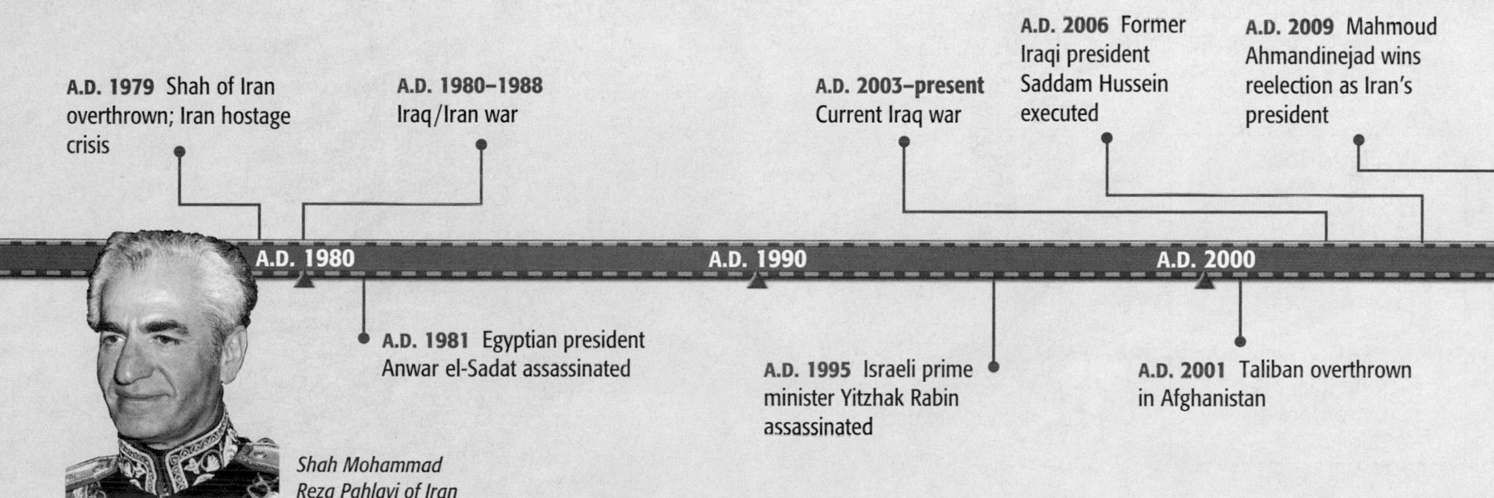

Shah Mohammad Reza Pahlavi of Iran

UNIT 6 REGIONAL ATLAS

PHYSICAL North Africa, Southwest Asia, and Central Asia

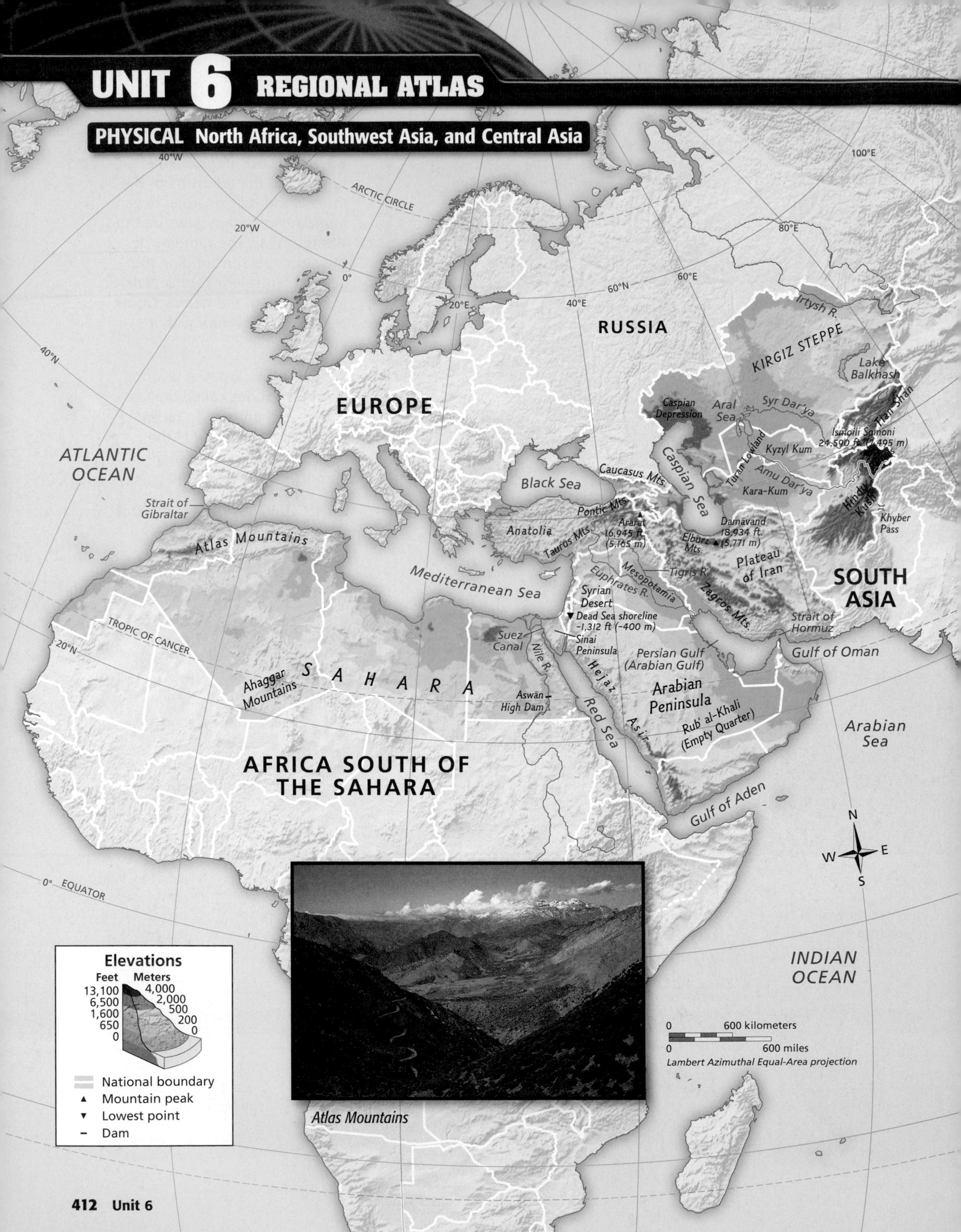

Atlas Mountains

A Difficult Land

The landscape of the region has a profound effect on the people who live there. As you study the maps and graphics on these pages, look for the geographical features that make the region unique. Then answer the questions below on a separate sheet of paper.

1. How would you describe the elevation of the region? What impact would that terrain have on people's ways of life?
2. How would the landforms and the deserts of the region affect settlement patterns?
3. Why are the Strait of Gibraltar, the Suez Canal, and the Strait of Hormuz important?

Oasis in the Sahara

Dunes in the Rub' al-Khali

Comparing Deserts

	Sahara	Rub' al-Khali (Empty Quarter)
Desert size:	3,500,000 sq. mi. (9,065,000 sq. km)	250,000 sq. mi. (650,000 sq. km)
Goes through:	11 countries	4 countries
Notable because:	World's largest desert	World's largest area of uninterrupted sand
Percentage sand:	25%	100%
People:	Over 2 million people live in the Sahara around many oases. Oases make up about 77,220 sq. mi. (200,000 sq. km) of the Sahara.	No permanent inhabitants. Bedouin tribes live on the outskirts of the desert. Oil workers sometimes take up temporary residence as the desert is rich in oil reserves.
Plant and animal life:	Contains about 150–230 species per 4,000 sq. mi. (10,360 sq. km). The Sahara contains grasses, shrubs and trees, more than 300 species of birds, many insects, rodents, and larger mammals such as gazelles, foxes, hyenas, and baboons. The Sahara is also home to several reptile species.	The Rub' al-Khali is relatively barren of life. It contains grasses, rodents, and insects, as well as migratory birds.

POLITICAL North Africa, Southwest Asia, and Central Asia

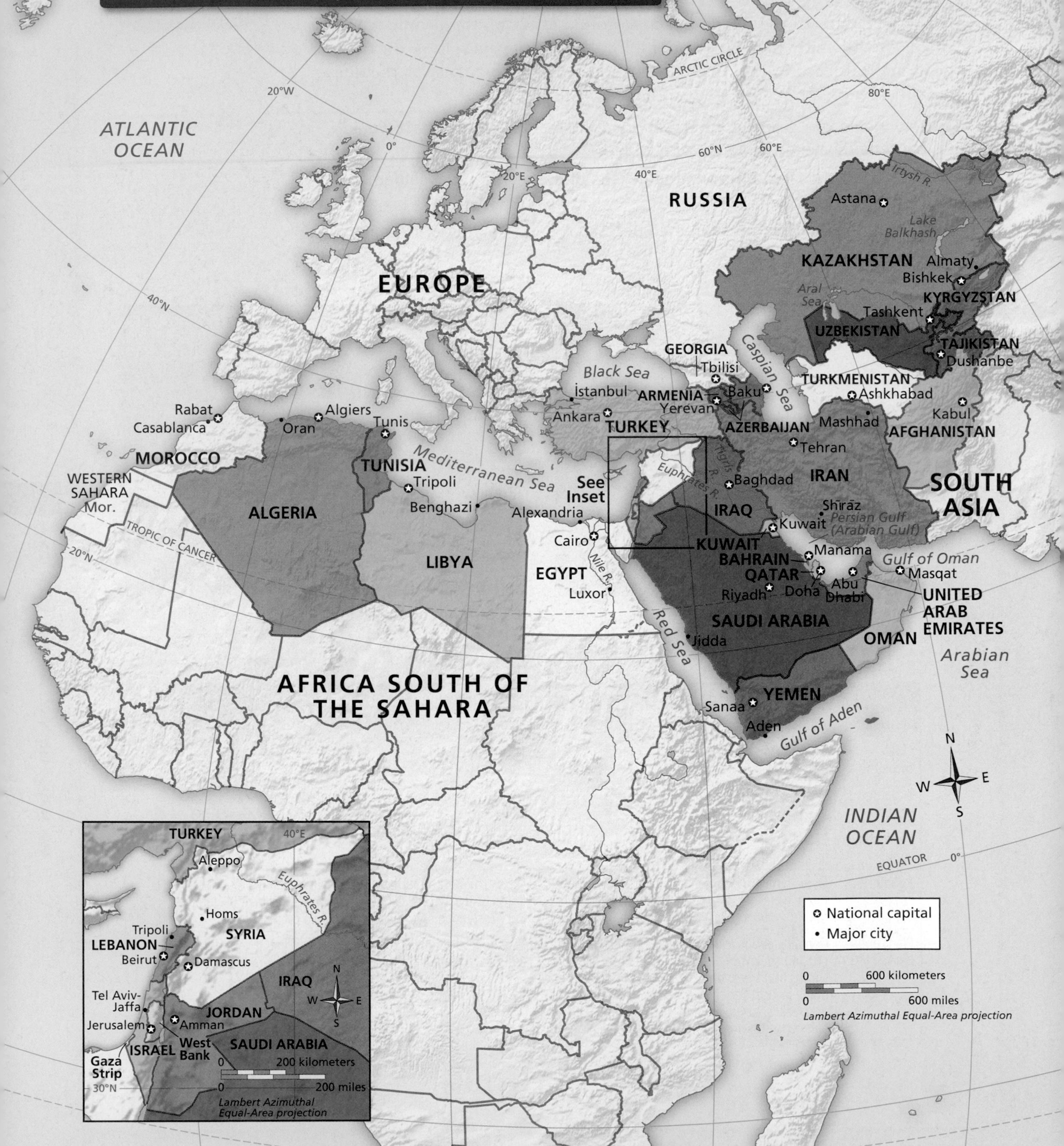

ATLANTIC OCEAN
RUSSIA
EUROPE
KAZAKHSTAN
Astana
Lake Balkhash
Irtysh R.
Almaty
Bishkek
KYRGYZSTAN
Aral Sea
Tashkent
UZBEKISTAN
TAJIKISTAN
Dushanbe
GEORGIA
Tbilisi
Caspian Sea
Black Sea
Istanbul
ARMENIA
Baku
Yerevan
TURKMENISTAN
Ashkhabad
Ankara
TURKEY
AZERBAIJAN
Mashhad
Kabul
AFGHANISTAN
Tehran
IRAN
Baghdad
Tigris R.
Euphrates R.
IRAQ
SOUTH ASIA
Shiraz
Kuwait
Persian Gulf (Arabian Gulf)
KUWAIT
Manama
BAHRAIN
QATAR
Doha
Abu Dhabi
Gulf of Oman
Masqat
Riyadh
UNITED ARAB EMIRATES
SAUDI ARABIA
Jidda
OMAN
Arabian Sea
YEMEN
Sanaa
Aden
Gulf of Aden
Red Sea
Rabat
Casablanca
Algiers
Oran
Tunis
MOROCCO
TUNISIA
Mediterranean Sea
WESTERN SAHARA Mor.
ALGERIA
Tripoli
Benghazi
Alexandria
See Inset
LIBYA
Cairo
EGYPT
Nile R.
Luxor
AFRICA SOUTH OF THE SAHARA
INDIAN OCEAN
TROPIC OF CANCER
ARCTIC CIRCLE
EQUATOR
20°W
0°
20°E
40°E
60°E
80°E
60°N
40°N
20°N
0°
N E S W

Inset:
TURKEY
40°E
Aleppo
Euphrates R.
Homs
SYRIA
Tripoli
LEBANON
Beirut
Damascus
IRAQ
Tel Aviv-Jaffa
JORDAN
Amman
Jerusalem
West Bank
ISRAEL
SAUDI ARABIA
Gaza Strip
30°N
0 200 kilometers
0 200 miles
Lambert Azimuthal Equal-Area projection

National capital
Major city

0 600 kilometers
0 600 miles
Lambert Azimuthal Equal-Area projection

Bonds and Barriers

There are a wide variety of cultural and religious differences in this region. As you study the maps on these pages, look for cultural features that make the region unique. Then answer the questions below on a separate sheet of paper.

1. How does the size of countries compare? What effect might those differences have on the relations between countries?
2. What is the dominant religion in the region?
3. What do language families suggest about linkages and differences among the people of the region?

Kurdish woman and children in Turkey

RELIGION

Christian
- Eastern Orthodox

Islam
- Shia
- Sunni

- ✡ Judaism
- Religions undifferentiated

Major religious minorities
- C Christian
- D Druze
- S Sufism

0 600 kilometers
0 600 miles
Lambert Azimuthal Equal-Area projection

RUSSIA
EUROPE
Black Sea
Caspian Sea
Mediterranean Sea
SOUTH ASIA
Red Sea
TROPIC OF CANCER
Arabian Sea
ATLANTIC OCEAN
AFRICA SOUTH OF THE SAHARA
60°N 40°N 20°N 40°W 20°W 0° 20°E 40°E 60°E

LANGUAGE

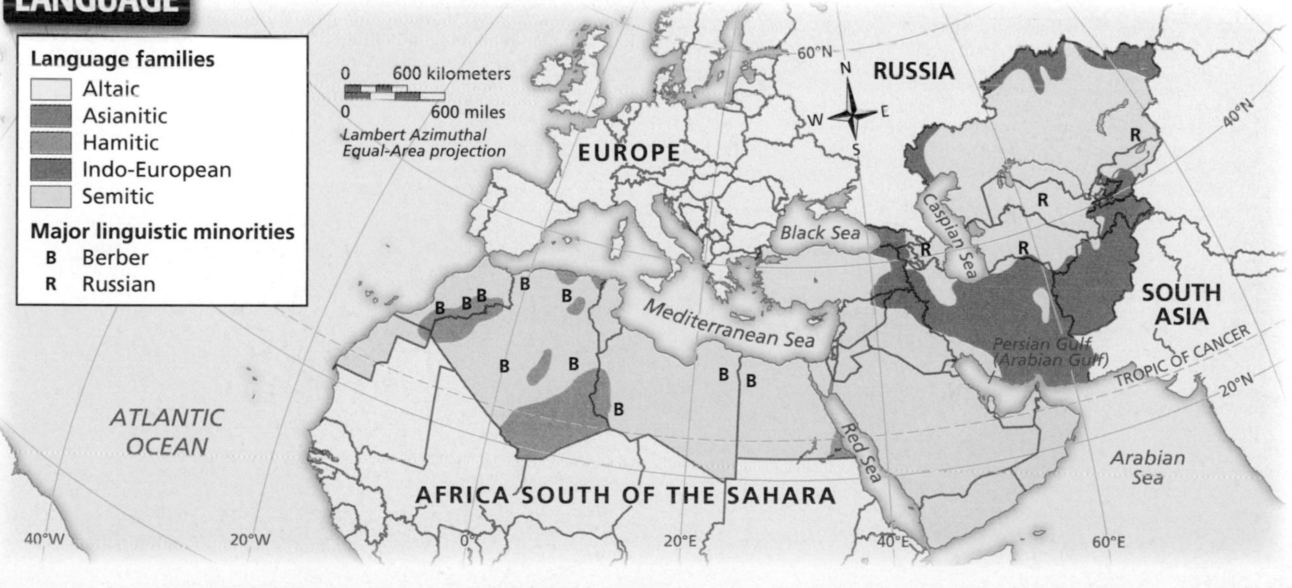

A Challenging Climate

This region is covered by dry climates. As you study the maps on these pages, look for environmental features that make the region unique. Then answer the questions below on a separate sheet of paper.

1. What climates cover most of the region? Why will those climates create challenges for people?
2. Which countries have a severe water problem? How can you tell?
3. Where do you think the water being used comes from? Why do you think so?

CLIMATE North Africa, Southwest Asia, and Central Asia

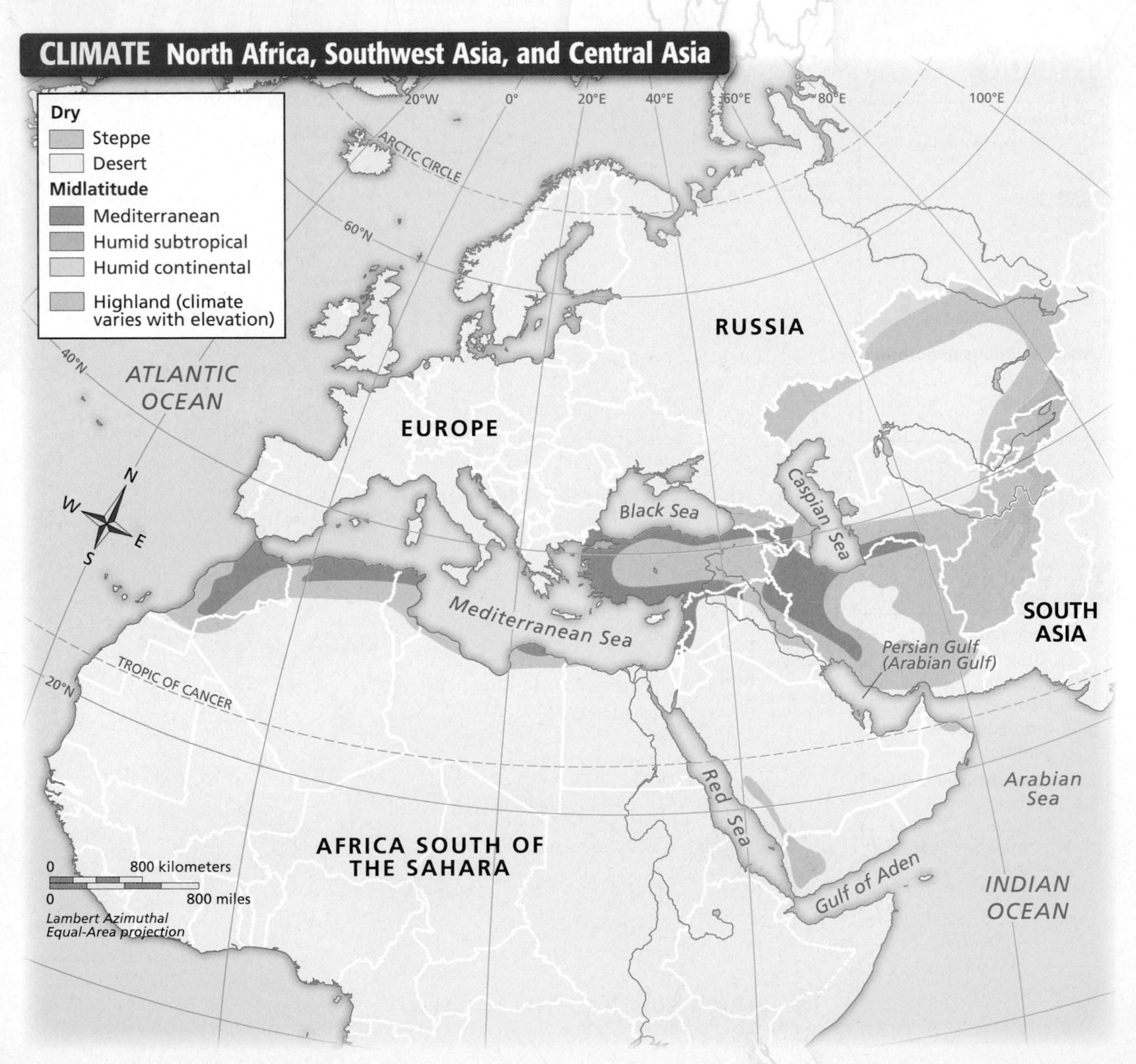

NATIONAL GEOGRAPHIC

WATER AVAILABILITY North Africa, Southwest Asia, and Central Asia

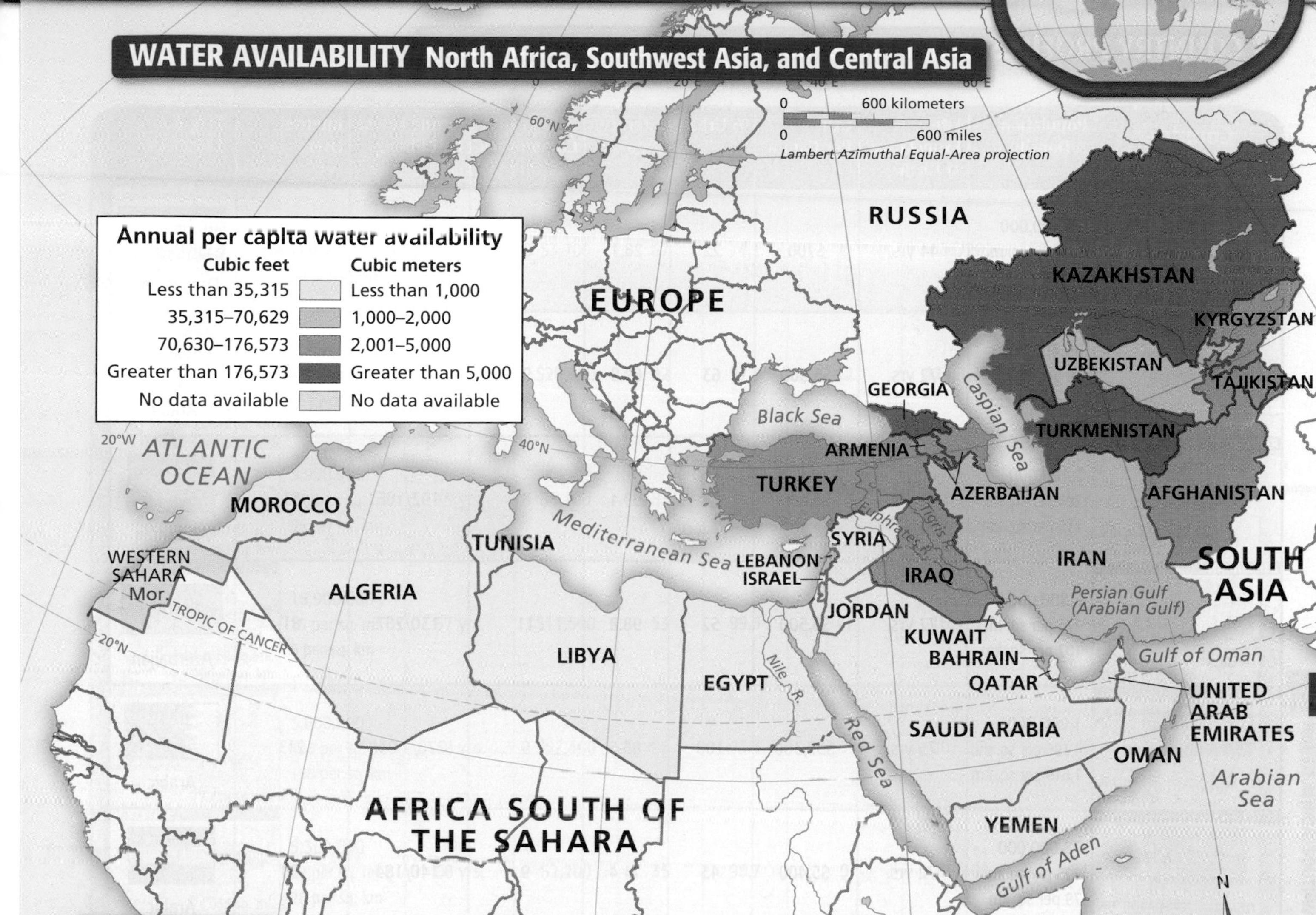

WATER USAGE

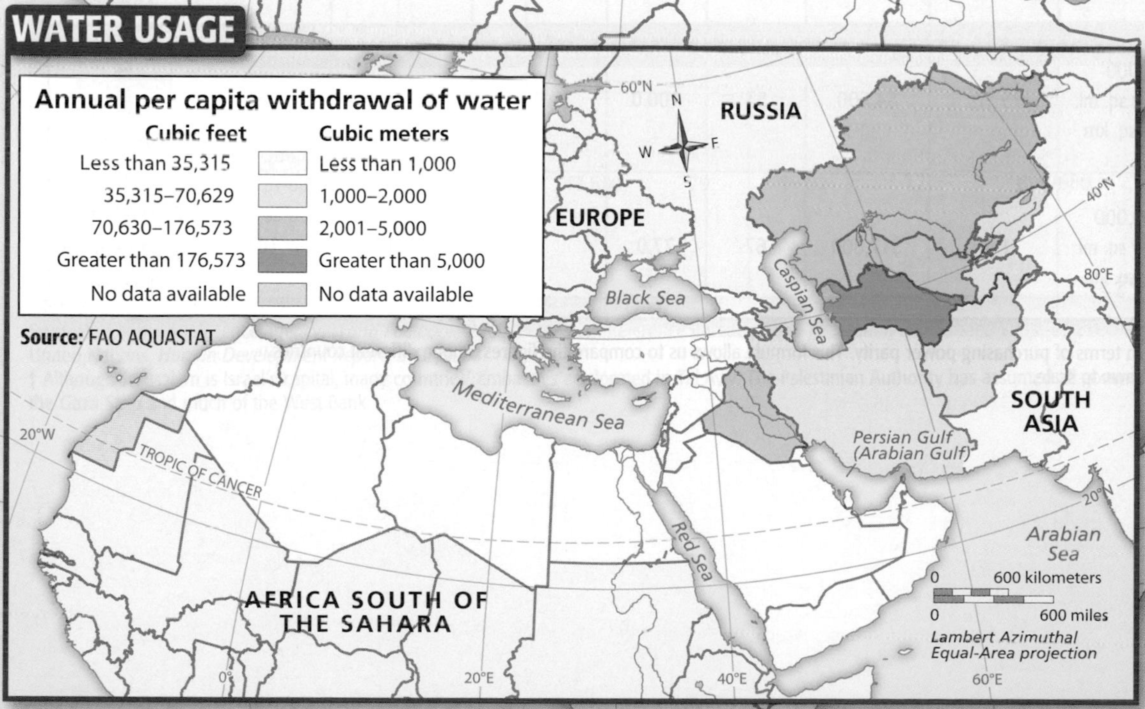

Source: FAO AQUASTAT

Source: FAO AQUASTAT

CHAPTER 17

PHYSICAL GEOGRAPHY OF North Africa, Southwest Asia, and Central Asia

BIG Idea

Places reflect the relationship between humans and the physical environment. A study of the physical geography of North Africa, Southwest Asia, and Central Asia reveals the relationship between humans and the physical environment, where seas and rivers help sustain life amid rugged mountain ranges, dry plateaus, and some of Earth's greatest deserts.

Essential Questions

Section 1: The Land

How has water played a role in defining the region of North Africa, Southwest Asia, and Central Asia?

Section 2: Climate and Vegetation

How does the climate of North Africa, Southwest Asia, and Central Asia affect natural vegetation and human activity?

Geography ONLINE

Visit glencoe.com and enter QuickPass™ code WGC9952C17 for Chapter 17 resources.

River valleys—such as the Dades Valley in Morocco—provide the region with fertile agricultural land.

Organizing Information Use a Layered-Look Book to gather and organize information about the region's land.

Reading and Writing As you read, write key ideas and examples associated with each heading under the appropriate flap of your Foldable.

SECTION 1

 section audio  spotlight video

Guide to Reading

Essential Question

How has water played a role in defining the region of North Africa, Southwest Asia, and Central Asia?

Content Vocabulary

- *kum* *(p. 425)*
- alluvial soil *(p. 426)*
- wadi *(p. 427)*
- phosphate *(p. 428)*

Academic Vocabulary

- shift *(p. 425)*
- complex *(p. 426)*
- revenue *(p. 428)*

Places to Locate

- Atlas Mountains *(p. 425)*
- Caucasus Mountains *(p. 425)*
- Arabian Peninsula *(p. 426)*
- Persian Gulf *(p. 426)*
- Sinai Peninsula *(p. 426)*
- Anatolia *(p. 426)*
- Dead Sea *(p. 426)*
- Caspian Sea *(p. 426)*
- Aral Sea *(p. 426)*
- Nile River *(p. 426)*
- Tigris River *(p. 426)*
- Euphrates River *(p. 426)*

Reading Strategy

Organizing As you read, use a chart like the one below to describe each body of water listed.

Body of Water	Description
Dead Sea	
Caspian Sea	
Aral Sea	

The Land

The vast region of North Africa, Southwest Asia, and Central Asia spans portions of Africa and Asia, where early civilizations thrived thousands of years ago in fertile river valleys. Today, ancient rivers such as the Nile remain vital to the people of the region, where water in an arid land continues to mean life.

NATIONAL GEOGRAPHIC Voices Around the World

"West of the emerald alfalfa fields and dusty green palm groves that flank the Nile, Saqqara rests atop a rocky escarpment the color of ripe wheat. Here the wind-rippled desert sand begins its sweep toward Libya. And here on the sunset bank of the Nile, the ancient Egyptians believed, was as close as mortal remains could get to the great beyond. In their view of the world, when the sun slipped beneath the desert horizon each evening, it traveled through the underworld ruled by Osiris, the god of the afterlife, until being reborn in the morning on the opposite side of the great river."

—A. R. Williams, "Death on the Nile," *National Geographic,* October 2002

An Egyptian man along the Nile River

Landforms

MAIN Idea In North Africa, Southwest Asia, and Central Asia, dramatic landforms can be found in a region dominated by deserts and mountains.

GEOGRAPHY AND YOU Do you live near a desert or in the mountains? Read to learn about the deserts and mountains that dominate the landscape in North Africa, Southwest Asia, and Central Asia.

Tectonic activity is responsible for shaping the landscape in many parts of North Africa, Southwest Asia, and Central Asia.

Earthquakes

The African, Arabian, Anatolian, and Eurasian plates come together in North Africa, Southwest Asia, and Central Asia. As the plates move, they build mountains, **shift** landmasses, and cause earthquakes. Tectonic activity built such mountains as the Atlas of Morocco and Algeria, the Zagros of southern Iran, and the Taurus of Turkey.

Earthquakes rumble throughout the region regularly. In 2005 Afghanistan, surrounded by active plate boundaries, experienced an earthquake measuring 6.4 on the Richter scale.

Mountains

As the physical map on page 412 of the Regional Atlas shows, Africa's longest mountain range, the **Atlas Mountains,** extends across Morocco and Algeria. Enough precipitation falls on the northern side of these mountains to water the coastal regions, making them ideal for farming.

In Southwest Asia, two mountain ranges, the Hejaz and Asir, stretch along the western coast of the Arabian Peninsula. The taller Asir Mountains receive more rainfall than the Hejaz, about 19 inches (48 cm) annually. This precipitation makes the Asir region the most agriculturally productive on the Arabian Peninsula.

The Pontic Mountains and the Taurus Mountains rise from the Turkish landscape. Between these ranges, the Anatolian Plateau stands 2,000 to 5,000 feet (610 to 1,524 m) above sea level. East of the Pontic range, camel-backed Mount Ararat, at almost 17,000 feet (5,182 m), overlooks the Turkish-Iranian border. The **Caucasus Mountains** rise north of Mount Ararat between the Black Sea and Caspian Sea.

Landforms and Tectonic Activity

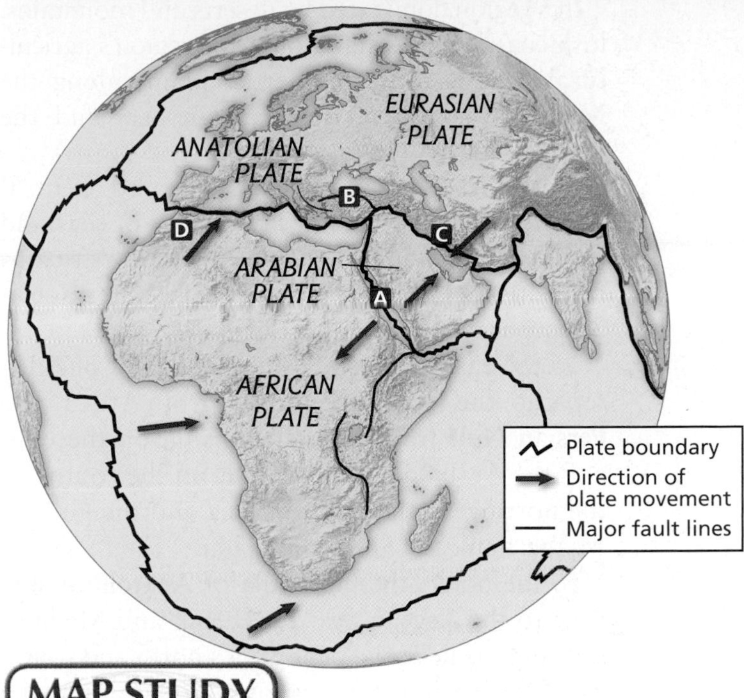

MAP STUDY

A RED SEA: Created as the African and Arabian plates moved away from each other. Continued widening may cause the Red Sea to become an ocean in millions of years.

B NORTH ANATOLIAN FAULT: Running the width of Turkey, it is one of the most active faults in the world, and has resulted in numerous earthquakes.

C ZAGROS MOUNTAINS: Formed by the collision of the Eurasian and Arabian plates, the mountains extend 932 miles (1,500 km).

D ATLAS MOUNTAINS: Located along a fault line, these mountains have a large amount of seismic activity.

1. **Location** Why does the region of North Africa, Southwest Asia, and Central Asia experience such a significant amount of tectonic activity?
2. **Regions** How was the creation of the Red Sea different from that of the Zagros Mountains?

West of the Tian Shan range, the Turan Lowland provides some irrigated farmland. To the south, dune-covered ***kums*** (KOOMZ), or deserts, offer a stark contrast to the cultivated fields of the lowland. The Kara-Kum, or black sand desert, covers most of Turkmenistan. The Kyzyl Kum, or red sand desert, blankets half of Uzbekistan. Farther west, the Ustyurt Plateau has salt marshes, sinkholes, and caverns.

Coastal Plains, Seas, and Peninsulas

In a region dominated by deserts and mountains, lush coastal plains stand out. The region's agricultural base is rooted in fertile plains along the Mediterranean Sea, the Caspian Sea, and the Persian Gulf.

North Africa, Southwest Asia, and Central Asia form an intricate jigsaw puzzle of seas and peninsulas. Edging the coast of North Africa as far as the Strait of Gibraltar, the Mediterranean Sea separates Africa and Europe.

To the east, the Red Sea and the Gulf of Aden separate the **Arabian Peninsula** from Africa. The **Persian Gulf** frames this peninsula on the east, and the Arabian Sea borders it on the south. To the northwest, the Gulf of Suez and the Gulf of Aqaba flank the **Sinai Peninsula.**

To the north, the peninsula of **Anatolia** points west to the Aegean Sea. The Black and Mediterranean Seas lie at the peninsula's north and south. The Dardanelles, the Sea of Marmara, and the Bosporus, which together connect the Aegean and Black Seas, also separate Europe and Asia.

Three landlocked bodies of salt water lie east of the Mediterranean Sea. The smallest of these, the **Dead Sea,** sits at the mouth of the Jordan River. In Central Asia, the **Caspian Sea** is the largest inland body of water on Earth. This sea laps the shores of both Asia and Europe.

East of the Caspian Sea is the **Aral Sea.** Until the 1960s, it was the world's fourth-largest inland sea. Now it is just a fraction of its former size. It began to dry up when the Soviet Union diverted water for irrigation from the rivers flowing into the sea. By the late 1980s, the continuing shrinkage split the sea into two separate bodies of water—the North Aral Sea and the South Aral Sea. By 2006, water levels in the North Aral Sea had increased after dams were built to ensure the flow of freshwater into the sea.

READING Check **Regions** What types of physical features dominate the region's landscape?

Geography ONLINE

Student Web Activity Visit glencoe.com, select the *World Geography and Cultures* Web site, and click on Student Web Activities–Chapter 17 for an activity about the Dead Sea.

Water Systems

MAIN Idea For thousands of years, people have depended on the region's rivers and fertile river valleys, where early civilizations thrived.

GEOGRAPHY AND YOU How did the Mississippi River influence human settlement in the United States? Read to learn about the relationship between major rivers and the establishment of early civilizations in the region of North Africa, Southwest Asia, and Central Asia.

Rivers are the lifeblood of North Africa, Southwest Asia, and Central Asia. Their lush and productive valleys have always welcomed travelers and provided food for local peoples.

Major Rivers: Cradles of Civilization

Egypt's **Nile River** is the world's longest river at 4,160 miles (6,693 km). The Nile Delta and the fertile land along the river's banks gave birth to one of the world's earliest civilizations. Today more than 90 percent of Egypt's people live in the Nile Delta or along the course of the river on only 3 percent of Egypt's land. The Aswān High Dam and other modern dams farther up the Nile now control the river's flow, reducing both flooding and deposits of **alluvial soil,** rich soil made up of sand and mud deposited by moving water.

The Aswān High Dam also provides water for agriculture and hydroelectric power for the country. Lake Nasser, a human-made reservoir created when the dam was built, stores water and helps regulate the Nile's flow. The water in Lake Nasser is used to bring additional land under irrigation and has also helped to convert flood land to irrigated farmland. Electricity is provided by generators powered by the dam.

Early civilizations also thrived in the Tigris-Euphrates river valley, a fertile farming valley in Southwest Asia. Known by ancient peoples as Mesopotamia, Greek for "land between two rivers," this valley owes its fertile character to the **Tigris** (TY•gruhs) **River** and the **Euphrates** (yu•FRAY•teez) **River,** which flow mainly through Iraq. A **complex** irrigation network has watered the valley and supported farming there for 7,000 years. Today the Tigris and Euphrates help irrigate Turkey, Syria, and Iraq.

Originating only 50 miles (80 km) from each other in eastern Turkey, the Tigris and Euphrates Rivers join in southern Iraq to form the Shatt al Arab, which empties into the Persian Gulf. The Euphrates is the longer river, flowing 2,335 miles (3,596 km) toward the sea. The Tigris extends about 1,180 miles (1,899 km). Dams control the flow of both rivers, and hydroelectric power plants provide electricity.

Streambeds

Many streams in arid North Africa and Southwest Asia flow only intermittently, appearing suddenly and disappearing just as quickly. In the region's deserts, runoff from infrequent rainstorms creates **wadis** (WAH•dees)—streambeds that remain dry until a heavy rain. Irregular rainstorms often produce flash flooding. During a flash flood, wadis fill with so much sediment that they can rapidly become mud flows, or moving masses of wet soil, which are a danger to humans and animals.

READING Check **Regions** Which river in the region is the world's longest river?

Natural Resources

MAIN Idea Some of North Africa, Southwest Asia, and Central Asia's most abundant resources are important to the world economy.

GEOGRAPHY AND YOU Why is oil important to the U.S. economy? Read to find out what role oil plays in making the United States and other countries dependent on this region of the world.

The lands of North Africa, Southwest Asia, and Central Asia contain many natural resources. These include petroleum and natural gas.

Oil and Natural Gas

As the graph on page 428 shows, over 60 percent of the world's known oil reserves lie beneath the region. About 50 percent of the world's known natural gas reserves are there as well. Unmeasured reserves include newly discovered gas fields in the Gaza Strip, in Egypt, and under the Caspian Sea.

MAP STUDY

1. **Place** How are the satellite image and the map similar in what they show? How are they different?
2. **Location** According to the map, where are most areas of oases and irrigated agriculture located?

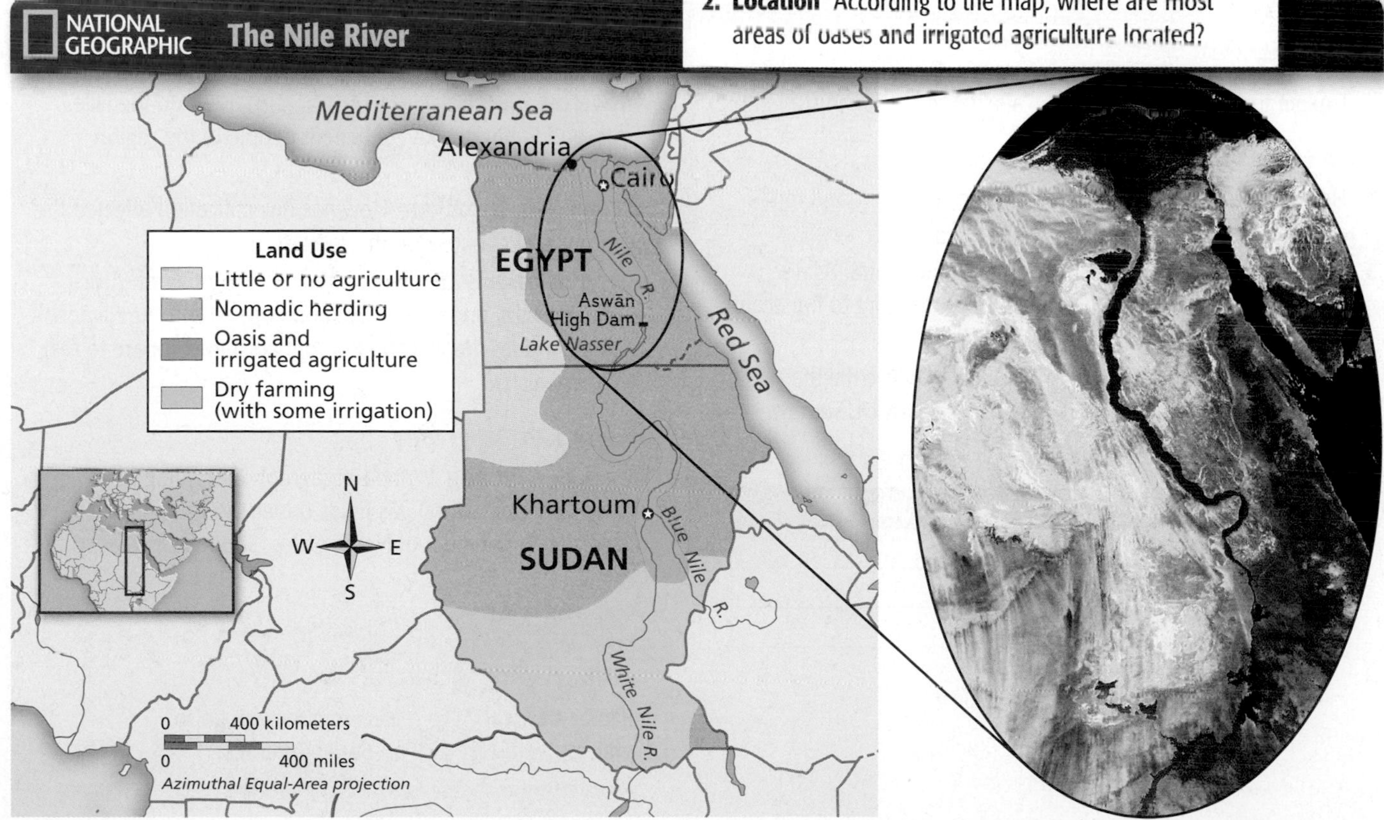

Petroleum exports have enriched the region, but heavy reliance on petroleum exports is risky. When oil prices fluctuate on world markets, as they did between 1997 and 1999, the region's economies suffer. By the time oil prices rose from a low of $7 per barrel to about $30 per barrel in early 2000, the economies of oil-exporting countries had been damaged.

Many factors can contribute to fluctuating oil prices. By 2008, with growing global demand for oil and the ongoing conflict in Iraq, along with other uncertainties in the region, oil prices rose to over $130 per barrel.

World Oil Reserves

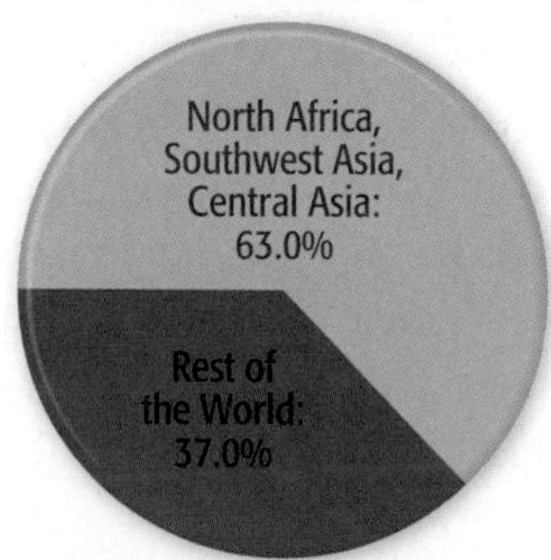

Source: www.eia.doe.gov, Energy Information Administration.

Minerals

Minerals also provide **revenue** for the region. Turkmenistan has one of the world's largest deposits of sulfur and large deposits of sulfate, which is used in paperboard, glass, and detergents. Morocco is one of the leading producers of **phosphate**—a chemical used in fertilizers. Deposits of chromium, gold, lead, manganese, and zinc are sprinkled across the region. Discoveries of iron ore and copper deposits indicate that the region may contain up to 10 percent of the world's iron ore reserves.

Building Diverse Economies

Some countries in the region are diversifying their economies to decrease their reliance on oil and mineral exports. The United Arab Emirates, for example, is investing oil earnings in banking, information technology, and tourism. Libya, which relies on oil for 95 percent of its export income, is investing in infrastructure, agriculture, and fisheries.

READING Check **Regions** On what natural resource do many of the region's economies depend?

SECTION 1 REVIEW

Vocabulary

1. Explain the significance of: *kum,* alluvial soil, wadi, phosphate.

Main Ideas

2. How have people depended on the region's rivers and fertile river valleys for thousands of years?
3. What are North Africa, Southwest Asia, and Central Asia's most valuable resources? How are they important to the economies of countries around the world?
4. Create a table like the one below to list and describe the varied and dramatic landforms found in North Africa, Southwest Asia, and Central Asia.

Region	Landforms
North Africa	
Southwest Asia	
Central Asia	

Critical Thinking

5. **Answering the Essential Question** How do the Nile, Tigris, and Euphrates Rivers benefit people in the region today?
6. **Making Generalizations** How has diversification affected the economies of countries in the region?
7. **Analyzing Visuals** Study the physical map of North Africa, Southwest Asia, and Central Asia on page 412 of the Regional Atlas. How does the elevation of Afghanistan compare to that of Turkmenistan?

Writing About Geography

8. **Descriptive Writing** Write a paragraph describing how development of natural gas fields under the Caspian Sea might affect the region of North Africa, Southwest Asia, and Central Asia.

Study Central™ To review this section, go to glencoe.com and click on Study Central.

SECTION 2

section audio   spotlight video

Guide to Reading

Essential Question

How does the climate of North Africa, Southwest Asia, and Central Asia affect natural vegetation and human activity?

Content Vocabulary
- oasis *(p. 430)*
- pastoralism *(p. 430)*
- cereal *(p. 432)*

Academic Vocabulary
- define *(p. 430)*
- annually *(p. 430)*
- exposure *(p. 431)*

Places to Locate
- Sahara *(p. 430)*
- Rub' al-Khali *(p. 430)*
- Kara-Kum *(p. 430)*

Reading Strategy

Categorizing As you read about the climates of North Africa, Southwest Asia, and Central Asia, complete a graphic organizer like the one below by identifying the region's three midlatitude climates.

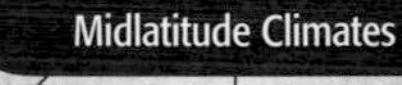

Climate and Vegetation

Large expanses of North Africa, Southwest Asia, and Central Asia have an average annual rainfall of 10 inches (25 cm) or less. As a result, much of the region contains arid areas and experiences desert climate.

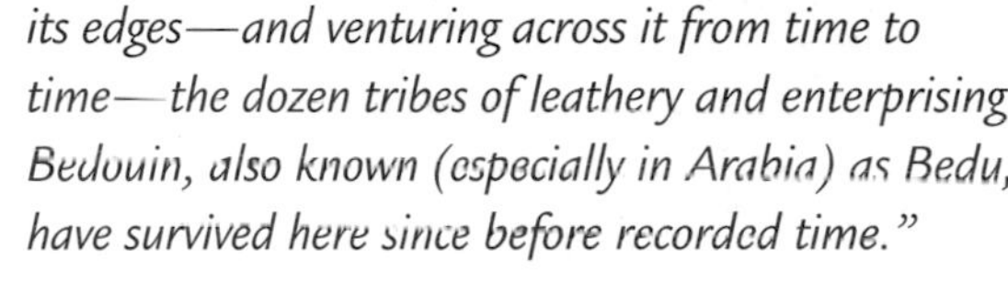

NATIONAL GEOGRAPHIC VOICES AROUND THE WORLD

"Welcome to the [Rub' al-Khali], or Empty Quarter—a world of harsh extremes that may rank as both the least, and most, hospitable place on Earth. . . . Taking up a fifth of the Arabian Peninsula, the [Rub' al-Khali] (literally, 'quarter of emptiness'), or the Sands for short, is the world's largest sand sea. . . . Because of these sandy expanses, not to mention its profound heat, the Sands have long been judged too unforgiving for all but the most resourceful humans, considered more a wasteland to cross than a landscape to settle in. Still, along its edges—and venturing across it from time to time— the dozen tribes of leathery and enterprising Bedouin, also known (especially in Arabia) as Bedu, have survived here since before recorded time."

—Donovan Webster, "The Empty Quarter," *National Geographic*, February 2005

A falconer in the Rub' al-Khali

Water and Climate

MAIN Idea Lack of water affects climate, natural vegetation, and human activities in North Africa, Southwest Asia, and Central Asia.

GEOGRAPHY AND YOU How much rain falls every year where you live? Read to find out how rainfall affects the dry climate regions of North Africa, Southwest Asia, and Central Asia.

Water scarcity defines the region's climates. The North African landscape, for example, is commonly associated with images of vast stretches of sand and the occasional watering hole. Yet, ancient cave paintings tell us that this part of the African continent was once wet and green. Differences and changes in climates across the region have affected and continue to affect natural vegetation and human activities.

Desert Climate

How much of the entire region is desert? Scientists **define** a desert climate as one in which precipitation averages 10 inches (25 cm) or less per year. By that definition, deserts encompass almost 50 percent of North Africa, Southwest Asia, and Central Asia. The **Sahara**, the largest desert in the world at about 3.5 million square miles (about 9.1 million sq. km), covers most of North Africa. In recent decades, droughts have expanded the extent of the Sahara.

NATIONAL GEOGRAPHIC *Grain and livestock—produced in Kazakhstan's steppe region—are important export commodities.*

Location What areas of Central Asia are characterized by steppe climate?

Weather patterns in the desert tend to be extreme. The deserts of Central Asia and northern parts of the Sahara and the Arabian Desert have relatively cold winters with freezing temperatures. Winters in the southern Sahara and the Arabian Desert are generally milder. Summers in these desert regions are long and hot.

A traveler crossing any of the region's deserts would probably see only a few *ergs,* or sandy, dune-covered areas. *Regs*—stony plains covered with rocky gravel called "desert pavement"—and an occasional *hamada,* a flat, sandstone plateau, would be more common. Sand covers less than 10 percent of the Sahara; desert pavement, mountains, and barren rock cover the rest.

The 250,000-square-mile (647,500-square-km) **Rub' al-Khali,** or Empty Quarter, has the largest area of sand in the region. One of several deserts on the Arabian Peninsula, the Rub' al-Khali covers almost the entire southern quarter of the peninsula.

Despite their arid conditions, the Sahara and other deserts in the region support vegetation such as cacti and drought-resistant shrubs. Nomadic herds of sheep, goats, and camels graze on brush in Central Asia's **Kara-Kum.** Small-scale farming is possible in an **oasis,** a place in the desert where underground water surfaces. Villages, towns, and cities developed around many Saharan oases.

Steppe Climate

Steppe is the second-largest climate region in North Africa, Southwest Asia, and Central Asia. The steppe borders the Sahara to the north and south. Steppe borders desert climate regions across Turkey to eastern Kazakhstan. Precipitation in this semi-arid climate region usually averages less than 14 inches (36 cm) **annually.** This amount is enough to support short grasses, providing pasture for sheep, goats, and camels, as well as shrubs and some trees. **Pastoralism,** the raising and grazing of livestock, is a way of life for people who live in a steppe climate.

READING Check **Regions** What type of natural vegetation grows in the deserts of the region?

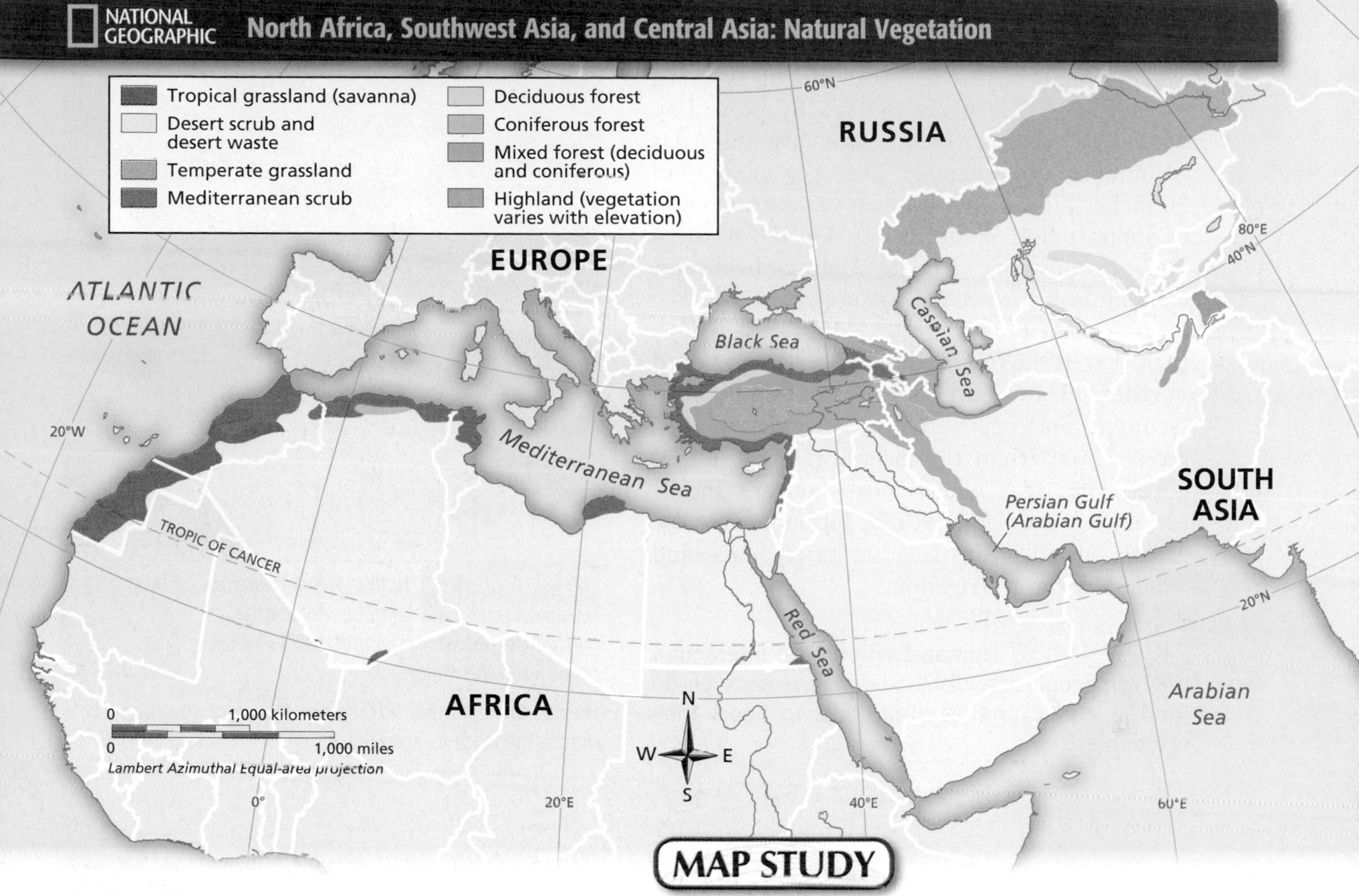

MAP STUDY

1. **Regions** Which type of vegetation dominates the region of North Africa, Southwest Asia, and Central Asia?
2. **Location** Compare the map above with the economic activity map on page 476. What type of vegetation supports livestock grazing?

Maps in MOtion Use **StudentWorks™ Plus** or glencoe.com.

Midlatitude Regions

MAIN Idea Countries within the region's midlatitudes benefit from rainfall in the Mediterranean, highland, and humid subtropical climates.

GEOGRAPHY AND YOU Describe the climate in U.S. coastal regions. Read to learn how coastal areas in North Africa, Southwest Asia, and Central Asia benefit from rainfall.

Mediterranean climates have cool, rainy winters and hot, dry summers. The climate map on page 416 of the Regional Atlas shows that this climate is common in the Tigris-Euphrates river valley, in highland areas, and on the coastal plains of the Mediterranean, Black, and Caspian Seas.

Exports and Tourists

Morocco, Tunisia, Syria, and other countries having Mediterranean climates boost their economies by exporting citrus fruits, olives, and grapes to Europe and North America. Some of these Mediterranean countries also benefit from tourism, as people from colder climates seek the sun and warmth of the Mediterranean region. The Moroccan city of Agadir, with more than 300 days of sunshine per year, attracts many of the country's 4 million tourists, who come mainly from Europe. Travelers to Morocco also visit the cultural attractions of ancient cities such as Fès, Marrakech, and Casablanca.

Higher areas, like the Caucasus Mountains, have a highland climate, which is generally wetter and colder than other climates in the region. The highland climate varies, however, with elevation and **exposure** to wind and sun.

Rainfall

Coastal and highland areas near mountain ranges usually receive the most rainfall, as moist, warm air is driven off the sea by prevailing winds. The North African coast near the Atlas Mountains, for example, averages more than 30 inches (76 cm) of rain each year, enough rain to support flourishing forests. On the coast of Oman, monsoons arrive in August, their rains creating lush forests and pastures.

More than 60 inches (152 cm) of rain falls each year at the foot of the Elburz Mountains in northern Iran. Batumi, in the Republic of Georgia, one of the region's wettest places, receives more than 100 inches (254 cm) of rain a year. In areas where more than 14 inches (36 cm) of rain falls yearly, farmers can raise **cereals**—food grains such as barley, oats, and wheat—without irrigation.

READING Check **Human-Environment Interaction** Why can people in midlatitude climate regions rely on tourism and exporting citrus fruits to boost their economies?

NATIONAL GEOGRAPHIC *With a Mediterranean climate, Morocco's agricultural potential is unmatched by other countries in the region.*

Human-Environment Interaction How else does Morocco's climate influence its economy?

SECTION 2 REVIEW

Vocabulary

1. Explain the significance of: oasis, pastoralism, cereal.

Main Ideas

2. How do countries within the region's midlatitudes benefit from rainfall in the Mediterranean, highland, and humid subtropical climates?
3. How does lack of water affect life in the dry climate regions of North Africa, Southwest Asia, and Central Asia? Use a table like the one below to identify and describe the climate, vegetation, and human activities in this part of the world.

Climate Region	Climate	Vegetation	Human Activities
Desert			
Steppe			

Critical Thinking

4. **Answering the**  Compare and contrast agriculture in steppe climate regions with that in Mediterranean climate regions. Which has the greater advantage?
5. **Drawing Conclusions** How did climate changes in the Sahara centuries ago affect its people?
6. **Analyzing Visuals** Study the natural vegetation map on page 431 and the physical map on page 412 of the Regional Atlas. What type of natural vegetation characterizes the lower elevations of Kazakhstan?

Writing About Geography

7. **Expository Writing** Write a paragraph explaining the possible effects of water and climate on settlement patterns in North Africa, Southwest Asia, and Central Asia.

Geography ONLINE

Study Central™ To review this section, go to glencoe.com and click on Study Central.

CHAPTER 17 VISUAL SUMMARY

STUDY TO GO — Study anywhere, anytime by downloading quizzes and flashcards to your PDA from glencoe.com.

A Atlas Mountains

- Africa's longest mountain range; reaches across Morocco and Algeria
- Northern slopes have Mediterranean climate and support farming

B Aral Sea

- Was the world's fourth-largest inland sea
- Began to dry up when the Soviet Union diverted water for irrigation
- Water levels increased slightly after dams were built, ensuring flow of freshwater

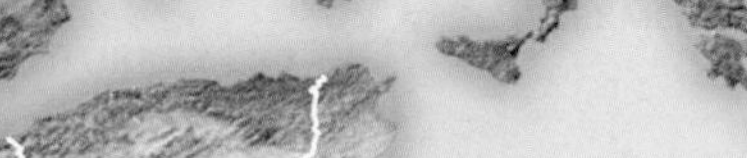

C Anatolia

- Surrounded by the Black Sea, Aegean Sea, and Mediterranean Sea
- Together, the Dardanelles, the Sea of Marmara, and the Bosporus connect the Black and Aegean Seas.
- Taurus Mountains located along southern part of peninsula

D Sahara

- World's largest desert; covers most of North Africa
- Droughts have expanded the Sahara in recent decades

E Arabian Peninsula

- Surrounded by Red Sea, Arabian Sea, and Persian Gulf
- Oil and natural gas reserves in eastern part of peninsula

F Nile River

- World's longest river
- Area is home to one of the world's earliest civilizations
- More than 90 percent of Egypt's population lives along the Nile Delta and Nile River on fertile land

G Tigris and Euphrates Rivers

- Known as Mesopotamia, Greek for "land between two rivers"
- Irrigation has supported farming for 7,000 years
- Two rivers join in Iraq to form Shatt al Arab

H Natural Resources

- Over 60 percent of the world's known oil reserves are located in the region.
- Some countries are investing in agriculture and fisheries to decrease their dependence on oil and mineral exports.
- Countries with Mediterranean climates export citrus fruits, olives, and grapes.

CHAPTER 18

CULTURAL GEOGRAPHY OF North Africa, Southwest Asia, and Central Asia

BIG Idea

Geography is used to interpret the past, understand the present, and plan for the future. The region of North Africa, Southwest Asia, and Central Asia has served as the crossroads for Asia, Africa, and Europe. As a result, it has been home to many ethnic groups and cultures.

Essential Questions

Section 1: North Africa
How have the Sahara and access to water affected the people of North Africa?

Section 2: The Eastern Mediterranean
How do ancient civilizations and cultures continue to influence the eastern Mediterranean today?

Section 3: The Northeast
What religious traditions have shaped the history of the Northeast?

Section 4: The Arabian Peninsula
What has affected the modern development of the Arabian Peninsula?

Section 5: Central Asia
How have geography and climate created challenges for Central Asia?

Geography ONLINE
Visit glencoe.com and enter *QuickPass*™ code WGC9952C18 for Chapter 18 resources.

An Egyptian man at the Fortress of Qaitbey, which offers views of Alexandria and the Mediterranean Sea.

Summarizing Information Make a Layered-Look Book to help you summarize information about each of the subregions discussed.

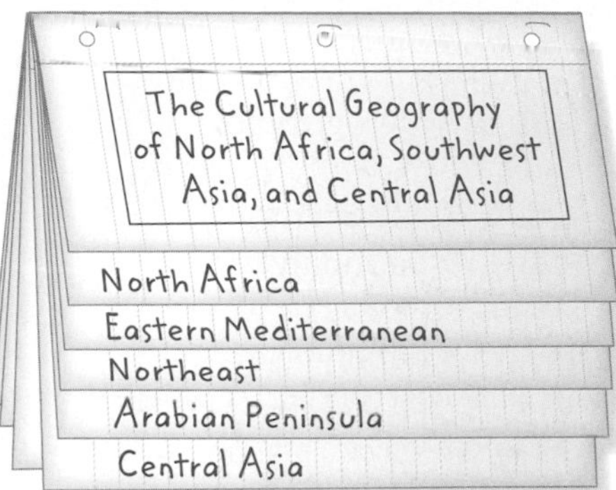

Reading and Writing As you read the chapter, make notes in your Foldable to summarize information about the cultures of each subregion.

SECTION 1

 section audio  spotlight video

North Africa

Guide to Reading

Essential Question
How have the Sahara and access to water affected the people of North Africa?

Content Vocabulary
- nomad *(p. 439)*
- bedouin *(p. 439)*
- infrastructure *(p. 440)*
- domesticate *(p. 440)*
- hieroglyphics *(p. 440)*
- geometric boundary *(p. 441)*
- nationalism *(p. 441)*

Academic Vocabulary
- principal *(p. 439)*
- medical *(p. 442)*

Places to Locate
- Egypt *(p. 439)*
- Morocco *(p. 439)*
- Algeria *(p. 439)*
- Tunisia *(p. 439)*
- Casablanca *(p. 440)*
- Algiers *(p. 440)*
- Tunis *(p. 440)*
- Tripoli *(p. 440)*
- Libya *(p. 440)*
- Cairo *(p. 440)*
- Suez Canal *(p. 441)*

Reading Strategy
Organizing Complete a chart similar to the one below by listing information about each of the major ethnic groups in North Africa.

	Berber	Arab	European
Country			
Religion			
Language			

As the birthplace of one of the world's earliest civilizations, North Africa has a rich cultural heritage. Although the traditions and cultures of some of the ethnic groups have changed with modernization and urbanization, some continue to hold on to their traditional ways of life.

NATIONAL GEOGRAPHIC Voices Around the World

"As we entered the village, children saw me and cried, 'Arrumi!' ('Roman!'), an offhand tribute to rulers 16 centuries gone and the name by which Berbers still refer to Westerners. Little appeared to have changed since the days of the Latins: Barefoot boys used sticks to prod sluggish cattle toward their pens; turbaned men sharpened scythes on whetstones; women trudged by, amphorae of sloshing water on their backs."

—Jeffrey Tayler, "Among the Berbers," *National Geographic*, January 2005

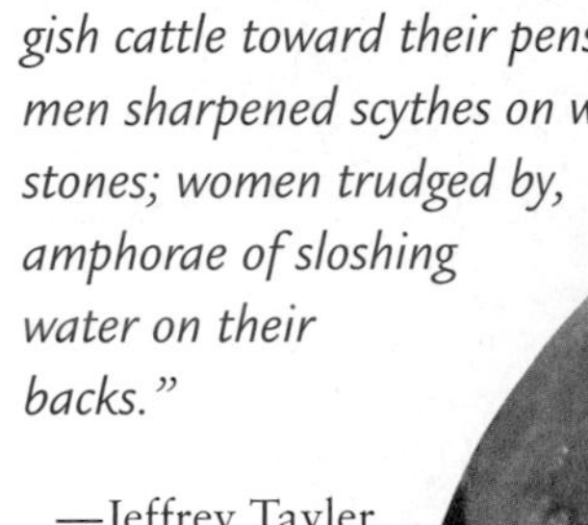

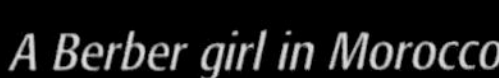
A Berber girl in Morocco

Population Patterns

MAIN Idea Indigenous ethnic groups, migrations, and the dramatic climate have shaped population patterns in North Africa.

GEOGRAPHY AND YOU How has climate affected your region of the United States? Read to learn how North Africa's water affects its population patterns.

Indigenous cultures in North Africa have mixed with those from the Arabian Peninsula and from Europe to form distinct cultures.

The People

While European influence remains in the coastal regions of North Africa, the primary influence on the subregion is a mix of indigenous and Arab cultures. The people indigenous to North Africa before Arab invasions are called Berbers. Most of the 15 million Berbers exist today as farmers, though previously they were pastoral **nomads,** groups of people who move from place to place depending on the season and availability of grass for grazing and water. The Berbers are most populous in the Atlas Mountains and in the Sahara.

The other **principal** ethnic group in North Africa is the Arab people. United by language, Arabs first migrated from the Arabian Peninsula to North Africa in the A.D. 600s. Nomadic **bedouin** (BEH•doo•ihn) are Arabic-speaking people who migrated to North Africa from deserts in Southwest Asia. Bedouin can be found herding animals in the desert where there is enough vegetation to support their herds or water for growing food in oases. **Egypt** was the primary gateway for Arabs migrating to North Africa.

North African peoples have had continuous contact with Europe for hundreds of years, most notably in **Morocco, Algeria,** and **Tunisia.** French, Spanish Muslims, Romans, and Jews have all influenced the culture in this subregion.

MAP STUDY

1. **Location** Where are North Africa's largest cities located?
2. **Regions** Which countries in the region are the most densely populated?

NATIONAL GEOGRAPHIC **North Africa, Southwest Asia, and Central Asia: Population Density**

Maps In MOtion Use **StudentWorks™ Plus** or glencoe.com.

POPULATION

Per sq. mi.	Per sq. km
1,250 and over	500 and over
250–1,249	100–499
63–249	25–99
25–62	10–24
2.5–24	1–9
Less than 2.5	Less than 1

Cities
(Statistics reflect metropolitan areas)
- ◇ Over 10,000,000
- ■ 5,000,000–10,000,000
- □ 2,000,000–5,000,000
- ⊙ 1,000,000–2,000,000

0 1,000 kilometers
0 1,000 miles
Lambert Azimuthal Equal-Area projection

NATIONAL GEOGRAPHIC *Open-air markets, such as this one in Tripoli, Libya, can be found in many cities throughout North Africa.*

Regions What are the major population centers in North Africa?

Density and Distribution

Geographic factors, especially the availability of water, have influenced settlement in the subregion. Because water is scarce, people have for centuries settled along seacoasts and in river deltas, such as along the Nile River in Egypt. Today more than 90 percent of Egypt's people live in the Nile Delta region, one of the world's most densely populated areas.

The major population centers in North Africa are urban: **Casablanca,** Morocco; **Algiers,** Algeria; **Tunis,** Tunisia; **Tripoli, Libya;** and **Cairo,** Egypt. As Egypt's primate city, Cairo dominates social and cultural life in the country. Cities such as Cairo have grown rapidly as people move there in search of a better life. Problems have arisen in some cities because urban growth has occurred too quickly to supply enough jobs and housing or to develop the **infrastructure**—basic urban necessities such as streets and utilities.

READING Check **Human-Environment Interaction** How has the Sahara affected settlement patterns in North Africa?

History and Government

MAIN Idea The Sahara, the Nile River valley, and multiple invasions influenced different cultures throughout North Africa's history.

GEOGRAPHY AND YOU How has the physical environment influenced events where you live? Read to learn how the physical environment has shaped the history of North Africa.

North Africa's location near Europe and Southwest Asia has made it vulnerable to numerous migrations and invasions over the centuries. As a result, European, Arab, and Berber influences exist in North Africa.

Early Peoples and Civilizations

Hunters and gatherers settled throughout North Africa by the end of the last Ice Age, about 10,000 years ago. By 6000 B.C., farming communities had arisen in areas along the Nile River and the Mediterranean Sea. The region's farmers were among the first to **domesticate** plants and animals, or adapt them from the wild for such uses as food, clothing, and transportation.

The Egyptian civilization developed in the fertile Nile River valley about 6,000 years ago. Annual floods from the Nile deposited rich soils on the floodplain. During dry seasons, Egyptians used sophisticated irrigation systems to water crops, enabling farmers to grow two crops each year. The Egyptians also developed a calendar with a 365-day year, built impressive pyramids as tombs for their rulers, and invented a form of picture writing called **hieroglyphics** (HY•ruh•GLIH•fihks).

Invasions

The seventh-century invasions of Arabs from the Arabian Peninsula heavily influenced the cultures of North Africa. The Berbers closely assimilated with Arab cultures in Algeria and Morocco, but less so in Tunisia. After brief Vandal and Byzantine invasions, Arab rule was established in Tunisia. The culture survived through Ottoman rule that lasted until 1922.

While Arab culture has persisted through the centuries in all North African countries, history has brought other cultural influences to the subregion. Internal invasions in the A.D. 600s from Arabs in the east brought Islamic religion and culture to Morocco. Muslim and Jewish exiles fleeing persecution by the Inquisition in Spain infused Morocco with Spanish culture in the 1400s. Early rule by Arab-Berber dynasties was eventually overturned by the Ottoman Empire in the 1500s in Algeria.

European colonial rule also affected people and cultures in North Africa. Algeria's short independence, followed by French invasion and conquest in the mid-1800s, imprinted French influence on the country. **Geometric boundaries**—which often follow straight lines and do not account for natural and cultural features—exist between Libya, Egypt, and Algeria. They were drawn by European colonial powers and often created conflict between the new countries because local practices of government were not the same as European ideas about governing.

During the 1800s a well-educated urban middle class developed in North Africa. Trained in European ways, this new middle class adopted European ideas about **nationalism,** or a belief in the right of an ethnic group to have its own independent country. This development stirred demands for self-rule that provided the basis for the modern countries that emerged in the subregion.

Independence

Egypt gained independence from the United Kingdom in 1922. The **Suez Canal,** an important shipping lane that connects the Mediterranean Sea to the Red Sea across Egypt, has made Egypt a key regional power. Egypt is also an important center of Arab nationalism.

Algeria gained independence from France when a strong nationalist movement led to a civil war in the mid-1900s. Since independence in 1962, Algeria has developed its resources and increased its standard of living. However, a civil war in the 1990s killed over 100,000 people.

Other countries in the region also gained independence in the 1950s and 1960s. Libya won independence from Italy in 1951, but was then ruled by a strong, Western-friendly monarchy. In 1969 a coup led by Colonel Muammar al-Qaddafi overthrew the monarchy. Al-Qaddafi has ruled the oil-rich country to this day. Tunisia separated from France in 1956. Morocco won independence from France in 1956 and today is a constitutional monarchy.

READING Check **Regions** What movement led to the demand for self-rule in North African countries?

Geography ONLINE
Student Web Activity Visit glencoe.com, select the *World Geography and Cultures* Web site, and click on Student Web Activities–Chapter 18 for an activity about Egypt.

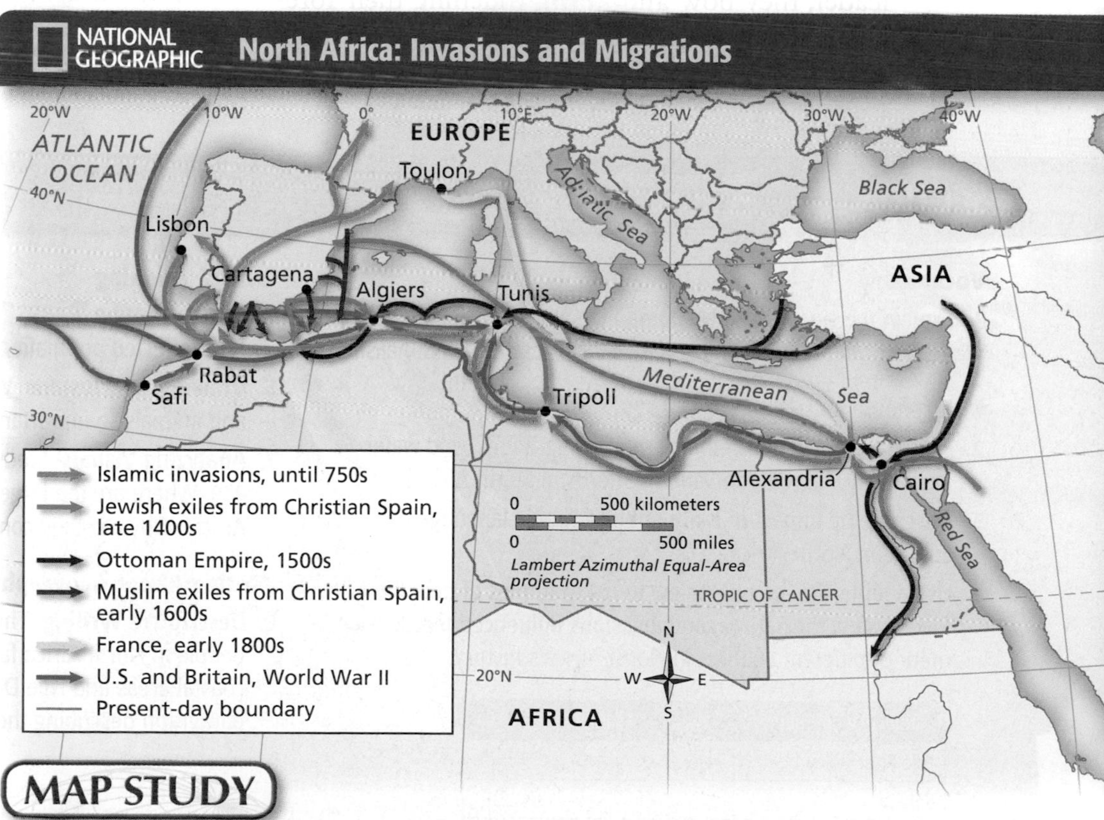

MAP STUDY

1. **Movement** Which cities were affected by Islamic invasions until the 750s?
2. **Movement** Which later migration route is similar to that of Islamic invasions?

CONFLICT: Israelis and Palestinians

The Problem:

Israelis and Palestinians each claim the same territory in Southwest Asia. After decades of uprisings, assassinations, attacks, failed peace agreements, and other struggles, conflict remains between these two groups attempting to coexist in the same area.

Separation These Israeli soldiers are patrolling the controversial barrier Israel constructed to protect itself from Palestinian terrorists.

Settler evicted from the Gaza Strip

Disengagement A woman cries in front of Israeli soldiers in the Gaza Strip settlement of Kfar Darom, where settlers have been evicted as part of Israel's disengagement plan.

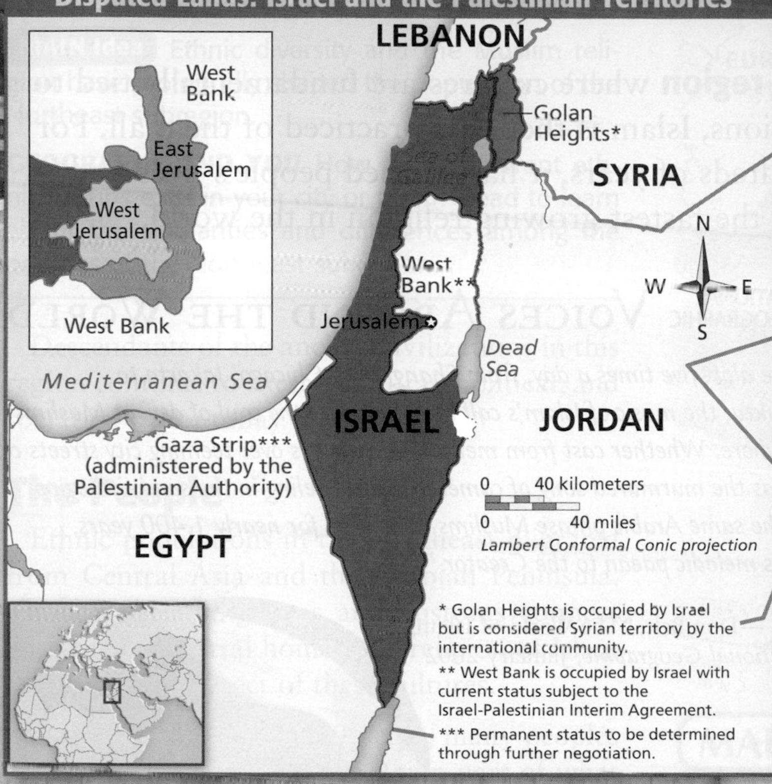

Term	Meaning
Hamas	a Palestinian Sunni Islamist terrorist organization founded in 1987
Palestinian National Authority	an interim administrative organization established in 1994 that nominally governs parts of the West Bank and all of the Gaza Strip, areas known as the Palestinian Territories
Road Map for Peace	a plan outlined by the "quartet"—the United States, European Union, United Nations, and Russia—that calls for an independent Palestinian state coexisting in peace with an Israeli state

Attempts at Peace

At the heart of the conflict lie the West Bank, the Gaza Strip, and East Jerusalem. East Jerusalem is the most disputed territory, as it is home to some of the holiest sites in the Jewish, Muslim, and Christian faiths. Israel claims a united Jerusalem as its capital, but this claim is not recognized by many nations. The Palestinians, on the other hand, want East Jerusalem to become the capital of their future country. Both sides have failed to reach agreement.

What are some of the attempts at peace? In 1993 a plan known as the Oslo Accords was enacted. This plan allowed for limited Palestinian self-rule in areas controlled by Israel, including the Gaza Strip and parts of the West Bank. The Palestinian National Authority (PNA) was established in 1994 to control security-related and civilian issues in the Palestinian urban areas. In 2005 Israel carried out a disengagement plan, removing all of its military and settlers from the Gaza Strip and some cities in the West Bank. The Annapolis Conference in 2007 was the first time a two-state solution was stated as the mutually agreed-upon strategy for achieving peace. However, the resolution was withdrawn after Israel had objections.

What is the current state of the region? The Palestinian Hamas-Fatah coalition government formed in 2006 ended amid conflicts between the two groups. Hamas later lost control of the West Bank. A truce between Israel and the group Hamas followed, but expired in December 2008 when conditions to extend the truce could not be agreed upon. Fighting continued into 2009.

One Solution:

Many feel that the only solution may be the creation of an independent country of Palestine. In 2009, U.S. president Barack Obama stated his support for an independent Palestinian state. The Road Map for Peace, originally presented in 2003, is being revisited.

President Obama (center) meets with Israeli prime minister Netanyahu (left) and Palestinian Authority president Abbas in New York City in September 2009.

THINKING GEOGRAPHICALLY

1. **Human Systems** Why is Jerusalem such an important city to both the Palestinians and the Israelis?
2. **Places and Regions** Look at the main picture. Several separation barriers have been built along Israeli and Palestinian territorial borders. Why do you think these barriers were built?

Population Patterns

MAIN Idea A shared religion, a common language, and rapid modernization have formed today's Arabian Peninsula.

GEOGRAPHY AND YOU Is the area where you live changing in noticeable ways? Read to learn how the Arabian Peninsula is a mix of tradition and modern cultures.

The harsh desert climate of the Arabian Desert has led most people to live along the coasts. Their lives are shaped by traditional Islamic culture and modernization driven by the oil industry.

The People

Most people in the subregion—about 65 million—are Arabs. Most Arabs are Muslims, but a small percentage follows Christianity or other religions. Islamic culture and the Arabic language have had a significant impact here.

Arabic-speaking peoples have lived on the Arabian Peninsula since before the spread of Islam. Many Arabic-speaking people today, however, descend from ancient groups such as the Egyptians, Phoenicians, Saharan Berbers, and peoples speaking other Semitic languages. In **Kuwait,** for example, the majority of people are Arab, but, instead of originating on the Arabian Peninsula, they migrated from other states in the region when oil was discovered in the first half of the 1900s.

Many people from South Asia also live in the larger cities of the eastern Arabian Peninsula. Muslims from Pakistan, India, Bangladesh, and Iran have immigrated to this area to find jobs.

Density and Distribution

As in North Africa, harsh desert and lush coastal climates dictate where the majority of people live on the Arabian Peninsula. Bedouin still roam the large Arabian Desert and settle in oases, but many have migrated to cities. This settlement pattern has become common in **Saudi Arabia,** where population densities can be as high as 2,600 people per square mile (1,000 per sq. km) in cities and some oases. Yet, because the country has over 829,999 sq. miles (2,149,690 sq. km), the density for all of Saudi Arabia is only 34 people per square mile (13 per sq. km). In **Bahrain,** 89 percent of the population lives in its two main cities, Manama and Al Muharraq. In **Oman,** more than half of the population lives along the coastal plain.

NATIONAL GEOGRAPHIC **The Arabian Peninsula: Citizens and Foreign Nationals**

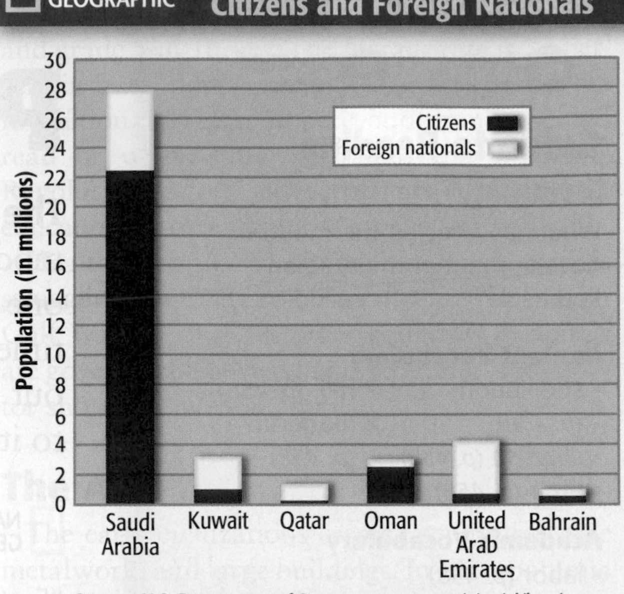

Source: U.S. Department of State, www.state.gov/r/pa/ei/bgn/

GRAPH STUDY

1. **Place** Which country has the largest population of foreign nationals?
2. **Place** Which country has about the same number of citizens as it does foreign nationals?

Graphs In MOtion Use **StudentWorks™ Plus** or glencoe.com.

Arab people in **Yemen** are historically a more settled group than Arab nomads. They have lived mostly in villages and small towns. The bedouin, however, have adapted to the desert as they roam for water and grazing areas for their herds. In contrast, 95 percent of the bedouin population of Saudi Arabia is now settled.

The discovery of oil in the early 1900s led to increased wealth, modernization, and immigration in many Arab countries. In many of these countries, such as the **United Arab Emirates,** these new arrivals make up a majority of the population. Only 15 to 20 percent of the people in the United Arab Emirates are citizens. Foreign workers make up more than 50 percent of **Qatar's** (KAH•tuhr) population and more than 60 percent in Kuwait.

READING Check **Movement** What discovery helped lead to modernization in some countries on the peninsula?

History and Government

MAIN Idea Conquering empires and unified governments have imposed cultures on the peoples of the Arabian Peninsula that remain influential to this day.

GEOGRAPHY AND YOU How do you think the culture of the United States might be shaping other cultures around the world? Read to learn about how the peninsula was shaped by various forces in its early and recent history.

Though some countries on the Arabian Peninsula are quite young, the histories of their peoples are long. Many of these histories through the 1800s are similar, as people of the region sought protection from Great Britain to fight off rule by other outsiders, such as the Ottoman Empire. This region also saw the birth of Islam, which has become a crucial part of each state's culture in the peninsula.

Early Cultures and Conquests

For 5,000 years vibrant cultures have existed on the Arabian Peninsula. One of the oldest centers of civilization in the area existed in Yemen between the 1100s B.C. and the A.D. 500s.

Because of the peninsula's harsh climate, it has been difficult for large settlements to exist anywhere but along the coast. An island near Kuwait was used by ancient Greeks as a huge trade center for about 2,000 years.

After a local powerful family gained control of territory on the Arabian Peninsula around 1750, it struggled against invasion by the Ottoman Empire and others. The Unified Kingdom of Saudi Arabia was finally established in 1932.

Also threatened by the Ottoman Empire, Kuwait, Bahrain, and Qatar signed treaties with Great Britain for protection in the 1800s and early 1900s. Parts of Yemen were under British control from the 1800s until 1967. Oman was independent for most of its history.

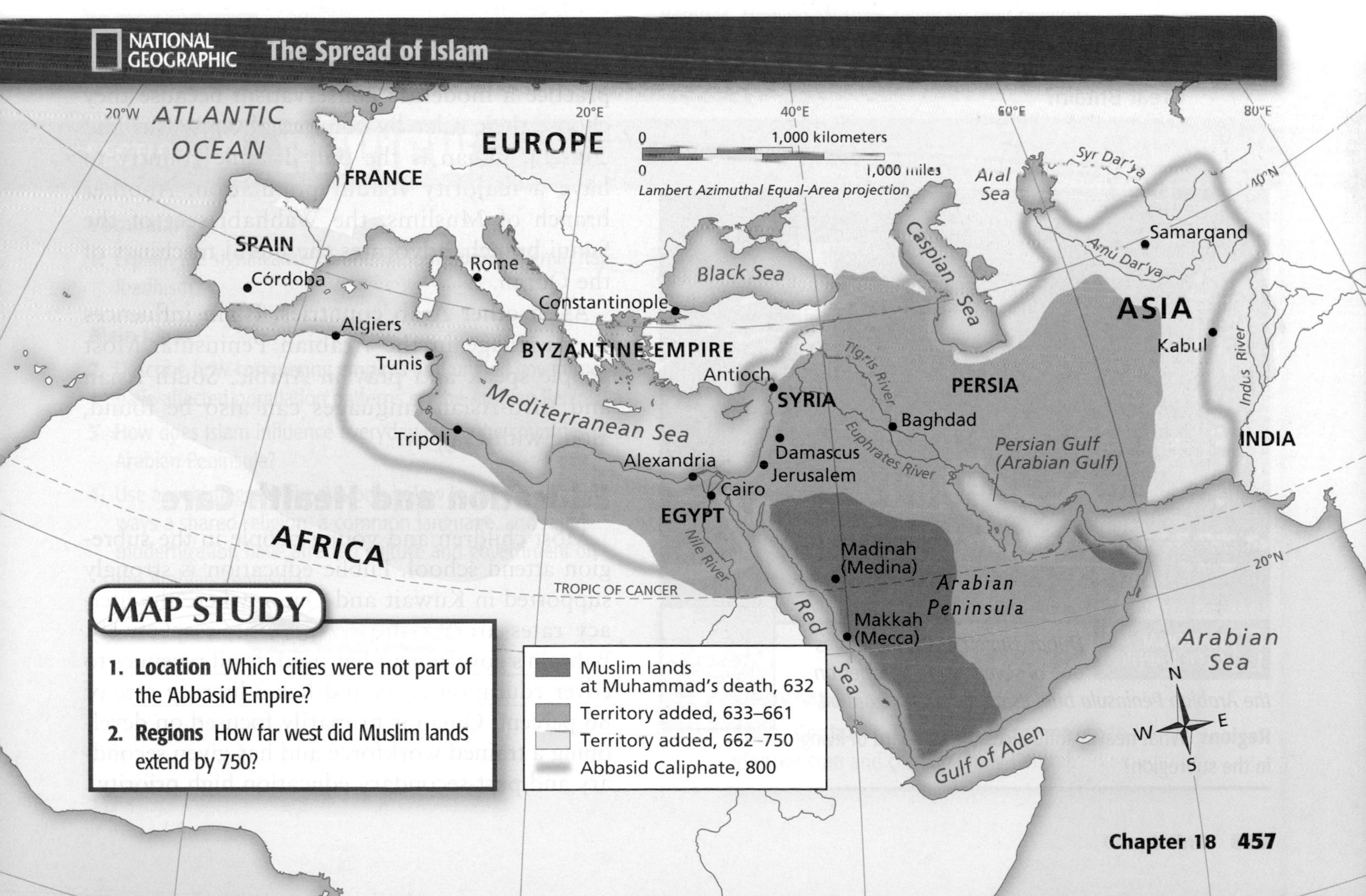

MAP STUDY

1. **Location** Which cities were not part of the Abbasid Empire?
2. **Regions** How far west did Muslim lands extend by 750?

Why Geography Matters

Afghanistan: A Troubled History

The Problem:

Afghanistan's inability to form a strong central government has led to terrorism and instability in the country. Obstacles such as warlords, a decentralized power base, and the lack of a democratic history stand in the way of the formation of a national democratic government.

Former warlord and now provincial governor Gul Agha Sherzai

Warlords Internal fighting along ethnic lines has given rise to warlords. These men command their own armed forces used to solidify control in the areas they occupy.

Voting These Afghan citizens are voting in a national election. Voters are required to dip a finger in ink when voting. An ink-stained finger signifies that the citizen has voted. This process helps prevent voters from voting more than once.

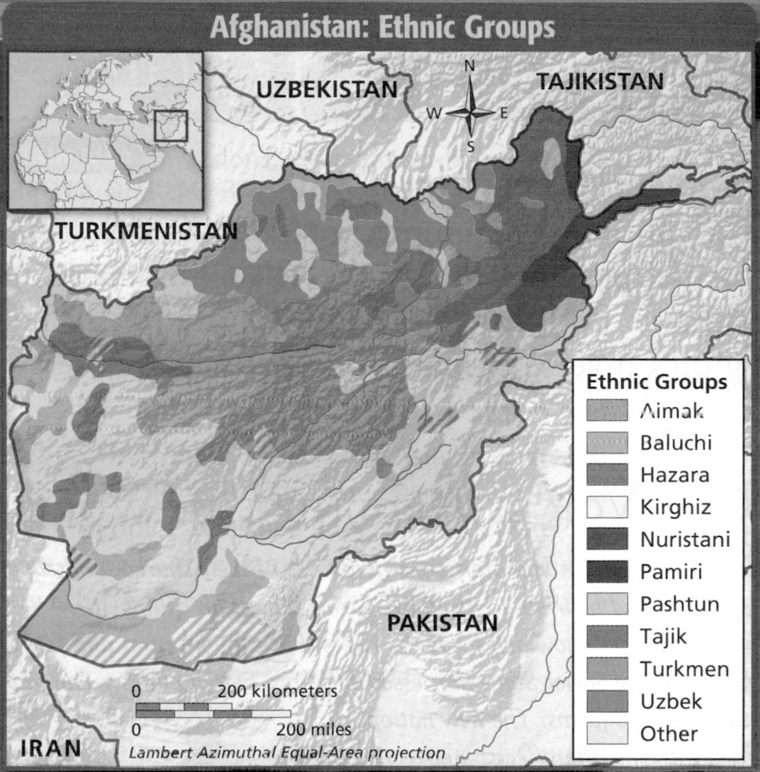

Afghanistan: Ethnic Groups

UZBEKISTAN
TAJIKISTAN
TURKMENISTAN
PAKISTAN
IRAN
N
S
E
W

Ethnic Groups
Aimak
Baluchi
Hazara
Kirghiz
Nuristani
Pamiri
Pashtun
Tajik
Turkmen
Uzbek
Other

0 200 kilometers
0 200 miles
Lambert Azimuthal Equal-Area projection

Challenges

Located on trade routes linking the Middle East and South Asia, Afghanistan has often fallen victim to invaders. In the last 30 years Afghanistan has again faced the external forces of invasion by other nations as well as the internal forces of ethnic diversity. Both threaten the country's stability.

How has the Soviet Union been involved? In 1979 the Soviet Union invaded Afghanistan, hoping to shore up the communist regime. Fearing an expansion of Soviet power, the United States, Britain, and China provided arms to guerilla rebels. In 1988 the United States, Pakistan, Afghanistan, and the Soviet Union signed agreements for the withdrawal of Soviet forces. Following the Soviet withdrawal, the rebel group Mujahadeen stormed the capital at Kabul and overthrew the government, creating the Islamic Republic of Afghanistan.

Who are the Taliban? In 1995 the Taliban, an Islamic military group promising to restore order, rose to power. They cracked down on rampant crime and drug trafficking, but they also severely limited the rights of women and used violence to enforce laws.

How has the U.S. been Involved? International terrorist Osama bin Laden was believed to be hiding in Afghanistan. When the Taliban refused to turn him over to stand trial for terrorism, the United Nations imposed trade restrictions. Following the 2001 attack on the World Trade Center and Pentagon, the United States attacked Afghanistan and ousted the Taliban. American forces worked with Afghans to try to stabilize the government, and in 2004 Hamid Karzai was named Afghanistan's president in the country's first successful election. After much controversy, Karzai was reelected in 2009.

One Solution:

The support of peace organizations combined with foreign aid may help strengthen and stabilize Afghanistan by helping to build an infrastructure, including a strong communications system. The economic development of the country would help in forming a national identity.

Accusations of fraud led to plans for a run-off election between Abdullah Abdullah (left) and Hamid Karzai (right) in 2009. Abdullah Abdullah later dropped out.

THINKING GEOGRAPHICALLY

1. **Human Systems** Research U.S. involvement in Afghanistan in more depth. Why do you think a stable Afghan government is important to the United States?
2. **Places and Regions** Look at the map of the various ethnic divisions in Afghanistan. Why has uniting the country under one centralized government been so difficult?

CONNECTING TO

THE UNITED STATES

Iraq's first parliamentary election since the overthrow of Saddam Hussein was held in December 2005. An estimated 240,000 Iraqis living in the United States, like this woman in Detroit, Michigan, were eligible to vote in the election by absentee ballot.

Just the Facts:

- Approximately half of the Arab population in the United States is concentrated in five states: California, Florida, Michigan, New Jersey, and New York.
- Outside of Israel, the United States is home to the largest Jewish population in the world.
- The cell phone was developed in Israel at the Israeli branch of Motorola.
- The alphabet was created in the land of the Phoenicians, which is now Lebanon.

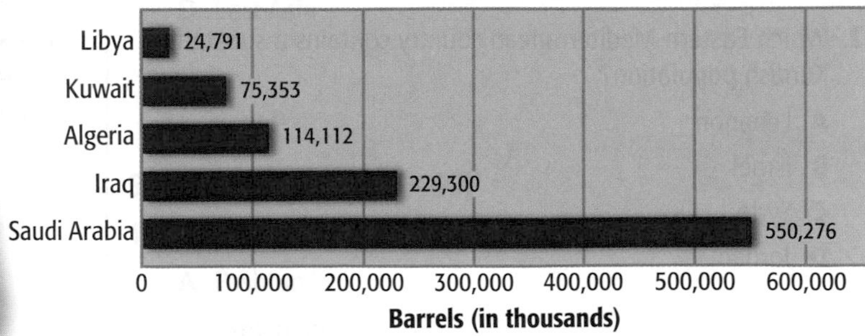

Importing oil The United States is the world's largest energy consumer. A large portion of U.S. oil is imported from countries in this region.

Making the Connection

World cultures are interconnected in many ways, and much of what we encounter in our daily lives finds its past and present in the countries of North Africa, Southwest Asia, and Central Asia.

Beef kabobs and rice pilaf

Food—Influencing Our Dinner Tables A wide variety of foods we eat daily comes from this region. Some of the most recognizable include melons, kebabs, yogurt, and rice pilaf.

Energy—Keeping Us on the Move The United States is the world's largest energy consumer, importing about half of all of its oil. The United States consumes about 21 million barrels of oil each day, most of which comes from this region. Currently, Saudi Arabia is the greatest supplier of oil to the United States.

Judaism, which began in this region, is one of the three most practiced religions in the United States.

Religion—A Melting Pot of Beliefs Christianity, Judaism, and Islam, the three most practiced religions in the United States, have deep connections to North Africa, Southwest Asia, and Central Asia. The region still holds great significance today for all three religions. The city of Makkah, in Saudi Arabia, is the holiest site in Islam. Jerusalem, the capital of Israel, is the holiest city of Judaism and is also of special importance to Christianity and Islam.

Paula Abdul, a well-known choreographer, singer, and television personality, is of Syrian descent.

Region's Immigration to the United States by Country (2008)

Country	Immigrants
Iran	13,852
Egypt	8,712
Israel	5,851
Iraq	4,795
Morocco	4,425
Lebanon	4,254
Turkey	4,210
Jordan	3,936
Afghanistan	2,813
Syria	2,641

Source: *U.S. Department of Homeland Security*

THINKING GEOGRAPHICALLY

1. **Human Systems** Research one of the religious sites mentioned in this feature. Explain its significance within the context of one or multiple religions.
2. **Environment and Society** Investigate some of the uses of oil other than gasoline for cars. What products are created using oil? How do these products impact your daily life?

CHAPTER 19

THE REGION TODAY

North Africa, Southwest Asia, and Central Asia

Big Idea

Countries are affected by their relationships with each other. North Africa, Southwest Asia, and Central Asia contain a large share of the world's oil and natural gas. A study of the region will explain how its natural resources influence the global economy and what challenges the region faces.

Essential Questions

Section 1: The Economy

How has the presence of oil affected the economies of the countries in North Africa, Southwest Asia, and Central Asia?

Section 2: People and Their Environment

How have technology and war impacted the environment in North Africa, Southwest Asia, and Central Asia?

Geography ONLINE

Visit glencoe.com and enter ***QuickPass***™ code WGC9952C19 for Chapter 19 resources.

Dubai, the largest city in the United Arab Emirates, is the main center of trade for the entire Persian Gulf region.

Identifying Information Create a Shutter Fold to identify the causes and effects of environmental problems in North Africa, Southwest Asia, and Central Asia today.

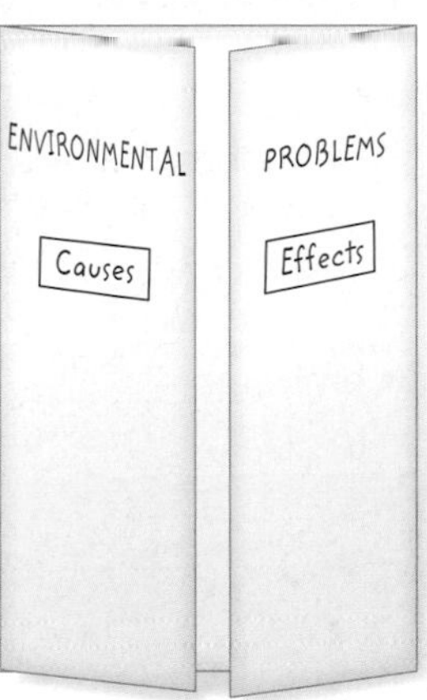

Reading and Writing As you read the chapter, identify and describe in the appropriate place in your Foldable the causes and effects of the region's environmental problems, such as the scarcity of freshwater in many parts of the region.

SECTION 1

 section audio  spotlight video

Guide to Reading

Essential Question
How has the presence of oil affected the economies of the countries in North Africa, Southwest Asia, and Central Asia?

Content Vocabulary
- arable *(p. 475)*
- commodity *(p. 475)*
- crude oil *(p. 476)*
- petrochemical *(p. 476)*
- landlocked *(p. 477)*
- embargo *(p. 479)*

Academic Vocabulary
- economic *(p. 475)*
- energy *(p. 475)*
- strategic *(p. 477)*

Places to Locate
- İstanbul *(p. 477)*
- Gulf of Aqaba *(p. 477)*
- Strait of Hormuz *(p. 477)*
- Baku *(p. 478)*
- Batumi *(p. 478)*

Reading Strategy
Taking Notes As you read about the region's economy, create an outline like the one below by using the major headings of the section.

I. Economic Activities
   A.
   B.
II. Transportation and Communications
   A.
   B.

The Economy

The oil-producing countries in North Africa, Southwest Asia, and Central Asia have experienced greater economic growth than other countries in the region. Tremendous wealth from oil and natural gas has brought many positive changes. For some countries, such as Saudi Arabia and Kuwait, vast oil reserves have defined their economic history.

NATIONAL GEOGRAPHIC Voices Around the World

"The fulcrum of Saudi history can be pinpointed exactly: the Persian Gulf city of Dammam on March 3, 1938, when American engineers unleashed the kingdom's first commercially viable oil gusher after 15 months of drilling. The joint venture between U.S. petroleum companies and Saudi Arabia's ruler, King Abdul Aziz ibn Saud, put the fledgling nation on the global economic map."

—Frank Viviano, "Saudi Arabia on Edge," *National Geographic*, October 2003

A Saudi man at an oil refinery

Economic Activities

MAIN Idea Economic activities in North Africa, Southwest Asia, and Central Asia are influenced by oil, natural gas, and water.

GEOGRAPHY AND YOU What natural resources are important to the U.S. economy? Read to learn how the vast oil and natural gas reserves in North Africa, Southwest Asia, and Central Asia affect global affairs and economic activities.

Oil and water are two key natural resources for **economic** activities in North Africa, Southwest Asia, and Central Asia. Those countries rich in oil generally have scarce water supplies; those countries with abundant water supplies generally lack oil resources.

Agriculture and Fishing

As the chart at the right shows, only a small part of the region's land is **arable,** or suitable for farming, yet a large percentage of the population works in agriculture. In Afghanistan, for example, where only 12 percent of the land is arable, 80 percent of the people farm for a living. Agriculture plays a smaller role in countries that have economies based on oil, such as Kuwait.

Areas of North Africa and Southwest Asia that have a Mediterranean climate are best suited for growing cereal crops, citrus fruits, grapes, olives, and dates. When rainfall is below normal, however, harvests of major crops such as wheat, barley, and corn seldom meet people's needs. Countries that grow these crops must often import additional grains to feed their people. Other crops like citrus fruits are important exports. For example, Georgia's humid subtropical climate is good for growing citrus fruits, grapes, and cotton.

The steppes of Central Asia provide fertile soil for growing crops and grasslands for grazing livestock. Uzbekistan is one of the world's largest cotton producers. Both Uzbekistan and Turkmenistan are important centers for raising silkworms. Wheat, cotton, potatoes, and tea earn Azerbaijan substantial export income, even though only 21 percent of the country's land is arable. Kazakhstan is a major grain producer.

Fish are an important food source in the region. Moroccan vessels bring in sardines and mackerel from the Atlantic Ocean. The majority of Israel's annual fish catch consists of freshwater fish raised in artificial ponds. Fishers from other countries harvest fish from the Persian Gulf, which is home to about 150 species. The size of fish catches has declined in the Caspian Sea because of overfishing and pollution. Still, Iran and several other countries bordering this sea have flourishing fishing industries.

Total Land Area vs. Arable Land (selected countries)

Country	Total Land Area	Arable Land	Percent Arable
Afghanistan	251,885 sq. mi. 652,230 sq. km	30,226 sq. mi. 78,267 sq. km	12%
Azerbaijan	31,903 sq. mi. 82,629 sq. km	6,699 sq. mi. 17,352 sq. km	21%
Georgia	26,911 sq. mi. 69,700 sq. km	3,229 sq. mi. 8,364 sq. km	12%
Iran	591,352 sq. mi. 1,531,595 sq. km	59,135 sq. mi. 153,159 sq. km	10%
Iraq	168,868 sq. mi. 437,367 sq. km	21,953 sq. mi. 56,858 sq. km	13%
Israel	8,356 sq. mi. 21,642 sq. km	1,253 sq. mi. 3,246 sq. km	15%
Kazakhstan	1,042,360 sq. mi. 2,699,700 sq. km	83,389 sq. mi. 215,976 sq. km	8%
Saudi Arabia	829,999 sq. mi. 2,149,690 sq. km	16,599 sq. mi. 42,994 sq. km	2%
Turkey	297,156 sq. mi. 769,632 sq. km	89,147 sq. mi. 230,889 sq. km	30%
Turkmenistan	181,441 sq. mi. 469,930 sq. km	9,072 sq. mi. 23,497 sq. km	5%

Source: www.cia.gov, The World Factbook 2009.

Industry

Petroleum and oil products are the main export **commodities,** or economic goods, of the region. It holds over 60 percent of the world's oil and is likely to continue to supply much of the world's oil. In addition to significant oil reserves, the region also holds about 50 percent of the world's natural gas reserves.

Oil has brought unimagined riches to the Persian Gulf countries. Trapped in pockets beneath the region's sandy soils are two-thirds of the world's known petroleum reserves. This "black gold" provides the raw material for everyday products, such as compact discs, crayons, and house paint. In addition, oil supplies more than half of the **energy** used worldwide. Almost overnight, oil profits transformed villages in Saudi Arabia, Kuwait, Bahrain, and other Persian Gulf countries into gleaming, modern cities.

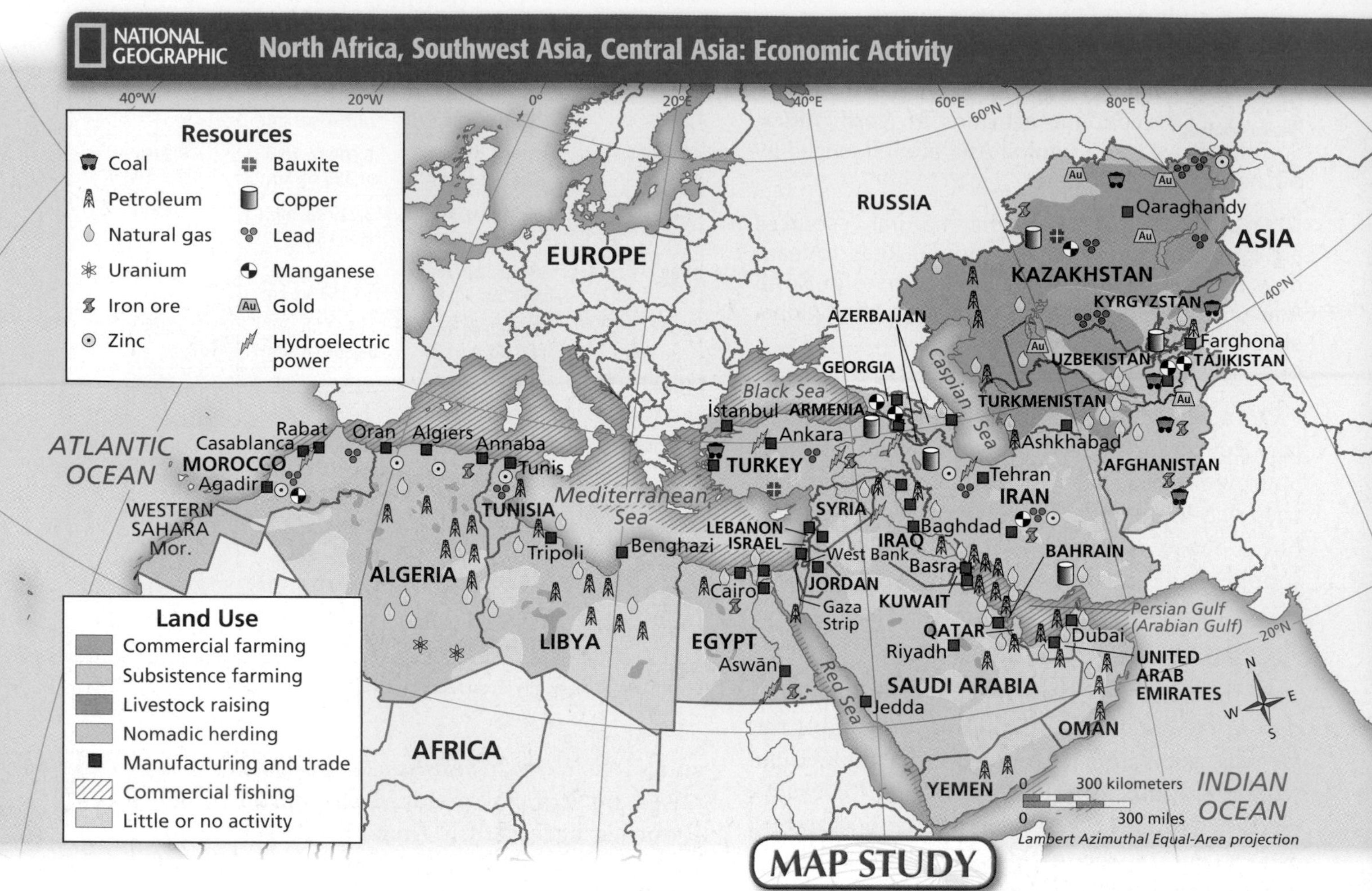

MAP STUDY

1. **Location** What is common about the location of most of the region's manufacturing and trading centers?
2. **Regions** What type of land use is most common in the region of North Africa, Southwest Asia, and Central Asia?

Maps In MOtion Use **StudentWorks™ Plus** or glencoe.com.

Oil, Natural Gas, and Mining Oil wealth has helped build industry in the region. Iran and Saudi Arabia operate oil-refining and oil-shipping facilities. Most other oil-producing countries export **crude oil**—petroleum that has not yet been refined—to industrialized countries. Natural gas has also advanced the region. Israel is exploring export opportunities for natural gas discovered offshore in the eastern Mediterranean.

Industries using **petrochemicals**—products derived from petroleum or natural gas—make fertilizers, medicines, plastics, and paints. They provide jobs and improve the standard of living.

Coal and copper mining and cement production are important in both Southwest Asia and Central Asia. In North Africa, Morocco is the largest exporter of phosphate, an essential ingredient in agricultural fertilizers.

Service Industries Service industries play significant roles in the region's economies. For example, Israel's growing high-tech sector accounts for the largest share of the country's overall manufacturing output.

North Africa and Southwest Asia are popular travel destinations because of their historical importance. Ancient monuments and religious sites attract followers of the three major religions that originated in the region. Christians and Jews visit Israel, Jordan, and other countries whose past is linked to the Bible. Muslims make a hajj to Makkah, Saudi Arabia. Other visitors come to enjoy sunny Mediterranean beaches.

Some countries, however, discourage visitors to limit foreign influences. After the Islamic revolution in 1979, the Iranian government placed restrictions on tourists from non-Muslim countries. Conflicts and instability in Algeria, Syria, Iraq, Israel, and Lebanon have also affected tourism.

READING Check **Regions** What economic activity has brought the greatest development to the region?

Transportation and Communications

MAIN Idea Advancements in transportation and communications are improving throughout the region, but the physical environment and government control have limited some development.

GEOGRAPHY AND YOU What physical features must be crossed to link major cities, oil fields, and ports in North Africa, Southwest Asia, and Central Asia? Read to learn how the physical environment has slowed the development of transportation and communications in the region.

Advances in transportation and communications systems in the region are bringing people closer together. Countries in the eastern Mediterranean have experienced the greatest expansion in transportation and communications.

Roads, Railroads, and Airlines

Road systems are unevenly distributed across the region. Extensive systems cross Iran, Turkey, and Egypt, connecting major cities with oil fields and seaports. In some countries, though, mountains and deserts make road building difficult and costly. In recent years, however, economic development and the growing number of vehicles demand the construction of more road systems.

In parts of the Caucasus area, roads provide the only access to the outside world. In Central Asia, the countries of Afghanistan, Kyrgyzstan, and Tajikistan are surrounded by formidable mountain ranges such as the Hindu Kush, Tian Shan, and Pamirs. **Landlocked** countries—those almost or entirely surrounded by land—such as these do not have access to the sea for transportation and trade.

To ease traffic congestion in crowded urban areas and to improve urban-rural connections, some governments have built rapid-transit systems and railroads. A new subway in **İstanbul,** Turkey, a city of more than 10 million people, carries commuters to and from the city's center. National rail lines also connect urban areas and seaports. In 2006 Tajikistan began working on the Anzob Highway Tunnel, which would provide a year-round link between the northern and southern parts of the country.

Since World War II, the growth of the air travel industry has benefited North Africa and Southwest Asia. In recent years, Central Asia has also benefited from increased air traffic. Before the breakup of the Soviet Union, Central Asian countries relied on the Soviet airline Aeroflot, but now some Central Asian countries have their own airlines.

Waterways and Pipelines

Water transportation is vital to the region. Ships load and unload cargo at ports on the Mediterranean and Black Seas. The Strait of Tiran—between the **Gulf of Aqaba** and the Red Sea—and the **Strait of Hormuz**—linking the Persian Gulf with the Arabian Sea—are of **strategic** and economic importance. Oil tankers entering and leaving the Persian Gulf must pass through the Strait of Hormuz. The Suez Canal, a major human-made waterway located between the Sinai Peninsula and the rest of Egypt, enables ships to pass from the Mediterranean Sea to the Red Sea.

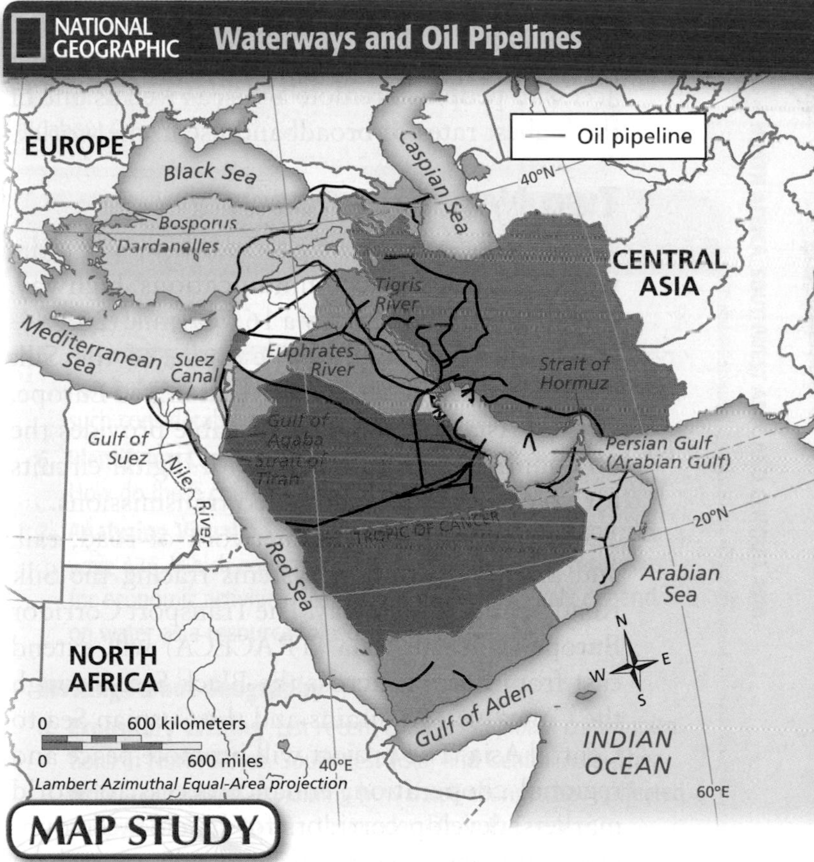

MAP STUDY

1. **Location** In which countries is the greatest concentration of oil pipelines located?
2. **Regions** Why are waterways such as the Strait of Hormuz and the Suez Canal of economic importance?

SECTION 2

 section audio

 spotlight video

People and Their Environment

Guide to Reading

Essential Question

How have technology and war impacted the environment in North Africa, Southwest Asia, and Central Asia?

Content Vocabulary

- aquifer *(p. 481)*
- desalination *(p. 481)*

Academic Vocabulary

- capacity *(p. 481)*
- project *(p. 482)*
- regime *(p. 483)*

Places to Locate

- Aswān High Dam *(p. 482)*
- Elburz Mountains *(p. 483)*
- Aral Sea *(p. 483)*

Reading Strategy

Organizing As you read about the environmental concerns of the region, complete a graphic organizer like the one below by describing the environmental challenges of the Caspian Sea, the Dead Sea, and the Aral Sea.

Body of Water	Challenges
Caspian Sea	
Dead Sea	
Aral Sea	

Like human actions in many places, those in North Africa, Southwest Asia, and Central Asia often threaten the environment. These actions take many forms—oil spills, urban sprawl, overuse of water supplies, and destructive wars. The dilemma people face is how to meet human needs while trying to restore the already damaged environment and protect it from further devastation.

NATIONAL GEOGRAPHIC Voices Around the World

"After 23 years of conflict Afghanistan is the neediest of all, a gutted shell of a state with millions of land mines embedded in its earth. . . .

Consider the arithmetic of Afghanistan: A million and a half people killed. Nearly four million living as refugees, including most of the veneer of educated men and women. Land mines preventing the use of thousands of acres of precious farmland. Kabul all but destroyed, the university in rubble. Highways, bridges—gone. Experts say it will take at least a decade to rebuild Afghanistan merely to its spare 1960s development level. And many more years to bring it into the 21st century."

—Mike Edwards, "Central Asia Unveiled," *National Geographic*, February 2002

An Afghan man wounded by a land mine

Managing Resources

MAIN Idea Growing populations in North Africa, Southwest Asia, and Central Asia severely strain the already scarce water resources.

GEOGRAPHY AND YOU What human actions impact water resources in the United States? Read to learn about solutions to the lack of freshwater in North Africa, Southwest Asia, and Central Asia.

According to the World Health Organization, more than 1 billion of the world's people cannot obtain clean drinking water. About 884 million people still use an unimproved source for drinking water.

Water Resources

Much freshwater in North Africa, Southwest Asia, and Central Asia comes from rivers, oases, and **aquifers**—underground layers of porous rock, gravel, or sand that contain water. As populations grow, demand for water taxes water resources.

The Nile, Tigris, Euphrates, Jordan, Amu Dar'ya, and Syr Dar'ya are the region's only major rivers, so only a few of the region's countries have enough freshwater for irrigation. Israel, for example, uses an elaborate system of human-made canals to funnel the freshwater of the Jordan River from north to south. In the rest of the region, people turn to smaller rivers and other sources for water.

Desalination

Limited water resources have prompted scientists to develop ways to remove salt from seawater, a process called **desalination.** As the world's population increases and becomes more concentrated in urban areas, desalination helps meet the need for more freshwater. The region now has about 75 percent of the world's freshwater-production **capacity**. Many countries, particularly those near the Persian Gulf, depend heavily on desalination plants. The costs to build and maintain such plants, however, are too much for some countries. These countries still face the challenge of acquiring enough freshwater for the needs of their people and industries.

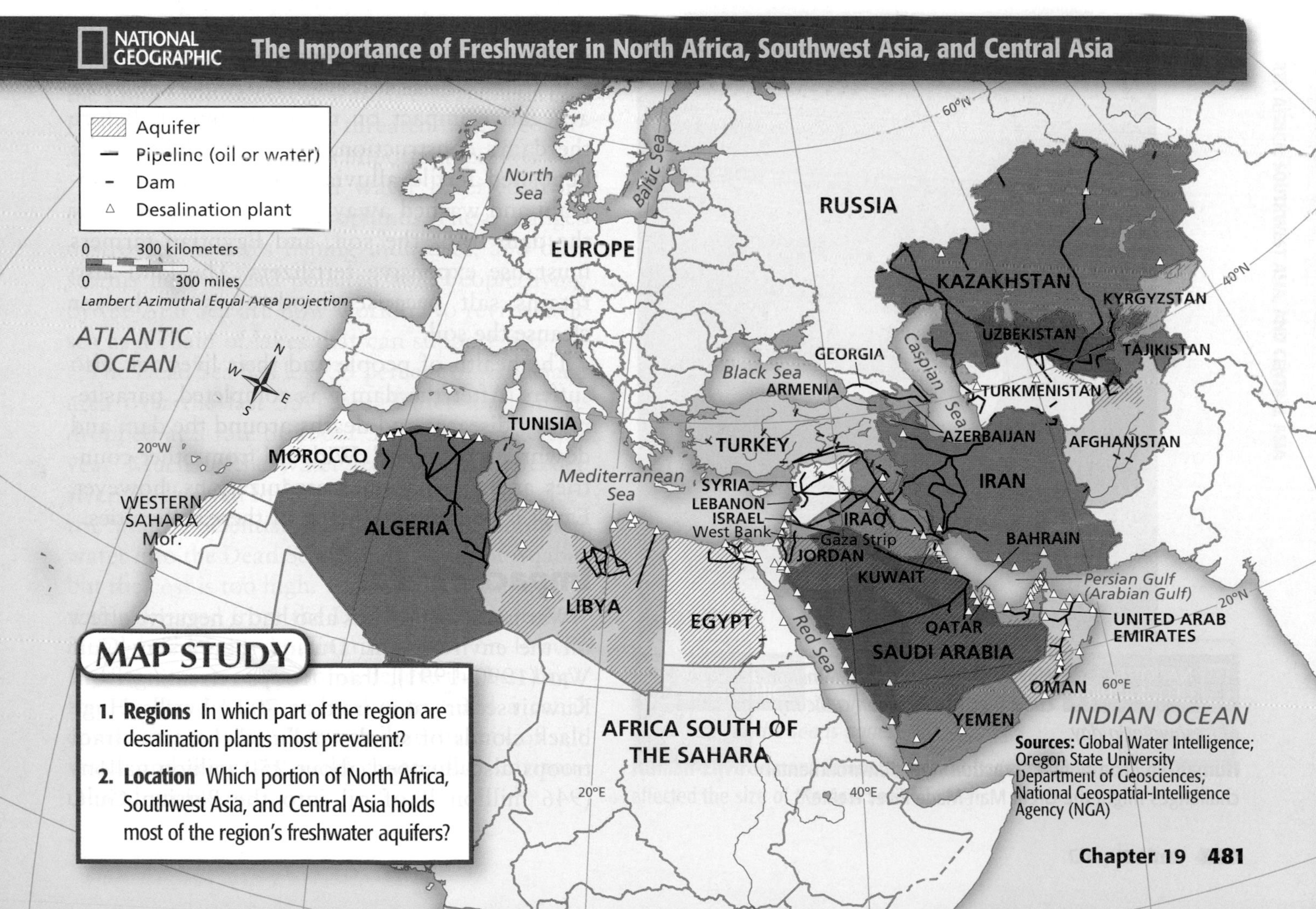

MAP STUDY

1. **Regions** In which part of the region are desalination plants most prevalent?
2. **Location** Which portion of North Africa, Southwest Asia, and Central Asia holds most of the region's freshwater aquifers?

NATIONAL GEOGRAPHIC **Case STUDY**

SUNNI AND SHIA MUSLIMS: Why are there two branches of Islam, and how do they relate to each other?

There are two main branches of Islam—Sunni and Shia. The main difference between the two is their belief about how the leader, or caliph, should be chosen. Sunni Muslims believe that the caliph should be chosen by the Islamic community. Shia Muslims believe that the caliph should be a descendant of Muhammad, the prophet of Islam.

The split dates from about A.D. 680. Both groups accept the first four caliphs who followed Muhammad. Sunnis believe that heirs of all four are legitimate leaders. Shias, however, believe that only heirs of the fourth caliph are legitimate. In A.D. 931 the twelfth caliph disappeared. This was an important event in Shia history, because the Shias believed they lost their divinely guided political leader.

Most Muslims in the world are Sunni. Only in Iran, Iraq, Bahrain, and Azerbaijan are the majority of the people Shia.

Understanding the Issue

The relations between Sunni and Shia Muslims can be understood from a variety of perspectives.

A Moral Issue Religions teach rules of conduct that one must follow in order to be in harmony with the world and to live a good life. Followers of many religions believe that they should work for the organization of their entire society according to these rules. When the rules of different groups are not the same, conflicts between the groups can become very pronounced, even violent.

A Cultural Issue Because religious belief deals with some of the deepest questions of human experience, it becomes a part of culture that people use to define their identity. Followers of one religion or religious branch may distrust people from other groups, based on their membership and identity.

A Political Problem Groups of people who identify with particular religious groups may form political factions. They want to see their particular group in power, or at least in an equal position with their rivals. In areas where one religious group has been favored and others disadvantaged, the disadvantaged group may become demanding of greater power if given the chance. For example, after Iraqi dictator Saddam Hussein (a Sunni) was driven from power, the Shia saw an opportunity to assume power. The Sunnis, fearing they would be disadvantaged, fought back in a violent insurgency.

Above right: Shrine of Imam Reza in Mashhad, Iran's holiest city
Above: Muslim men attend Friday prayer.

World Muslim Population

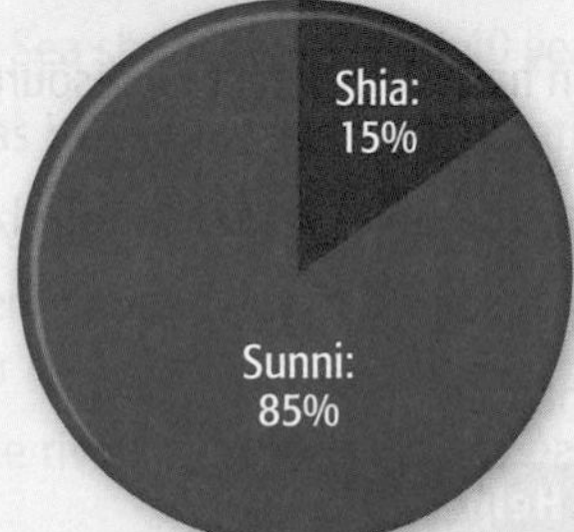

Source: www.cia.gov, The World Factbook 2006; www.adherants.com.

Possible Solutions

The division of Muslims between Sunni and Shia has become an important issue as political and religious factions emerge.

Democratic Government In countries where there are many religious and ethnic groups, democratic political institutions help ensure that all groups have a say in the government. Groups may form political parties, which work to win votes at the ballot box instead of through violence.

Human Rights Laws Along with democratic government, where the will of the majority prevails, laws protect the interests of the minority. If members of the minority feel they have been wrongly treated, they can take their problems to an impartial court, where the issue is decided on the basis of laws.

Economic Development People whose lives are improving economically are less likely to become frustrated and violent. They see that their lives are getting better and that fighting will simply destroy everything they have worked for. They are less likely to follow extremist leaders, especially when they see that they can retain the most important parts of their religious beliefs along with economic development.

Muslim women in France march during a demonstration in support of wearing head coverings.

Understanding the Case

The primary resources listed below provide information about Sunni and Shia Muslims. Use these resources, along with what you have learned in Unit 6, to complete the activities listed on the next page.

Conflict Between Sunnis and Shias in Iraq

Primary Source ①

Excerpt from "Sunni-Shia schism 'threatening to tear Iraq apart,' says conflict group," by Michael Howard, www.guardian.co.uk, February 27, 2006.

People in Iraq identify strongly with the branch of Islam to which they belong. These groups are forming the basis for conflicting groups struggling for power.

Iraq is on the verge of breaking up along religious, ethnic and tribal lines—a process bloodily amplified by the Shia versus Sunni violence in the wake of last week's bomb attack on the gold-domed shrine in Samarra, the International Crisis Group [ICG] says in a report out today.

[ICG] warns that, left unchecked, the widening fissures in Iraqi society that have been exposed since the removal of the Ba'athist regime in 2003 could bring further "instability and violence to many areas, especially those with mixed populations". . . .

Five days of violence in the wake of the Samarra bombing, have left more than 200 dead and many mosques smashed. . . .

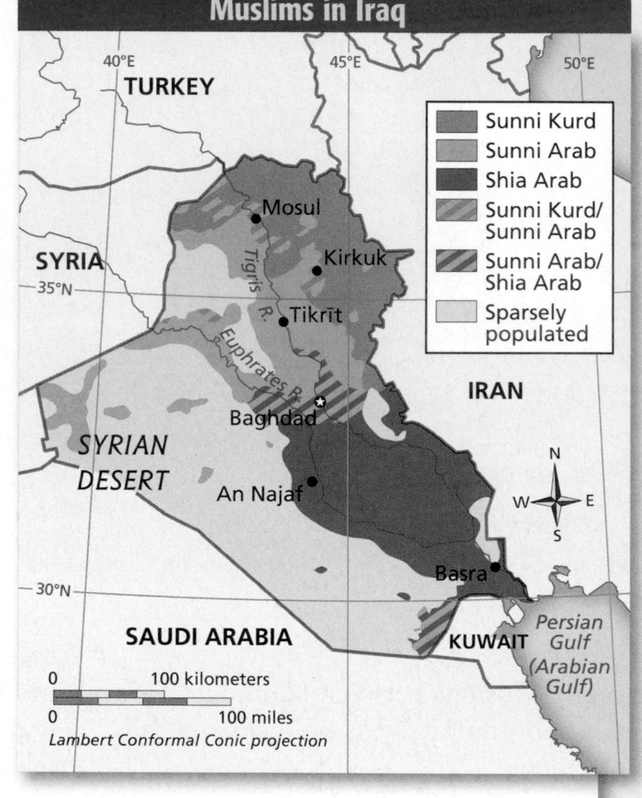

There were further ominous signs of the "cleansing" of once mixed neighbourhoods in and around Baghdad. Scores of Shia families were reported to have fled homes in the restive western Sunni suburb of Abu Ghraib. Shia community leaders said they were being housed temporarily in schools and other buildings in Shia areas. In the latest round of attacks, a bomb destroyed a minibus as it was leaving a bus station in the mostly Shia town of Hilla, 60 miles south of Baghdad, killing five people and wounding three.

In Baghdad at least 18 people were killed and more than 50 injured when mortar rounds slammed into houses in two mainly Shia neighbourhoods. Also, two US soldiers were killed by a roadside bomb. Iraq's political leadership staged a show of unity by appearing on TV on Saturday night.

The prime minister Ibrahim al Jaafari said that all or most of the leaders "expressed the importance of accelerating the political process without any delay". Sunni leaders who pulled out of talks to form a national unity government hinted they may soon rejoin the process. . . .

Joost Hiltermann, the director of the ICG's Middle East Project . . . said it was encouraging that Shia and Sunni religious leaders had called for unity and calm.

Primary Sources

Attempts to Control Violence

Primary Source 2

U.S. military forces, sent to Iraq to topple Saddam Hussein's government and set the country on a path to democracy, find themselves caught between the fighting Sunni and Shia factions. U.S. forces try to keep the violence contained and the factions separate from each other.

Source: Steve Sack, Minnesota, *The Minneapolis Star-Tribune* 5/5/06

A Turning Point?

Primary Source 3

Excerpt from "National Unity Is Rallying Cry in Iraq Elections," by Steven Lee Myers, *The New York Times*, October 1, 2009.

A truce between Sunni and Shia Muslims in Iraq seems possible.

Iraqi politics has a new catchphrase, the "yes, we can" of the country's coming parliamentary elections. It is "national unity," and while skepticism abounds, it could well signal the decline of the religious and sectarian parties that have fractured Iraq since 2003.

Across the political spectrum—Sunni and Shiite, secular and Islamic—party leaders have jettisoned explicit appeals to their traditional followers and are now scrambling to reach across ethnic or sectarian lines. . . .

Prime Minister Nuri Kamal al-Maliki, a conservative Shiite whose party has deep Islamic roots, has enlisted support from Sunni tribal leaders in areas that once were—and might again be—the heartland of opposition to the central government. . . .

"I do believe that there is genuine opportunity for restoring our coexistence, our historical coexistence," said Vice President Tariq al-Hashimi, who broke with the main Sunni party, the Iraqi Islamic Party, this year. "I mean, in the past, we used to live together here. What we need, in fact, is real and genuine reconciliation."

. . . [M]any people view the apparent transformations of some parties cynically. Even as Iraq's political leaders all pledge national unity, Parliament remains so paralyzed by infighting that [it is] unable to pass any significant legislation. . . .

Analyzing the Case

1. **Drawing Conclusions** Review the information in the primary sources above. How do religious traditions and beliefs combine with political upheaval to increase the conflict between Sunnis and Shias in Iraq?
2. **Making Predictions** How might the conflict between Sunnis and Shias eventually be resolved?
3. **Forming a Peace Team.** Divide the class into two teams, one representing Sunnis and one representing Shias. Teams should research the following questions:
   - What is the origin of the split between Sunni and Shia Muslims?
   - Where in the world is each group dominant?
   - Do the groups have different degrees of political power? Why?

   Teams should then conduct a negotiation session to come to an understanding of how to begin to resolve their differences.
4. **Writing About the Case** Write a one-page essay in which you discuss one of the issues addressed in #3 above.

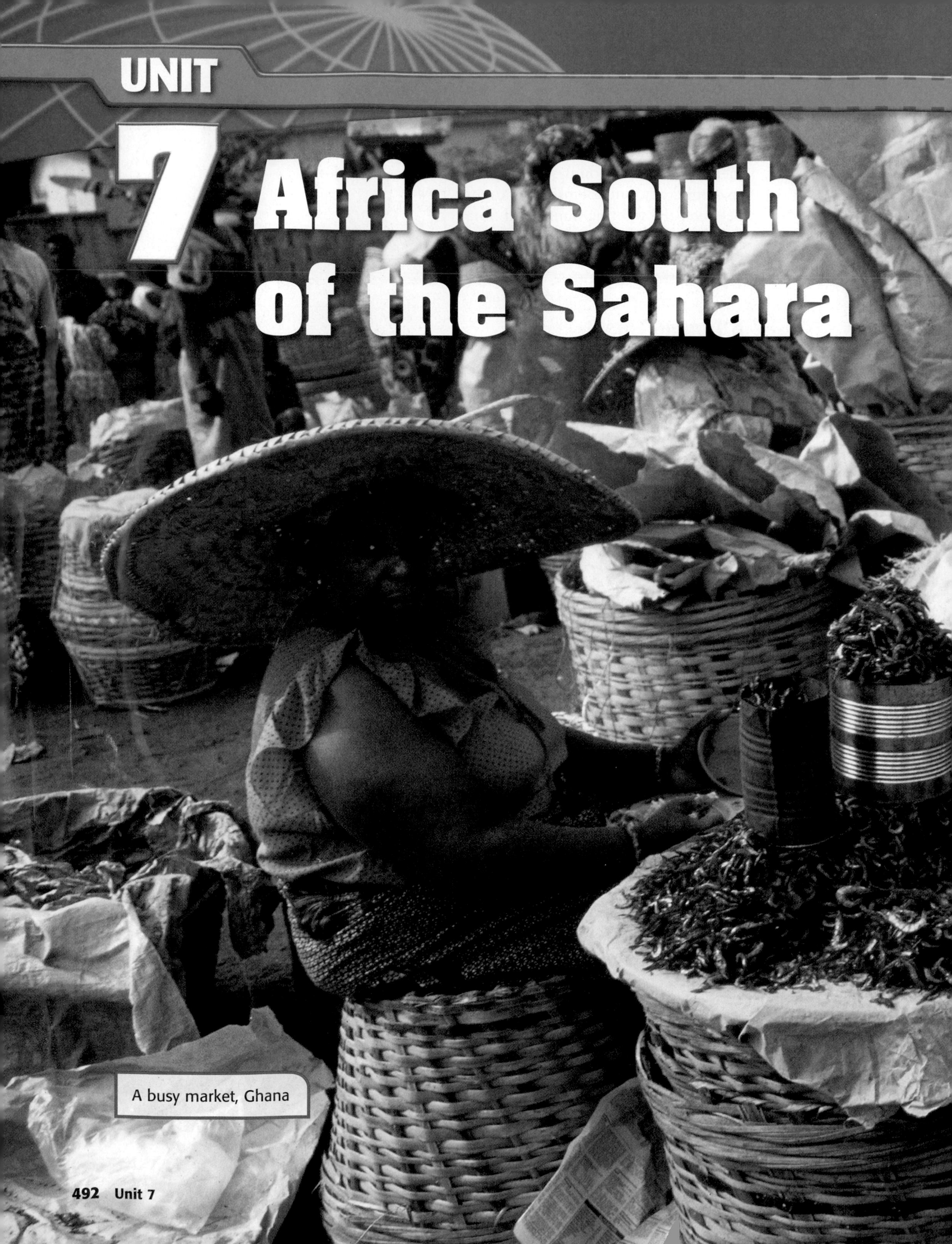

UNIT 7 Africa South of the Sahara

A busy market, Ghana

Why It Matters

Africa south of the Sahara presents a rich mosaic of ethnic groups who speak hundreds of languages. Over the past 60 years, a number of countries in the region have gained independence. Today they are working toward greater political and economic unity. They are also strengthening their voice in global affairs through such international organizations as the United Nations.

Africa South of the Sahara

PHYSICAL GEOGRAPHY Africa south of the Sahara occupies the southern three-fourths of the continent of Africa. The area has a generally low elevation compared to other regions. Plateaus and low mountains tend to be found to the east and south. Some of these highlands end abruptly and plunge to low coastal plains that are very narrow. More extensive lowland areas lie to the north and west.

Two deep cuts in the land run from north to south in the eastern part of the region. Called the Great Rift Valley and the Western Rift Valley, these deep depressions show the effect of two of Earth's tectonic plates pulling apart. Volcanic mountains are found nearby, and many low areas of the rifts have filled with water, forming a chain of lakes.

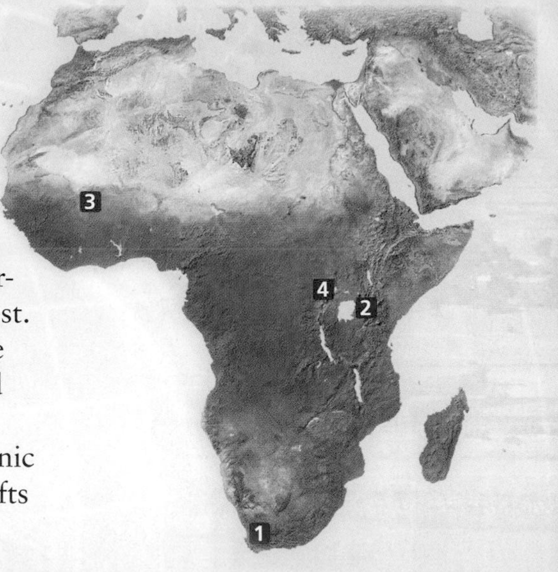

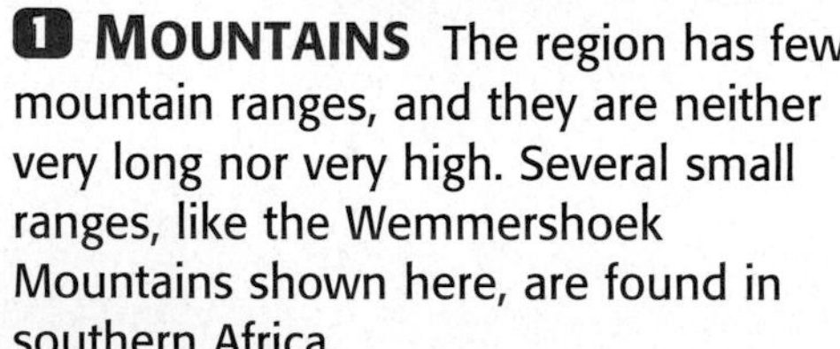

1 MOUNTAINS The region has few mountain ranges, and they are neither very long nor very high. Several small ranges, like the Wemmershoek Mountains shown here, are found in southern Africa.

2 PLAINS AND PLATEAUS Seas of grass cover some of the lowland areas of Africa, like this part of the Masai Mara National Reserve in Kenya.

3 LAKES AND RIVERS The city of Djenné, Mali, sits along the banks of the Bani River, one of the tributaries of the Niger River, which winds through much of West Africa.

4 NATURAL RESOURCES Miners pan for gold in the Democratic Republic of the Congo. Africa is rich in precious gemstones, coal, and metals like gold, iron, and uranium.

UNIT 7 WHAT MAKES THIS A REGION?

Africa South of the Sahara

CULTURAL GEOGRAPHY Africa south of the Sahara is a region of great cultural diversity. The region is made up of several thousand different ethnic groups with different languages and religions. These cultural differences do not always follow national boundaries. Intermarriage between members of different groups has lessened group differences, but in some cases ethnic rivalries have remained strong.

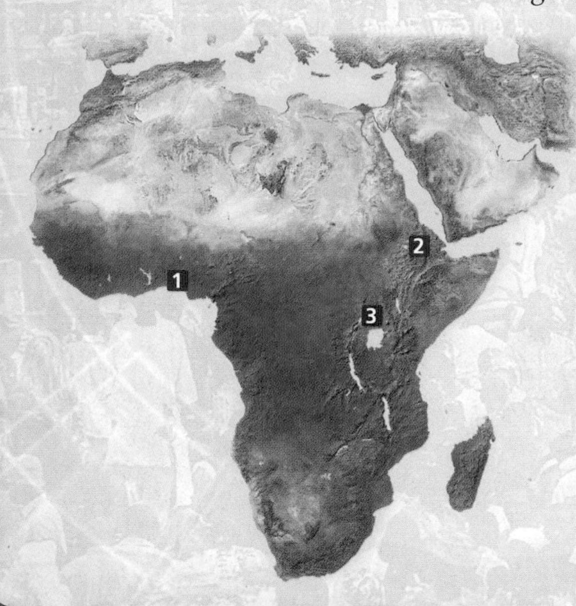

REGIONAL TIME LINE

Glass goblet from the Axum Empire

- **1700–1500 B.C.** Kush Empire flourishes
- **500 B.C.–A.D. 8** Axum Empire flourishes
- **A.D. 750–1240** Ghana trading empire flourishes
- **A.D. 1652** Cape Town founded on southern tip of Africa
- **A.D. 1881** Scramble by European countries for African colonies begins
- **A.D. 1907** Mosque at Djenné built

1500 B.C. | A.D. 1 | A.D. 1000 | A.D. 1300 | A.D. 1600 | A.D. 1900

1 ECONOMY Most people in Africa south of the Sahara live by subsistence farming, but bustling cities like Lagos, Nigeria, are centers of business and industry.

2 CULTURE Ethiopian Christians march in a procession during a religious festival. Christianity was adopted in Ethiopia in the A.D. 300s.

3 PEOPLE There are a variety of ethnic groups in the countries of Africa south of the Sahara, including Uganda. In the past, many Africans identified with one ethnic group. Recently intermarriage and immigration have contributed to increased ethnic diversity in parts of Africa.

Nelson Mandela

A.D. 1948 Alan Paton publishes *Cry, the Beloved Country*

A.D. 1957 Ghana becomes the first nation south of the Sahara to gain its independence

A.D. 1950 — A.D. 1970 — A.D. 1990 — A.D. 2010

A.D. 1984–85 Drought and civil war cause catastrophic famine in Ethiopia

A.D. 1994 Fall of apartheid in South Africa; Nelson Mandela elected president of South Africa

A.D. 1994 Hutu militias kill more than 800,000 Tutsis and moderate Hutus in Rwanda

A.D. 2006 Islamic courts take control of Mogadishu, Somalia

A.D. 2008 Kenyans hold riots to protest alleged rigged elections

PHYSICAL Africa South of the Sahara

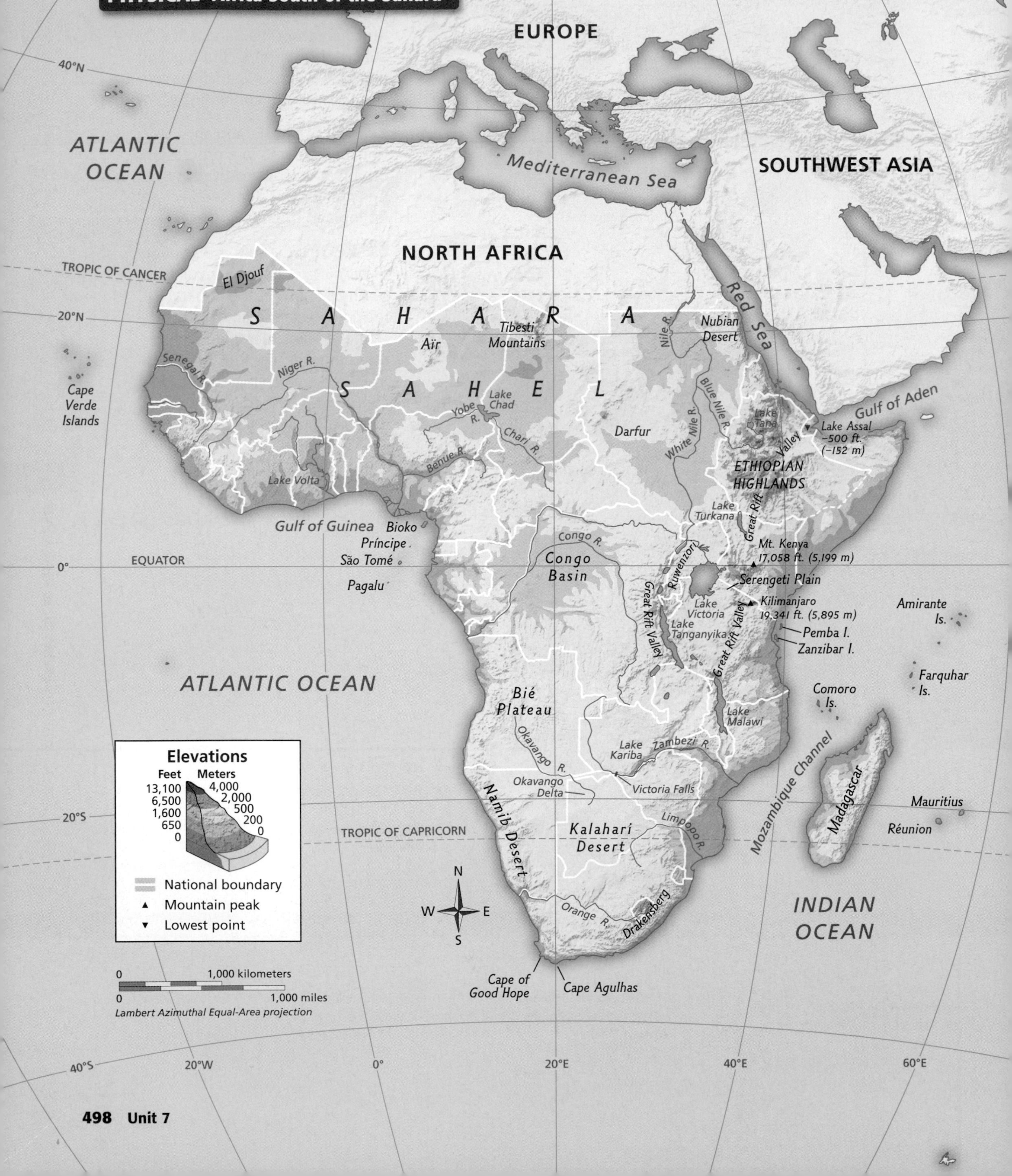

EUROPE

ATLANTIC OCEAN

Mediterranean Sea

SOUTHWEST ASIA

NORTH AFRICA

TROPIC OF CANCER

El Djouf

S A H A R A

Tibesti Mountains

Aïr

Nubian Desert

Red Sea

Nile R.

Senegal R.

Niger R.

Cape Verde Islands

S A H E L

Lake Chad

Yobe R.

Chari R.

Blue Nile R.

White Nile R.

Lake Tana

Darfur

Gulf of Aden

Lake Assal –500 ft. (–152 m)

Great Rift Valley

ETHIOPIAN HIGHLANDS

Benue R.

Lake Volta

Lake Turkana

Gulf of Guinea

Bioko

Príncipe

São Tomé

Pagalu

EQUATOR

Congo R.

Congo Basin

Mt. Kenya 17,058 ft. (5,199 m)

Ruwenzori

Serengeti Plain

Great Rift Valley

Lake Victoria

Lake Tanganyika

Kilimanjaro 19,341 ft. (5,895 m)

Great Rift Valley

Pemba I.

Zanzibar I.

Amirante Is.

Farquhar Is.

ATLANTIC OCEAN

Bié Plateau

Comoro Is.

Lake Malawi

Okavango R.

Lake Kariba

Zambezi R.

Mozambique Channel

Madagascar

Okavango Delta

Victoria Falls

Namib Desert

Mauritius

Réunion

Limpopo R.

TROPIC OF CAPRICORN

Kalahari Desert

Orange R.

Drakensberg

INDIAN OCEAN

Cape of Good Hope

Cape Agulhas

N S E W

Elevations

Feet	Meters
13,100	4,000
6,500	2,000
1,600	500
650	200
0	0

National boundary

▲ Mountain peak

▼ Lowest point

0 1,000 kilometers

0 1,000 miles

Lambert Azimuthal Equal-Area projection

40°N 20°N 0° 20°S 40°S

20°W 0° 20°E 40°E 60°E

Powerful Waterways

The region's major rivers move quickly through rapids and waterfalls or broaden into marshy inland deltas in lowland areas. As you study the maps and graphics on these pages, look for the geographical features that make the region unique. Then answer the questions below on a separate sheet of paper.

1. How might rapids, waterfalls, and marshes affect the way people can use the region's rivers?
2. Compare Victoria Falls to Niagara Falls. Which has the greater potential to generate energy?
3. Would you expect the Niger River to have as many waterfalls as the Zambezi River? Why or why not?

Comparing Waterfalls

Water Volume at Victoria Falls

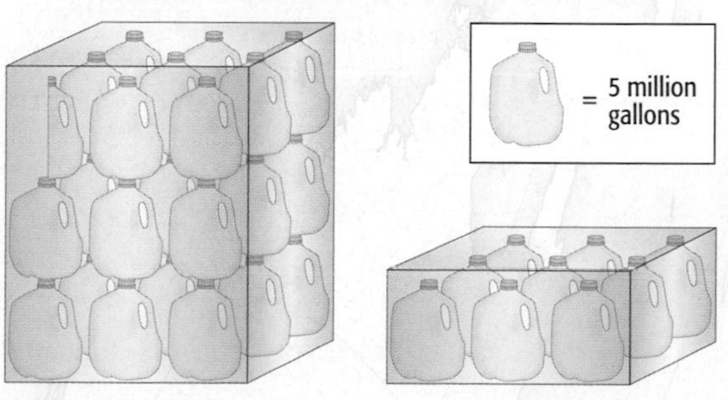

Victoria Falls
132 million gallons
(500 million l per minute)

Niagara Falls
45 million gallons
(170 million l per minute)

AFRICA SOUTH OF THE SAHARA

UNIT 7 REGIONAL ATLAS

POLITICAL Africa South of the Sahara

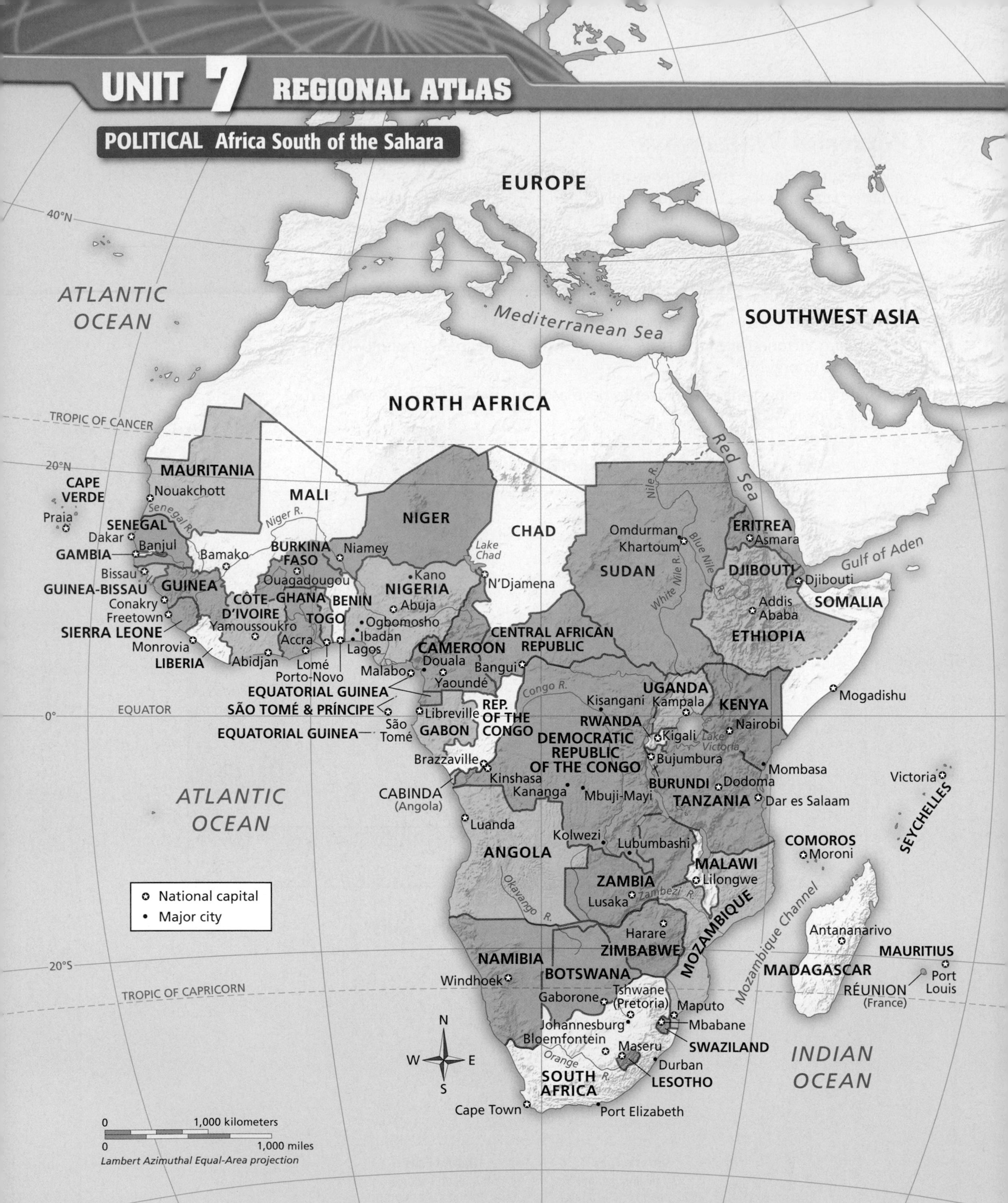

EUROPE
ATLANTIC OCEAN
Mediterranean Sea
SOUTHWEST ASIA
NORTH AFRICA
TROPIC OF CANCER
40°N
20°N
Red Sea
Nile R.
MAURITANIA
Nouakchott
Senegal R.
CAPE VERDE
Praia
MALI
Niger R.
NIGER
CHAD
Lake Chad
Omdurman
Khartoum
Blue Nile R.
White Nile R.
SUDAN
ERITREA
Asmara
Gulf of Aden
DJIBOUTI
Djibouti
SOMALIA
Addis Ababa
ETHIOPIA
SENEGAL
Dakar
Banjul
GAMBIA
Bamako
BURKINA FASO
Niamey
Kano
Bissau
GUINEA-BISSAU
GUINEA
Ouagadougou
NIGERIA
N'Djamena
Conakry
Freetown
SIERRA LEONE
CÔTE D'IVOIRE
GHANA
BENIN
TOGO
Abuja
Ogbomosho
Ibadan
Lagos
Yamoussoukro
Accra
Monrovia
LIBERIA
Abidjan
Lomé
Porto-Novo
CAMEROON
Douala
Yaoundé
Malabo
CENTRAL AFRICAN REPUBLIC
Bangui
Congo R.
EQUATORIAL GUINEA
SÃO TOMÉ & PRÍNCIPE
São Tomé
EQUATORIAL GUINEA
EQUATOR
0°
Libreville
GABON
REP. OF THE CONGO
Brazzaville
Kinshasa
Kisangani
UGANDA
Kampala
KENYA
Nairobi
Mogadishu
RWANDA
Kigali
Lake Victoria
DEMOCRATIC REPUBLIC OF THE CONGO
Bujumbura
BURUNDI
Mombasa
Dodoma
TANZANIA
Dar es Salaam
Kananga
Mbuji-Mayi
CABINDA (Angola)
Luanda
ATLANTIC OCEAN
Victoria
SEYCHELLES
COMOROS
Moroni
Kolwezi
Lubumbashi
ANGOLA
MALAWI
Lilongwe
ZAMBIA
Lusaka
Zambezi R.
Okavango R.
MOZAMBIQUE
Mozambique Channel
Antananarivo
Harare
ZIMBABWE
MAURITIUS
NAMIBIA
BOTSWANA
MADAGASCAR
Port Louis
RÉUNION (France)
Windhoek
Tshwane (Pretoria)
Gaborone
Maputo
Mbabane
Johannesburg
Bloemfontein
SWAZILAND
Maseru
Orange R.
Durban
LESOTHO
SOUTH AFRICA
INDIAN OCEAN
Cape Town
Port Elizabeth
TROPIC OF CAPRICORN
20°S
40°S
20°W
0°
20°E
40°E
60°E
N
S
E
W

National capital
Major city

0 1,000 kilometers
0 1,000 miles
Lambert Azimuthal Equal-Area projection

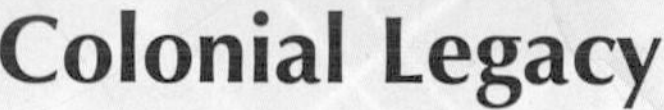

Colonial Legacy

In the late 1800s, much of Africa was colonized by European powers. As you study the maps and graphics on these pages, look for political features that make the region unique. Then answer the questions below on a separate sheet of paper.

1. What patterns of colonial rule do you see? How many countries were colonized by the French?
2. When did most countries in the region gain independence? How stable would you expect governments in the region to be? Why?
3. How many countries are landlocked? How might that characteristic affect their economies?

COLONIZATION Africa South of the Sahara

Colonial Powers, 1913
- Belgian
- British
- French
- German
- Italian
- Portuguese
- Spanish
- Independent

0 1,000 kilometers
0 1,000 miles
Lambert Azimuthal Equal-Area projection

DATES OF INDEPENDENCE

0 1,000 kilometers
0 1,000 miles
Lambert Azimuthal Equal-Area projection

AFRICA SOUTH OF THE SAHARA

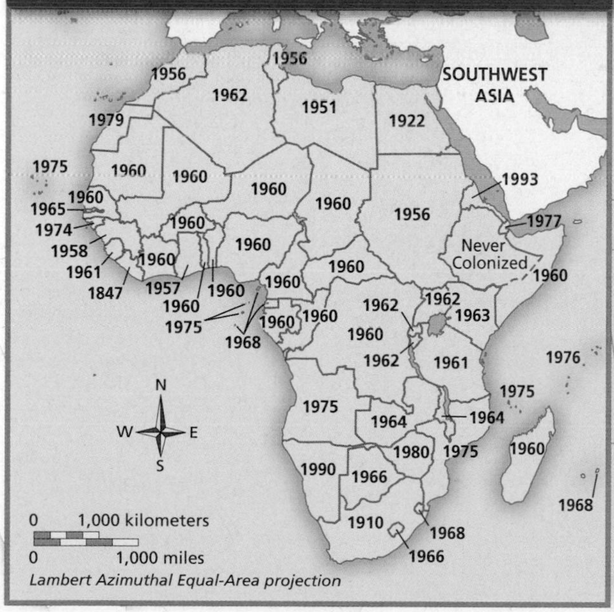

Growing Crisis in the Land

Though famous for its tropical rain forests, the region has many areas with dry climates and poor soils. As you study the maps and graphics on these pages, look for environmental features that make the region unique. Then answer the questions below on a separate sheet of paper.

1. What type of vegetation dominates the northern portion of the region?
2. Which parts of Africa south of the Sahara are most vulnerable to desertification?
3. Compare the two maps. What relationship can you see between vegetation and risk of desertification?

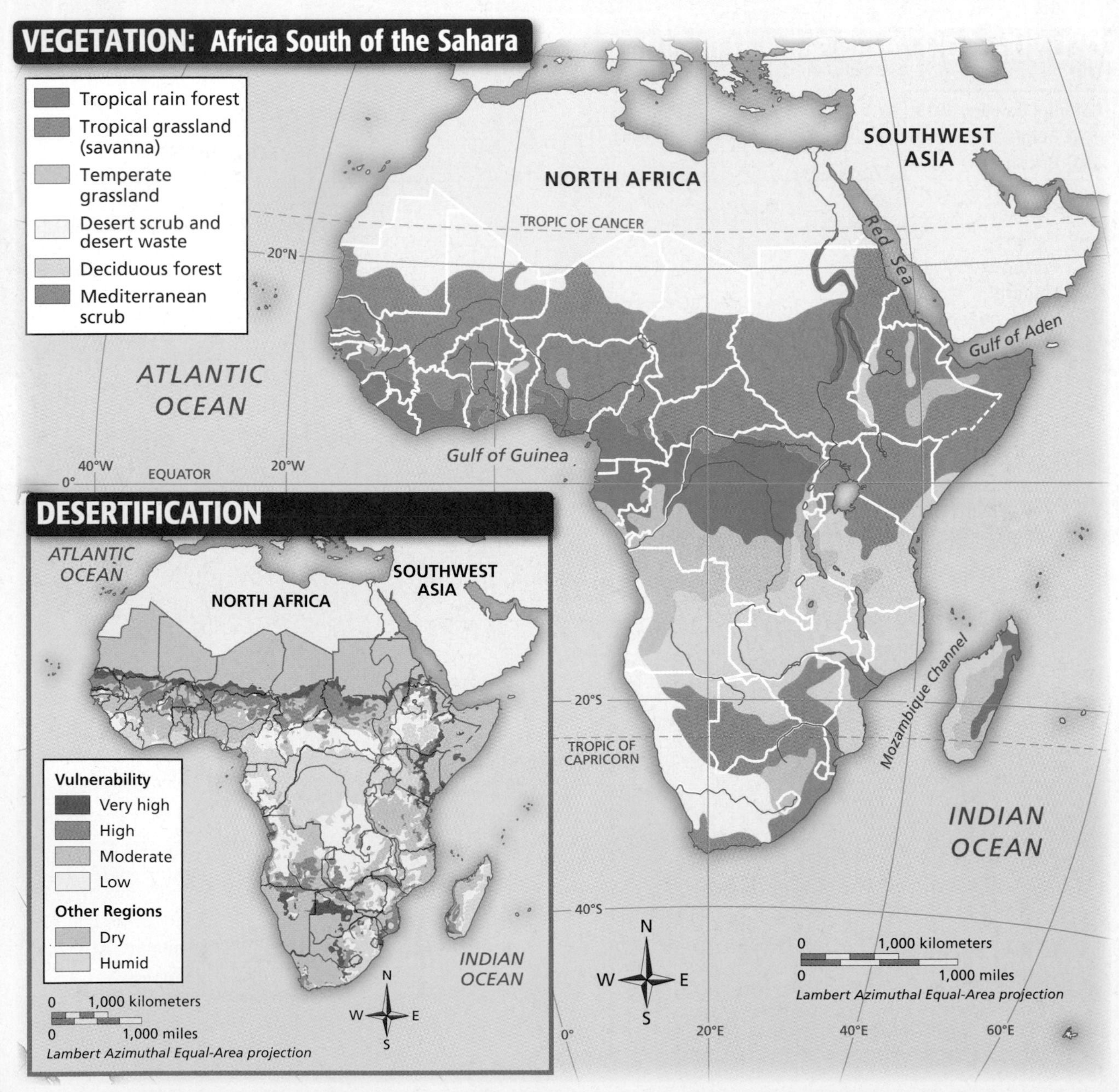

Country, Capital, & Area	Population & Density	Life Expectancy at Birth	GDP Per Capita*	% Urban	Literacy Rate (%)	Years of Compulsory Education	Phone Lines/ Cell Phones (per 1,000 people)	Internet Users (per 1,000 people)	Flag & Language
ANGOLA Luanda 481,353 sq. mi. 1,246,700 sq. km	17,100,000 36 per sq. mi. 14 per sq. km	46 yrs.	$8,800	57	67.4	6	6/69	11	Portuguese
BENIN Porto-Novo 42,711 sq. mi. 110,622 sq. km	8,900,000 208 per sq. mi. 80 per sq. km	56 yrs.	$1,500	41	34.7	6	9/89	50	French
BOTSWANA Gaborone 218,816 sq. mi. 566,730 sq. km	2,000,000 9 per sq. mi. 4 per sq. km	49 yrs.	$13,900	60	81.2	10	75/466	34	English, Setswana
BURKINA FASO Ouagadougou 105,715 sq. mi. 273,800 sq. km	15,800,000 149 per sq. mi. 58 per sq. km	57 yrs.	$1,200	16	21.8	11	7/43	5	French
BURUNDI Bujumbura 9,915 sq. mi. 25,680 sq. km	8,300,000 837 per sq. mi. 323 per sq. km	49 yrs.	$400	10	59.3	6	4/20	5	Kirundi, French
CAMEROON Yaoundé 182,514 sq. mi. 472,710 sq. km	18,900,000 104 per sq. mi. 40 per sq. km	52 yrs.	$2,300	57	67.9	6	6/138	15	English, French
CAPE VERDE Praia 1,557 sq. mi. 4,033 sq. km	500,000 321 per sq. mi. 124 per sq. km	71 yrs.	$3,800	59	76.6	6	141/161	49	Portuguese
CENTRAL AFRICAN REPUBLIC Bangui 240,535 sq. mi. 622,984 sq. km	4,500,000 19 per sq. mi. 7 per sq. km	45 yrs.	$700	38	48.6	10	2/25	3	French
CHAD N'Djamena 486,179 sq. mi. 1,259,200 sq. km	10,300,000 21 per sq. mi. 8 per sq. km	47 yrs.	$1,600	27	25.7	6	1/22	4	French, Arabic
COMOROS Moroni 863 sq. mi. 2,235 sq. km	700,000 811 per sq. mi. 313 per sq. km	64 yrs.	$1,100	28	56.5	8	28/27	33	Arabic, French

*The CIA calculates per capita GDP in terms of purchasing power parity. This formula allows us to compare the figures among different countries.

Note: Countries and flags are not drawn to scale.

Sources: Central Intelligence Agency, *World Factbook,* 2009; Population Reference Bureau, *World Population Data Sheet,* 2009; UNESCO Institute for Statistics; United Nations, *Human Development Report,* 2007/2008.

Country, Capital, & Area	Population & Density	Life Expectancy at Birth	GDP Per Capita*	% Urban	Literacy Rate (%)	Years of Compulsory Education	Phone Lines/ Cell Phones (per 1,000 people)	Internet Users (per 1,000 people)	Flag & Language
CONGO, DEMOCRATIC REPUBLIC OF THE Kinshasa 875,312 sq. mi. 2,267,048 sq. km	68,700,000 78 per sq. mi. 30 per sq. km	53 yrs.	$300	33	67.2	8	NA/48	2	French
CONGO, REPUBLIC OF THE Brazzaville 131,854 sq. mi. 341,500 sq. km	3,700,000 28 per sq. mi. 11 per sq. km	53 yrs.	$3,900	60	83.8	10	4/123	13	French
CÔTE D'IVOIRE Yamoussoukro 122,782 sq. mi. 318,003 sq. km	21,400,000 174 per sq. mi. 67 per sq. km	52 yrs.	$1,700	48	48.7	10	14/121	11	French
DJIBOUTI Djibouti 8,949 sq. mi. 23,180 sq. km	900,000 101 per sq. mi. 39 per sq. km	55 yrs.	$3,700	87	67.9	10	14/56	13	French, Arabic
EQUATORIAL GUINEA Malabo 10,831 sq. mi. 28,051 sq. km	700,000 64 per sq. mi. 25 per sq. km	59 yrs.	$37,200	39	87.0	5	20/192	14	Spanish, French
ERITREA Asmara 38,996 sq. mi. 101,000 sq. km	5,100,000 131 per sq. mi. 50 per sq. km	58 yrs.	$700	21	58.6	8	9/9	16	Afar
ETHIOPIA Addis Ababa 386,102 sq. mi. 1,000,000 sq. km	82,800,000 214 per sq. mi. 83 per sq. km	53 yrs.	$800	16	42.7	6	9/6	2	Amharic
GABON Libreville 99,846 sq. mi. 257,667 sq. km	1,500,000 15 per sq. mi. 6 per sq. km	59 yrs.	$14,200	84	63.2	11	28/470	48	French
GAMBIA Banjul 3,861 sq. mi. 10,000 sq. km	1,600,000 414 per sq. mi. 160 per sq. km	55 yrs.	$1,300	54	40.1	6	29/163	33	English
GHANA Accra 87,851 sq. mi. 227,533 sq. km	23,800,000 271 per sq. mi. 105 per sq. km	59 yrs.	$1,500	48	57.9	9	15/129	18	English
GUINEA Conakry 94,872 sq. mi. 245,717 sq. km	10,100,000 106 per sq. mi. 41 per sq. km	56 yrs.	$1,100	33	29.5	6	3/20	5	French

*The CIA calculates per capita GDP in terms of purchasing power parity. This formula allows us to compare the figures among different countries.
Note: Countries and flags are not drawn to scale.

Country, Capital, & Area	Population & Density	Life Expectancy at Birth	GDP Per Capita*	% Urban	Literacy Rate (%)	Years of Compulsory Education	Phone Lines/ Cell Phones (per 1,000 people)	Internet Users (per 1,000 people)	Flag & Language
GUINEA-BISSAU Bissau 10,857 sq. mi. 28,120 sq. km	1,600,000 147 per sq. mi. 57 per sq. km	46 yrs.	$600	30	42.4	6	7/42	20	Portuguese
KENYA Nairobi 219,746 sq. mi. 569,140 sq. km	39,100,000 178 per sq. mi. 69 per sq. km	54 yrs.	$1,600	19	85.1	8	8/135	32	English, Kiswahili
LESOTHO Maseru 11,720 sq. mi. 30,355 sq. km	2,100,000 179 per sq. mi. 69 per sq. km	40 yrs.	$1,500	24	84.8	7	27/137	24	Sesotho, English
LIBERIA Monrovia 37,189 sq. mi. 96,320 sq. km	4,000,000 108 per sq. mi. 42 per sq. km	56 yrs.	$500	58	57.5	6	NA/49	NA	English
MADAGASCAR 224,534 sq. mi. 581,540 sq. km Antananarivo	19,500,000 87 per sq. mi. 34 per sq. km	59 yrs.	$1,000	30	68.9	5	4/27	5	French, Malagasy
MALAWI 36,324 sq. mi. 94,080 sq. km Lilongwe	14,200,000 391 per sq. mi. 151 per sq. km	46 yrs.	$800	17	62.7	8	8/33	4	Chichewa
MALI 471,118 sq. mi. 1,220,190 sq. km Bamako	13,000,000 28 per sq. mi. 11 per sq. km	48 yrs.	$1,200	31	46.4	9	6/64	4	French
MAURITANIA 397,955 sq. mi. 1,030,700 sq. km Nouakchott	3,300,000 8 per sq. mi. 3 per sq. km	57 yrs.	$2,100	40	51.2	9	13/243	7	Arabic
MAURITIUS Port Louis 784 sq. mi. 2,030 sq. km	1,300,000 1,658 per sq. mi. 640 per sq. km	72 yrs.	$12,800	42	84.4	9	289/574	146	Creole, French
***Comparing Lands:** The region of Africa south of the Sahara is about three times the size of the contiguous United States.*									
	306,800,000 87 per sq. mi. 33 per sq. km	78 yrs.	$47,800	79	99.0	12	606/680	630	English

Sources: Central Intelligence Agency, *World Factbook,* 2009; Population Reference Bureau, *World Population Data Sheet,* 2009; UNESCO Institute for Statistics; United Nations, *Human Development Report,* 2007/2008.

CHAPTER 20

PHYSICAL GEOGRAPHY OF

Africa South of the Sahara

BIG Idea

Physical processes shape Earth's surface. Africa south of the Sahara is a region of dramatic landforms and great natural resources. A study of its physical geography will explain some of the processes that have shaped and continue to shape the diverse landscapes, climates, and vegetation of the region.

Essential Questions

Section 1: The Land

What kinds of physical processes have shaped the African landscape?

Section 2: Climate and Vegetation

What factors influence climate in Africa south of the Sahara?

Visit glencoe.com and enter *QuickPass*™ code WGC9952C20 for Chapter 20 resources.

A hungry elephant strips bark from an acacia tree in the Great Rift Valley, Tanzania.

FOLDABLES™ Study Organizer **Comparing Information** Use a Three-Tab Book to make a Venn diagram for comparing two major river basins in Africa south of the Sahara.

Reading and Writing As you read the chapter, gather information about the Niger and Zambezi Rivers, including location, length, direction of flow, and uses. In the appropriate place in your Foldable, note what is unique to each river and what they have in common.

Mountains and Plateaus

Africa south of the Sahara is a series of plateaus that rise in elevation from the coast inland and from west to east. Ranging in elevation from 500 feet (152 m) in the west to 8,000 feet (2,438 m) or more in the east, the plateaus are outcroppings of the solid rock that makes up most of the continent. The edges of the plateaus are marked by **escarpments**—steep, often jagged cliffs. Most are located less than 20 miles (32 km) from the coast. Rivers crossing the plateaus plunge down the escarpments in **cataracts,** or waterfalls.

Most African mountains dot the Eastern Highlands, an area that stretches from Ethiopia almost to the Cape of Good Hope. These highland areas include the Ethiopian Highlands as well as volcanic summits, such as Kilimanjaro and Mount Kenya. West of the Eastern Highlands, the **Ruwenzori Mountains** divide Uganda and the Democratic Republic of the Congo. Covered with snow and cloaked in clouds, they are also called the "Mountains of the Moon." Moist air from the Indian Ocean creates the clouds that wrap around the Ruwenzoris.

Farther south is the **Drakensberg Range** in South Africa and Lesotho. These mountains rise to more than 11,000 feet (3,353 m) and form part of the escarpment along the southern edge of the continent.

READING Check **Place** Where are most of Africa's mountains located?

NATIONAL GEOGRAPHIC *A road hugs the coastal escarpment near Cape Town, South Africa.*

Regions What type of landform dominates most of Africa?

Water Systems

MAIN Idea Landforms and physical processes have influenced the region's water systems, which include deep lakes, spectacular waterfalls, and great rivers.

GEOGRAPHY AND YOU In what ways have landforms affected the Mississippi River in the United States? Read to learn how the land has influenced the water systems in Africa south of the Sahara.

The land has influenced the **region's** water systems in important ways. Lakes and rivers are located in huge basins formed millions of years ago by the uplifting of the land. The great rivers of Africa originate high in the plateaus and eventually make their way to the sea. Escarpments and ridges break the rivers' paths to the ocean with rapids and cataracts. The broken landscape makes it impossible to navigate most of the region's rivers from mouth to source.

Land of Lakes

As the map on page 513 shows, most of the region's lakes, including Lakes Tanganyika and Malawi, are near the Great Rift Valley. **Lake Victoria,** the largest lake in Africa, is located between the eastern and western branches of the Great Rift. It is the world's second-largest freshwater lake, after Lake Superior in North America. Lake Victoria is the source of the White Nile River. Despite its large size, Lake Victoria is comparatively shallow with a depth of only 270 feet (82 m).

Lake Chad, outside the Great Rift Valley in west-central Africa, is threatened with extinction. Although fed by three large streams, landlocked Lake Chad is shrinking. Droughts in the 1970s completely dried up the northern portion of the lake, and the water level continues to be shallow even during years when rainfall is normal. Because of the climate, much of the lake's water evaporates. It also seeps into the ground. Other factors contributing to the shrinkage of Lake Chad include global warming, irrigation, and desertification. **Desertification** occurs when long periods of drought and land use destroy the vegetation. The land is left dry and barren, unable to support life. As Lake Chad shrinks, the desert expands on the dry lake bottom.

A Human-Made Lake

Lake Volta in West Africa ranks among the largest human-made lakes in the world. The lake was created in the 1960s by damming the Volta River south of Ajena, Ghana. The new lake flooded more than 700 villages, forcing more than 70,000 people to find new homes.

Although the dam was originally built to provide hydroelectric power to an aluminum plant, the people of Ghana today benefit from the lake in many ways. It supplies irrigation for farming in the plains below the dam and is well stocked with fish. The hydroelectric plant now also generates electricity used throughout Ghana.

River Basins

The **Niger** (NY•juhr) **River** is known by many names along its course, but all of them have roughly the same meaning—"great river." The Niger is the main artery in western Africa, extending about 2,600 miles (4,183 km) in length. Originating in the highlands of Guinea, the river forms a great arc. It flows northeast and then curves southeast to the Nigerian coast. In addition to being important to agriculture, the Niger River is a major means of transportation.

This great river does not flow as one well-defined stream into the Atlantic Ocean. At Aboh in southern Nigeria, the Niger splits into a vast inland **delta,** a triangular section of land formed by sand and silt carried downriver. The Niger Delta stretches 150 miles (241 km) north to south and extends to a width of about 200 miles (322 km) along the Gulf of Guinea.

The **Zambezi River** of south-central Africa also meets the ocean in a delta. The Zambezi flows 2,200 miles (3,540 km) from its source near the Zambia-Angola border in the west to the Indian Ocean in the east, where it fans out in a delta that is 37 miles (60 km) wide. The Zambezi's course to the sea is interrupted in many places by waterfalls. At **Victoria Falls,** on the border of Zambia and Zimbabwe, the Zambezi plummets a sheer 355 feet (108 m).

Unlike most African rivers, the **Congo River** reaches the sea through a deep **estuary** (EHS•chuh•WEHR•ee), or passage where freshwater meets seawater. The Congo's estuary is 7 miles (10 km) wide, and ships can navigate the deep water. The remaining 2,700 miles (4,344 km) of the Congo form a large **network** of navigable

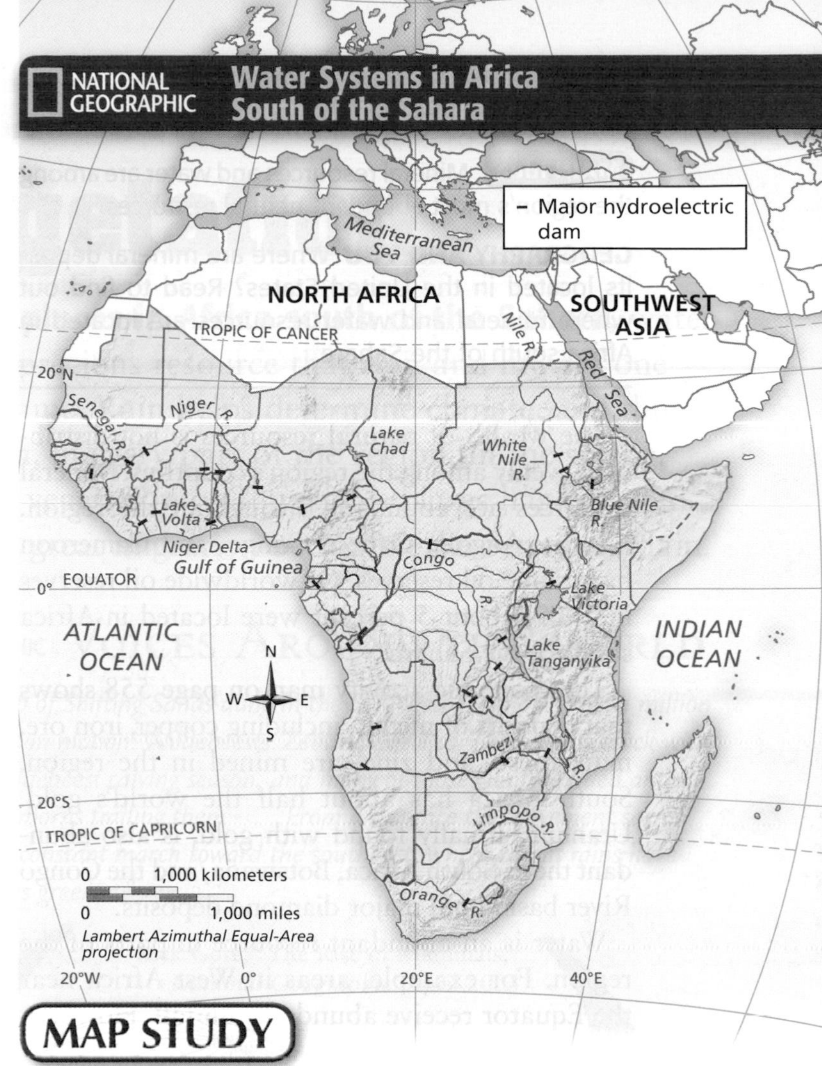

MAP STUDY

1. **Location** Which part of the region has the largest concentration of hydroelectric dams?
2. **Place** Compare this map to the physical map on page 498. Why do northern and southern areas of the region have few rivers?

Maps In MOtion Use **StudentWorks™ Plus** or glencoe.com.

waterways for smaller boats. Some parts of the river, however, have rapids and waterfalls that present serious obstacles to traffic. The river plunges almost 900 feet (274 m) in numerous cataracts not far from where it meets the Atlantic Ocean.

READING Check **Regions** Which lake is the largest in Africa?

Geography ONLINE

Student Web Activity Visit glencoe.com, select the *World Geography and Cultures* Web site, and click on Student Web Activities—Chapter 20 for an activity about the Great Rift Valley.

CHAPTER 21

CULTURAL GEOGRAPHY OF Africa South of the Sahara

Big Idea

Places reflect the relationship between humans and the physical environment. Current events in Africa south of the Sahara can best be understood by knowing how the region's diverse peoples, histories, and cultures are influenced by the physical environment.

Essential Questions

Section 1: The Sahel

How has the physical environment of the Sahel influenced ways of life in this subregion?

Section 2: East Africa

How has East Africa's location along the Red Sea and the Indian Ocean influenced its culture?

Section 3: West Africa

How has the history of West Africa been affected by its physical location and influence by outside peoples?

Section 4: Central Africa

How have different peoples responded to the physical environment of Central Africa?

Section 5: Southern Africa

How has colonial rule affected the peoples of southern Africa?

Geography ONLINE

Visit glencoe.com and enter *QuickPass*™ code WGC9952C21 for Chapter 21 resources.

The Masai of Kenya and Tanzania maintain some of their cultural traditions while engaging in global economic activities.

FOLDABLES™ Study Organizer

Summarizing Information Make a Folded Table to help you summarize information about each of the subregions in Chapter 21.

	The Sahel	East Africa	West Africa	Central Africa	Southern Africa
Population Patterns					
History and Government					
Culture					

Reading and Writing As you read the chapter, summarize information about the population patterns, history and government, and culture for each subregion of Africa south of the Sahara.

History and Government

MAIN Idea West Africa's history has been shaped by indigenous and outside forces, each with their own cultures.

GEOGRAPHY AND YOU How do you think religions in the United States were shaped by the religions of Europe? Read to learn how the cultures of West Africa were influenced by different outsiders.

Powerful West African empires gave way to colonial rule, which later dissolved into independent countries. The resources that helped build the empires also lured European powers **seeking** wealth. The colonial period, combined with continuing economic problems, has left West Africa with many challenges.

Early Empires

As shown on the map on page 526, trading empires grew strong in West Africa around the A.D. 700s. Today the countries of **Ghana** and **Mali** are named after two of these ancient empires. Ghana grew rich by trading gold for salt brought by camel caravans across the Sahara. Peoples south of the Sahara highly valued salt for use as a food preservative.

Gold was plentiful in Ghana. The Spanish-Arab geographer al-Bakri, who traveled to West Africa in the 1000s, reported, for example, that even the king's dogs wore collars of gold and silver. Ghana's wealth was reflected in its large capital, Kumbi. This prosperous empire flourished for almost 500 years.

NATIONAL GEOGRAPHIC *Kwame Nkrumah, the first president of independent Ghana, worked for African unity and development.*

Regions How did colonialism create ethnic and political divisions within Africa?

The Colonial Era

By the 1400s, the Portuguese, who were seeking gold and later, enslaved Africans, had set up trading posts along the African coast. Foreign travelers who reached the West African trading centers of Timbuktu, Kano, Gao, and Wangara were impressed with the bustling markets and vibrant cultures.

By the time Europeans were actively trading with Africans in the 1600s and 1700s, African chiefs and kings had enslaved and traded prisoners of war for centuries. Arab traders had brought enslaved Africans to the Islamic world since the A.D. 800s. The slave trade greatly increased when Europeans began shipping Africans to the Americas to work on plantations.

While Europeans and Americans were shipping enslaved Africans to colonies in America, some French colonizers were trying to end slavery and interethnic strife within Africa. For example, the **legal** prohibitions the French placed on slavery helped encourage black Moors to return to areas of **Mauritania** they had previously left for fear of being captured and enslaved. Such ethnic conflict created difficulties for the country in addition to those created by colonialism.

Nigeria: A Colonial Legacy

In 1914 the British formed the colony of Nigeria from several smaller ethnic territories. As a result, many different ethnic and religious groups lived within Nigeria's boundaries. In the north, cultures were based on Islam. Southern cultures were based on African religions or on Christianity. After Nigeria gained independence in 1960, these ethnic and religious differences erupted in civil war. Ethnic and religious divisions continue to plague Nigeria today.

READING Check **Regions** What attracted European colonizers to West Africa?

Culture

MAIN Idea West African culture has been shaped by hundreds of years of European and Arab influences.

GEOGRAPHY AND YOU Why do you speak the language you do? Read to learn about the different languages spoken in West Africa.

West Africa's diversity is seen in the many languages spoken there. Various religions are practiced as well. The recent popularity of West African music in the United States is evidence of the many art forms of the region.

Language and Religion

Hundreds of languages are spoken in West Africa. Some languages are widespread, such as English and French. Arabic is common in the northern areas of West Africa. Yoruba, part of the Congo-Kordofanian language group, is widely spoken in many dialects. It is printed in books, newspapers, and pamphlets, and taught in television and radio broadcasting schools. Yoruba is also taught in primary and secondary schools, as well as at universities in the southern part of West Africa.

Religion plays an integral role in everyday life in West Africa. The most common religions are Islam, Christianity, and traditional African religions. Although many followers of different religions live together peacefully, conflict sometimes occurs between competing religious groups.

Education and Health Care

Free, universal education in West Africa remains inconsistent throughout the subregion. Literacy rates range from 29 percent in Niger to 68 percent in Nigeria. Education has been a high priority in Ghana, where government spending on education has steadily increased since the 1960s. Primary and secondary education is mandatory and free, as stated in the constitution. About 3.7 million students attended primary schools in Ghana in 2008. Students who complete high school in Ghana usually attend universities, training colleges, or polytechnic schools to learn trades. Sierra Leone has done a poorer job, however, and its support of education has steadily declined over the last 30 years.

Teen Life in Nigeria

Nigerians take great pride in their heritages. Nigerians identify first with their ethnicity, next with their religion, and then with their nationality. There are 250 different ethnic groups, speaking 250 different languages and having 250 distinct cultural heritages.

In 1960 Nigeria became an independent country. However, because of the widely varying cultural differences within the country, it has been difficult to unite the population.

Did you know . . .

- With 250 languages spoken in Nigeria, English was named the national language, but only about half of the population uses English.
- October 1 is National Day, a national holiday celebrating the day Nigeria became an independent country.
- Most Nigerian students wear uniforms to school.
- After school, children in Nigerian villages help tend the crops and prepare meals with their families.
- School is often difficult for children. After the third grade, teachers speak mainly English during class and not the ethnic languages that students speak at home.
- The voting age is 21.

CONNECTING TO

THE UNITED STATES

U.S. Immigration from Africa, 2008	
Ethiopia	12,917
Somalia	10,745
Ghana	8,195
Nigeria	8,172
Kenya	6,998

Source: U.S. Department of Homeland Security

Graduates wearing stoles made of Kente cloth

Just the Facts:

- About 12 percent of Americans identify themselves as African American.
- Modern-day blues music evolved from spirituals—the songs of enslaved Africans.
- More than half of African Americans in the United States live in the South.
- Since 1995 Nigeria has accounted for a large number of immigrants from Africa to the United States.

Singers today, like Usher, have been greatly influenced by African rhythms and African American musicians of the past.

African art is very distinctive and has influenced the work of artists such as Picasso and Matisse.

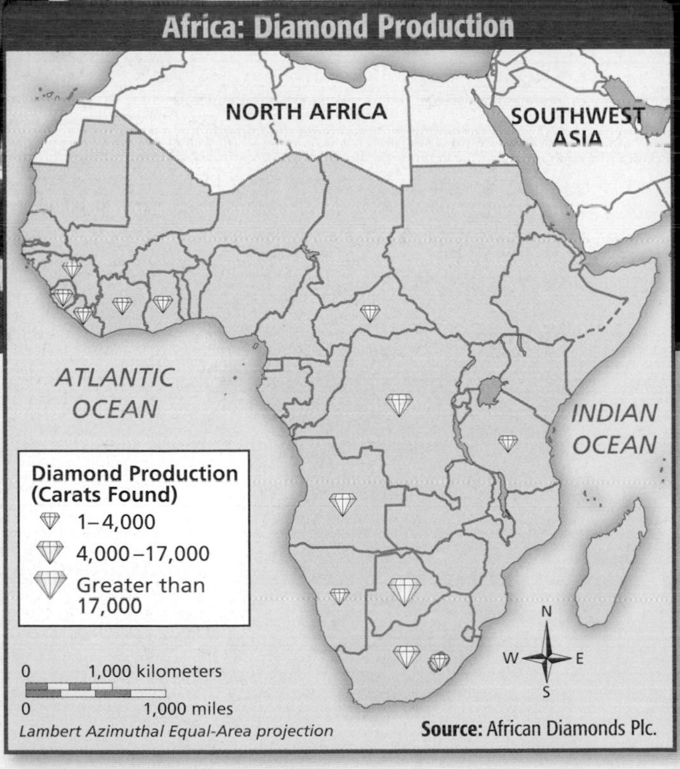

Making the Connection

The history of Africans in America dates back to before the founding of the nation. Unlike many who settled in North America, most Africans were forced to come to the region as enslaved persons. Despite this painful start, the influences of African culture have been melded into our national culture.

Crafts—Fabric of Pride Kente cloth has its origins with the Ashanti people of modern-day Ghana. They began weaving Kente around the seventeenth century for use at prestigious events by Ashanti royalty. The cloth is made by weaving individual strips of cloth that have meaningful colors and patterns. Today Kente cloth can be seen adorning many modern items such as neckties, hats, and umbrellas.

Music—The Beat Goes On The strong rhythms that form the core of African music are part of the musical heritage of the United States. Spirituals were the earliest American music with an African rhythm. These religious songs were first developed in the southern United States by enslaved Africans. Later these same rhythms were used in other forms of American music, including blues, jazz, and rock 'n' roll.

Diamonds The demand for diamonds has not ceased since their discovery thousands of years ago. Today most commercial diamonds come from Africa. The most notable deposits are located in South Africa, Namibia, Botswana, the Democratic Republic of the Congo, Angola, Tanzania, and Sierra Leone.

Food The vegetable okra originated near present-day Ethiopia. It came to the United States with enslaved persons. Okra continues to be popular in the ethnic cuisine of African Americans in the southern United States.

Africa: Diamond Production

Source: African Diamonds Plc.

THINKING GEOGRAPHICALLY

1. **Human Systems** The cultural contributions of African Americans are numerous. Research one aspect such as art, music, or food. Show how these contributions can be traced back to Africa.
2. **Environment and Society** Conduct research to find out how U.S. and international demand for diamonds has affected diamond-producing countries like Sierra Leone.

CHAPTER 22

THE REGION TODAY

Africa South of the Sahara

BiG Idea

The characteristics and distribution of human populations affect human and physical systems. A study of Africa south of the Sahara today will explain the immense challenges the region faces in gaining economic independence and stability while finding the best use of its natural resources, which are key to the region's development.

Essential Questions

Section 1: The Economy

How might economic activities in Africa south of the Sahara be affected by the environment?

Section 2: People and Their Environment

How do human actions contribute to environmental problems like desertification and deforestation?

Table Mountain rises behind Cape Town, South Africa, an economic and industrial center in the region.

Geography ONLINE

Visit glencoe.com and enter *QuickPass*™ code WGC9952C22 for Chapter 22 resources.

Identifying Information Use a Four-Door Book to identify and describe four problems facing Africa south of the Sahara today.

Reading and Writing As you read the chapter, identify and describe in the appropriate place in your Foldable four problems facing the region today. After you name and describe each problem on the front of each door, present possible solutions to each problem inside the Foldable.

Wildlife Conservation

The Problem:

Within ten years, rhinos could be extinct in the wild. After 50 million years only five species of rhinoceros remain, and all are threatened with or on the verge of extinction.

Rhino skulls left by poachers

Poaching Poaching–the illegal killing of animals–is profitable. Rhinoceros horn is a valuable ingredient in traditional Asian medicine.

Refugees in Mozambique

Humans Refugees fleeing civil wars and floods seek shelter and food, which can destroy wildlife and wildlife habitats.

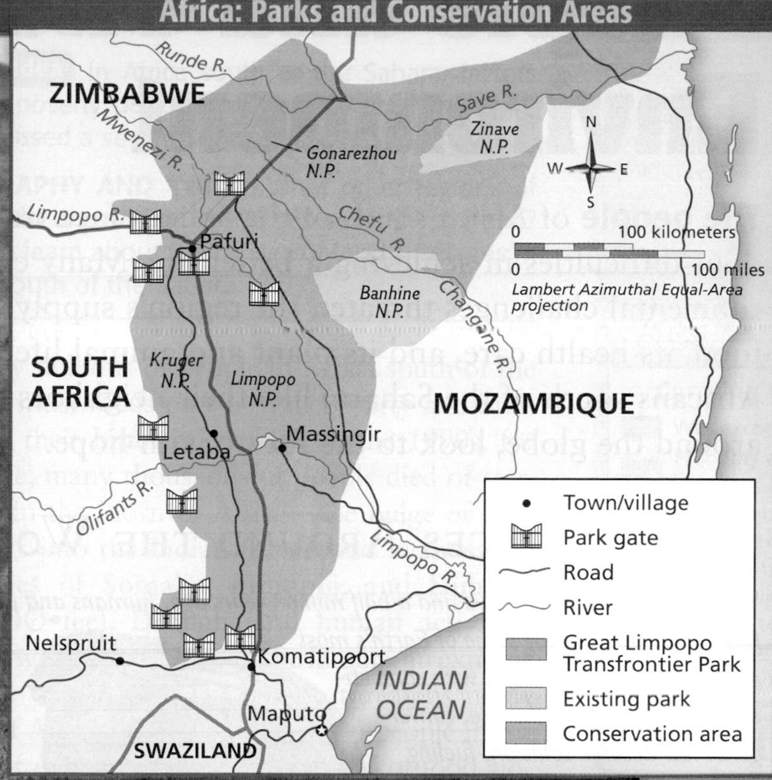

NATIONAL GEOGRAPHIC

Great Limpopo Transfrontier Park

At almost 13,500 square miles (35,000 sq. km), the Great Limpopo Transfrontier Park is the largest park of this type, spanning the borders of Mozambique, South Africa, and Zimbabwe. The GLTP joined five national parks to provide the diverse ecosystems needed for a variety of species to survive.

What is a Transfrontier Park? Park nations form transfrontier parks by joining their national parks. Recognizing that ecosystems do not end at national borders, the countries remove fences and other barriers to allow wildlife to follow ancient migration routes.

Why are they called peace parks? Transfrontier parks are called *peace parks* because nations must cooperate to create and manage the parks. Besides protecting wildlife and the environment, the parks promote regional peace through socioeconomic development. The local people living in or near the parks are being taught how to benefit from wildlife conservation. If tourism, game farming, and controlled hunting produce jobs and income, the outcome should be economic stability and peace.

How does the creation of the GLTP help the rhino? The Great Limpopo Transfrontier Park will bring rhinos together in a safe environment with ample grazing, allowing them to repopulate and avoid extinction.

One Solution:

Transfrontier parks, or *peace parks,* bring together formerly warring countries. These neighbors cooperatively manage the parks to protect the biodiversity of the region.

Transfrontier parks, like Great Limpopo, protect animals like the white rhinoceros and help the economy by bringing tourism to the area.

THINKING GEOGRAPHICALLY

1. **Environment and Society** Research Africa's rhinos in more depth. What factors have caused them to become endangered?
2. **Places and Regions** Study other transfrontier parks. What benefits do they offer the environment and wildlife?

Challenges for the Future

MAIN Idea As Africa south of the Sahara faces the future, human activities continue to have both positive and negative impacts on the environment.

GEOGRAPHY AND YOU What environmental challenges does the United States face? Read to find out about the future outlook for Africa south of the Sahara.

People in Africa south of the Sahara are working to overcome some of the region's serious challenges. The region has already taken important steps, however, toward preserving the environment. Democratic reforms are taking root in countries such as Ghana, Nigeria, and Liberia. Efforts to encourage private enterprise have also had **positive** results. New ranching laws, for example, have allowed people to engage in crocodile farming, a highly profitable business that has brought this species back from low numbers due to trapping and hunting. Rhinoceroses and elephants are also beginning to thrive again as their habitats are protected and as poaching is discouraged by stricter laws.

To save endangered species, some countries have created huge game reserves. These reserves—which include Tanzania's Serengeti National Park, Kenya's Masai Mara, and Ghana's Kakum National Park—have helped some animals make a comeback. The parks also attract millions of tourists each year. **Ecotourism,** or tourism based on concern for the environment, has become a big business in parts of the region, bringing millions of dollars into African economies. Governments give rural peoples an economic stake in the reserves. Some train to work in the reserves as trail guides or become involved in development planning.

Increasingly, the protection of tropical forests is a priority in the region. In 1999 leaders from six central African countries signed an agreement to preserve the forests. The effects of this and similar efforts have yet to be seen, but they are a strong signal that Africans today are moving toward a more positive future.

READING Check **Human-Environment Interaction** What has helped bring crocodiles back from low numbers due to trapping and hunting in the region?

SECTION 2 REVIEW

Vocabulary

1. Explain the significance of: carrying capacity, habitat, extinction, poaching, ecotourism.

Main Ideas

2. How have poverty, population growth, war, and drought caused severe strain on the environment in Africa south of the Sahara?
3. What kinds of human activities have destroyed large areas of the region's tropical forests and threatened wildlife?
4. Create a diagram like the one below to show the environmental challenges facing Africa south of the Sahara. List the causes and effects of each challenge. Be sure to consider the positive and negative impacts of each challenge.

Critical Thinking

5. **Answering the Essential Question** How do you think war has contributed to desertification in the region?
6. **Summarizing Information** What is the central issue in the debate over the renewal of the ivory trade?
7. **Analyzing Visuals** Study the map of carrying capacity in Africa south of the Sahara on page 565. In what parts of the region has the land's carrying capacity not been exceeded?

Writing About Geography

8. **Expository Writing** Think about the challenges Africa south of the Sahara faces today. Choose one problem in the region that might have an impact on the rest of the world. Write a paragraph explaining why the problem is a global issue.

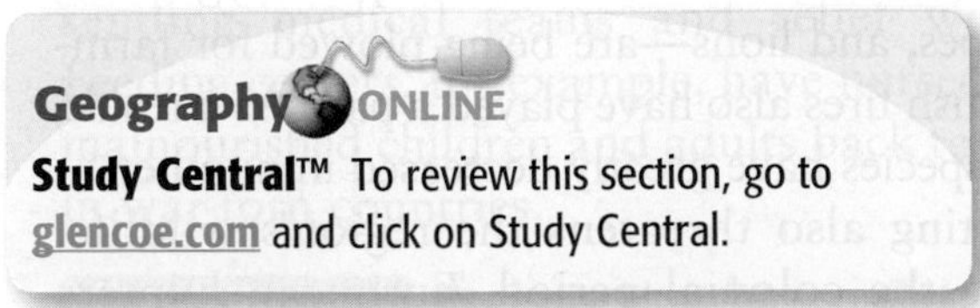

Geography ONLINE
Study Central™ To review this section, go to glencoe.com and click on Study Central.

Study anywhere, anytime by downloading quizzes and flashcards to your PDA from glencoe.com.

THE REGION TODAY

Friday — Section A

Slowly Industrializing

AGRICULTURAL LIFE

- Most African farmers are subsistence farmers that provide food only for their family or village.
- If there is a drought and crops fail, subsistence farmers do not have food or money to survive.

MAKING THE CHANGE

- Many countries are trying to shift from an agricultural economy to an industrial economy.
- Although Africa is rich in minerals it lacks the infrastructure to process and manufacture the minerals.
- Some manufacturing has taken place, but continued expansion will be a challenge due to a lack of skilled workers, frequent power outages, and political conflicts.

Protecting Endangered Species

- Most African countries have set up large parks and game reserves to protect endangered species.
- These parks and reserves also serve as a source of income for the countries as tourists come from around the world to see Africa's wildlife.
- Private enterprise has also helped save endangered species. For example, crocodile farmers earn money selling crocodile leather. However, they also release crocodiles into the wild, increasing the population.

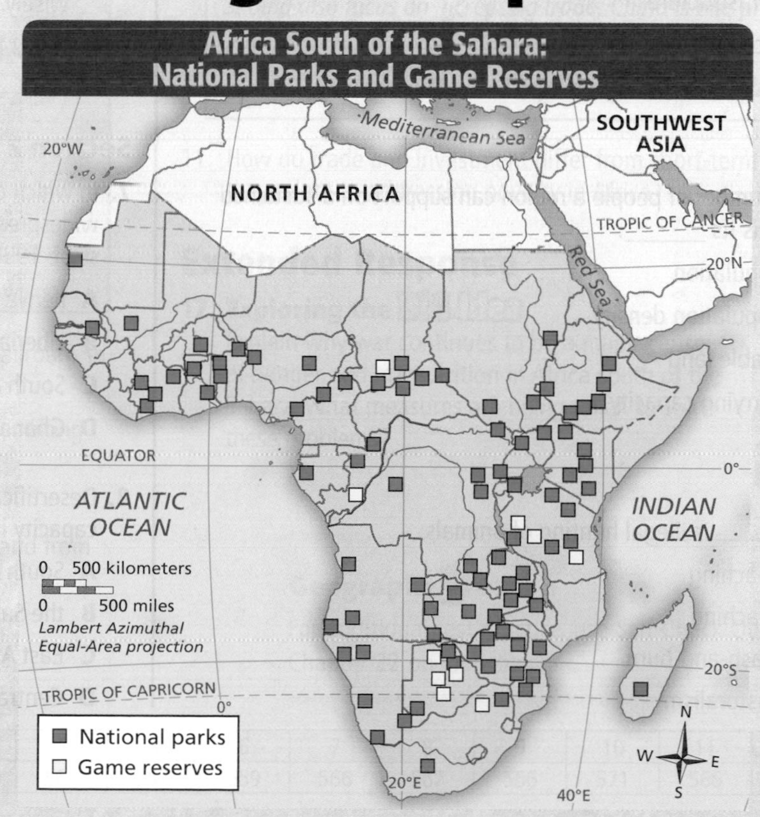

UNIT

8 South Asia

Taj Mahal, across from the Jamuna River, Agra, India

NGS ONLINE To learn more about South Asia visit www.nationalgeographic.com/education.

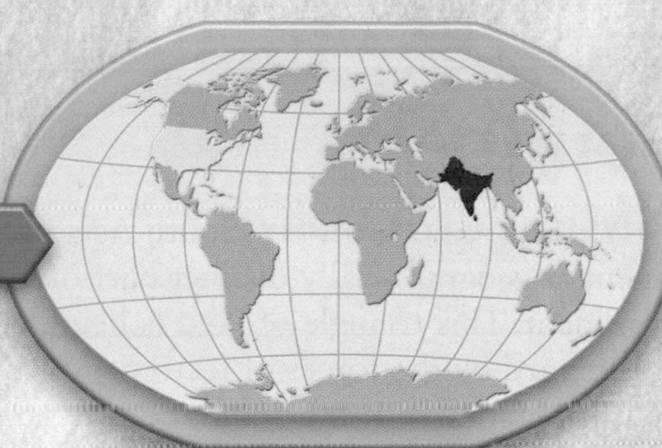

Why It Matters

Many of the countries of South Asia have earned their independence relatively recently, but they have their roots in very ancient civilizations. The rich culture, minerals, and spices of the area have attracted foreign invaders for hundreds of years. Since the subcontinent shook off the cloak of British colonial rule in the twentieth century, political and religious rivalries within the region have threatened its peace and stability. The governments of South Asia are struggling to overcome their differences and increase the region's role in trade and technological development.

STORY OF A Tsunami

Tsunami Glossary

Richter scale	a way of measuring the magnitude of an earthquake
epicenter	the point directly above an earthquake's focus

Path of Destruction

Disaster The earthquake that caused the 2004 Indian Ocean tsunami was the fourth most powerful earthquake recorded since 1900. The ensuing waves caused by the earthquake resulted in nearly 200,000 deaths, some as far away as Africa. Entire villages were swept away when the tsunami waves crashed on the shorelines of over a dozen different countries. The economic impact of the catastrophe is still being felt.

The Indian Ocean tsunami hits Thailand.

The Environmental Impact The 2004 Indian Ocean tsunami destroyed numerous coral reefs, mangrove forests, and coastal wetlands and contaminated many wells used for drinking water. Large salt deposits, left behind by the tsunami's receding ocean water, ruined much of the farmland in the affected area.

Nature's Fury

On the morning of December 26, 2004, an earthquake measuring over 9.0 on the Richter scale caused the worst tsunami in recent history. With its epicenter located just west of Sumatra, the earthquake caused a massive displacement of ocean water, resulting in nearly 100-foot (30-m) waves that crashed over the nearby islands. The lack of a tsunami warning system left the people of the region unprepared for the death and destruction they faced.

An Indonesian man surrounded by the rubble that was his home.

2004 Indian Ocean Tsunami: Human Toll

Country	Fatalities	Missing	Total
Indonesia	130,736	37,000	167,736
Sri Lanka	35,322	–	35,322
India	12,405	5,640	18,045
Thailand	8,212	–	8,212
Maldives	82	26	108
Somalia	78	211	289
Malaysia	69	6	75
Myanmar (Burma)	61	–	61
Tanzania	13	–	13
Bangladesh	2	–	2
Seychelles	2	–	2
Kenya	1	–	1
Total	**186,983**	**42,883**	**229,866**

Source: UN Office of the Special Envoy for Tsunami Recovery.

An Indonesian woman receives medical treatment in a Korean-run clinic.

The World Responds In the days following the 2004 Indian Ocean tsunami, the global community came together to provide aid to the affected countries. As of 2005, nearly $13 billion had been pledged to help the recovery effort.

THINKING GEOGRAPHICALLY

1. **Environment and Society** What advantages might a tsunami warning system have provided people who lived in the affected area?
2. **Places and Regions** Research the 2004 Indian Ocean tsunami. Use the knowledge you gain to create a map of the area. On your map, show the epicenter of the earthquake and where the waves hit the shore.

Dry Regions

Along the lower Indus River in the northwestern part of South Asia, a desert climate keeps the land arid and windswept. The **Thar Desert** lies to the east of the Indus River. The vegetation here is desert scrub, low, thorny trees, and grasses. Livestock graze in some areas, and irrigation makes it possible to grow wheat near the Indus River. Much of this area, however, remains sandy with little vegetation.

Surrounding this desert, except on the coast, is a steppe climate. Few trees grow in this semiarid grassland. In northwestern India, rainfall averages less than 20 inches (51 cm) per year. Another steppe area runs through the center of the Deccan Plateau between the Eastern and Western Ghats in southern India. This area receives less rainfall than the coasts. As the summer monsoons approach the west coast of India, they rise up the Western Ghats and the air cools, releasing moisture as rainfall. However, as the winds make their way over the mountains, they lose most of their moisture. This reduced rainfall—a result of the rain shadow effect—makes the area relatively dry. Scrub and deciduous forests cover vast stretches of India's interior.

Regions How is vegetation related to climate in the northern part of South Asia?

Seasonal Weather Patterns

MAIN Idea Seasonal weather patterns bring much-needed rainfall to South Asia, but monsoon winds, as well as other natural disasters, can also bring devastating hardships.

GEOGRAPHY AND YOU In what locations in the United States have hurricanes recently caused severe damage? Read to learn about the effects of monsoon rains and other natural disasters that frequently occur in South Asia.

Both the high temperatures of the hot season and the heavy rains of the wet season deliver mixed blessings to South Asia, a region prone to natural disasters.

Monsoon Rains

Much of South Asia experiences three distinct seasons—hot (from late February to June), wet (from June or July until September), and cool (from October to late February). These **periods**

GRAPH STUDY

1. **Regions** During which months do Karachi and Miami receive approximately the same amount of rainfall?
2. **Place** According to the charts, which months make up Karachi's rainy season?

NATIONAL GEOGRAPHIC **Comparing Climates**

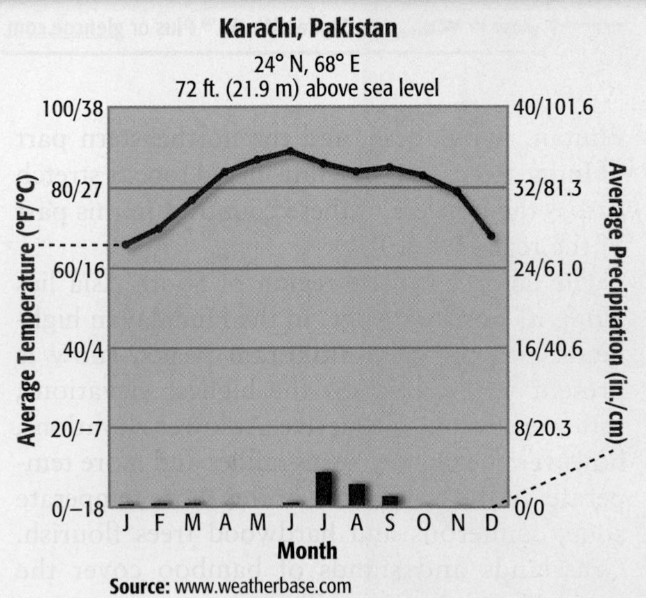

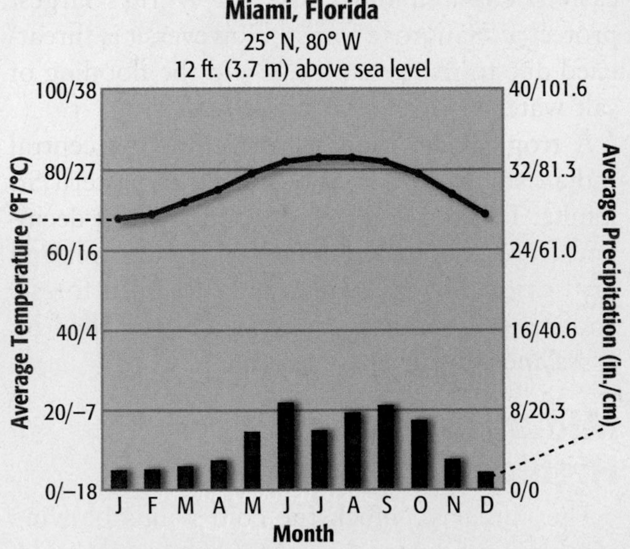

depend on seasonal winds called **monsoons.** In the hot season, air across South Asia is heated and becomes unstable. It rises and **triggers** a change in wind direction. The winds carry moist ocean air from the south and southwest, bringing monsoon rains. The winds carrying moisture-laden air from the Indian Ocean cause heavy rainfall and flooding across the subcontinent. During the cool winter season, this pattern is reversed as air in the Asian interior cools and becomes more stable and blows south across the subcontinent toward the ocean. The air is dry and cool compared to the wet season.

Rains from the monsoons are heaviest in eastern South Asia. When the rains sweep over the Ganges-Brahmaputra delta, the Himalaya block them from moving north. As a result, the rains move west to the Gangetic Plain, bringing rainfall needed for crops. People celebrate the arrival of the monsoon rains because without them, the farmers will not produce as much food. India's 700 million farmers depend on the monsoon rains.

Natural Disasters

Temperature and rainfall impact agriculture in the region. High temperatures and water allow farmers to produce crops, including the rice that many people living in Bangladesh and India depend on year-round. The extreme heat, however, can result in evaporation, and without rainfall there is drought.

The monsoon winds also have benefits and drawbacks. Rainfall waters crops, but areas outside the path of the monsoon, such as western Pakistan, may receive little or no rainfall during the year. When the people of Bangladesh are planting rice, and those on the Gangetic Plain are planting winter crops, other areas may suffer from drought. Too much rain can also be a problem. In the low-lying lands of Bangladesh, monsoons may cause flooding that kills people and livestock, leaves thousands homeless, and ruins crops.

Cyclones, the same as hurricanes in the Atlantic Ocean, are a natural hazard in South Asia. They have high winds and heavy rains. A 2007 cyclone struck Orissa, India, with winds of about 155 miles per hour (249 km per hour) and waves over 10 feet (3 m) high. The storm killed over 3,200 people and caused more than $450 million in damage.

NATIONAL GEOGRAPHIC *The same land in South Asia often looks quite different during the wet and dry seasons.*

Regions From what direction do the winter monsoon winds come? What kind of weather do they bring?

Tectonic activity in the Himalaya and beneath the Indian Ocean affects countries such as India, Bangladesh, and Sri Lanka. Earthquakes and **tsunamis,** huge waves caused by underwater earthquakes, are also natural hazards. In October 2005, an earthquake of magnitude 7.6 devastated northern Pakistan and Kashmir. The earthquake killed more than 70,000 people within minutes and left millions of other people without food and shelter. In December 2004, a tsunami in the Indian Ocean struck parts of South Asia and Southeast Asia. In Sri Lanka, more than 30,000 people were killed and many fishing villages were destroyed.

READING Check **Location** In what part of South Asia are monsoon rains heaviest?

Student Web Activity Visit glencoe.com, select the *World Geography and Cultures* Web site, and click on Student Web Activities–Chapter 23 for an activity on the formation of the Himalaya and attempts to reach the summit of Mount Everest.

NATIONAL GEOGRAPHIC *Severe cyclones are common in coastal Bangladesh, causing loss of life, injury, and property damage.*

Human-Environment Interaction What other weather events affect the countries of South Asia?

SECTION 2 REVIEW

Vocabulary

1. Explain the significance of: monsoon, cyclone, tsunami.

Main Ideas

2. Describe the seasonal weather patterns that bring much-needed rain to South Asia.
3. How can monsoon winds and other natural disasters bring devastating hardships to South Asia?
4. Although much of South Asia has tropical climates, the region does have a variety of other types of climates and vegetation. On a chart like the one below, list and describe South Asia's climate regions and their corresponding natural vegetation.

Climate Region	Climate Conditions	Natural Vegetation
Tropical		

Critical Thinking

5. **Answering the Essential Question** What impact do the summer monsoon winds have on the environment and people of South Asia?
6. **Analyzing Information** When do the three seasons occur in much of South Asia, and how would you describe each?
7. **Analyzing Visuals** Compare the climate map on page 587 of the Regional Altlas and the vegetation map on page 597. In what parts of the region are climates and vegetation most varied?

Writing About Geography

8. **Descriptive Writing** Think about the climate and vegetation of a country in South Asia you would like to visit. Write a paragraph describing the natural features of the country.

Study Central™ To review this section, go to glencoe.com and click on Study Central.

VISUAL SUMMARY

STUDY TO GO

Study anywhere, anytime by downloading quizzes and flashcards to your PDA from glencoe.com.

A THE HIMALAYA

- Created by collision of tectonic plates; extend more than 1,500 miles (2,414 km) across northern edge of South Asia
- Includes Mount Everest, the world's highest peak at 29,028 feet (8,848 m)

B GANGETIC PLAIN

- World's longest alluvial plain
- Watered by the Ganges, Brahmaputra, and Indus Rivers; agriculturally productive area
- India's most densely populated area

C VINDHYA AND SATPURA RANGES

- Mountains in central India created by collision of tectonic plates
- Separate the distinct cultures of northern and southern India

E DECCAN PLATEAU

- Plateau region in southern India; located between Western Ghats and Eastern Ghats
- Rich soil with wet and dry seasons

D INDUS RIVER

- Flows mainly through Pakistan; empties into Arabian Sea
- Known as the cradle of ancient India
- Remains an important transportation route

G BRAHMAPUTRA RIVER

- Flows through India and Bangladesh
- Joins the Ganges River to form a delta; empties into Bay of Bengal
- Major inland waterway; also provides hydroelectric power

F GANGES RIVER

- Flows east from the Himalaya; empties into Bay of Bengal
- Drainage basin covers about 400,000 square miles (about 1 million sq. km)
- Named for Hindu goddess Ganga; sacred to Hindus

H MONSOONS

- Hot season is from late February to June
- Wet season is from June or July to September
- Cool season is from October to late February

SOUTH ASIA

CHAPTER 23

STANDARDIZED TEST PRACTICE

When you have finished, check your work to be sure you have answered all the questions.

Reviewing Vocabulary

Directions: Choose the word or words that best complete the sentence.

1. A(n) _______ is a large, distinct landmass that is joined to a continent.

A island
B peninsula
C country
D subcontinent

2. An area of fertile soil deposited by river floodwaters is called a(n) _______.

A glacier
B alluvial plain
C plateau
D loess

3. _______ are winds that change direction with the season.

A Monsoons
B Foehns
C Chinooks
D Mistrals

4. A storm like a hurricane in South Asia is called a _______.

A monsoon
B typhoon
C tornado
D cyclone

Reviewing Main Ideas

Directions: Choose the best answers to complete the sentences or to answer the following questions.

Section 1 *(pp. 590–593)*

5. _______ border South Asia on the north.

A Rivers
B Plains
C Mountains
D Plateaus

6. What is the source of the major rivers in the northern part of South Asia?

A Karakoram Range
B Western Ghats
C Eastern Ghats
D the Himalaya

Section 2 *(pp. 596–600)*

7. Why is there a band of humid subtropical climate across Nepal, Bhutan, Bangladesh, and the northeastern part of India?

A This region is in the low latitudes.
B This region is only a few feet above sea level.
C The Himalaya mountain ranges block the coldest winds blowing from Central Asia.
D This region is near the sea.

8. Where are the dry regions of South Asia?

A in the north-central part
B in the lower Indus River valley
C in the lower Ganges River valley
D in the far south

Critical Thinking

Directions: Choose the best answers to complete the sentences or to answer the following questions.

9. Why are dams especially important in South Asia?

A They provide employment to the large population.

B They clean the water of silt.

C They control flooding and store irrigation water for the dry season.

D They provide sources of national pride.

Base your answer to question 10 on the map and on your knowledge of Chapter 23.

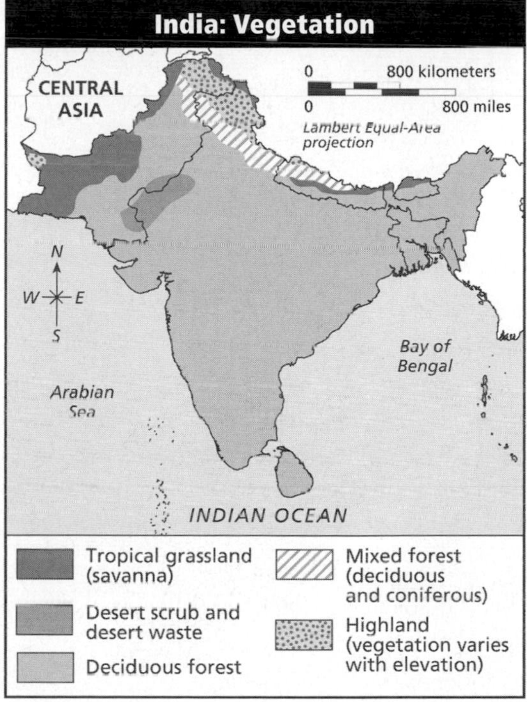

10. What kind of vegetation is found throughout most of India?

A mixed forest

B deciduous forest

C tropical grassland

D highland

Document-Based Questions

Directions: Analyze the document and answer the short-answer questions that follow the document.

In September 2008, flooding struck Bangladesh near the end of the summer monsoon.

> *Bangladesh is bracing for possible major flooding in the next two days with heavy rain forecast for neighbouring India set to swell its rivers to dangerous levels, an official said Thursday.*
>
> *More than 70,000 people were marooned by flooding five days ago after rivers in northern and central Bangladesh burst their banks. . . .*
>
> *. . . [O]fficials were on high alert with downpours expected in northeastern India likely to cause problems for parts of Bangladesh.*
>
> *[Flood Forecasting and Warning Centre director Saiful Hossain] said Bangladesh's three major rivers—the Ganges, the Brahmaputra and the Meghna—were all rising, particularly the Brahmaputra, the country's biggest.*
>
> *"All three systems will rise which will affect the flooding situation, particularly in the southern central part of the country, as well as parts of the north," Hossain said.*
>
> *Bangladesh is criss-crossed by a network of 230 rivers and suffers annual floods, with at least a fifth of the country submerged each year.*
>
> —"Bangladesh braces for floods as heavy rain forecast," www.terradaily.com, September 4, 2008

11. Why do rivers in India contribute to flooding in Bangladesh?

12. What options might be available to Bangladesh to combat this chronic problem?

Extended Response

13. Exploring the BIG Idea
Describe the importance of rivers to the livelihood of South Asia's population.

STOP

Geography ONLINE
For additional test practice, use Self-Check Quizzes—Chapter 23 on glencoe.com.

Need Extra Help?

If you missed questions. . .	1	2	3	4	5	6	7	8	9	10	11	12	13
Go to page. . .	591	592	598	600	591	592	592	598	594	603	603	603	593

CHAPTER 24 CULTURAL GEOGRAPHY OF South Asia

BiG Idea

Cultures are held together by shared beliefs and common practices and values. A study of the cultural geography of South Asia will reveal how the region's history is tied to its ancient past through language and social structures.

Essential Questions

Section 1: India

What things can one learn about a country by studying its people?

Section 2: Pakistan and Bangladesh

In what ways can conflict shape a country?

Section 3: Nepal, Bhutan, Maldives, and Sri Lanka

How might the location of a country influence its culture?

Geography ONLINE

Visit glencoe.com and enter QuickPass™ code WGC9952C24 for Chapter 24 resources.

Women celebrate Holi, the Hindu festival of colors, in Allahabad, India.

FOLDABLES™ Study Organizer

Summarizing Information Make a bound Trifold Book to help you summarize information about the cultural geography of the countries of South Asia.

Religions

Ethnic Conflicts

Art

Reading and Writing As you read the chapter, use your Foldable to summarize the cultural histories of each country in South Asia.

SECTION 1

section audio

spotlight video

Guide to Reading

Essential Question

What things can one learn about a country by studying its people?

Content Vocabulary

- *jati* (p. 607)
- megalopolis (p. 607)
- dharma (p. 608)
- reincarnation (p. 608)
- karma (p. 608)
- mercantilism (p. 608)
- imperialism (p. 608)
- raj (p. 608)
- guru (p. 609)

Academic Vocabulary

- identify (p. 607)
- facility (p. 607)
- requires (p. 608)

Places to Locate

- Mumbai (Bombay) (p. 607)
- Kolkata (Calcutta) (p. 607)
- Delhi (p. 607)
- Khyber Pass (p. 608)

Reading Strategy

Categorizing Complete a chart similar to the one below by describing India's major cities.

City	Description
Mumbai	
Kolkata	
Delhi	

India

The culture of today's India is making an impact on the global community. While most Westerners have never seen a Bollywood film, the name given to Mumbai's thriving center of film production, India's film industry is the largest in the world.

NATIONAL GEOGRAPHIC Voices Around the World

"Bollywood has become a globally recognized brand; like Darjeeling tea or the Taj Mahal, it has become an emblem of India. Its films are popular in the Middle East, Central Asia, Africa, Latin America—and now the U.S. and Europe, where immigrants from Bollywood-loving countries make up most of the audiences and provide more than 60 percent of overseas revenues. With the recent buzz surrounding Bollywood-inspired films like Bride and Prejudice *and* Monsoon Wedding, *and the nomination of one—*Lagaan*—for a 2001 Oscar, even Hollywood is starting to take notice of its rival."*

—Suketu Mehta, "Welcome to Bollywood," *National Geographic,* February 2005

A young fan in Mumbai with movie postcards

Population Patterns

MAIN Idea Population density and distribution, as well as urbanization, continue to shape India's population patterns.

GEOGRAPHY AND YOU What has influenced the growth of cities in your region of the United States? Read to learn how India's largest cities are growing.

One of the most significant characteristics of India's population is its size. Over 1.1 billion people—more than 15 percent of the world's population—live in the country. Besides being large, India's population is also ethnically diverse.

The People

The largest number of Indians are descended from the Dravidians, who have lived in the south of India for 8,000 years, and the Aryans, who entered from Central Asia more than 3,000 years ago. Many Indians traditionally **identify** themselves by their religion—as Hindus, Muslims, Buddhists, Sikhs, Jains, or Christians. Hindus also identify themselves by a ***jati***, a group that defines one's occupation and social position.

Density and Distribution

With 1,020 people per square mile (394 people per sq. km), India's average population density is about seven times the world average. Although population densities are generally high throughout India, the distribution of population varies from place to place. Factors such as climate, vegetation, and physical features affect the number of people the land can support. The Thar Desert is sparsely populated, while the highest population concentrations are found on the fertile Gangetic Plain and along the monsoon-watered coasts of southern India. Within parts of these agriculturally productive areas, densities exceed more than 2,000 people per square mile (772 people per sq. km).

Most of India's population is rural—about 70 percent of people live in villages. They farm and struggle to grow enough food for their families. Part of their crops often goes to the owners of the fields they farm.

In recent years growing numbers of Indians have been migrating to urban areas, drawn by the hope of better jobs and higher wages. As urban populations grow, however, they strain public resources and **facilities**. India's cities are among the world's most densely populated. **Mumbai (Bombay)** is India's main port on the Arabian Sea as well as its largest city, with a population of more than 20 million. **Kolkata (Calcutta)**, a thriving port city on a branch of the Ganges River, is the center of India's iron and steel industries. **Delhi** (DEH•lee), India's second-largest city, is part of a **megalopolis**, or chain of closely linked metropolitan areas.

READING Check **Human-Environment Interaction** Why is population density high on the Gangetic Plain?

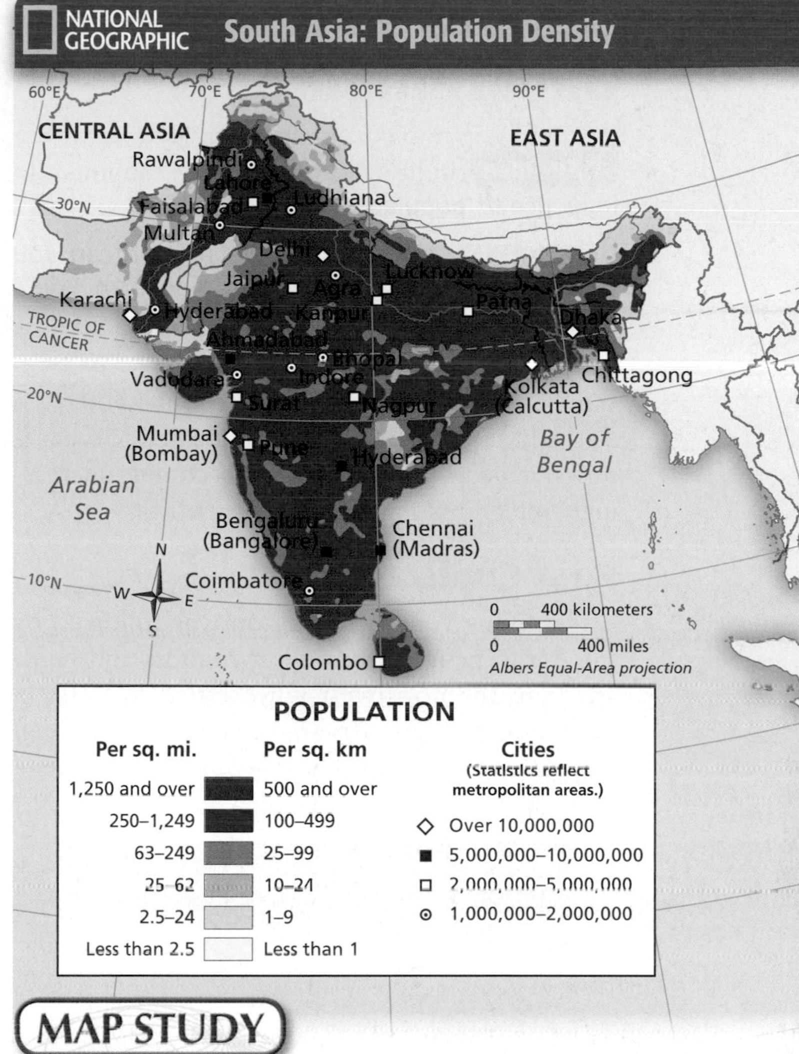

MAP STUDY

1. **Place** Which parts of India have the highest population density?
2. **Place** How do population densities in India compare to those in Pakistan?

Maps In MOtion Use **StudentWorks™ Plus** or glencoe.com.

History and Government

MAIN Idea India's ancient history continues to influence its populations today.

GEOGRAPHY AND YOU How far back can you trace your family's history? Read to learn how India's past continues to influence its people.

India's history dates back more than 4,500 years to the Indus Valley civilization, located in what is now Pakistan. Today India is becoming increasingly integrated into the modern world.

First Civilizations

The Indus Valley civilization was followed by the Aryan people, a group of hunters and herders from the northwest who settled in India in the 2000s B.C. The Aryans created a rigid social structure that Europeans called the caste system.

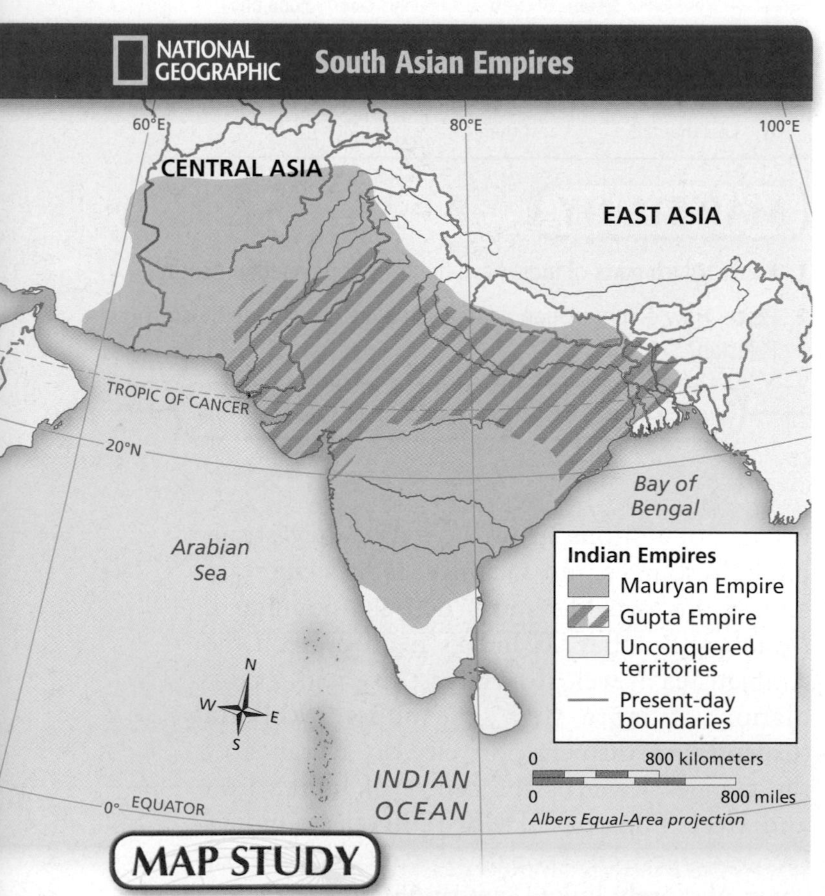

MAP STUDY

1. **Location** Which empire was limited to the Ganges River valley?
2. **Human-Environment Interaction** Which geographic factors may have aided the cultural advances of the Gupta Empire?

A caste, or *jati,* is the position in life one is born into. The Aryans' sacred writings, the Vedas, outline Aryan ideas about social structure and religion. Understanding India's religions is key to understanding its history and culture.

Hinduism Hindu belief **requires** every person to carry out his or her **dharma** (DUHR•muh), or moral duty. Hindus also believe that after death people undergo **reincarnation,** or rebirth as another living being until overcoming personal weaknesses and earthly desires. In the law of **karma,** good deeds—actions in accord with one's dharma—move one toward this point.

Buddhism Siddhartha Gautama became known as the Buddha, or the Awakened One, after perceiving the true nature of human existence. The Buddha taught that people suffer because they are too attached to material things. Buddhism teaches people to think clearly, work diligently, and show compassion for all living things.

Invasions and Empires

After the Aryans, other groups with new cultures entered northwestern India through the **Khyber Pass** in the Hindu Kush mountains. The Mauryan Empire maintained control from about 320 B.C. to 180 B.C. The Gupta Empire ruled from about A.D. 320 to A.D. 500 and became one of the most advanced civilizations in the world. After this empire came the Muslim-led Mogul Empire, during which many Indians converted to Islam.

The final invaders were Europeans. The British employed a policy of **mercantilism,** an economic system of using colonies for supplying materials and markets to the colonizing country. The British practiced **imperialism,** or political and economic domination. They called their Indian empire the British **raj,** the Hindi word for "empire." The British introduced the English language, restructured the educational system, built railroads, and developed a civil service.

Independence

India's fight for independence was led by Mohandas K. Gandhi. Using nonviolent methods, he inspired people to seek self-rule. Indians won freedom in 1947. Britain divided the land into Hindu India and Muslim Pakistan. Today India is the world's largest democracy.

READING Check **Place** Which group brought the caste system to India?

Culture

MAIN Idea India's people share an amazingly diverse culture rooted in religious traditions.

GEOGRAPHY AND YOU What has influenced the language and religion of the United States? Read to learn about the influences on India's culture.

India's rich culture can be seen in its many religious expressions, its native art forms, and its wholehearted embrace of movies.

Education and Health Care

Children are required to attend school for 9 years. India's average literacy rate is 61 percent, with literacy rates in rural areas somewhat lower. The government is committed to extending educational opportunities to women and members of the lower social classes.

India's state-run hospital system has improved in recent years. Diseases such as malaria—which were once widespread—have been brought under control. Other health problems continue, such as HIV infection and AIDS.

Language and Religion

The people of India speak 22 official languages and hundreds of local dialects, with Hindi the most widely spoken. English, the common language of international business and tourism, is also widely spoken in parts of India that were once under British rule.

Most people in India are Hindus. Other religions practiced in the country include Islam, Buddhism, Christianity, and Sikhism. Sikhism, founded in the early A.D. 1500s by a **guru**, or teacher, named Nānak, teaches that there is one God and that good deeds and meditation bring release from the cycle of reincarnation. Most of South Asia's Sikhs live in northwestern India, and many want an independent Sikh state there.

The Arts

Artistic expression is as much a part of Indian life as religious practice. Two great epic poems—the *Mahābhārata* (muh•hah•BAH•ruh•tuh) and the *Rāmāyana* (rah•MAH•yah•nuh)—combine Hindu social and religious beliefs. India has numerous classical dance styles, most of which are based on themes from Hindu mythology.

India

India is a country of great diversity and rich cultural heritage. The lives teens live are largely influenced by the caste or social class into which they are born. Those in the lowest classes have less opportunity for education and recreation. They often work long days on farms or in shops. In spite of this, most teens enjoy sports, games, and spending time with friends, just like teens in the United States.

Did you know . . .

- Nearly 400 languages are spoken in India. Hindi and English are used in government communications.
- One of the most popular sports is cricket, a game similar to baseball.
- The diet of many Indians is largely vegetarian. Most people in India follow either the Hindu or Muslim religions. The Hindu religion forbids the eating of beef, and Muslims do not eat pork.
- Indian children in urban areas wear uniforms to school and often attend school at least two Saturdays a month in addition to weekly attendance.
- American movies are very popular with Indian teens and adults. They also enjoy Bollywood films produced by the Indian film industry.
- Unlike drivers in the United States, Indians drive on the left side of the road.

Since 1896, when motion pictures first arrived in India, movies have been a popular form of entertainment. India's film industry is the world's largest, producing more full-length feature films each year than any other country. It is centered in Mumbai and is nicknamed "Bollywood," a combination of Bombay and Hollywood.

Family Life and Leisure

Family is the most important social unit for most people in India. Extended families live together, sharing household chores and finances. Arranged marriages based on caste, economic status, and education have been the traditional path to marriage for most Indians. However, this is slowly changing. In many instances, a bride will go to live with her husband and his extended family. Within the family, there is a clear order of influence based on gender and age, and in the case of women, the number of male children.

Leisure time in India is spent in various ways. For the middle class, it means going to the movies and watching television. In rural areas, a break during the agricultural season allows families to attend weddings and other family celebrations.

READING Check **Place** How has Hinduism influenced Indian literature?

NATIONAL GEOGRAPHIC *India is home to the world's largest motion-picture industry, producing 1,091 films in 2006 compared to 485 produced in the United States.*

Location Where is the center of India's film industry located?

SECTION 1 REVIEW

Vocabulary

1. Explain the significance of: *jati,* megalopolis, dharma, reincarnation, karma, mercantilism, imperialism, raj, guru.

Main Ideas

2. How does India's ancient history influence its people today? Give examples.
3. Describe aspects of India's culture that have been influenced by religious beliefs and traditions.
4. Complete a table like the one below by listing examples of how population density, population distribution, and urbanization shape India's population patterns.

Influences	Examples
Population density	
Population distribution	
Urbanization	

Critical Thinking

5. **Answering the Essential Question** How is life in urban areas in India different from life in rural areas? How is it similar?
6. **Making Generalizations** How has the physical geography of South Asia contributed to the development of diverse cultures?
7. **Analyzing Visuals** Study the population density map on page 607. In which parts of South Asia is population density the highest?

Writing About Geography

8. **Descriptive Writing** How might Gandhi's promotion of nonviolence and self-rule have helped the country achieve independence from the British? Describe how Gandhi might have affected the people's drive toward independence.

Study Central™ To review this section, go to glencoe.com and click on Study Central.

SECTION 2

 section audio

 spotlight video

Guide to Reading

Essential Question

In what ways can conflict shape a country?

Content Vocabulary

- total fertility rate *(p. 612)*
- Sikh *(p. 613)*

Academic Vocabulary

- technique *(p. 612)*
- overseas *(p. 613)*
- confirmed *(p. 613)*

Places to Locate

- Dhaka *(p. 612)*
- Islamabad *(p. 612)*
- Karachi *(p. 612)*
- Mohenjo Daro *(p. 613)*
- Harappa *(p. 613)*

Reading Strategy

Taking Notes As you read about the history of Pakistan and Bangladesh, use the major headings of the section to create an outline similar to the one below.

I. Population Patterns
   A.
   B.
II. History and Government
   A.
   B.
   C.

Pakistan and Bangladesh

Pakistan and Bangladesh share a similar history of Muslim influence, British colonialism, and the pursuit of independence. Today, sweeping events like the global war on terrorism have put countries like Pakistan in the world spotlight, revealing a people accustomed to hardship throughout history. The people of Pakistan live at an international crossroads between two cultural regions: Islam and Hindu India.

NATIONAL GEOGRAPHIC Voices Around the World

"Mohmand is not a friendly place either. Nearly every house is a castle built on steeples of rock, and every farmer toiling in his field has a rifle strapped to his back. Still, the rules of hospitality apply, and one day a local chieftain named Iftikhar Chandar invites us, and our police escort, to his house for lunch. Rope-strung cots are set out in a courtyard under the shade of an ancient grapevine arbor, and we feast on roasted goat and okra. Bees drone lazily around us."

—Tim McGirk, "Tracking the Ghost of bin Laden in the Land of the Pashtun," *National Geographic*, December 2004

A Pakistani soldier along the Pakistan-Afghanistan border

SOUTH ASIA

THE UNITED STATES

World Spice Production, 2007	
India	72%
Bangladesh	8%
Pakistan	3%
Other countries	17%

Source: Food and Agricultural Organization of the United Nations.

Women in Los Angeles celebrate a springtime Hindu New Year.

Just the Facts:

- 4.9 percent of medical doctors in the United States are Indian.
- The single largest group of Muslim immigrants to the United States is from South Asia.
- The first Hindu temple in the United States was established in Penn Hills, Pennsylvania, in 1976.
- California, New York, and New Jersey boast the largest Indian American populations in the United States.

Actors and actresses of the Indian film industry are becoming more recognizable in the United States. Actress Aishwarya Rai (right), pictured with American actresses Kerry Washington and Andie MacDowell, enjoys increasing popularity in mainstream cinema.

Making the Connection

South Asia is one of the most populous regions in the world. The region, which includes Bangladesh, Bhutan, India, Maldives, Nepal, Pakistan, and Sri Lanka, is home to almost five times as many people as the United States. In recent years immigration, outsourcing, and the entertainment industry have helped South Asia and the United States become more closely connected than ever before.

Food—Spices Many spices native to this region are exported to countries around the world. Cinnamon is made from the ground bark of the cinnamon tree, which is native to Sri Lanka and southern India. Other regional spices include nutmeg, cardamom, and cloves.

Cinnamon is used in the United States to flavor cereals and desserts.

Religion Recent immigration to the United States from countries such as India has made Hinduism one of the fastest-growing religions in the country. Hindu temples can be found throughout the United States, including prominent temples in Malibu, California, and Lanham, Maryland, located just outside of Washington, D.C.

Cinema—Hollywood and Bollywood The Indian film industry annually produces more films than the United States. These films usually feature song-and-dance segments, a style not seen in U.S. films. Still, the industries in both countries affect each other. As Indian films become more Westernized, including some Western actors and plotlines, the style and music of Indian cinema, in turn, influence American films.

U.S. Immigration from South Asia, 2008

Country	Number
India	63,352
Pakistan	19,719
Bangladesh	11,753
Nepal	4,093
Sri Lanka	1,935
Bhutan	42
Maldives	NA

Source: U.S. Department of Homeland Security.

International College Students in the United States

Country of Origin	Number of Students
India	94,563
China	81,127
South Korea	69,124
Japan	33,974
Canada	29,051

Source: www.iie.org, Institute of International Education, "Open Doors Report of International Educational Exchange, 2008."

The highest number of international college and university students in the United States comes from India. India has held the top position since it surpassed China in 2002.

THINKING GEOGRAPHICALLY

1. **Human Systems** Compare the two charts on this page. What effect might higher education have upon Indian immigration to the United States?
2. **Places and Regions** What are some of the major similarities and differences between the Indian and the American film industries?

SOUTH ASIA

CHAPTER 25

THE REGION TODAY

South Asia

BiG Idea

Economic systems shape relationships in society. South Asian countries are working to increase trade, industrialization, and technology, but ongoing political and religious conflicts threaten the region's stability, economic development, and environment.

Essential Questions

Section 1: The Economy

What does it mean for countries to be economically interdependent?

Section 2: People and Their Environment

What environmental challenges might countries with large populations face?

Visit glencoe.com and enter QuickPass™ code WGC9952C25 for Chapter 25 resources.

Ricksha drivers maneuver the crowded streets of Dhaka, Bangladesh, one of the fastest-growing cities in South Asia.

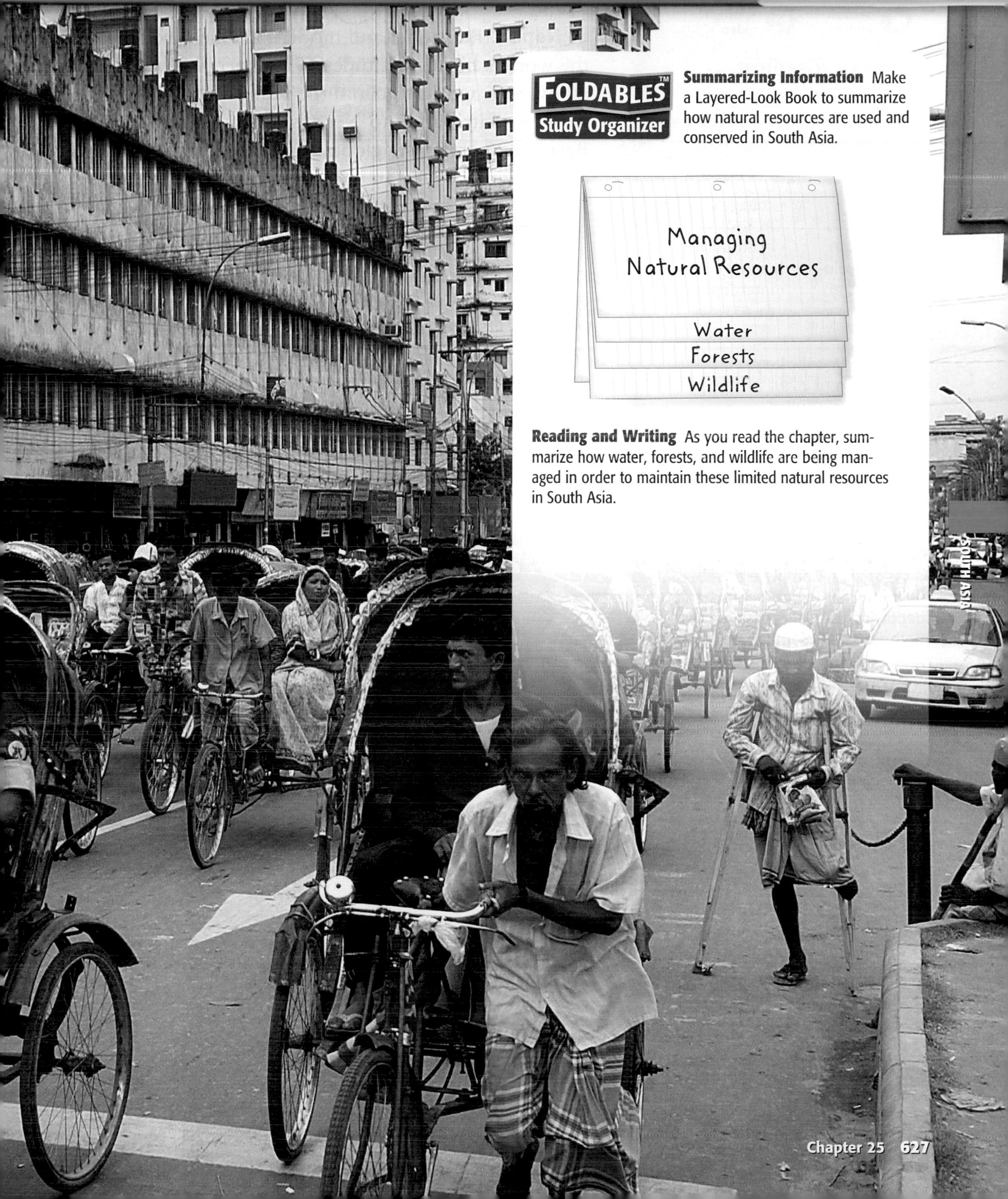

FOLDABLES™ Study Organizer

Summarizing Information Make a Layered-Look Book to summarize how natural resources are used and conserved in South Asia.

Reading and Writing As you read the chapter, summarize how water, forests, and wildlife are being managed in order to maintain these limited natural resources in South Asia.

WHY GEOGRAPHY MATTERS

India: Skilled Laborers Needed

The Problem:

The rapid growth in the number of India's private colleges has resulted in education programs that turn out underprepared employees who lack the basic skills required by industry.

Demand The number of Indian students enrolling in universities and colleges outnumbers the space available. These students attend St. Stephen's College, one of the colleges within the University of Delhi system.

Growth of Recognized Indian Educational Institutions

Year	Professional Colleges	Universities
1995–1996	1,354	226
2000–2001	2,223	254
2001–2002	2,409	272
2005–2006	17,625	348

Rapid Increase The number of colleges in India has exploded in the last ten years to satisfy the demand for education.

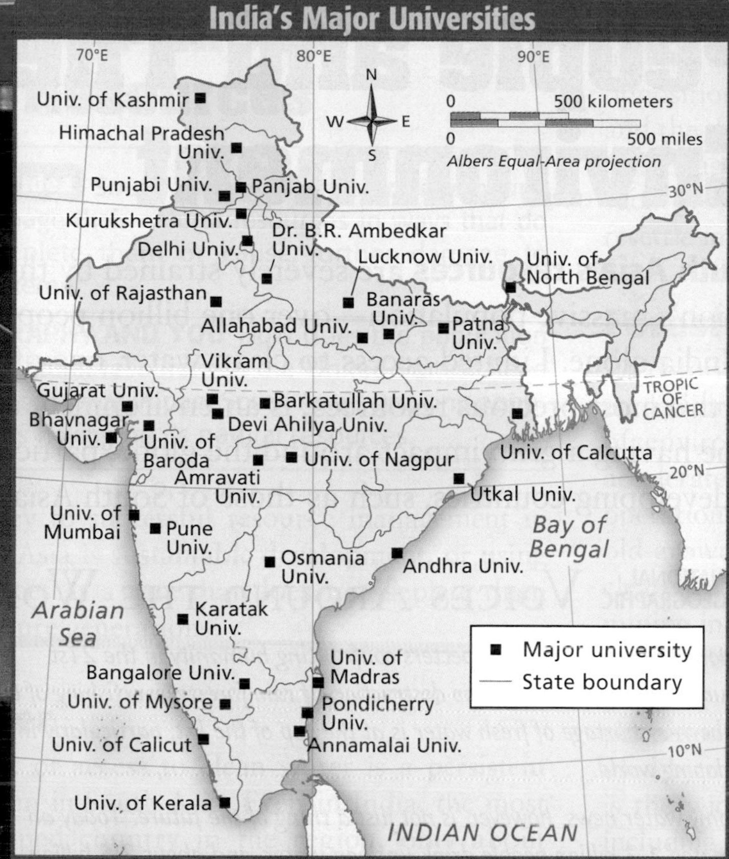

Education System

The need for highly skilled engineers to fill positions in India's growing high-tech industries is soaring. Although India's universities and colleges continue to graduate approximately 400,000 engineers a year, many are under-trained.

Why have skill levels of graduates decreased? The universities in India have not been able to graduate enough students to meet the need. Quotas and space limit the numbers and kinds of students that universities are able to accept. As a result, a number of private colleges have sprung up to educate a larger segment of the population. Unfortunately these new institutions often graduate students lacking in both English and technical skills.

How are private companies responding? To prepare new employees for the workforce, many companies have established internal training centers. These training centers specialize in providing graduates with job-specific skills, particularly communication and technical skills. New employees participate in these programs for anywhere from two to six months.

What does the future hold? The need for skilled engineers is unlikely to decrease soon. Private companies are lobbying for private investment, including foreign investment, in India's higher education system. Without better training, India may be faced with a workforce unable to fill the needs of high-tech industry.

One Solution:

Private companies are establishing training centers to fill the education gaps in India's workforce.

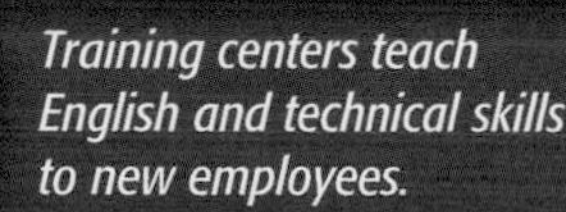

Training centers teach English and technical skills to new employees.

THINKING GEOGRAPHICALLY

1. **Human Systems** What action could the Indian government take to improve and standardize the education of university and college students?
2. **Environment and Society** What positive effects could internal and foreign investment have on India's higher education system? What negative effects could it have?

UNIT 9 East Asia

Traditional drummers, Seoul, South Korea

Why It Matters

East Asia and the United States are important trading partners. Many American companies manufacture goods in East Asia, and East Asia exports a variety of its own products to the United States. When you go shopping, notice the many items, ranging from cars and computers to clothing and furniture, that have been produced in East Asia or that are made of products exported from the region.

CHAPTER 26

PHYSICAL GEOGRAPHY OF East Asia

BIG Idea

Certain processes, patterns, and functions help determine where people settle. East Asia is a region of contrasts—from towering mountains in the west to fertile plains in the east, from subarctic forests in the north to tropical rain forests in the south. A study of the physical geography of East Asia will explain how these patterns and processes have affected human settlement in the region.

Essential Questions

Section 1: The Land

How has East Asia's location on the Ring of Fire affected the region's physical and human systems?

Section 2: Climate and Vegetation

What factors influence climate in East Asia?

Visit glencoe.com and enter *QuickPass*™ code WGC9952C26 for Chapter 26 resources.

Chinese farmers use terraced fields to grow crops such as rice.

Identifying Information Use a Six-Tab Book to identify definitions of the content vocabulary terms in the chapter.

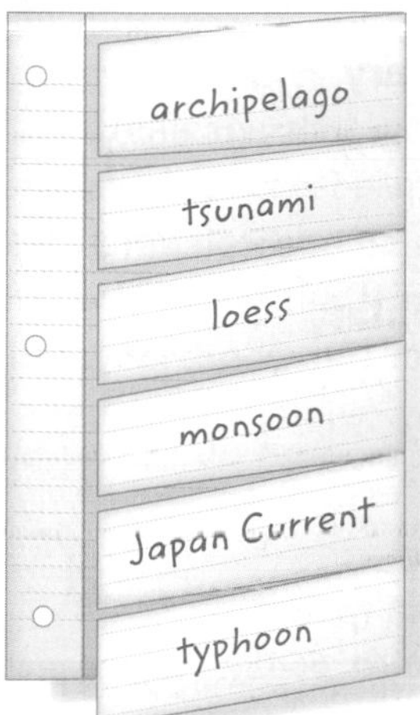

Reading and Writing As you read the chapter, write definitions and sentences using each term under the appropriate tab in your Foldable. Include a small sketch to help illustrate each word.

Peninsulas, Islands, and Seas

The physical map on page 654 shows that many peninsulas and islands dot the coast of East Asia. The **Korean Peninsula** juts southeast from China's Northeast Plain, separating the Sea of Japan (East Sea) from the Yellow Sea. The peninsula, home to North Korea and South Korea, consists mainly of mountains surrounded by coastal plains.

Tectonic activity is responsible for the creation of many of the region's islands and seas. For example, the process of subduction created the Sea of Japan millions of years ago. Tectonic activity also helped create other bodies of water, including the Yellow Sea, the East China Sea, and the **South China Sea.**

Four large, mountainous islands and thousands of smaller ones form the **archipelago** (AHR•kuh•PEH•luh•GO), or island chain, of Japan. Most of these islands were created by volcanic activity over millions of years. Honshū is the central and largest island, with Hokkaidō to the north and Kyūshū and Shikoku to the south. Most of Japan's major cities are located on the island of Honshū. Surrounding Japan are the Sea of Okhotsk on the north, the Sea of Japan and the East China Sea on the west, and the Philippine Sea on the south. To the east and southeast is the Pacific Ocean.

NATIONAL GEOGRAPHIC The Gobi

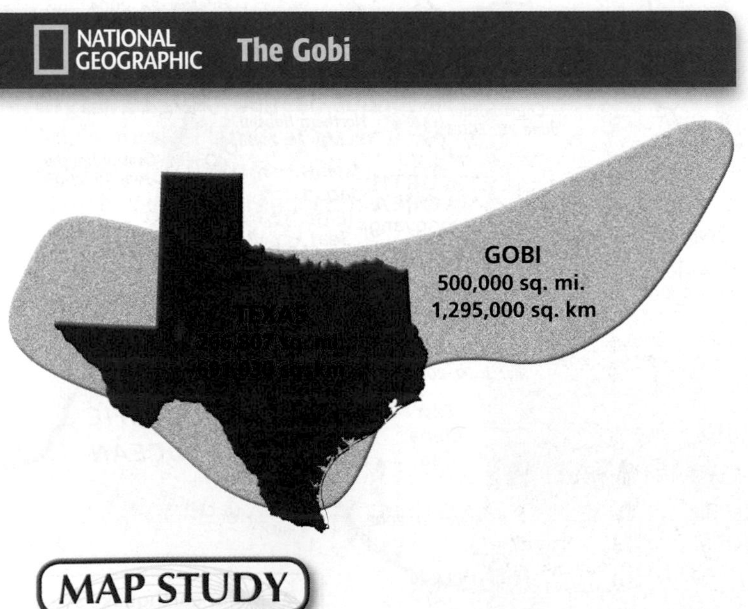

MAP STUDY

1. **Regions** How does the Gobi compare to the size of Texas?
2. **Place** Compare the Gobi to the Sahara, discussed in Chapter 17. How are these deserts similar? How are they different?

Mountains, Highlands, and Lowlands

Numerous mountain ranges fan out from an area of high peaks and deep valleys called the **Pamirs** in western China. The ranges that begin in this remote interior region include the Kunlun Shan and Tian Shan. (*Shan* is Chinese for "mountains.") Farther north, the Altay Shan form a natural barrier between Mongolia and China. To the south and west, the world's highest mountains, the **Himalaya,** separate China from South Asia. They include Mount Everest, the world's tallest peak at 29,028 feet (8,848 m), which spans the border between China and Nepal.

The Kunlun Shan bends to become the Qinling (CHIHN•LIHNG) Shandi, crossing central China from west to east. To the east, the lower Changbai Shan of Manchuria extend into the Korean Peninsula, to become the Northern Mountains.

Coastal plains surround the mountain interiors of Japan and Taiwan. Mount Fuji, at 12,388 feet (3,776 m), is a **dramatic,** cone-shaped volcano rising above the plains of Japan's island of Honshū. Also called Fujiyama, Mount Fuji is an important spiritual symbol to Japan's people.

The **Plateau of Tibet,** or Plateau of Xizang (SHEE•ZAHNG), in China's southwest corner, is East Asia's highest plateau region. Its average elevation is about 15,000 feet (4,600 m). Other rugged highlands stretch north and eastward at lower elevations. In the far north, the Mongolian Plateau's extensive highlands are mostly grassy pasture. The region's only extensive lowland areas are China's **Northeast Plain** and **North China Plain.**

Broad expanses of wasteland, including the deserts and salt marshes of the Tarim Basin, lie between the Kunlun Shan and Tian Shan. West of the Tarim Basin is the dry, sandy **Taklimakan.** To the northeast is another desert, the **Gobi,** whose frequent dust storms make life difficult in southern Mongolia and north-central China. Less than three inches of rain fall annually there.

READING Check **Regions** What physical process formed the islands of Japan and Taiwan?

Geography ONLINE

Student Web Activity Visit glencoe.com, select the *World Geography and Cultures* Web site, and click on Student Web Activities–Chapter 26 for an activity about tsunamis in East Asia.

Water Systems

MAIN Idea Landforms and physical processes have shaped East Asia's rivers, which provide transportation, water, and rich mineral deposits for fertile soils.

GEOGRAPHY AND YOU Can you name some of the major river systems in the United States? Read to learn what East Asia's river systems provide for the region and its people.

East Asia's rivers serve densely populated urban centers as transport routes. They provide hydroelectric power for energy, and the fertile soil in their basins is used for farming.

China's Rivers

China's major rivers begin in the Plateau of Tibet and flow eastward to the Pacific Ocean. The **Huang He** (HWAHNG HUH), or Yellow River, is northern China's major river system. This river is called "yellow" because it carries tons of fine, yellowish-brown topsoil called **loess** (LEHS), eroded from the western regions by wind and water. When deposited, the rich soil—along with water from the river—makes the North China Plain a major wheat and soybean farming area. Also called "China's sorrow," the Huang He often floods its banks, killing hundreds of thousands of people.

Central China's **Chang Jiang** (CHAHNG JYAHNG), or Yangtze River, is Asia's longest river at 3,965 miles (6,380 km). It flows through spectacular gorges and broad plains and empties into the ocean near Shanghai. The Chang Jiang, a major transport route, provides water for a large agricultural area where more than half of China's rice and other grains are grown. Many hope that construction of the Three Gorges Dam, a hydroelectric dam on the Chang Jiang, will put an end to flooding along the lower portions of the river. The dam will also make more water available for irrigation and provide hydroelectric power for China's growing population and economy.

The **Xi** (SHEE), also called the West River, is southern China's most important river system. Near the ports of Guangzhou (GWAHNG•JOH) and Macau (muh•KOW), the soil deposits of the Xi form a huge, fertile delta, one of China's fast-developing areas.

The world's longest artificial waterway, China's Grand Canal, was begun in the 400s B.C. Over the centuries, the canal has been expanded and rebuilt. Today, the Grand Canal moves people and goods along a 1,085-mile (1,746-km) course from Beijing to Hangzhou (HAHNG•JOH).

Rivers in Japan and Korea

In contrast to China's long rivers, the rivers of Japan and Korea are short and swift. They flow through mountainous terrain, often forming spectacular waterfalls. During the wet season, they provide hydroelectric power. The courses of many of Japan's rivers—such as the Shinano and Tone Rivers—have been altered for irrigation and to regulate water flow. Korea's chief rivers flow from inland mountains westward toward the Yellow Sea. The Han River flows through South Korea's capital, Seoul. In North Korea, the Yalu (or Amnok) River flows west, forming the border with China.

READING Check **Regions** How do the rivers in Japan and Korea compare with those in China?

The Three Gorges Dam is the world's largest hydroelectric dam.

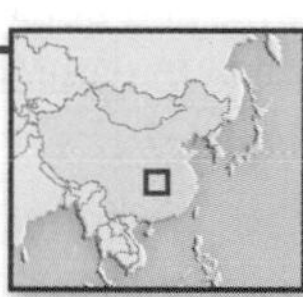

Location How will the Three Gorges Dam aid China?

EAST ASIA

CHAPTER 27 CULTURAL GEOGRAPHY OF East Asia

BIG Idea

Geography is used to interpret the past, understand the present, and plan for the future. East Asia's increasing participation in the global community, and the resulting diffusion of the region's cultures, continue to have a profound effect on the world. East Asia's growing and aging populations bring challenges for the future.

Essential Questions

Section 1: China

How might rural-to-urban migration affect the physical landscape in China?

Section 2: Japan

How might culture influence people's perceptions of Japan?

Section 3: North Korea and South Korea

How can two countries with similar histories move in very different directions today?

Visit glencoe.com and enter *QuickPass*™ code WGC9952C27 for Chapter 27 resources.

The Kiyomizu-dera Buddhist temple is one of the best-known sights of Kyōto, Japan.

FOLDABLES™ Study Organizer

Summarizing Information Make a Three-Tab Book to help you organize information about the culture of the countries in East Asia.

East Asian Culture

China	Japan	North Korea and South Korea

Reading and Writing As you read the chapter, use your Foldable to organize the main ideas about the cultural geography of each country under the appropriate tab. Be sure to include information about population patterns, history and government, and culture.

CONNECTING TO

THE UNITED STATES

New York City is home to one of the country's many Chinatowns.

Freedom Festival, Detroit, Michigan

Just the Facts:

- There are 13.1 million Asian Americans living in the United States today.
- In 2006 Asia supplied about 60 percent of the technology imports to the United States.
- There are approximately 25 Chinatowns located throughout the United States in various cities.
- In 2005 China surpassed Canada as the number 1 exporter of goods to the United States.
- In 2005 basketball player Yao Ming received a record 2,558,278 votes for the NBA All-Star team.

Asian Americans by Origin

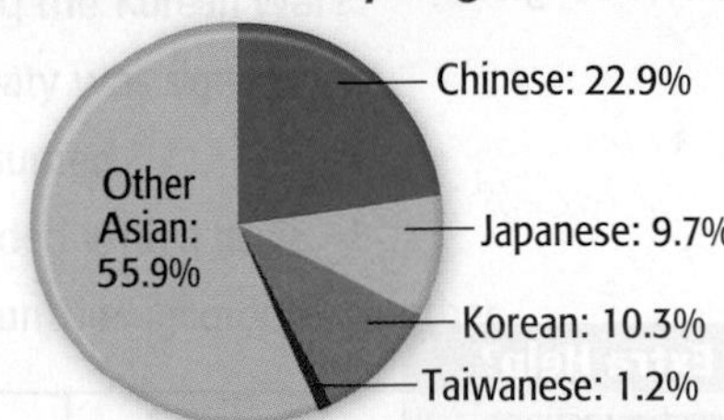

Source: www.census.gov

NATIONAL GEOGRAPHIC

Food	
sushi	Japan
green tea	China
Kimchi	Korea
Kung Pao chicken	China

Making the Connection

The connections between East Asia and the United States are obvious in everything from the cars we drive to the food we eat. Immigration and trade continue to bring the countries of this region in closer contact with the United States.

Food Immigrants brought East Asian cuisine to the United States. The food traditionally includes lots of vegetables and fish and is frequently served with steamed rice. The cuisine served in U.S. restaurants is often Americanized, making use of fewer vegetables and more red meat and sauces.

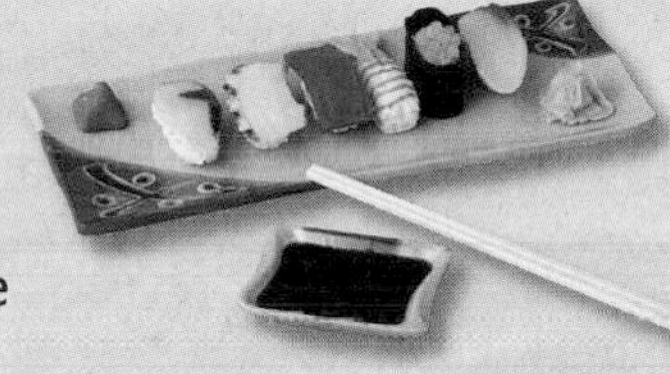

Sushi, a traditional Japanese dish, is enjoyed by many Americans.

Chinatowns During the 1840s and 1850s, the United States experienced its first major Chinese immigration. Most newcomers worked in the California gold mines and railroad construction. These immigrants were often discriminated against by other American citizens because of differences in culture and language. As a result, the Chinese often settled together in what have become known as Chinatowns. Many of these were located on the West Coast, especially in California.

Going Out With a Bang Simple fireworks were made as early as the Chinese Han dynasty (206 B.C.) by roasting bamboo, which made a loud sound. Daoist monks later played with gunpowder, creating more modern fireworks. Today, across the United States, large fireworks displays are used to commemorate holidays such as Independence Day.

Several different types of high-tech devices such as computers and MP3 players are produced in East Asian countries for sale in the United States and throughout the world.

THINKING GEOGRAPHICALLY

1. **Human Systems** What are some of the technologies and products coming from East Asia that you encounter every day? How important are these things to your life?
2. **Places and Regions** Why do you think more Asian Americans in the United States come from China than from any other East Asian country?

CHAPTER 28 THE REGION TODAY

East Asia

BiG Idea

Economic systems shape relationships in society. East Asia is experiencing rapid economic changes. Agriculture continues to play a major role in the region's economy, but industry and trade are becoming more important, especially in China, which is undergoing a dramatic economic transformation. The impact of this transformation is global, as China becomes more involved in the global economy and faces enormous environmental problems.

Essential Questions

Section 1: The Economy

How might the move toward a global economy impact countries in East Asia?

Section 2: People and Their Environment

How has rapid industrialization affected the environment in East Asia?

Geography ONLINE

Visit glencoe.com and enter *QuickPass*™ code WGC9952C28 for Chapter 28 resources.

Victoria Harbor—home to most of Hong Kong's port facilities—holds traces from the past, such as the traditional Chinese junk at the right.

Summarizing Information Make a Folded Chart to summarize current environmental issues in East Asia and to make predictions about future environmental issues in the region.

Environmental Issues	Present	Future
Land		
Water		
Air		

Reading and Writing As you read, summarize the present environmental problems and their impact on East Asia's land, water, and air. Then choose a specific point in the future and make predictions about the condition of the region's environment at that point in time. Use details from the chapter to support your predictions.

CHINA'S GROWING ENERGY DEMANDS

The Problem:

The industrialization of China has led to an increased need for energy resources. Access to affordable energy is required for China's continued economic growth.

China's Economy These workers are making computer chips in one of China's high-tech factories. Economic activities, such as manufacturing, agriculture, and construction, are important to China's economy but strain energy resources.

Traffic jam in Shanghai

Energy Consumption Automobile ownership in China has increased with the rising incomes of Chinese workers, intensifying the demand for gasoline. Car ownership grew by 34 percent between 2005 and 2006.

Daily Oil Production and Consumption in China, 1993–2007

Thousands of barrels: 0, 1,000, 2,000, 3,000, 4,000, 5,000, 6,000, 7,000, 8,000

Oil Production
Oil Consumption

Year: 1993, 1995, 1997, 1999, 2001, 2003, 2005, 2007

Source: www.eia.doe.gov, Energy Information Administration.

A Growing Strain on Energy Resources

China is the world's second-largest energy consumer after the United States. China's demand for energy is expected to grow at a rate of about 5 percent per year, effectively doubling the country's energy consumption by 2020. Even though it is currently the third-largest oil-producing country in the world, consumption has long outgrown its production. Now China must look for new ways to fuel its growing energy needs.

What has caused China's increase in energy consumption? China's economy continues to grow rapidly as the country industrializes. With an increase in manufacturing comes an increase in energy consumption. Growing car ownership and increased air travel across China also contribute to the problem. As China's consumption and needs have grown, the cost of oil on the world market has increased dramatically. Finding a less expensive supply of oil has become imperative for China.

What is being done to cope with increased energy demand? China has initiated efforts to meet increasing energy needs. Some of the policies undertaken include: increasing exploration to find new oil reserves in China; expanding the use of other energy resources, such as nuclear power and renewable resources; promoting energy conservation; and investment in energy-efficient technologies.

One Solution:

China is heavily dependent on imported energy resources and could increase exploration for new oil reserves within the country's own borders.

Oil rigs drill for oil in China.

THINKING GEOGRAPHICALLY

1. **Environment and Society** Use the Internet to research China's use of nuclear power and renewable sources of energy. What percentage of China's electricity is currently generated by nuclear power?
2. **Places and Regions** What are some of the potential social, economic, and environmental problems of rapid industrialization, such as is happening in China?

UNIT

10 Southeast Asia

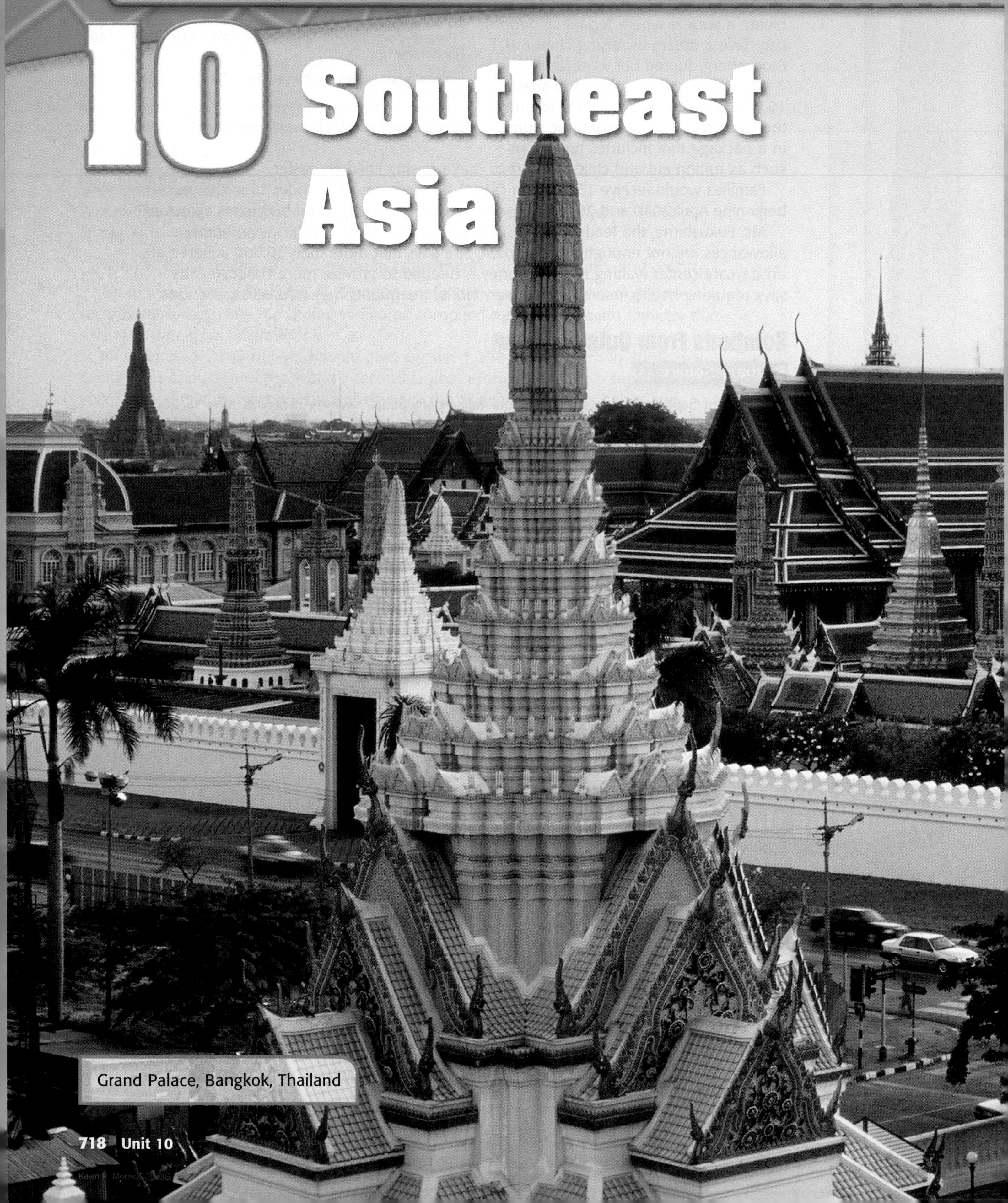

Grand Palace, Bangkok, Thailand

Why It Matters

Southeast Asia is a vital crossroads of trade and commerce. The region is rich in natural resources such as tin, petroleum, rubber, tea, spices, and valuable woods. In recent years, many Southeast Asians have migrated to the United States, bringing their own religions and cultures with them. You are probably familiar with the flavors of many Southeast Asian dishes, available now in restaurants in the United States.

Southeast Asia

PHYSICAL GEOGRAPHY Southeast Asia has two areas, which are distinct physically. The larger section, the mainland, is wedged on the edges of South Asia and East Asia. It is a region of several mountain ranges that run from north to south with plateaus or valleys between. The land descends to coastal plains that include the rich deltas of five rivers.

The islands of Southeast Asia form an arch curving from the southwest to the southeast of the mainland. Most of its hundreds of islands belong to the countries of Indonesia or the Philippines. Formed by the collision of some of Earth's tectonic plates, these islands have many active volcanoes.

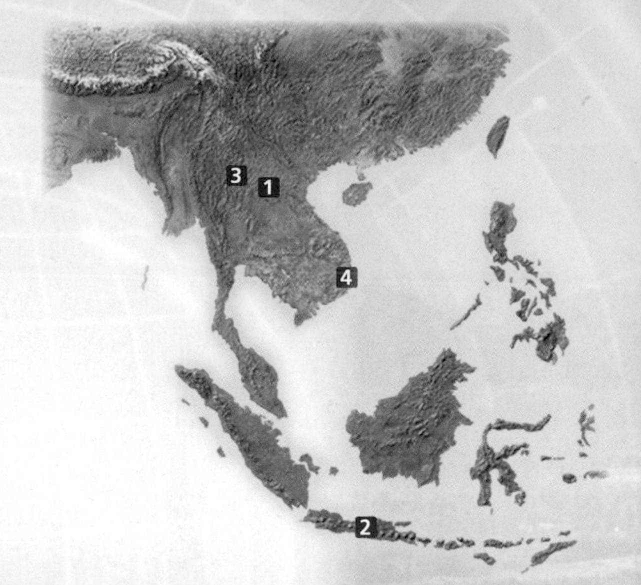

1 PLAINS AND PLATEAUS
Hollowed-out stones, once used for storage, litter a plateau in Laos. The uplands of this region are often lined by mountains.

2 MOUNTAINS Java's Gunung Merapi is aptly named—the name means "Fire Mountain." This volcano is among the most active in the world.

3 LAKES AND RIVERS The Mekong River flows along the border of Laos and Thailand before entering Vietnam and forming a large delta. It is the longest of the region's five major rivers.

3

4

4 NATURAL RESOURCES Workers in Vietnam use rakes to gather salt that has precipitated from seawater.

SOUTHEAST ASIA

CHAPTER 29

PHYSICAL GEOGRAPHY OF Southeast Asia

BIG Idea

The characteristics and distribution of ecosystems help people understand environmental issues. A study of the physical geography of Southeast Asia will explain its beginnings, its natural barriers of mountains and water, its tempestuous volcanoes, and its abundant natural resources.

Essential Questions

Section 1: The Land

How has Southeast Asia's location affected the region's physical features?

Section 2: Climate and Vegetation

How do you think climate influences ecosystems in Southeast Asia?

Visit glencoe.com and enter ***QuickPass***™ code WGC9952C29 for Chapter 29 resources.

Temple ruins dot the plains at Bagan (Pagan), Myanmar.

Identifying Information Create a Trifold Book to identify information about the physical geography of Southeast Asia. On one side of the Foldable, label the sections *Mountains, Volcanoes,* and *Rivers.*

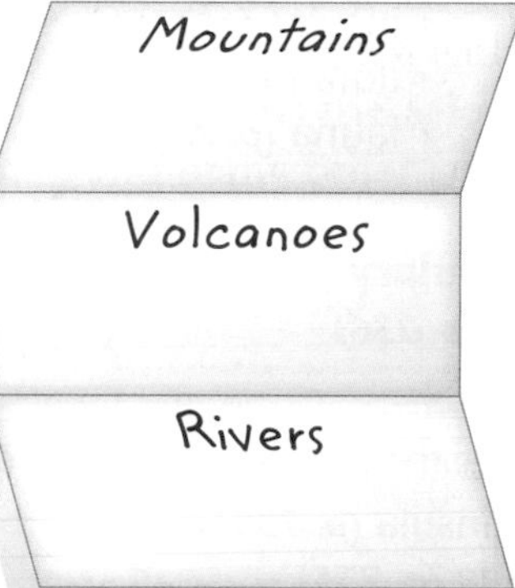

Reading and Writing As you read the chapter, write descriptions of these physical features and name specific examples of each.

CHAPTER 30

CULTURAL GEOGRAPHY OF Southeast Asia

BIG Idea

The movement of people, goods, and ideas causes societies to change over time. Southeast Asia's location at the crossroads of vital trade routes, as well as its culturally diverse population, make it an important part of the global community.

Essential Questions

Section 1: Mainland Southeast Asia
What leads people to divide areas of land to create new countries?

Section 2: Island Southeast Asia
How might outside cultures influence the indigenous culture of a region or country?

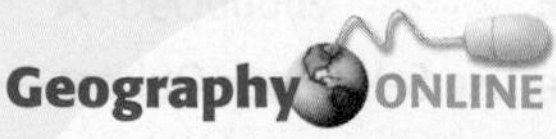

Geography ONLINE
Visit glencoe.com and enter *QuickPass*™ code WGC9952C30 for Chapter 30 resources.

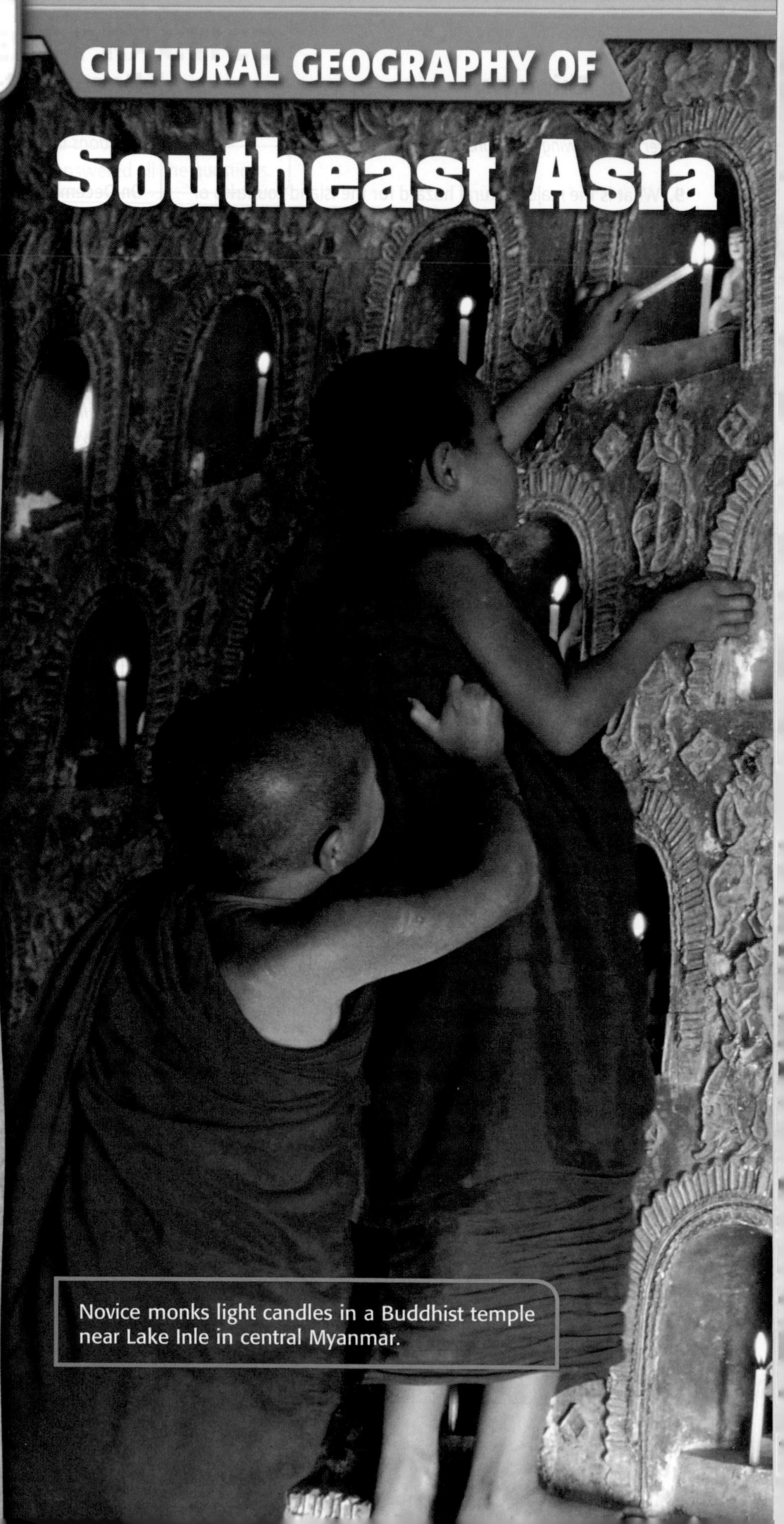

Novice monks light candles in a Buddhist temple near Lake Inle in central Myanmar.

FOLDABLES™ Study Organizer

Summarizing Information Make a Two-Tab Book to help you summarize information about the cultures of mainland and island Southeast Asia.

Southeast Asia

Mainland Countries | Island Countries

Reading and Writing As you read the chapter, write notes in your Foldable about the population patterns, history and government, and culture of each subregion.

CONNECTING TO

THE UNITED STATES

Dancers at a Philippine Independence Day celebration in New York City

Just the Facts:

- Filipino Americans make up the third-largest Asian American group in the United States after Indian Americans and Chinese Americans.
- St. Paul, Minnesota, has the largest urban Hmong population in the world.
- California, Texas, and Minnesota are the states with the greatest Southeast Asian populations.
- Southeast Asian immigration is recent, the majority occurring after the Vietnam War.
- Tagalog, one of the languages spoken in the Philippines, is the sixth-most-spoken language in the United States.

Southeast Asian American Population	
Filipino Americans	2,366,501
Vietnamese Americans	1,464,611
Cambodian Americans	203,719
Hmong Americans	192,575
Laotian Americans	181,963
Thai Americans	147,599
Indonesian Americans	59,304
Malaysian Americans	13,693

Source: U.S. Census Bureau.

Making the Connection

Southeast Asia has had a long and often tumultuous history with the United States. The 1960s and 1970s largely consisted of conflict between the two regions. Today trade, diplomacy, and immigration to the United States have established an alliance that has been strengthened since the early 1990s.

War Between 1959 and 1975, the Vietnam War pitted the United States and the South Vietnamese against the North Vietnamese and the National Liberation Front (NLF). Many lives were lost on both sides.

Immigration—Little Cities After the end of the Vietnam War, immigration from Southeast Asia increased. Little Saigon is the name given to a number of Vietnamese communities in the United States, named after the capital of the former South Vietnam. The largest Little Saigons in the United States are in Orange County, California; Houston, Texas; and San Jose, California. Cambodians have also built strong communities in California, Massachusetts, Minnesota, Washington, Texas, and Florida.

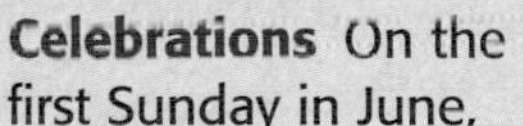

Pad Thai, a dish from Thailand, is popular in the United States.

Celebrations On the first Sunday in June, New York City holds a large parade celebrating Philippine Independence. Los Angeles's Vietnamese community holds a Tet Festival yearly to celebrate the Vietnamese New Year.

Food Southeast Asian cuisine is growing in popularity in the United States. Thai food is perhaps the most popular.

Little Saigons bring Southeast Asian culture to heavily populated Vietnamese communities in several large U.S. cities.

Actress and singer Vanessa Anne Hudgens is of Filipino and Chinese descent on her mother's side.

THINKING GEOGRAPHICALLY

1. **Human Systems** Describe how the Vietnam War affected Southeast Asian immigration to the United States.
2. **The Uses of Geography** How might California's location and history account for its large Southeast Asian population?

CHAPTER 31

THE REGION TODAY

Southeast Asia

BIG Idea

Geography and the environment play an important role in how a society is shaped over time. Rich in natural resources, Southeast Asia is a vital crossroads of trade. As the region becomes more urbanized and uses its natural resources to industrialize, it faces a variety of environmental problems.

Essential Questions

Section 1: The Economy

How do Southeast Asia's mountainous terrain, volcanic soil, and tropical climate influence economic activities in the region?

Section 2: People and Their Environment

How might humans affect their physical environment?

Geography ONLINE

Visit glencoe.com and enter **QuickPass**™ code WGC9952C31 for Chapter 31 resources.

Singapore Harbor is the world's busiest port in terms of total shipping tonnage.

Organizing Information Use a Shutter Fold to list the causes and effects of pollution and environmental problems in Southeast Asia.

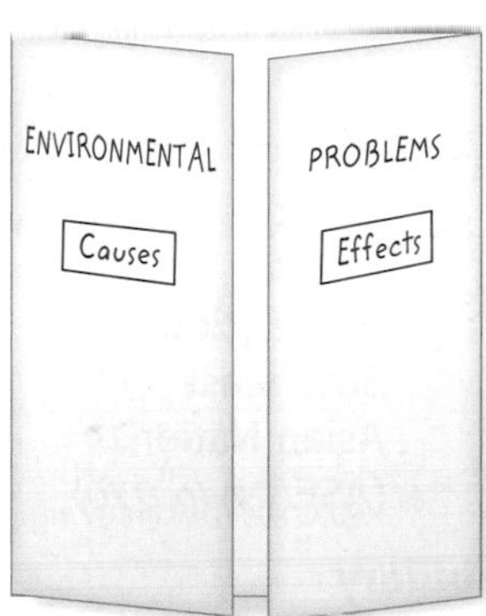

Reading and Writing As you read the chapter, record notes in your Foldable describing land, air, and water pollution in urban and rural Southeast Asia. Use this information to discuss how pollution spreads throughout an environment.

Story of a Volcano Mount Pinatubo

Path of Destruction

Disaster In June 1991, the eruption of Mount Pinatubo, located in the Philippines, was one of the most destructive and violent volcanic eruptions of the twentieth century.

The eruption of Mount Pinatubo caused incineration and destruction up to 11 miles (17 km) away, and 300 people were killed as a result. The summit of Mount Pinatubo was 5,725 feet (1,745 m) above sea level. Following the eruption, which created a large, basin-like depression in the summit, the peak of Mount Pinatubo was reduced to 4,872 feet (1,485 m).

A school covered in volcanic ash months after the eruption

What were the regional effects? The fallout from the eruption of Mount Pinatubo affected more than 212,511 acres (86,000 ha) of farmland and fishponds, rendering much of the land infertile. Approximately 100,000 people were left homeless. Health-care facilities, airports, schools, and a nearby U.S. air base were damaged and closed.

NATIONAL GEOGRAPHIC

Mount Pinatubo: Ashfall

110°E
120°E
CHINA
20°N
Gulf of Tonkin
Luzon
Philippine Sea
South China Sea
Mt. Pinatubo
PHILIPPINES
10°N
Ash fall
N
W
E
S
0 300 kilometers
0 300 miles
Miller Cylindrical projection
Sulu Sea
Celebes Sea

A Successful Evacuation

The Mount Pinatubo evacuation was the most successful in history. Prior to its eruption, effective monitoring of the volcano enabled authorities to evacuate 60,000 Filipinos and 18,000 American military workers and their families at a nearby air base. The evacuation saved thousands of lives.

This village near Mount Pinatubo was destroyed by a mud slide. Mud slides are common in the aftermath of a volcanic eruption.

Eight months after the eruption, the air quality was still bad, and much work was still being done to clean up the area.

What were the global effects? When Mount Pinatubo erupted, a gas of fine particles called aerosols was released into the upper atmosphere. The haze lasted for three years following the eruption. Global temperatures dropped by an estimated 0.9°F (0.5°C), and the amount of sunlight reaching Earth was reduced.

THINKING GEOGRAPHICALLY

1. **Environment and Society** Think about the devastating effects of natural disasters on a region—particularly the effects on agricultural systems, health-care facilities, and education. How can the loss of an infrastructure devastate a country for years following a disaster?
2. **Human Systems** Research the Mount Pinatubo evacuation in more depth. What methods were used to safely evacuate people?

Understanding the Case

The primary sources below provide information about the environmental effects of Agent Orange used during the Vietnam War. Use these resources, along with what you have learned in Unit 10, to complete the activities on the next page.

Primary Sources

Health Impacts

Primary Source 1

Excerpt from "Agent Orange: Birth defects plague Vietnam; U.S. slow to help," by Jason Grotto, *Chicago Tribune*, December 8, 2008.

The people of Vietnam blame many health problems on exposure to Agent Orange.

"I was about 16 when I saw the planes flying overhead, and I saw the spraying [of Agent Orange and other defoliants] until I was married," said [Dao Thi] Kieu, 58. "It smelled like ripe guava. No trees could survive. It made my clothes wet."

Her vivid memories are supported by data from spraying missions analyzed by the Tribune, which show at least seven sorties that dispensed nearly 13,000 gallons of defoliants passed over Kieu's fields.

Since then, the story of Kieu's life can be told with simple, heartbreaking math. She had eight children. Seven of them were born with severe deformities. Of those, five died before age 8. She also lost her husband, who served in the U.S.-backed South Vietnamese army, to cancers associated with herbicide exposure.

Decades after the Vietnam War ended, the most contentious question surrounding the use of defoliants by the U.S. military is the impact on the health of untold numbers of Vietnamese.

At the heart of the controversy is the suspected link between the herbicides and birth defects in Vietnam, where more than 5 out of every 100 children are born with some form of physical or mental abnormality, a fourfold increase since the start of the war, according to Vietnamese scientists.

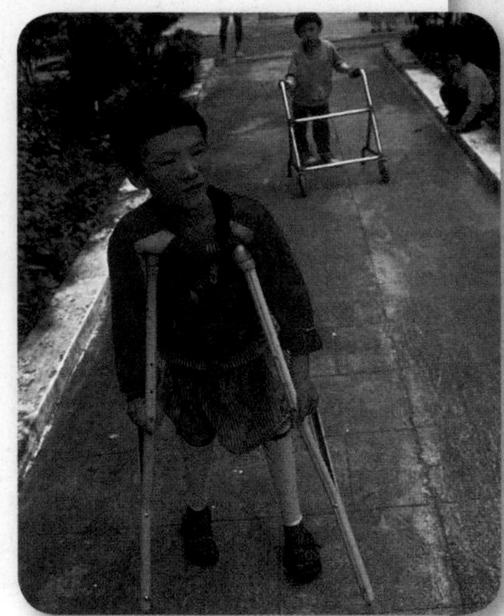

A young Vietnamese victim of Agent Orange

The Environment

Primary Source 2

Excerpt from "US, Vietnam face Agent Orange legacy," by Ian Timberlake, Agence France-Presse, September 23, 2009.

Agent Orange was used to kill vegetation in order to deprive the enemy of a hiding place and food supplies during the war. However, its effects lasted long after the war ended.

. . . US and Vietnamese officials have identified the old US bases in Danang, Bien Hoa—near the former Saigon—and Phu Cat as significant "hotspots" where spillage, washing of aircraft and loading of the herbicides contributed to contamination.

At Danang airport now, dioxin levels are still 300–400 times higher than internationally accepted levels. . . .

Almost two years ago Vietnamese officials, assisted by the US, installed a concrete cap over the former Agent Orange mixing and loading area and improved drainage and filtering of lake sediment inside the Danang airbase.

Authorities also banned people from eating fish or other foods from lakes on the property.

These temporary measures have prevented contamination from spreading, officials from both sides said. . . .

Without further action, contaminated material at the hotspots will continue to be dispersed through soil particles as well as water currents, wildlife and air, [Koos Neefjes, an adviser on dioxin to the United Nations in Hanoi] told annual US-Vietnamese Agent Orange talks this month.

Dioxin can be passed through the food chain via fish or foul.

Health of American Veterans

Primary Source 3

Excerpt from "Report Sees Agent Orange Link to More Illnesses," by Janie Lorber, *The New York Times,* July 25, 2009.

Like the Vietnamese, American soldiers who were exposed to Agent Orange have suffered long-enduring effects, as have their families.

An expert panel reported on Friday that two more diseases may be linked to exposure to Agent Orange, a defoliant used by the American military during the Vietnam War.

People exposed to the chemical appear, at least tentatively, to be more likely to develop Parkinson's disease and ischemic heart disease, according to the report. . . .

The results, though not conclusive, are an important first step for veterans groups working to get the government to help pay for treatment of illnesses they believe have roots on the battlefield. Some other conditions linked to Agent Orange already qualify.

Agent Orange Registry Statistics

Agent Orange Registry Examinations	Number of Vietnam Veterans
Initial Exams	454,069
Follow-Up Exams	52,115
Total Female Exams	8,791
Total Exams	506,184

Note: Female examinations number included in initial and follow-up exam numbers.
Source: Agent Orange Review (August 2008), U.S. Department of Veterans Affairs.

Claud Tillman, a 61-year-old veteran from Knoxville, Tenn., who lost his job repairing guns after he received a diagnosis of Parkinson's disease, said those benefits could help dig him out of tens of thousands of dollars in debt.

Mr. Tillman has not worked since March 2007 and now lives on loans from relatives, including his son. "It sure has messed my life up," said Mr. Tillman, who said he was sure he became ill after exposure to Agent Orange while serving in Vietnam. "I don't know how to explain it. It won't be long till I'm living under a bridge. I am confident that that's where it came from, but there's no way to prove it."

Analyzing the Case

1. **Comparing and Contrasting** Review the information in the primary sources above. Compare and contrast Agent Orange and nuclear weapons.
2. **Making Predictions** Can the environment in Vietnam be cleaned up so that Agent Orange is no longer a threat to the health of the Vietnamese people? Explain your answer.
3. **Conducting a Debate** Use the following questions to initiate a class debate:
   - How would chemical herbicides help military forces to fight a war?
   - What effects do they have on the soldiers?
   - What effects do they have on civilians?
4. **Writing About the Case** Write a one-page essay in which you answer one of the questions in #3 above.

UNIT 11 Australia, Oceania, and Antarctica

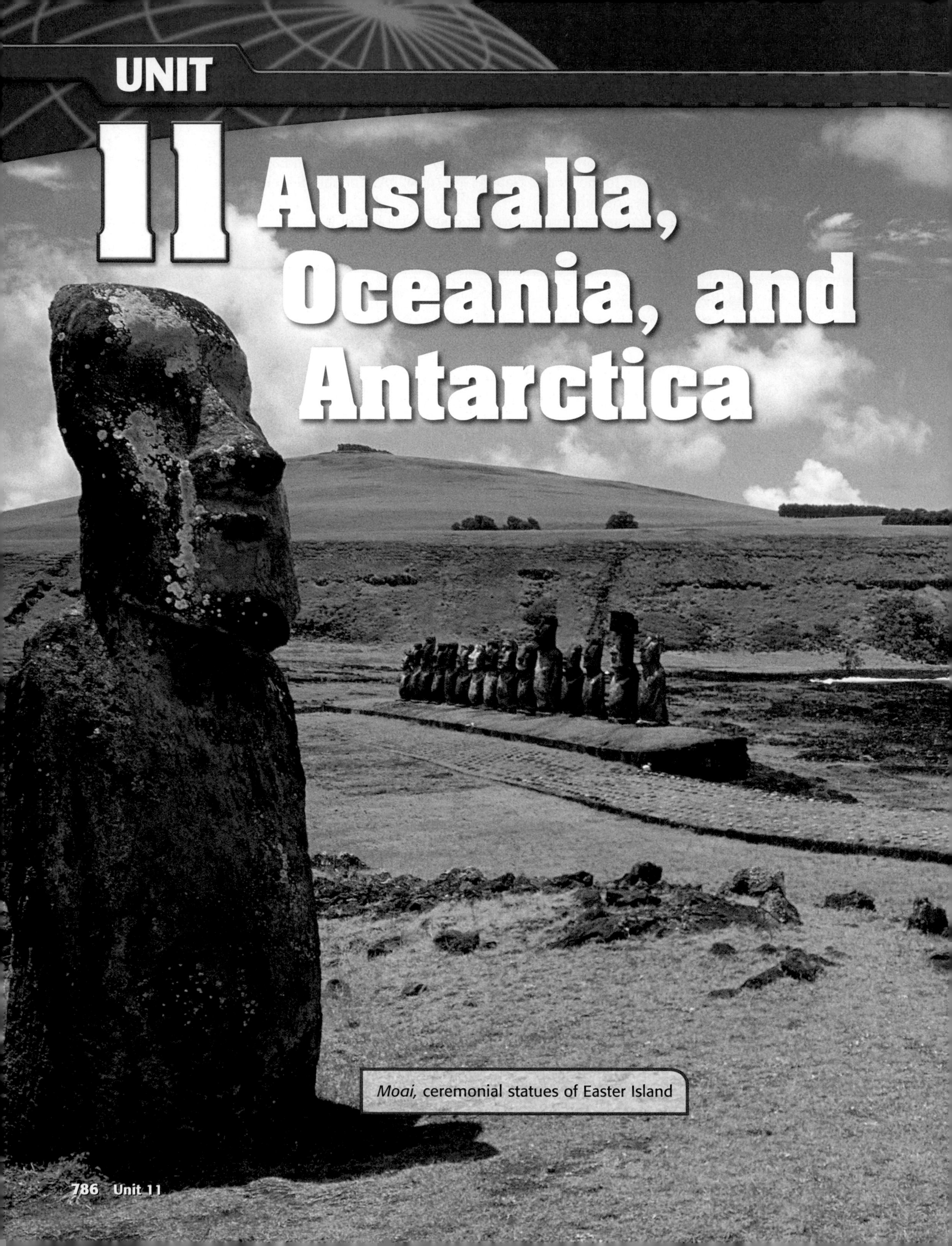

Moai, ceremonial statues of Easter Island

NGS ONLINE To learn more about Australia, Oceania, and Antarctica visit www.nationalgeographic.com/education.

Why It Matters

Vast and sparsely populated, the region of Australia, Oceania, and Antarctica is perhaps the most diverse of the world's regions. Parts of the region—Australia and Oceania—are developing close economic ties to other countries in the Pacific Rim, the area bordering the Pacific Ocean. Such ties to prosperous Pacific Rim nations will influence global trade and trading networks for decades to come. Cold, icy Antarctica lacks a permanent human population, but the data being gathered there by scientists will broaden your understanding of the world's climates and resources in the years ahead.

PHYSICAL Australia and Oceania

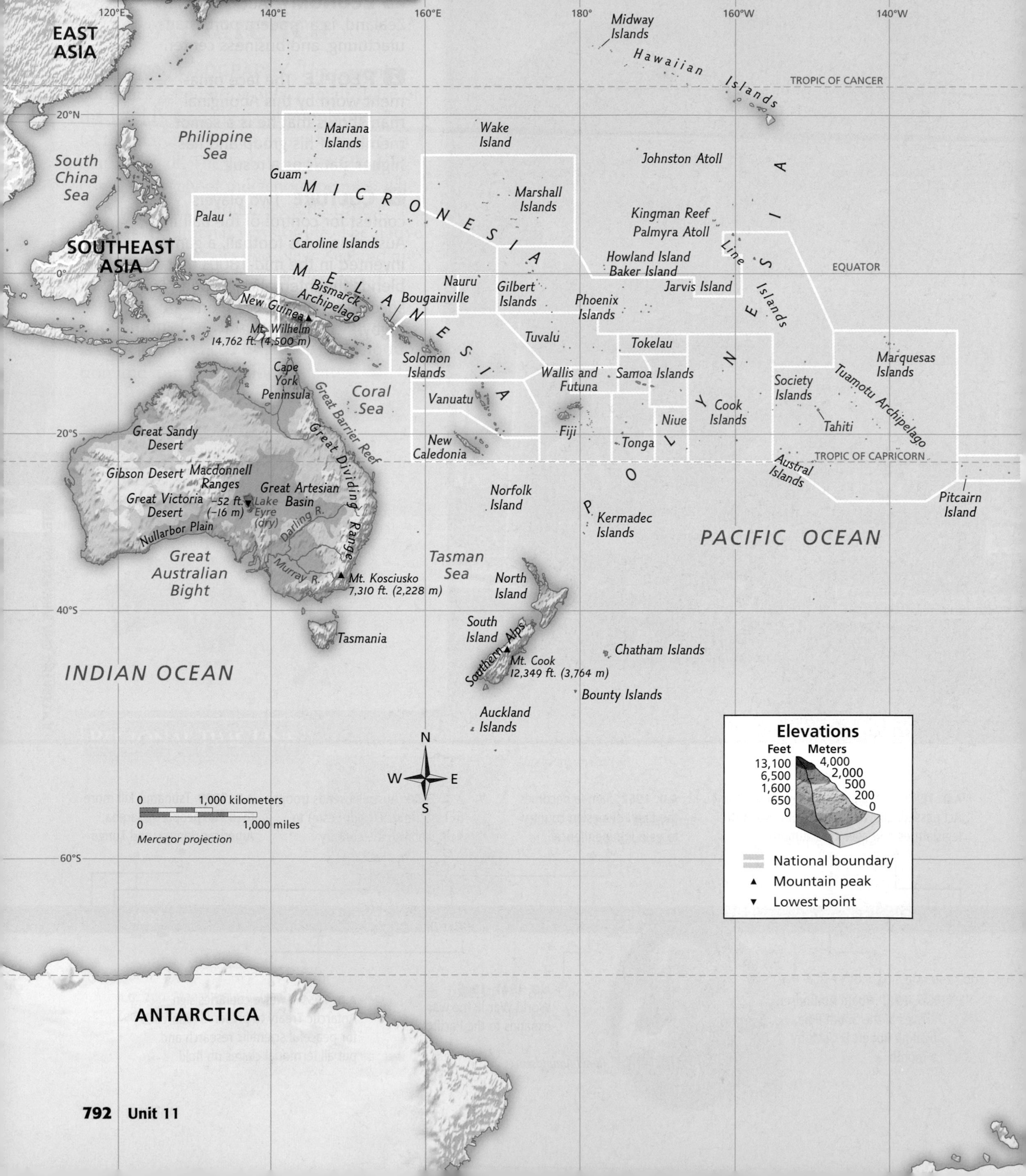

120°E
140°E
160°E
180°
160°W
140°W
20°N
0°
20°S
40°S
60°S
TROPIC OF CANCER
EQUATOR
TROPIC OF CAPRICORN
EAST ASIA
SOUTHEAST ASIA
South China Sea
Philippine Sea
Mariana Islands
Guam
Palau
Caroline Islands
MICRONESIA
MELANESIA
POLYNESIA
Midway Islands
Hawaiian Islands
Wake Island
Johnston Atoll
Marshall Islands
Kingman Reef
Palmyra Atoll
Howland Island
Baker Island
Jarvis Island
Line Islands
Nauru
Gilbert Islands
Phoenix Islands
Bismarck Archipelago
New Guinea
Bougainville
Mt. Wilhelm 14,762 ft. (4,500 m)
Tuvalu
Tokelau
Solomon Islands
Wallis and Futuna
Samoa Islands
Marquesas Islands
Society Islands
Tuamotu Archipelago
Tahiti
Cape York Peninsula
Coral Sea
Vanuatu
Fiji
Niue
Cook Islands
Tonga
New Caledonia
Austral Islands
Pitcairn Island
Great Barrier Reef
Great Sandy Desert
Gibson Desert
Macdonnell Ranges
Great Artesian Basin
Great Victoria Desert
–52 ft. (–16 m)
Lake Eyre (dry)
Darling R.
Great Dividing Range
Nullarbor Plain
Great Australian Bight
Murray R.
Mt. Kosciusko 7,310 ft. (2,228 m)
Norfolk Island
Kermadec Islands
PACIFIC OCEAN
Tasman Sea
North Island
South Island
Southern Alps
Mt. Cook 12,349 ft. (3,764 m)
Tasmania
Chatham Islands
Bounty Islands
Auckland Islands
INDIAN OCEAN
ANTARCTICA
N
W
E
S
0 1,000 kilometers
0 1,000 miles
Mercator projection

Elevations
Feet: 13,100; 6,500; 1,600; 650; 0
Meters: 4,000; 2,000; 500; 200; 0
National boundary
▲ Mountain peak
▼ Lowest point

Low, Dry, and Isolated

Each part of the region is isolated from other regions of the world, and Australia and Antarctica are the two driest continents. As you study the maps and graphics on these pages, look for the geographical features that make the region unique. Then answer the questions below on a separate sheet of paper.

1. What features on the map show that Australia has a dry climate? What parts of that continent might be less dry?
2. Use the drawings below to explain how many of the Pacific islands were formed.

The Birth of an Island

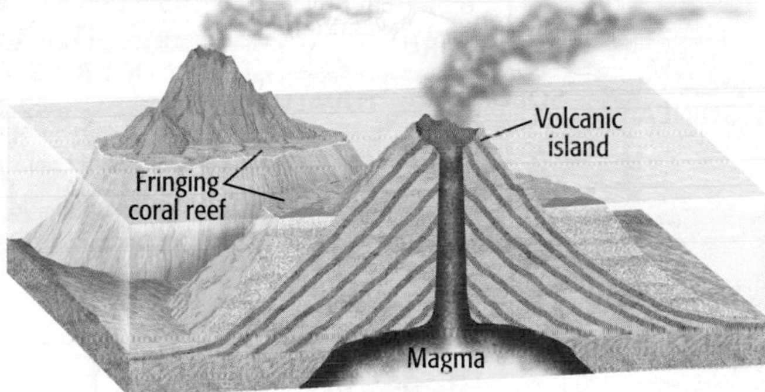

1. Coral atolls begin as volcanoes surrounded by coral reefs.

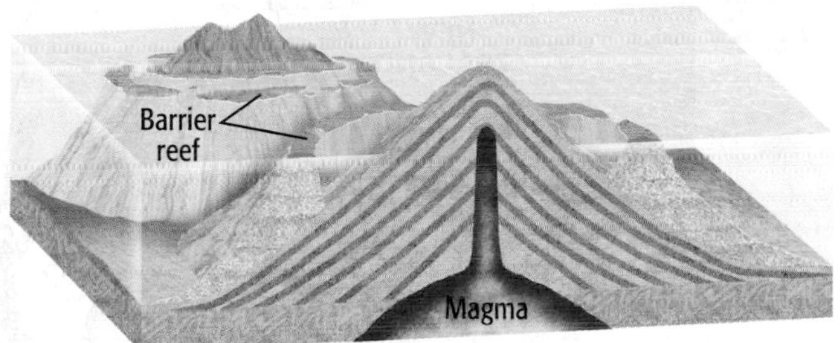

2. The extinct volcano erodes away. The coral reef expands to become a larger barrier reef.

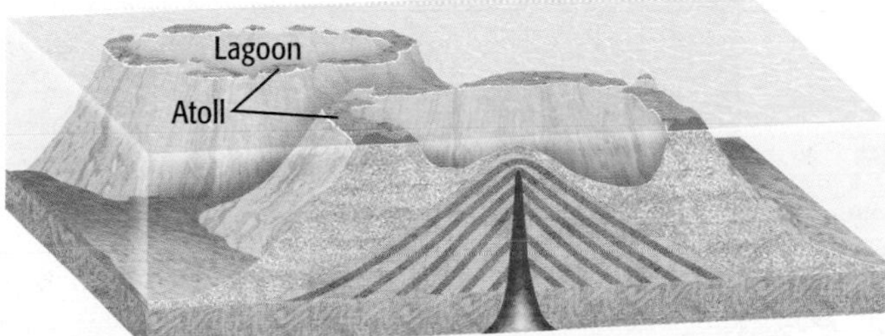

3. Eventually, all that remains is a coral atoll surrounding a lagoon.

CHAPTER 32

PHYSICAL GEOGRAPHY OF Australia, Oceania, and Antarctica

BIG Idea

Geography and the environment play an important role in how a society is shaped over time. The physical geography of Australia, Oceania, and Antarctica includes extreme differences—from the dry Australian Outback to the volcanic islands of Oceania to the cold ice caps of Antarctica. Each of these subregions offers unique opportunities for economic growth, tourism, and scientific research.

Essential Questions

Section 1: The Land

How might the physical geography contribute to the unique character of the region of Australia, Oceania, and Antarctica?

Section 2: Climate and Vegetation

How does the climate of the region affect the lives of people in Australia and Oceania?

Visit glencoe.com and enter QuickPass™ code WGC9952C32 for Chapter 32 resources.

Palau's Rock Islands are limestone coral reefs lifted above sea level.

FOLDABLES™ Study Organizer

Summarizing Information Create a Three-Tab Book to summarize information about the physical geography of Australia, Oceania, and Antarctica. Write the name of a subregion on each of the three tabs.

Physical Geography

Australia | Oceania | Antarctica

Reading and Writing As you read the chapter, write notes under the appropriate tab about the physical geography of the three subregions discussed in this chapter.

CHAPTER 33

CULTURAL GEOGRAPHY OF Australia and Oceania

BIG Idea

The movement of people, goods, and ideas causes societies to change over time. The geography and climates of Australia and Oceania have drawn people from great distances. Migrations of people from island to island made it a varied and fascinating region. Later, European and Japanese colonization reshaped the region. Today, international travel has made the region more accessible than ever.

Essential Questions

Section 1: Australia and New Zealand

How did the migration and settlement of foreign cultures affect indigenous cultures in Australia and New Zealand?

Section 2: Oceania

How did European colonization affect the island countries of Oceania?

Visit glencoe.com and enter QuickPass™ code WGC9952C33 for Chapter 33 resources.

Women use face paint to celebrate their indigenous culture in Papua New Guinea.

Organizing Information Make a Three-Pocket Book to help you organize information about each of the areas in the chapter: Australia and New Zealand and Oceania.

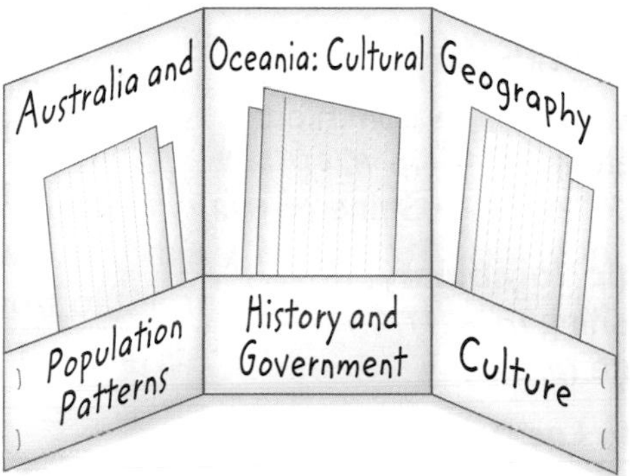

Reading and Writing As you read the chapter, write information about the cultural geography of each subregion on note cards and then file them in the Three-Pocket Book.

SECTION 2

 section audio  spotlight video

Guide to Reading

Essential Question

How did European colonization affect the island countries of Oceania?

Content Vocabulary

- horticulture *(p. 825)*
- trust territory *(p. 827)*
- subsistence farming *(p. 827)*
- pidgin English *(p. 828)*

Academic Vocabulary

- generation *(p. 826)*
- temporary *(p. 827)*

Places to Locate

- Melanesia *(p. 825)*
- Papua New Guinea *(p. 825)*
- Micronesia *(p. 825)*
- Kiribati *(p. 825)*
- Guam *(p. 825)*
- Mariana Islands *(p. 825)*
- Polynesia *(p. 825)*
- Samoa *(p. 825)*
- Tonga *(p. 825)*
- Tuvalu *(p. 825)*
- Tahiti *(p. 825)*
- Fiji *(p. 825)*

Reading Strategy

Organizing On a chart like the one below, list the different ways of life of the peoples of Oceania.

Ways of Life	
Government	Culture

Oceania

Hundreds of indigenous cultures peopled the islands of Oceania in the South Pacific when European explorers arrived. Thousands of islands had been their home for thousands of years. Many of these cultures shared religious beliefs that tied them to the land and sea.

NATIONAL GEOGRAPHIC VOICES AROUND THE WORLD

"Samoa itself is said to mean 'sacred center.' . . . [T]his is where the world began as the creator, Tagaloagagi, first called forth earth, sea, and sky from rock. . . . Language links and artifacts suggest that the first distinctly Polynesian culture may have developed here some 3,000 years ago. Over the centuries that followed, seafarers in double-hulled sailing vessels stocked with pigs, dogs, and fruits spread that culture across much of the Pacific."

—Douglas Chadwick, "The Samoan Way," *National Geographic,* July 2000

A Samoan man and his granddaughter

Population Patterns

MAIN Idea Migration of people among the islands in Oceania has shaped life on the islands today.

GEOGRAPHY AND YOU What would it be like to live on a small island in the Pacific? Read to learn about the people who live in Oceania.

Oceania spreads across thousands of miles in the Pacific Ocean. Its diverse peoples lead lives that are closely tied to water.

Many Peoples

The islands of Oceania were probably first settled by peoples from Asia more than 30,000 years ago. Waves of migrants from Asia continued to arrive over many centuries. Meanwhile, people already living there moved from island to island and settled into three major groups—Melanesians, Micronesians, and Polynesians.

Melanesia Located in the southwestern Pacific Ocean, **Melanesia** includes independent island countries, such as **Papua New Guinea,** Fiji, and the Solomon Islands, as well as French-ruled New Caledonia. Melanesian cultures differ greatly, even among groups living on the same island. One of Papua New Guinea's largest indigenous populations is the Chimbu, notable for their egalitarian social structure.

Micronesia Micronesia is situated in the western Pacific east of the Philippines. Among the independent countries of Micronesia are the Federated States of Micronesia, Nauru, and **Kiribati.** The area also includes the U.S. territories of **Guam** and the **Mariana Islands.** Micronesians have several languages and cultures.

Polynesia Polynesia is located in the central Pacific Ocean. Three independent countries—**Samoa, Tonga,** and **Tuvalu**—are found in Polynesia. Other island groups, known as French Polynesia, are under French rule and include **Tahiti,** Polynesia's largest island.

The largest population of Polynesians lives in the Samoan Islands. In the past, they practiced **horticulture,** or the raising of plants and fruit on small plots of land. Women gathered wild plants and were weavers. Today most Polynesians share similar languages and culture.

NATIONAL GEOGRAPHIC *Many people in Oceania have blended elements of their traditional culture (brightly colored printed fabrics) with those of Western cultures (clothing).*

Place How does Melanesian culture differ from that of Polynesia?

Asians Asian communities also exist in the South Pacific area. Chinese traders and South Asian workers settled parts of Oceania during the 1800s, and today their descendents live in such places as French Polynesia and **Fiji.**

Density and Distribution

Oceania spans a vast area. However, a higher percentage of the islands are unsuited for human habitation. The area's population, therefore, is divided unequally among the island countries. Papua New Guinea leads with about 6.6 million people, whereas Nauru—the world's smallest republic—has a population of only 10,000. Most islanders live on their countries' coasts rather than in the often-rugged interiors.

Oceania's population is growing at a higher average rate than the United States because it has a relatively young population. The land area of Oceania's 25,000 islands totals only 551,059 square miles (1,427,246 sq. km), and the population density varies greatly. Because Papua New Guinea has a large area, its population density is only 38 people per square mile (15 per sq. km).

READING Check **Regions** What are the three island groups in Oceania?

CONNECTING TO

THE UNITED STATES

Dancers at Hawaii's Polynesian Culture Center

Just the Facts:

- Both the United States and Australia were colonized by Great Britain. Both countries have English city names such as Sydney and Melbourne.
- New Zealand is a popular locale for shooting U.S. films because of its landscape variety.
- Hawaii is a state in the United States but is culturally connected to Oceania.
- Pacific Islander Americans were the smallest racial group counted in the 2000 U.S. Census, making up 0.3 percent of the United States population.

Troy Polamalu is one of several players in the NFL of Samoan descent.

Making the Connection

The United States, Australia, New Zealand, and some of the islands in Oceania were originally British colonies, so the language, foods, culture, and many of the customs are shared. However, the Pacific Islands have many unique cultural attributes as well.

The coconut

Sports—Samoans in the NFL Samoa produces a high proportion of NFL players, considering more than 200 of the 500,000 Samoans in the world play Division I college football. Players of Samoan descent are estimated to be 40 times more likely to make it to the NFL than any other player.

Arts—The Film Industry Many recent movies have been filmed in New Zealand, including *King Kong* and *The Lord of the Rings* trilogy. New Zealand is also the birthplace of several world-famous directors and actors, including Peter Jackson, Sam Neill, Anna Paquin, and Russell Crowe. Actors from Australia include Cate Blanchett, Hugh Jackman, Nicole Kidman, Heath Ledger, and Naomi Watts.

Academy Award–winning actor Russell Crowe was born in Wellington, New Zealand.

Food—Similarities and Differences The diets of Australians and Americans are very similar largely due to British influence. Many U.S. fast-food chains have franchises in Australia. Instead of calling it fast food, Australians often refer to this food as take-away.

The food of Oceania is quite different. Commonly referred to as Polynesian, this food has an Asian influence. It often makes use of foods readily available on the islands such as coconut and fish.

The Summer Olympics were held in Sydney, Australia, in 2000.

THINKING GEOGRAPHICALLY

1. **Human Systems** Research the contributions of an American sports figure of Pacific Islander descent. Relay your findings to the class.
2. **Physical System** Research the landscape of New Zealand. What features make the country so popular with the film industry?

CHAPTER 34

THE REGION TODAY

Australia and Oceania

BIG Idea

Places reflect the relationship between humans and the physical environment. Australia and New Zealand are among the world's leaders in exports of agricultural products. The region's unique ecosystems have led to the growth of tourism. Although physical barriers and long distances can be obstacles for people in Australia and Oceania, improved transportation and communications have helped make the region more interdependent.

Essential Questions

Section 1: The Economy
How might the landscape affect economic activities in Australia and Oceania?

Section 2: People and Their Environment
How can human activity have a negative impact on the environment?

Geography ONLINE
Visit glencoe.com and enter *QuickPass*™ code WGC9952C34 for Chapter 34 resources.

As the largest urban area in New Zealand, Auckland plays a dominant role in the region's economy.

Organizing Information Create a Four-Door Book to identify environmental problems and possible solutions in Australia and Oceania.

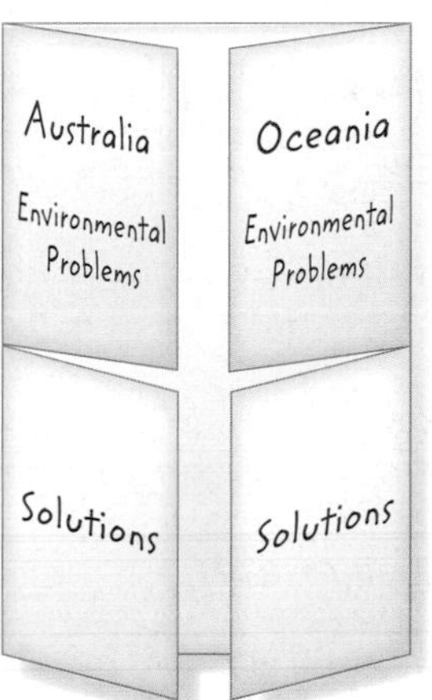

Reading and Writing As you read, identify at least two environmental problems and possible solutions in Australia and Oceania. Use this information to predict which solutions might be most effective and why.

SECTION 1

 section audio  spotlight video

The Economy

Guide to Reading

Essential Question

How might the landscape affect economic activities in Australia and Oceania?

Content Vocabulary

- station *(p. 837)*
- copra *(p. 837)*
- grazier *(p. 837)*

Academic Vocabulary

- devoted *(p. 837)*
- involved *(p. 838)*

Places to Locate

- Fiji *(p. 837)*
- Papua New Guinea *(p. 837)*
- Nauru *(p. 838)*

Reading Strategy

Organizing Complete a web diagram like the one below by identifying the service industries that have emerged in South Pacific countries.

Service Industries in the South Pacific

Australia and Oceania contain striking scenery composed of snowy mountain glaciers, ancient rock formations, coral reefs and their carnivals of sea life, unspoiled beaches, and bubbling geothermal fields. With natural wonders such as these, it is easy to understand why despite its remote geographic locations, tourism is a growing part of the region's economies.

NATIONAL GEOGRAPHIC Voices Around the World

"New Zealand is red-hot right now. Blazing onto the world stage in 2001 with the Lord of the Rings *movies, which were made here, followed in 2005 by* King Kong, *this island nation surged to the top of travel lists as word of its epic scenery, high-test adventure sports, award-winning wines, and innovative art scene got out. . . . 'New Zealand is the most beautiful country in the world,' says actor Grant Roa, who plays Uncle Rawiri in the film* Whale Rider. *'It has spiritual places where you can get away from civilization, including beaches without roads or cell phone coverage.'"*

—Carrie Miller,
"Insiders New Zealand,"
National Geographic Traveler,
March 2006

A hiker in New Zealand

Economic Activities

MAIN Idea Agriculture is the most important economic activity in Australia and Oceania, but new industries are contributing to national economies.

GEOGRAPHY AND YOU In what parts of the United States do ranching and raising livestock take place? Read to find out why raising sheep and cattle is important to Australia's economy.

The remote geographic locations and challenging environments of Australia and Oceania influence how people earn their livings.

Agriculture

Agriculture is by far the most important economic activity in the South Pacific. For example, Australia and New Zealand export large quantities of farm products. Australia is the world's leading producer of wool.

Although less than 5 percent of Australians work in agriculture, much of the country's vast land area is **devoted** to raising livestock. Because of the generally dry climate, animals need large areas to find enough vegetation to eat. As a result, some Australian ranches, called **stations,** are gigantic—as large as 6,000 square miles (15,540 sq. km), about the size of Connecticut.

As a result of Australia's dry climate, less than 10 percent of its land is arable, or suitable for growing crops. Irrigation, fertilizers, and modern technology help Australian farmers make the best use of the limited agricultural land.

More than half of New Zealand's land is used for agriculture. New Zealand ranchers, known as **graziers,** raise sheep, cattle, and red deer. Surprisingly, the country has nearly 20 times more livestock than people! New Zealand's soils, among the most fertile in the region, allow farmers to grow wheat, barley, potatoes, and fruits.

Throughout Oceania, the lack of arable land limits agriculture. Island farmers sometimes practice subsistence farming and fishing.

Some islands, however, have areas of rich, often volcanic soil and ample rainfall. The major cash crop is **copra** (KOH•pruh), or dried coconut meat. Among the island countries that export crops are **Fiji,** a producer of sugarcane, copra, and ginger; and **Papua New Guinea,** a supplier of coffee, copra, and cacao.

Mining and Manufacturing

A variety of mineral deposits exists in some parts of the South Pacific. Australia is a leading exporter of diamonds, gold, bauxite, opals, and iron ore. Extracting these minerals, however, is hampered by high transportation costs. In addition, public debate about Aboriginal land rights limits where mining can occur.

With some exceptions, few significant mineral resources are found in other areas of the South Pacific. New Zealand has a large aluminum smelting industry, and Papua New Guinea's rich deposits of gold and copper have only recently been exploited.

Australia and New Zealand are the South Pacific's major producers of manufactured goods. Since agriculture is important in these two countries, food processing is their most important manufacturing activity. Relatively isolated geographically, Australia and New Zealand must import costly machinery and raw materials to set up manufacturing industries capable of producing exports. As a result, consumer product industries generally manufacture products such as appliances for home consumption.

The rest of the South Pacific is less industrialized than Australia and New Zealand. Manufacturing in Oceania is limited to small-scale enterprises such as apparel production.

NATIONAL GEOGRAPHIC *In Australia's Outback, ranchers often use helicopters to herd cattle over thousands of acres of rough terrain.*

Human-Environment Interaction Why are some Australian ranches so large?

Service Industries

Throughout Australia and Oceania, a range of service industries have emerged as major contributors to national economies. In contrast, few countries in Oceania are large enough to support extensive service industries other than tourism. **Nauru,** however, has become **involved** in international banking and investment companies. In recent decades, the expansion of air travel has boosted tourism in Australia and Oceania. Among the South Pacific's attractions are its indigenous cultures, unique wildlife, and contrasting physical features.

READING Check **Movement** What often limits agriculture in Australia and Oceania?

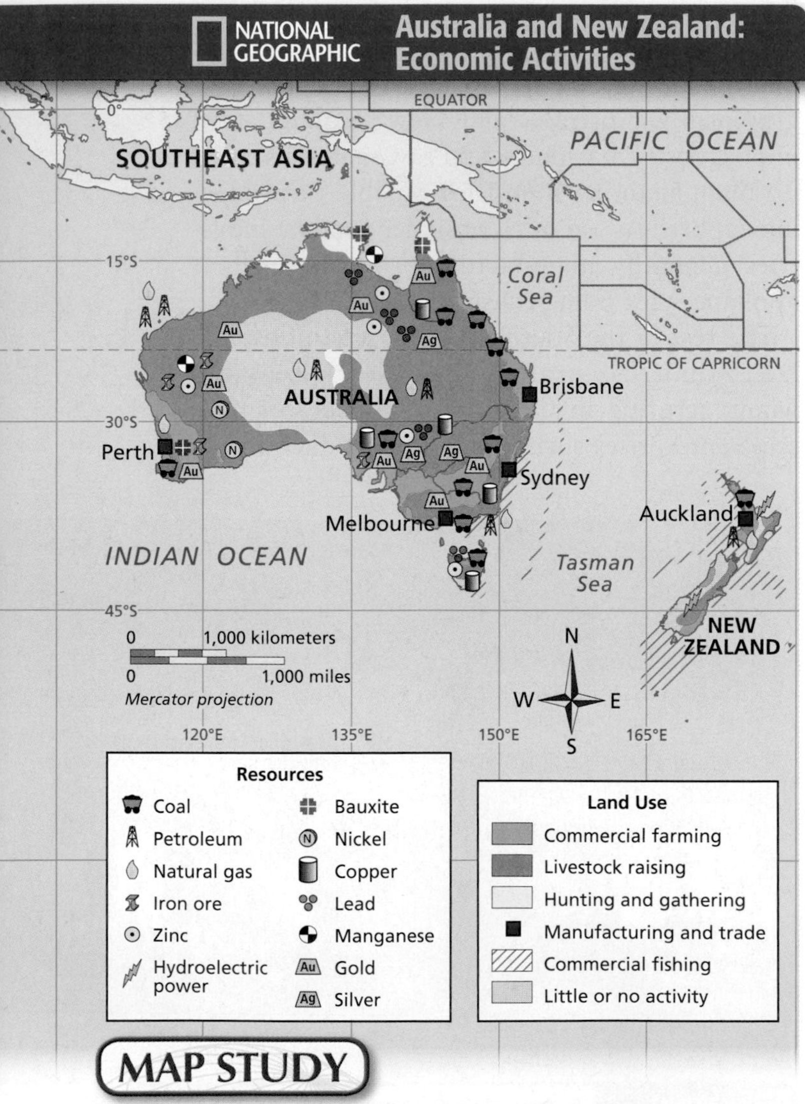

MAP STUDY

1. **Place** Where are most of Australia's coal deposits located?
2. **Regions** What type of land use dominates much of Australia?

Maps In MOtion Use **StudentWorks™ Plus** or glencoe.com.

Transportation and Communications

MAIN Idea The physical environment creates obstacles to transportation and communications in Australia and Oceania, but new technologies are helping to conquer these obstacles.

GEOGRAPHY AND YOU How might vast expanses of land affect the forms of transportation used to move people and goods? Read to learn how people overcome physical barriers and long distances in Australia and Oceania.

Australia and Oceania contain thousands of miles of coastlines, barren desert, rain forests, and vast expanses of rugged terrain. Physical barriers and long distances challenge the movement of people and goods in the region.

Transportation

Australia and New Zealand have the most developed road and rail systems in the region. In the coastal areas of these countries, highways are well maintained, and subways provide public transportation in urban areas. Few roads, however, exist in the isolated Australian Outback.

In Oceania many island countries are too small, too poor, or too rugged to have well-developed road or rail systems. Some governments, however, are improving these systems.

Long distances, harsh climates, or obstacles to land travel make air and water travel important. Cargo ships and planes move imports and exports to and from far-flung Pacific territories. Commercial airlines and cruise ships bring travelers.

Communications

In the South Pacific region, the same geographic obstacles that hinder land travel also make communications difficult. In the Australian Outback, some cattle stations are large enough to maintain their own post offices and telephone exchanges. The development of modern technology, however, has helped increase contacts within Australia and Oceania and with the rest of the world. Cellular, digital, and satellite communications and the Internet are becoming common in some areas.

READING Check **Regions** Why are air and water travel especially important to the region?

Trade and Interdependence

MAIN Idea Trade between Australia, Oceania, and other parts of the world has increased due to improvements in transportation and communications, as well as to the creation of trade agreements.

GEOGRAPHY AND YOU Why would transportation and communications links play a key role in boosting trade in a remote region? Read to learn how the South Pacific's remote geographic location has affected its trade relations.

In recent decades, improved transportation and communications links have increased trade between the South Pacific region and other parts of the world. The region's agricultural and mining products are its greatest sources of export income. Countries in Oceania export copra, timber, fish, vegetables, many varieties of spices, and handicrafts.

During most of the 1900s, Australia and New Zealand traded mainly with the United Kingdom and the United States. In recent years, however, these South Pacific countries have increased trade with their neighboring Asian countries of Japan, Taiwan, and China. Australia, for example, is a member of the Asia-Pacific Economic Cooperation forum (APEC) and is pursuing free trade agreements with China and the Association of Southeast Asian Nations (ASEAN).

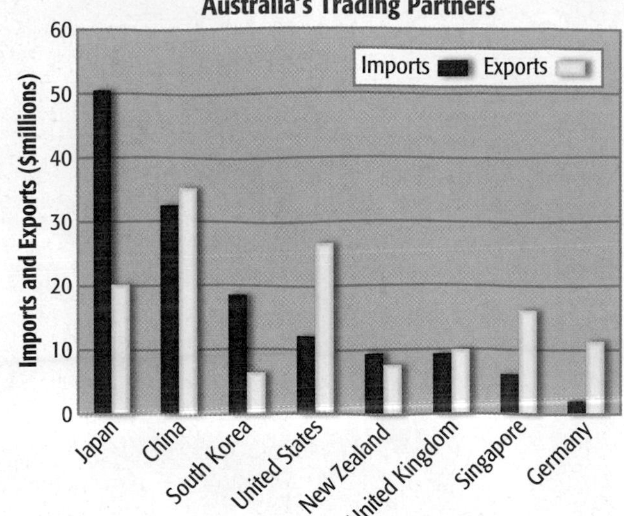

Source: Australian Bureau of Statistics.

READING Check **Regions** What are the South Pacific region's greatest sources of export income?

SECTION 1 REVIEW

Vocabulary

1. Explain the significance of: station, grazier, copra.

Main Ideas

2. What is the most important economic activity in the South Pacific region? What new industries are contributing to national economies in the region?
3. Describe the factors that have increased trade between Australia, Oceania, and other parts of the world. Give examples.
4. Use a table like the one below to identify ways in which the physical environment creates obstacles to transportation and communications in Australia and Oceania. Then describe the new technologies that are helping the region overcome these obstacles.

	Obstacles	Technologies
Transportation		
Communications		

Critical Thinking

5. **Answering the Essential Question** How has the region's physical environment influenced the growth of tourism?
6. **Summarizing Information** How have changes in transportation and communications affected economic activities in the South Pacific region?
7. **Analyzing Visuals** Study the physical map on page 792 of the Regional Atlas and the economic activity map on page 838. Which parts of Australia are the most developed? Least developed?

Writing About Geography

8. **Expository Writing** Create a table that shows major economic activities for five countries in the region. Then write a paragraph explaining why the economies of some countries focus on one major product.

Study Central™ To review this section, go to glencoe.com and click on Study Central.

The Great Barrier Reef

The Problem:

Increased water temperatures and environmental pollutants threaten the Great Barrier Reef. The survival of the reef and its marine life are at stake if this deterioration continues at its current rate.

Bleached coral on the Great Barrier Reef

Coral Bleaching Extended exposure to unusually high water temperatures and pollution can kill coral reefs.

Changing Climate Global warming has led to increased water temperatures, which are killing coral, altering the underwater ecosystem, and shrinking the Great Barrier Reef.

NATIONAL GEOGRAPHIC

Saving a Natural Wonder

Located in the Coral Sea off the coast of northeast Australia, the Great Barrier Reef is the world's largest coral reef system. It is home to a variety of plant and animal species. Currently, this natural wonder is threatened by an array of forces.

What is coral bleaching? Coral bleaching is a natural response to environmental stress. Stressing agents, including increased water temperatures and environmental pollution, cause the coral to lose its vibrant color. Sustained exposure to these elements can result in the death of the coral organism. The most widespread coral bleaching event to affect the Great Barrier Reef occurred in 2002. Continued monitoring of coral bleaching provides important information about the health of the Great Barrier Reef.

What factors can we control? Several measures have been taken to address reef destruction. The Australian government has placed restrictions on fishing and has created sanctuaries to preserve the unique biodiversity of the area. *The Reef Water Quality Protection Plan,* another government measure, reduces the amount of land-based pollution such as sediment and pesticide pollution from reaching the ocean.

Can the Great Barrier Reef be saved? With tourism making up a large portion of the region's economy, the reef's survival is of paramount importance. The continued conservation efforts of the government and those who depend on the reef for their livelihood will go a long way toward restoring the resilience of the reef.

One Solution:

Environmental programs that protect and monitor the condition of the Great Barrier Reef are key to its survival.

Increased observation, monitoring, and environmental protection efforts will help sustain the ecosystem of the Great Barrier Reef.

THINKING GEOGRAPHICALLY

1. **Environment and Society** Why are conservation efforts at the Great Barrier Reef important for the environment and the economy?
2. **Places and Regions** Investigate coral reefs throughout the world. Where are they located? Is coral bleaching a problem in these areas?

SECTION 2

 section audio

 spotlight video

Guide to Reading

Essential Question

How can human activity have a negative impact on the environment?

Content Vocabulary

- marsupial *(p. 843)*
- introduced species *(p. 843)*
- food web *(p. 844)*
- ozone layer *(p. 845)*
- El Niño-Southern Oscillation (ENSO) *(p. 845)*
- diatom *(p. 846)*

Academic Vocabulary

- restore *(p. 843)*
- poses *(p. 844)*

Places to Locate

- Tasmania *(p. 843)*
- Murray-Darling River Basin *(p. 844)*
- Great Barrier Reef *(p. 844)*

Reading Strategy

Categorizing Use the major headings of the section to create an outline like the one shown below.

I. Human Impact on Resources
   A.
   B.
   C.
II. Challenges for the Future

People and Their Environment

Australia is home to one of the most diverse and complex ecosystems on Earth—the Great Barrier Reef. The Australian government has taken steps to protect this complex ecosystem, but human activities on land impact what happens in the sea.

NATIONAL GEOGRAPHIC Voices Around the World

"The Great Barrier Reef covers 135,000 square miles, an expanse greater than Poland. . . . In 1975 virtually the whole offshore nation was declared Great Barrier Reef Marine Park, one of the world's first national marine sanctuaries and still the biggest. Intended mainly to prevent oil drilling and mining on the reef, the park remains open to many other uses. Commercial fishing, sportfishing, spearfishing, and the collecting of aquarium fish and shells are regulated, however, and some segments are set aside as no-take zones, research zones, or special reserves for troubled species. . . . On the other hand, the coral realm is not immune to the changes taking place in ecosystems on land. Cane fields, other croplands, and development along Queensland's coastal plain have replaced many seaside wetlands, the natural filters for fresh water coming from the continent."

—Douglas H. Chadwick, "Great Barrier Reef," *National Geographic,* January 2001

Croplands near Cape York, Australia, impact the reef

Human Impact on Resources

MAIN Idea Australia and Oceania have many natural resources, but the region's environment is threatened by human activities.

GEOGRAPHY AND YOU In what other regions of the world are freshwater sources threatened by human activity? Read to learn why the protection of freshwater resources is a major concern in the South Pacific.

Australia and Oceania hold some of the planet's richest and most diverse natural resources. Unfortunately, these resources have not always been well managed, and today the region faces many environmental issues. Conservation efforts, however, are gaining recognition in the region.

Unusual Animals

The continent of Australia, separated for so long from other landmasses, is home to many unique animal species. Kangaroos, koalas, and wallabies are just some of Australia's 144 species of **marsupials**—mammals whose young must mature in a pouch after they are born. The Australian island of **Tasmania** gave its name to the Tasmanian devil, a powerful meat-eating marsupial about the size of a badger. Australia's strangest wildlife may be the duck-billed platypus and the echidna—a spiny anteater—the only mammals in the world that lay eggs.

These unusual wildlife species, however, have been seriously threatened by the human introduction of various nonnative animals. These **introduced species** include the hunting dogs called dingoes brought from Asia by migrating Aborigines. European settlers brought sheep, cattle, foxes, cats, and rabbits to Australia. In the absence of natural predators, these animals have multiplied and taken over the habitats of Australia's native species. Some of Australia's native species have become extinct, and at least 16 kinds of marsupials are now endangered. Efforts to **restore** Australia's ecological balance include the use of electric fencing to keep out nonnative animals, hunting and trapping programs, the introduction of natural predators, and native wildlife reserves.

Because of its isolated, remote location surrounded by ocean, New Zealand is also home to many unique animal species—specifically seabirds, which flourished without mammalian predators. Among New Zealand's surprising and unlikely species is the penguin.

Many factors, including human settlement, threaten New Zealand's animal species. However, as in Australia, the most serious threat to New Zealand's native animals comes from introduced species, such as cats, rats, and ferrets.

To protect their endemic species from introduced predators, New Zealand is taking steps such as implementing predator-control techniques and establishing island sanctuaries. Island sanctuaries, some of which are still free of introduced species, have been key to New Zealand's conservation efforts. Of the hundreds of small islands and islets along New Zealand's coast, more than 200 are managed by the Department of Conservation as protected reserves.

Forest, Soil, and Water

The protection of forest, soil, and freshwater resources is a major concern throughout the South Pacific region. In Australia, many woodlands have been cleared for farms and grazing lands.

NATIONAL GEOGRAPHIC *Although koalas are protected by law, their habitat, which is often cleared for the construction of houses and roads, is not.*

Human-Environment Interaction What are Australians doing to help restore the country's ecological balance?

This leaves little protection against erosion. When the soil surface is exposed to wind and water, erosion occurs. In Australia, soil erosion is a major problem, which has been compounded by overgrazing in arid areas and by the country's worst drought in over a century. Soil conservation in the region is closely linked to reducing deforestation. Countries with valuable timber resources, such as New Zealand, Papua New Guinea, and Vanuatu, are developing plans to use forest resources without damaging the environment.

Drought, salt, irrigation, and agricultural runoff threaten Australia's freshwater sources. In the fertile **Murray-Darling River Basin,** one of the world's largest drainage basins, the use of water for agriculture and growing city populations has dramatically reduced the rivers' flow. Large areas within the basin are also at risk from increasing soil salinity, which is one of Australia's most pressing environmental problems today. One of the major causes of increasing salinity in Australia's water and soil has been the replacement of native vegetation with pastures and nonnative shallow-rooted crops.

Oceania also faces challenges in managing its freshwater resources. Many small coral atolls and volcanic islands hold only limited supplies of freshwater. Agricultural runoff and inadequate sanitation cause pollution that further threatens these freshwater supplies. The lack of clean drinking water keeps the standard of living low and **poses** barriers to economic growth in some countries. Improvement will come with better management of runoff, construction of additional sanitation facilities, and development of less expensive ways of removing salt from ocean water.

Agricultural runoff, chemical fertilizers, and organic waste also threaten the region's oceans. Toxic waste in particular endangers Australia's **Great Barrier Reef** and other Pacific coral reefs. Coral environments are increasingly stressed by tourists, boaters, divers, oil shale mining, and increasing water temperatures. Coral bleaching—the breakdown of the relationship between coral and the algae that provide these organisms with nutrients—is often the result of such stress.

Pollution also affects all kinds of marine life, including the tiny organisms that make up coral reefs. Algae—on which these organisms thrive—and plankton are key parts of the ocean's **food web,** the interlinking chains of predators and their food sources in an ecosystem. As these tiny living things are destroyed, the larger plants and animals that rely on them for food also die off.

NATIONAL GEOGRAPHIC

Levels of Coral Bleaching in the Great Barrier Reef

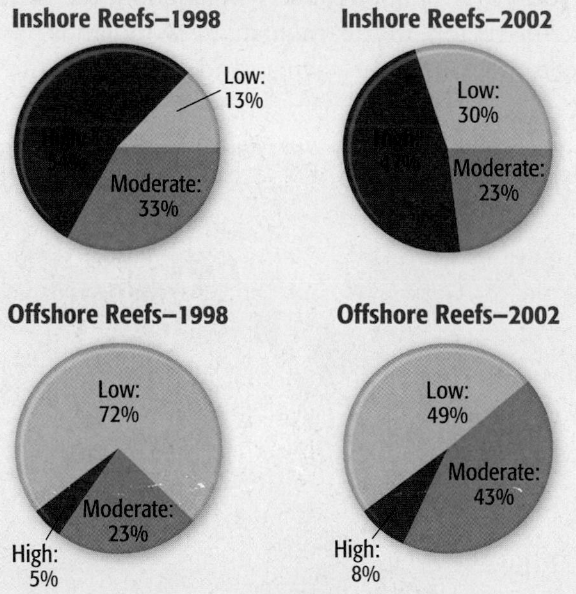

Inshore Reefs–1998

Low: 13%
Moderate: 33%

Inshore Reefs–2002

Low: 30%
Moderate: 23%

Offshore Reefs–1998

Low: 72%
Moderate: 23%
High: 5%

Offshore Reefs–2002

Low: 49%
Moderate: 43%
High: 8%

Note: Inshore = less than 6.2 miles (10 km) from the coast
Source: Australian Institute of Marine Science.

GRAPH STUDY

1. **Human-Environment Interaction** In which year did inshore reefs suffer higher levels of coral bleaching?
2. **Place** Which type of reef has suffered higher levels of coral bleaching?

Graphs In MOtion Use **StudentWorks™ Plus** or glencoe.com.

The Nuclear Legacy

The testing of nuclear weapons has had major effects on the region's environment. In the late 1940s and 1950s, the United States and other countries with nuclear capability carried out aboveground testing of nuclear weapons in the South Pacific. The dangers of such testing were gravely underestimated at the time. In 1954 the United States exploded a nuclear device on Bikini Atoll in the Marshall Islands. The people of Bikini Atoll had been moved to safety, but those living on Rongelap Atoll, downwind of the explosion, were exposed to massive doses of radiation that resulted in deaths, illnesses, and genetic abnormalities.

Although the U.S. testing was stopped, the effects of radiation exposure and environmental damage have continued through several generations. Today the atolls affected by the testing remain off-limits to human settlement. Recent studies, however, offer hopeful signs of eventual environmental recovery. In the 1990s the United States government provided $90 million to help decontaminate Bikini Atoll and set up a $45 million trust fund for blast survivors and their offspring from Rongelap Atoll.

The nuclear legacy also has had political effects. Antinuclear activism is a major factor in regional politics. In 1985 New Zealand banned nuclear-powered ships and those with nuclear weapons from entering its waters. Because of this ban, which remains in effect today, the United States withdrew from a defense agreement with New Zealand. In the mid-1990s, French plans to conduct nuclear tests on an atoll in French Polynesia aroused antinuclear demonstrations. The international outcry led to an early halt to the tests.

READING Check **Human-Environment Interaction** What contributes to increasing soil and water salinity in Australia?

Challenges for the Future

MAIN Idea Environmental concerns related to atmosphere and climate changes threaten Australia and Oceania, as well as other world regions.

GEOGRAPHY AND YOU Why might global warming have devastating effects around the world? Read to find out how climate changes are affecting Australia and Oceania.

Like other world regions, Australia and Oceania are threatened by global atmospheric and climate changes. In the 1970s scientists found a hole in the **ozone layer** over Antarctica. The ozone layer's protective gases prevent harmful solar rays from reaching Earth's surface. The ozone hole over Antarctica grew dramatically between 1975 and 1993, when it covered more than 9 million square miles (23 million sq. km). In October 2006, the average area of the ozone hole was the largest ever observed at 10.6 million square miles (27.4 million sq. km).

The loss of protective ozone may be behind the global rise in the rates of skin cancer and cataracts, conditions caused by overexposure to the sun's ultraviolet rays. Increased solar radiation that reaches Earth through the ozone hole may also contribute to global warming, the gradual rise in Earth's temperatures.

Climate and weather in the South Pacific are highly sensitive to changes in the El Niño weather pattern called **El Niño-Southern Oscillation (ENSO).** This seasonal weather event can cause both droughts and powerful cyclonic storms in the South Pacific. These ENSO-related weather patterns are believed to be increasing in frequency and severity and may also be linked to global warming.

NATIONAL GEOGRAPHIC **The Ozone Hole**

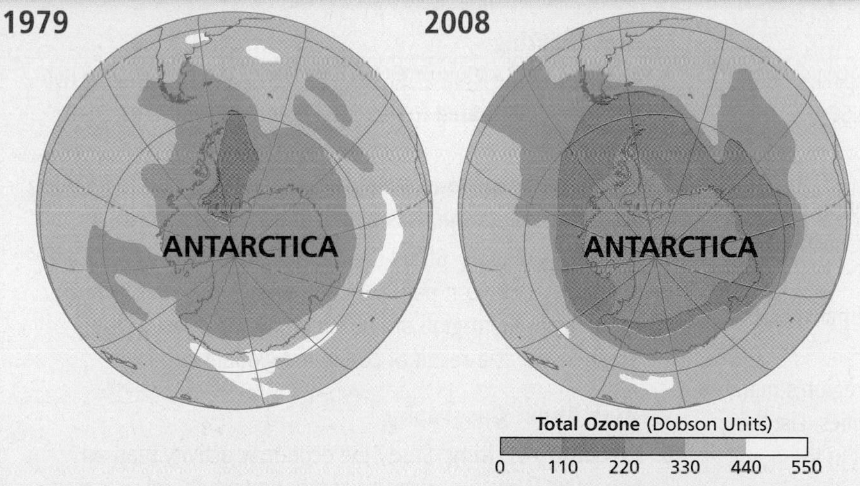

Note: Dobson units are the standard way to express amounts of ozone in the atmosphere. The ozone hole is defined as the area having less than 220 Dobson units.

MAP STUDY

1. **Regions** How did the size and shape of the ozone hole change between 1979 and 2008?
2. **Place** Over which parts of Antarctica did the ozone hole not extend in 1979?

Maps In MOtion Use **StudentWorks™ Plus** or glencoe.com.

NATIONAL GEOGRAPHIC *Rising ocean temperatures have affected the endangered hawksbill turtle, which makes its home in coral reefs.*

Human-Environment Interaction How else might rising ocean temperatures affect life on Earth?

Some scientists claim that continued rises in Earth's temperatures could be devastating. If polar ice caps were to melt and a rise in the level of ocean waters occurred, many of Oceania's islands would be flooded. In addition, rising ocean temperatures affect certain types of plankton and algae that grow in warm waters, causing overgrowth and the choking out of other life-forms. **Diatoms**—plankton that flourish in cold ocean waters—would die if temperatures rose, affecting life-forms that feed on them. Scientists in the region, especially in Antarctica, are studying global warming and are hoping to discover causes, predict consequences, and provide solutions.

READING Check **Human-Environment Interaction** How might the ozone hole over Antarctica affect people around the world?

Student Web Activity Visit glencoe.com, select the *World Geography and Cultures* Web site, and click on Student Web Activities–Chapter 34 for an activity about introduced species in the region.

SECTION 2 REVIEW

Vocabulary

1. Explain the significance of: marsupial, introduced species, food web, ozone layer, El Niño-Southern Oscillation (ENSO), diatom.

Main Ideas

2. Describe the environmental concerns related to atmosphere and climate changes that threaten Australia and Oceania and other world regions.
3. Why are introduced species a threat to native animal species in Australia and New Zealand?
4. Use a table like the one below to show how the region's many natural resources are threatened by human activities. List the resources and examples of their mismanagement in the region. Then list possible solutions.

Resource	Example of Mismanagement	Possible Solution

Critical Thinking

5. **Answering the Essential Question** What human activities have contributed to increased wind and soil erosion in the region?
6. **Drawing Conclusions** What steps would you take to increase awareness about the risks of global warming? Explain.
7. **Analyzing Visuals** Study the physical map on page 792 and the political map on page 794 of the Regional Atlas. Which countries in the region are at the greatest risk from rising ocean levels as a result of continued global warming?

Writing About Geography

8. **Expository Writing** Study the economic activity map on page 838. Write a paragraph comparing a mineral-rich area shown on the map to a mineral-rich area in another region. Explain the effects of mining on both environments.

Study Central™ To review this section, go to glencoe.com and click on Study Central.

STUDY TO GO

Study anywhere, anytime by downloading quizzes and flashcards to your PDA from glencoe.com.

The Region Today

Tuesday Section B

Island Economies

- Australia and New Zealand export large quantities of agricultural products.
- South Pacific island farmers practice subsistence farming and raise pigs and chickens.
- Most people in Australia and New Zealand work in service industries.
- Tourism in the region is growing and its attractions include indigenous cultures, unique wildlife, and interesting geographic features.

Cleaning Up

Challenges

- The testing of nuclear weapons in the 1940s, 1950s, and 1960s has resulted in lasting environmental effects in the region.
- Pollution is resulting in global warming and climate change, leading to drought in Australia and storms in the South Pacific.
- Agricultural runoff and inadequate sanitation pollute water in Oceania.

Australia and Oceania: Nuclear Testing Sites

120°E 140°E 160°E 180° 160°W 140°W
TROPIC OF CANCER
20°N
Johnston Atoll
PACIFIC OCEAN
Enewetak Atoll
Bikini Atoll
MARSHALL ISLANDS
Pacific Ocean test sites
EQUATOR
0°
Kiritimati
KIRIBATI
Malden Island
Fangataufa Atoll
FRENCH POLYNESIA
Mururoa Atoll
Montebello Islands
20°S
TROPIC OF CAPRICORN
AUSTRALIA
Emu Field
Maralinga (Woomera Prohibited Area)
• Nuclear test site
40°S
INDIAN OCEAN
N S E W
0 1,000 kilometers
0 1,000 miles
Mercator projection

Solutions

- In the 1990s, the United States gave $90 million to help decontaminate Bikini Atoll.
- Scientists are studying global warming to discover causes and solutions.
- Sanitation and runoff management programs hope to solve Oceania's water pollution problems.
- Environmental protection programs aim to save endangered wildlife, including the Great Barrier Reef.

CHAPTER 34

STANDARDIZED TEST PRACTICE

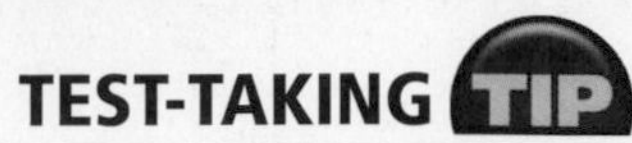

When you have finished, check your work to make sure you have answered all the questions.

Reviewing Vocabulary

Directions: Choose the word or words that best complete the sentence.

1. Australian ranches are called ________.

A plantations
B the Outback
C stations
D reserves

2. ________ is dried coconut meat.

A Cane
B Jerky
C Ginger
D Copra

3. The young of ________ must mature in the mother's pouch after they are born.

A marsupials
B mammals
C birds
D reptiles

4. The ________ is the interlinking chains of predators and their food sources in an ecosystem.

A food supply
B food chain
C food web
D food pyramid

Reviewing Main Ideas

Directions: Choose the best answers to complete the sentences or to answer the following questions.

Section 1 *(pp. 836–839)*

5. The dominant economic activity in Australia and Oceania is ________.

A agriculture
B mining
C industry
D fishing

6. Raising livestock takes up a large land area in Australia because ________.

A most Australians are ranchers
B the livestock take up a lot of space
C rich station owners like to own lots of property
D the livestock must roam large areas to find enough to eat

Section 2 *(pp. 842–846)*

7. A major threat to Australia's native wildlife is ________.

A global warming
B other native wildlife
C introduced species
D plants

8. In 1985 ________ banned nuclear-powered ships and those with nuclear weapons from its waters.

A Australia
B New Zealand
C Fiji
D Papua New Guinea

GO ON

Critical Thinking

Directions: Choose the best answers to complete the sentences or to answer the following questions.

9. How has distance to Europe and North America influenced Australia and New Zealand?

A It has made them poor.

B It has made imported goods expensive.

C It has left them unexplored and unsettled by Europeans.

D It has made them technologically backward.

Base your answer to question 10 on the map and on your knowledge of Chapter 34.

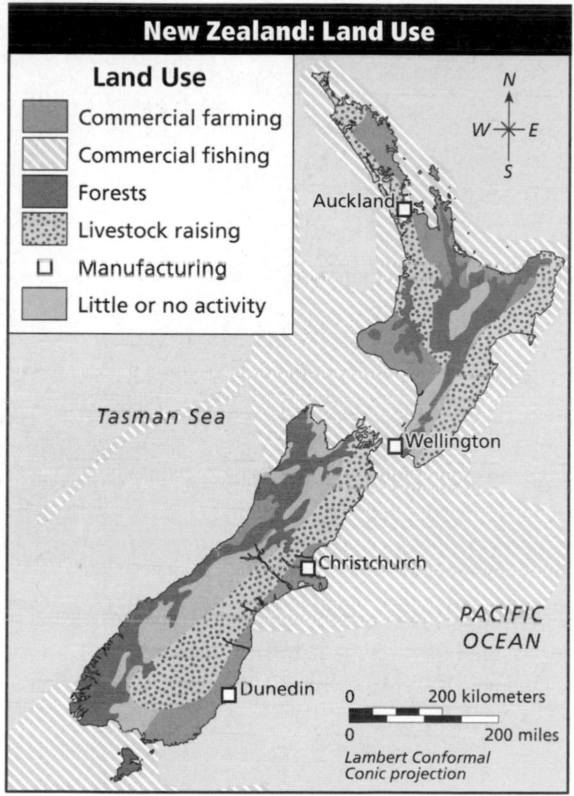

10. Where is most of New Zealand's cropland?

A in the mountains

B in the central plateaus of the islands

C along the coasts

D New Zealand has no cropland.

Document-Based Questions

Directions: Analyze the document and answer the short-answer questions that follow the document.

New Zealand takes a somewhat independent attitude in world affairs. For example, it has declared itself a nuclear-free zone. A university professor from Auckland describes his country's action.

> *In July 1984 a newly elected Labour government implemented a policy which dissociated New Zealand's military establishment from the nuclear component of the then-Cold War confrontation. In 1987* the New Zealand Nuclear Free Zone, Disarmament and Arms Control Act *was passed, giving legal force to the policy. Although the passage of the Act was regarded by the United States as a near-hostile action on the part of New Zealand, none of its three explicit purposes would have been expected intrinsically to evoke international controversy. These purposes are:*
>
> *(i) to establish a Nuclear Free Zone in New Zealand, (ii) to promote and encourage an effective contribution by New Zealand to international disarmament and arms control, and (iii) to implement locally a number of treaties to which New Zealand is party, specifically, the Partial Test Ban Treaty, the Nuclear Non-Proliferation Treaty, the Sea-bed Treaty, the Biological Weapons Convention and the Treaty of Rarotonga (which created the South Pacific Nuclear Free Zone).*
>
> —Peter R. Wills, "New Zealand's Nuclear Free Status"

11. Besides avoiding the dangers of nuclear material in New Zealand, what did the country hope to accomplish by becoming a nuclear-free zone?

12. How does location help make it possible for New Zealand to take this step?

Extended Response

13. Exploring the BIG Idea

Describe how human activity has affected the unusual animals of Australia and New Zealand. What measures have these countries taken to protect these unique species?

For additional test practice, use Self-Check Quizzes—Chapter 34 on glencoe.com.

Need Extra Help?

If you missed questions...	1	2	3	4	5	6	7	8	9	10	11	12	13
Go to page...	837	838	843	844	837	837	843	845	838	849	849	849	843

NATIONAL GEOGRAPHIC

Case STUDY

CLASH OF CULTURES: What is the relationship between the indigenous peoples and Europeans in Australia and New Zealand?

People were living in Australia and New Zealand long before the Europeans arrived. The indigenous peoples of Australia are known as Aborigines. They have lived in Australia for at least 50,000 years. The Maori are the indigenous people of New Zealand. They originally came from Polynesia between A.D. 950 and 1350.

Understanding the Problem

Aborigines and Maori suffer from lower incomes, educational levels, and health conditions. In both countries, there are conflicts over landownership.

A Moral Dilemma European settlers came to Australia and New Zealand beginning around 1800 and developed very prosperous societies. Many of the indigenous peoples suffered from disease, warfare, and cultural destruction as a result of European settlement. The Australian government tried to reform the Aborigines' culture to make them European. The government also took Aborigine children and placed them in boarding schools. In New Zealand, the Maori and the Europeans dealt with each other more as equals, but the Maori were still disadvantaged. Even today the Aborigines and the Maori suffer varying degrees of disadvantage and discrimination.

An Economic Problem Aborigines and Maori have higher unemployment rates and lower levels of education. Thus, they do not achieve their highest potential in the workforce. Their rates of illness are higher, requiring more health services.

A Political Issue Aborigines and Maori have become a political force in their countries. They have organized to press their governments for more rights and more aid. Both countries face the political problem of landownership. The Aborigines and the Maori claim that the land in their countries belongs to them and that European settlers took it illegally. The Maori, for example, point out that the Treaty of Waitangi in 1840 provided that the Maori would continue to own their land even as they were subject to the British government. The Aborigines have reservations, but most are in desert and tropical areas where few non-Aborigines want to live and where there are few services and markets for their products.

Above right: A woman from the Eastern Arrernte Aboriginal people sits amid carvings made by her ancestors.
Above: The new Maori king Tuheitia Paki was coronated on August 21, 2006.

Indigenous Education and Employment

Characteristic	Australia		New Zealand	
	Non-Aborigine	Aborigine	Non-Maori	Maori
Population	19,447,590	407,698	4,027,947	565,329
Number of unemployed	481,158	22,644	78,627	27,873
Number of university graduates	818,682	2,782	297,942	17,907

Source: Census of Australia, 2006; Census of New Zealand, 2006.

Possible Solutions

Raising levels of well-being of Aborigines and Maori requires a combination of different strategies.

Health and Social Services The destruction of their cultures by European settlers left the Aborigines and Maori with economic and health problems. Both groups are still trying to find a way to live successfully in a European society. To do so, they need social services that will improve their quality of life and their ability to participate fully in a modern society.

Education Education can also help Aborigines and Maori to participate in their traditional culture and to reap economic benefits. New Zealand has established schools in which lessons are taught in the Maori language. Both countries have programs to encourage Aborigine and Maori youth to pursue more education. Educating nonindigenous people about indigenous cultures is also important to creating better intercultural relations.

Government In both countries, indigenous people can vote. Australia has elected a few Aborigines to Parliament, and New Zealand has elected a few Maori. Electing more members of these groups will help them win the resources they need to improve their status.

Access to Health Care in Aboriginal Communities

	In community	Less than 6.2 miles (10 km)	6.2–14.9 miles (10–24 km)	15.5–30.4 miles (25–49 km)	31.1–61.5 miles (50–99 km)	62.1–154.7 miles (100–249 km)	More than 155.3 miles (250 km)
Nearest hospital	14,090	7,743	5,634	4,766	7,968	21,080	30,912
Aboriginal Primary Health Care Center	41,450	7,743	3,402	3,572	6,464	12,552	12,934

Source: Australia Bureau of Statistics.

Contents

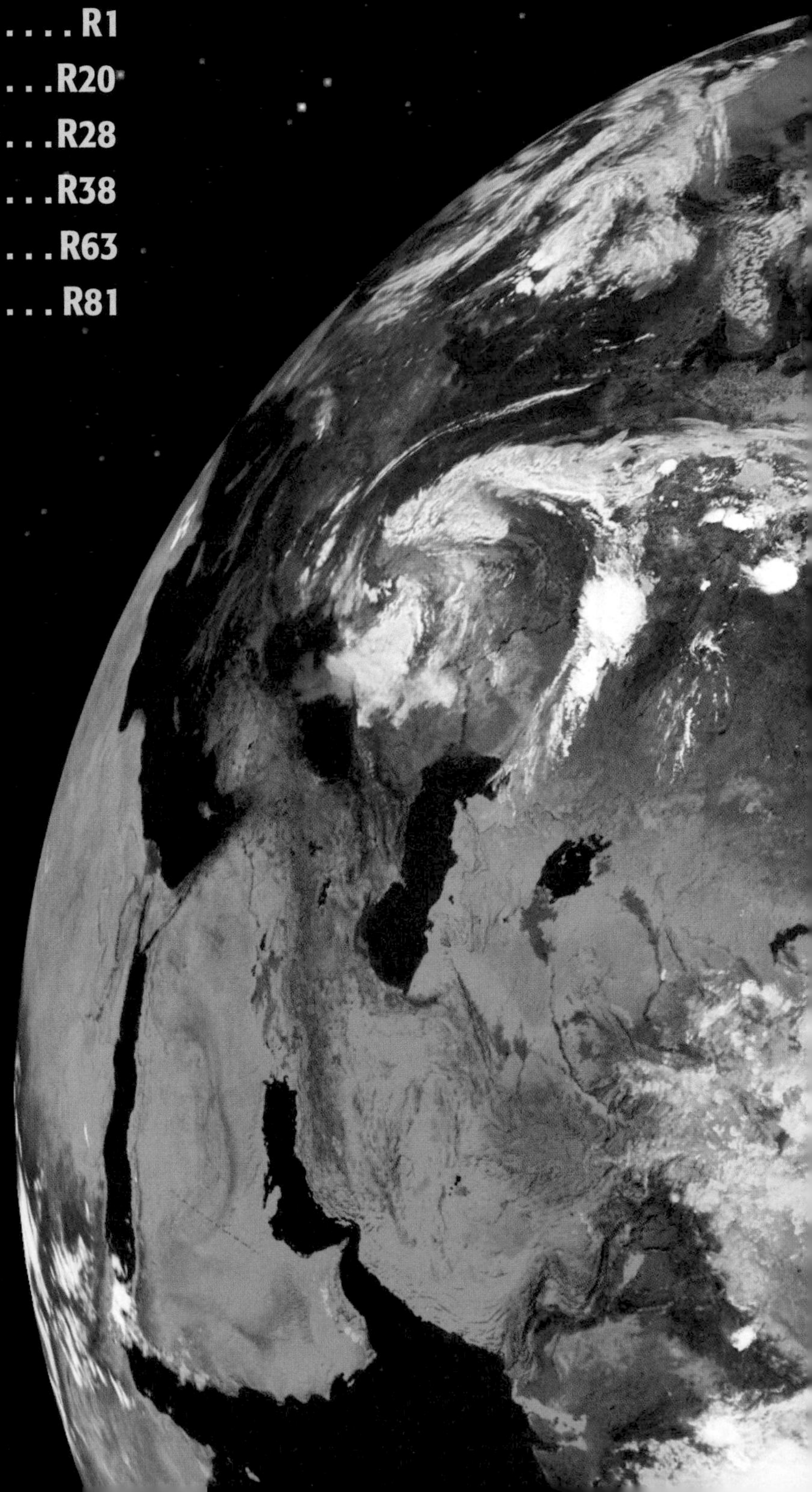

Skills Handbook

Table of Contents

Critical Thinking Skills

Social Studies Skills

Comparing and Contrasting

Why Learn This Skill?

When you make comparisons, you determine similarities among ideas, objects, or events. When you contrast, you are noting differences between ideas, objects, or events. Comparing and contrasting are important skills because they help you choose among several possible alternatives.

Learning the Skill

Follow these steps to learn how to compare and contrast. Then answer the questions below.

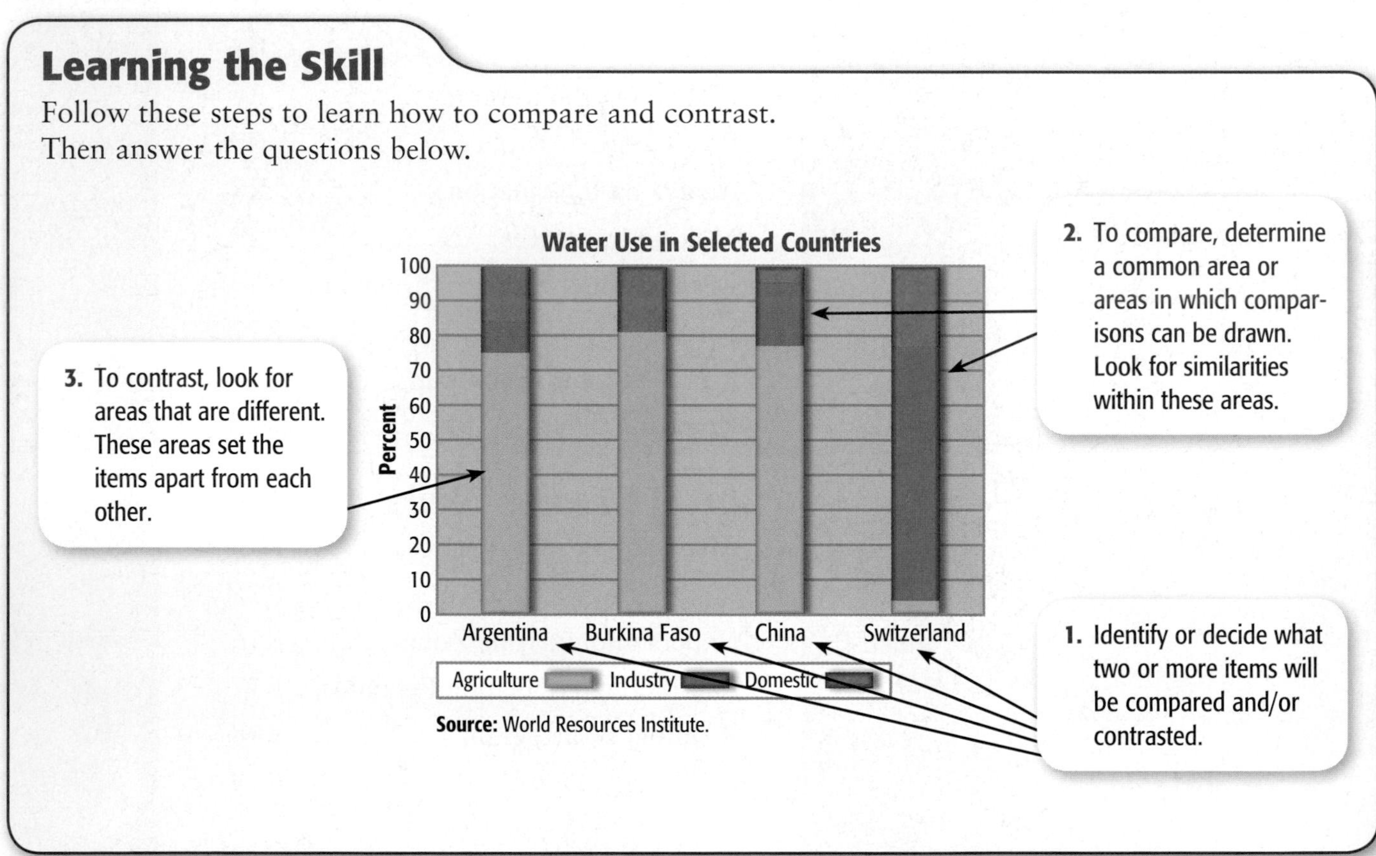

Practicing the Skill

1. What characteristics does the graph use to compare and contrast the different countries?
2. How are Argentina, Burkina Faso, and China similar?
3. How is Switzerland different from the other countries?
4. Which country might you infer has the fewest cities? Explain your answer.

APPLYING THE SKILL

With a partner, research four U.S. cities. Choose two characteristics, such as population and average annual rainfall, to compare and contrast. Use your information to create a bar graph. Develop three questions based on your graph. Exchange your work with another pair of students and answer the questions based on their graph.

Drawing Conclusions

Why Learn This Skill?

A conclusion is a logical understanding that you reach based on details or facts that you read or hear. When you draw conclusions, you use stated information to figure out ideas that are unstated.

Learning the Skill

Follow these steps to draw conclusions.
Then answer the questions below.

1. Read carefully for stated facts and ideas.
2. Summarize the information and list important facts.
3. Apply related information that you may already know.
4. Use your knowledge and insight to develop some logical conclusions.

Darfur Facts

2.8 million people currently live in camps or makeshift settlements in Darfur or in other countries.	The International Criminal Court issued an arrest warrant for Sudanese president Omar Al Bashir in 2009.
290,000 Sudanese have fled to Chad because of continuing violence.	300,000 people in Darfur have been killed or died as a result of the conflict.
More than 10 aid agencies were expelled from Sudan in March 2009.	Displaced women from Darfur face the risk of attack if they leave their camps.
China and Russia are major suppliers of arms to Sudan and permanent members of the UN Security Council.	The UN Security Council agreed to extend an existing arms embargo to the Sudanese government in March 2005.

Source: Amnesty International, Fall 2009.

Practicing the Skill

1. Which facts from the chart support the conclusion "the Sudanese government is trying to drive out the people of Darfur"?
2. What conclusion might you draw about why the UN Security Council waited so many years before extending an existing arms embargo to Sudan?

APPLYING THE SKILL

Find an article describing a current conflict in Africa. Use the steps on this page to draw conclusions about the causes of the conflict. Summarize your conclusions in a paragraph.

Reading a Cartogram

Why Learn This Skill?

Maps that distort country size and shape in order to show certain types of data are called cartograms. In a cartogram, country size reflects some value other than land area, such as population or gross national product. The cartogram is a tool for making visual comparisons.

Learning the Skill

Follow these steps to learn how to read a cartogram. Then answer the questions below.

1. Read the map title and key to identify the kind of information presented in the cartogram.
2. Look for relationships among the countries. Determine which countries are largest and smallest.
3. Compare the cartogram with a standard land-area map. Determine the degree of distortion of particular countries.
4. Study these relationships and comparisons. Identify the most important information presented in the cartogram.

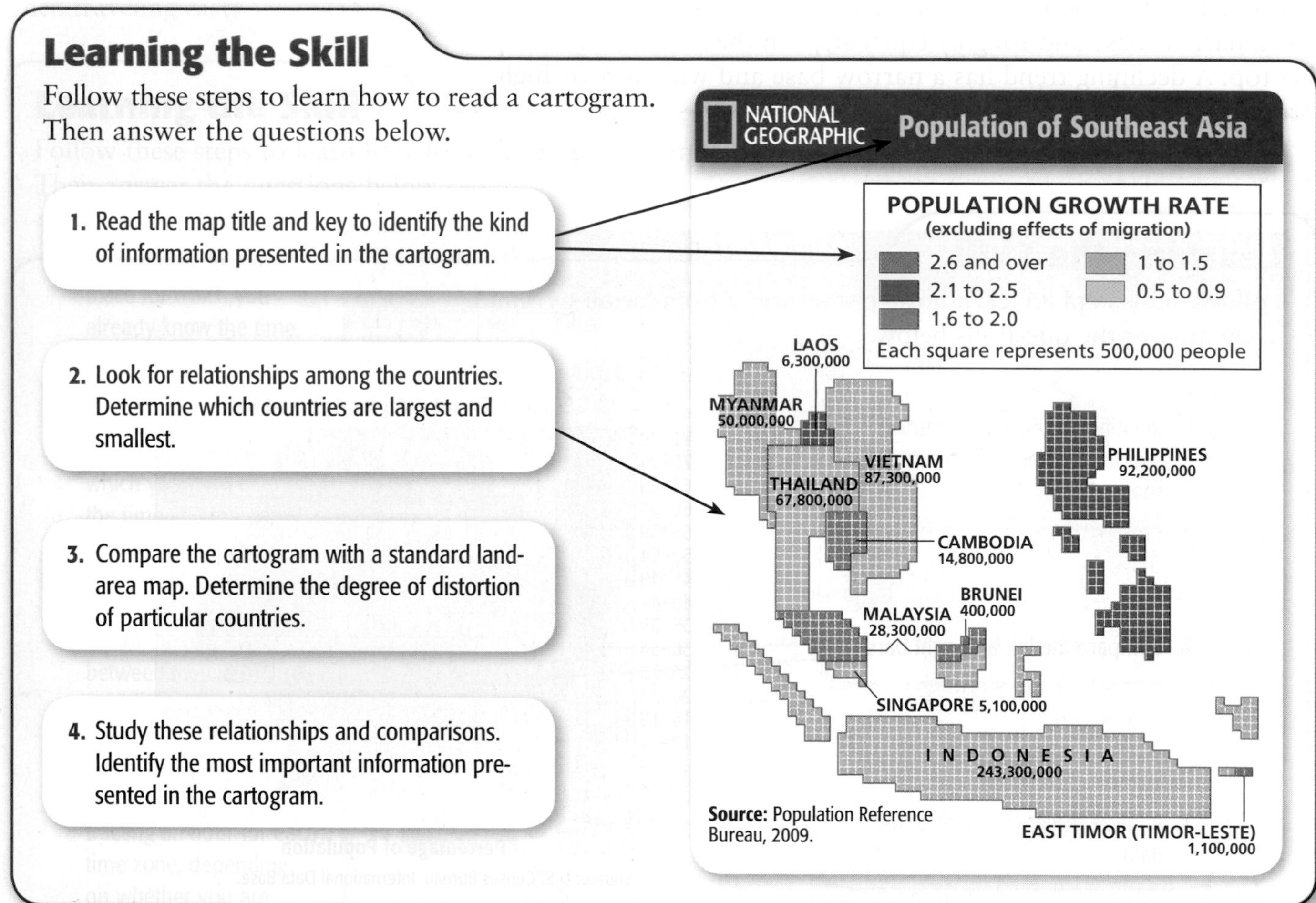

Source: Population Reference Bureau, 2009.

Practicing the Skill

1. What data determine the relative sizes of countries on this cartogram?
2. What characteristics determine the color of the squares on this cartogram?
3. Compare the cartogram with a standard land-area map. How has the relative size of Singapore been changed on the cartogram? How would you explain this change?

APPLYING THE SKILL

Research the gross domestic product (GDP) of each country in Southeast Asia. Then create a cartogram that compares the GDP of these countries. Include a key from the symbols you use.

Comparing Data

Why Learn This Skill?

Economists compare data in order to identify economic trends, draw conclusions about the relationships of sets of economic information, analyze the effectiveness of economic programs, or perform other types of analysis. It is often easiest to compare data that is organized in charts, tables, or graphs.

Learning the Skill

Follow these steps to compare and contrast data:

- Look at each set of data separately to understand what each one means on its own.
- Look for relationships among the sets of data. Ask yourself: How are these sets of information connected to each other?
- Note similarities and differences among the sets of data.
- Draw conclusions about what the sets of data, taken together, might mean.

Practicing the Skill

Study the following table and then answer the questions below.

1. What topics about Ireland are covered in the chart?
2. What characteristics make up a picture of the Irish economy?
3. By how many percentage points did inflation change from 2000 to 2008?
4. What trend does this indicate?
5. Study the data about the Irish people. What two trends can you identify?

Ireland Data Profile

	2000	2005	2008
People			
Population, total	3.8 million	4.2 million	4.5 million
Life expectancy at birth, total (years)	76.4	79	79
Mortality rate, under 5 (per 1,000 live births)	NA	5	4 (2007)
Ratio of girls to boys in primary and secondary education (%)	102.8	103	103 (2007)
Environment			
Surface area (sq. km)	68,883	68,883	68,883
Forest area (sq. km)	6,090.0	6,690.0	NA
Agricultural land (% of land area)	64	61.4	NA
Energy use (kg of oil equivalent per capita)	3,727	3,647	NA
Economy			
GDP (current US$)	95.0 billion	196.4 billion	281.78 billion
GDP growth (annual %)	9.9	4.7	-2.3
Inflation, GDP deflator (annual %)	4.8	3.1	4.0
Exports of goods and services (% of GDP)	97.7	82	NA
Imports of goods and services (% of GDP)	84.5	70	NA
States and markets			
Market capitalization of listed companies (% of GDP)	86.2	58.1	17.5
Military expenditure (% of GDP)	0.7	0.6	0.6
Mobile phone subscribers (per 100 people)	65	103	113
Internet users (per 100 people)	17.8	36.9	63.5

Sources: World Development Indicators database, April 2009; World Population Data Sheet 2009; CIA World Factbook 2009.

APPLYING THE SKILL

Use an almanac or the Internet to find climate data on your state. Compare temperatures and precipitation across the last several decades and identify the trends the data indicates.

ENGLISH

industrialization transition from an agricultural society to one based on industry (p. 108)

Industrial Revolution the rapid major change in the economy with the introduction of power-driven machinery (p. 297)

infrastructure the basic urban necessities like streets and utilities (p. 559)

***institute** to establish in a position or office (p. 751)

insular constituting an island, as in Java (p. 736)

intelligentsia intellectual elite (p. 374)

***intensive** marked by special effort (p. 296)

***interaction** the action or influence of people, groups, or things on one another (p. 670)

interdependent relying on one another for goods, services, and ideas (p. 633)

intermediate directions the courses or routes of northeast, northwest, southeast, and southwest (p. 10)

***internal** existing or lying within (p. 217)

interrupted projection a map of the Earth in which the Earth's surface appears cut along arbitrary lines, each section projected separately (p. 7)

introduced species plants and animals placed in areas other than their native habitat (p. 843)

Inuit a member of the Arctic native peoples of North America (p. 156)

***involve** to include or take part in as a participant (p. 838)

***isolate** to set or keep apart from others (p. 203)

J

Japan Current a warm-water ocean current that adds moisture to the winter monsoons (p. 670)

jati in traditional Hindu society, a social group that defines a family's occupation and social standing (p. 607)

jazz musical form that developed in the United States in the early 1900s, blending African rhythms and European harmonies (p. 153)

***job** a specific duty, role, or function (p. 687)

jute plant fiber used to make string and cloth (p. 629)

K

karma in Hindu belief, the sum of good and bad actions in one's present and past lives (p. 608)

key a map legend (p. 10)

kolkhoz in the Soviet Union, a small farm worked by farmers who shared in the farm's production and profits (p. 392)

kum term for a desert in Central Asia (p. 425)

L

***labor** those who do work for wages (p. 458)

ESPAÑOL

industrialización transición de una sociedad agrícola a una industrial (p. 108)

Revolución Industrial el cambio fundamental y rápido que tuvo la economía con la introducción de maquinaria que funciona con energía (p. 297)

infraestructura las necesidades básicas urbanas tales como calles y servicios (p. 559)

***instituir** establecerse en un cargo u ocupación (p. 751)

insular constituye una isla, como en Java (p. 736)

intelectualidad elite intelectual (p. 374)

intensivo que exige un esfuerzo especial (p. 296)

***interacción** la acción o influencia que se ejerce recíprocamente sobre personas, grupos o cosas (p. 670)

interdependiente la dependencia recíproca para bienes, servicios e ideas (p. 633)

direcciones intermedias direcciones de noreste, noroeste, sureste, y suroeste (p. 10)

***interno** que ocurre o existe en el interior (p. 217)

proyección interrumpida una proyección que corta la superficie de la Tierra a través de líneas arbitrarias y proyecta cada sección por separado (p. 7)

especies introducidas plantas y animales colocadas en áreas diferentes a su hábitat natural (p. 843)

inuit un miembro de los pueblos indígenas del Ártico de Norteamérica (p. 156)

***involucrar** tomar parte en un asunto como participante (p. 838)

***aislar** mantener algo aparte y separado de otros (p. 203)

J

Corriente de Japón corriente oceánica de agua templada que proporciona humedad a los monzones de invierno (p. 670)

jati en la sociedad tradicional hindú, un grupo social que define una ocupación tradicional de la familia y su posición social (p. 607)

jazz forma musical que se desarrolló en los Estados Unidos a principios de los 1900, mezclando ritmos africanos con armonías europeas (p. 153)

***trabajo** una tarea específica, labor o función (p. 687)

yute fibra que se usa para hacer cordón y tela (p. 629)

K

karma en la creencia Hindú, la suma de las acciones buenas y malas en la vida presente y las vidas pasadas de una persona (p. 608)

leyenda simbología del mapa (p. 10)

koljoz en la Unión Soviética, una granja pequeña trabajada por granjeros que comparten la producción y las utilidades (p. 392)

kum término regional para un desierto en Asia Central (p. 425)

L

***obrero** persona que trabaja a cambio un salario o sueldo (p. 458)

ENGLISH

lagoon shallow pool of water at the center of an atoll (p. 804)

lama Buddhist religious leader (p. 618)

landlocked enclosed or nearly enclosed by land (p. 477)

language family group of related languages that have all developed from one earlier language (p. 76)

latifundia in Latin America, large agricultural estates owned by families or corporations (p. 243)

latitude distance north or south from the equator measured in degrees (p. 8)

***layer** one thickness or fold over or under another (p. 663)

leach to wash nutrients out of the soil (p. 516)

leeward being in or facing the direction toward which the wind is blowing (p. 59)

***legal** of or relating to law (p. 536)

light industry manufacturing aimed at making consumer goods such as textiles or food processing rather than heavy machinery (p. 325)

lingua franca a common language used among people with different native languages (p. 532)

***link** to join or connect (p. 358)

literacy rate the percentage of people in a given place who can read and write (p. 153)

lithosphere surface land areas of the earth's crust, including continents and ocean basins (p. 32)

llano (LAH•noh) fertile grassland in inland areas of Colombia and Venezuela (p. 204)

***locate** to state and fix exactly the place, limits, or position (p. 281)

location a specific place on the Earth (p. 8)

lode deposit of minerals (p. 768)

loess (LEHS) fine, yellowish-brown topsoil made up of particles of silt and clay, usually carried by the wind (p. 281)

longitude distance measured by degrees or time east or west from the prime meridian (p. 8)

Loyalist a colonist who remained loyal to the British government during the American Revolution (p. 156)

M

Maastricht Treaty a 1992 meeting of European governments in Maastricht, the Netherlands, that formed the European Union (p. 328)

magma molten rock that is pushed up from the Earth's mantle (p. 35)

***maintain** to keep in an existing state (p. 465)

***major** greater in dignity, rank, importance, or interest (p. 78)

malnutrition faulty or inadequate nutrition (p. 220)

mantle thick middle layer of the Earth's interior structure, consisting of dense, hot rock (p. 35)

ESPAÑOL

laguna estanque de poca profundidad en el centro de un atolón (p. 804)

lama líder religioso budista (p. 618)

país sin costas marinas rodeado completamente por tierra (p. 477)

familia de lenguajes grupo de lenguajes relacionados que se desarrollaron de un lenguaje anterior (p. 76)

latifundios en Latinoamérica, extensas propiedades de cultivo que pertenecen a familias o corporaciones (p. 243)

latitud distancia norte y sur desde la línea ecuatorial medida en grados (p. 8)

***estrato** capa o grosor que se sobrepone a otras o se extiende por debajo de ellas (p. 663)

lixiviar lavar o sacar los nutrientes de la tierra (p. 516)

sotavento la parte opuesta a la dirección que sopla el viento (p. 59)

***legal** de o relativo a la ley (p. 536)

industria ligera industria orientada a la producción de bienes de consumo como textiles o procesamiento de alimentos en vez de maquinaria pesada (p. 325)

lengua franca un lenguaje común usado entre la gente con diferentes lenguas nativas (p. 532)

***unir** conectar o enlazar (p. 358)

índice de alfabetización el porcentaje de personas en un lugar dado que puede leer y escribir (p. 153)

litosfera áreas de superficie terrestre de la corteza de la Tierra, incluye los continentes y las cuencas de los océanos (p. 32)

llano fértiles praderas o llanos en el interior del territorio de Colombia y Venezuela (p. 204)

***localizar** establecer y fijar exactamente el lugar, límite o posición (p. 281)

ubicación un lugar específico en la Tierra (p. 8)

veta depósito de minerales (p. 768)

loes capa superficial del suelo, fina, amarillenta y marrón compuesta de partículas de limo y arcilla, usualmente arrastrada por el viento (p. 281)

longitud la distancia medida en grados o tiempo al este u oeste del Meridiano de Greenwich (p. 8)

lealista un colonizador que se mantuvo leal al gobierno británico durante la Revolución de las Colonias Norteamericanas (p. 156)

M

Tratado de Maastricht una reunión en 1992 de los gobiernos europeos en Maastricht, Holanda, en la que se formó la Unión Europea (p. 328)

magma roca fundida que emerge desde el manto de la Tierra hacia arriba a través de conductos volcánicos (p. 35)

***mantener** conservar en un mismo estado (p. 465)

***mayor** trascendental en dignidad, rango, importancia o interés (p. 78)

malnutrición nutrición pobre o inadecuada (p. 220)

manto gruesa capa mediana de la estructura interior de la Tierra que consiste de densa piedra caliente (p. 35)

D

E

W

X

Y

Z

Acknowledgments **RA45** From *Geography for Life: National Geography Standards 1994* Copyright © 1994 National Geographic Research & Exploration; **47** From "Earthquake dramatizes human ecological assault on the Himalayas," The Associated Press, October 22, 2005. Copyright © 2005, The Associated Press. Reprinted by permission; **113** From "Urban Poverty and Health in Developing Countries," by Mark R. Montgomery. Population Bulletin 64, No. 2 (2009); **145** From "Mountaintop Removal" by Tim Thornton. *The Roanoke Times*, July 2, 2006. Reprinted by permission; **182** From "Offshoring" *The Economist*, August 27, 2009. Copyright © 2009 The Economist Newspaper Limited. Reprinted by permission; **182** From "A Grand Goal for More U.S. Manufacturing Jobs" by Jessie Scanlon, *BusinessWeek*, August 31, 2009. Reprinted from August 31, 2009 issue of Bloomberg BusinessWeek by special permission, copyright © 2009 by Bloomberg L.P.; **183** from "Furniture makers mull moving factories back home" by Emily Kaiser. All rights reserved. Republication or redistribution of Thomson Reuters content, including framing or similar means, is expressly prohibited without the prior written consent of Thomson Reuters. Thomson Reuters and its logo are registered trademarks or trademarks of the Thomson Reuters group of companies around the world. © Thomson Reuters 2009. Thomson Reuters journalists are subject to an Editorial Handbook which requires fair presentation and disclosure of relevant interests; **257** From "4 Giants in Cattle Industry Agree to Help Fight Deforestation" by Alexei Barrioneuvo, from *The New York Times*, © October 7, 2009 The New York Times. All rights reserved. Used by permission and protected by the Copyright Laws of the United States. The printing, copying, redistribution, or retransmission of the Material without express written permission is prohibited; **260** From "The Other Mexico" by *BusinessWeek*, April 9, 2009. Reprinted from April 9, 2009 issue of Bloomberg BusinessWeek by special permission, copyright © 2009 by Bloomberg L.P.; **260** From "Latin American Economies on the Rebound," by Tyler Bridges, *The Miami Herald*, September 28, 2009 Copyright © 2009, The Miami Herald. Reprinted by permission; **261** From "Nicaragua: Emerging Economy, Ideal BPO Destination" by Brendan B. Read, Technology Marketing Corporation, June 18, 2009. Copyright © 2009, Technology Marketing Corporation. Reprinted by permission; **291** From "Southeast Europe Hit by Heatwave," by David Chance, from reuters.com, August 16, 2007. All rights reserved. Republication or redistribution of Thomson Reuters content, including by framing or similar means, is expressly prohibited without the prior written consent of Thomson Reuters. Thomson Reuters and its logo are registered trademarks or trademarks of the Thomson Reuters group of companies around the world. © Thomson Reuters 2007. Thomson Reuters journalists are subject to an Editorial Handbook which requires fair presentation and disclosure of relevant interests; **319** From "Leipzig: A city with many identities, and none" by Jeffrey Fleishman, *The Los Angeles Times*, November 17, 2006; **340** From "Why does Europe need the Treaty of Lisbon?" from Europa. Reprinted by permission of the Publications Office of the European Union; **340** From "Czech president attacks acolytes of EU integration" by Ingrid Melander, from reuters.com, Motoring Section, February 19, 2009. All rights reserved. Republication or redistribution of Thomson Reuters content, including by framing or similar means, is expressly prohibited without the prior written consent of Thomson Reuters. Thomson Reuters and its logo are registered trademarks or trademarks of the Thomson Reuters group of companies around the world. © Thomson Reuters 2009. Thomson Reuters journalists are subject to an Editorial Handbook which requires fair presentation and disclosure of relevant interests; **341** from "The Future's Lisbon" *The Economist*, October 8, 2009. Copyright © 2009 The Economist Newspaper Limited. Reprinted by permission; **404** From "Threat of oil spill menaces Russian Pacific Island" by Alexander Osipovich, www.exaptica.com, October 29, 2009. Copyright © 2009 Expatica Communications BV. Reprinted by permission; **404** From "Environmental Groups Urge Exxon Executive to Re-Route Sakhalin I Oil Pipeline," Pacific Environment, July 9, 2008. Copyright © 2008, Pacific Environment. Reprinted by permission; **405** From "Putin's Annual Message Boosts Infrastructure", *Executive Intelligence Review*, May 4, 2007 Copyright © 2007 EIRNS. Reprinted by permission; **490** From "Sunni-Shia schism 'threatening to tear Iraq apart' says conflict group," by Michael Howard, *The Guardian*, February 27, 2006. Reprinted by permission of the author; **491** From "National Unity is Rallying Cry in Iraq Elections" by Steven Lee Myers, from *The New York Times*, © October 1, 2009, The New York Times. All rights reserved. Used by permission and protected by the Copyright Laws of the United States. The printing, copying, redistribution or retransmission of the Material without express written permission is prohibited; **551** From *Out of Africa* by Isak Dinesen, copyright 1937 by Random House, Inc. and renewed 1965 by Rungstedlundfonden. Used by permission of Random House, Inc. and The Rungstedlund Foundation; **571** from "Why Foreign Aid Is Hurting Africa" by Dambisa Moyo. Reprinted by permission of *The Wall Street Journal*. Copyright © 2009 Dow Jones & Company, Inc. All Rights Reserved Worldwide. License number 2338941421254; **574** From *Voices of Women Entrepreneurs in Kenya* by IFC Gender Entrepreneurship Markets & Foreign Investment Advisory Services. Copyright ©, the International Bank for Reconstruction and Development, The World Bank. Reprinted by permission; **574** From "Record number of women contest Malawi elections" Agence France-Presse, May 17, 2009; **575** From "In Africa, women turn to microfinance" by Anne-Laure Buffard, *The Washington Times*, November 20, 2008; **603** From "Bangladesh braces for floods as heavy rain forecast" Agence France-Presse, September 4, 2008; **623** From "Is a New Form of Democracy Evolving in India?" by Bradford Smith. American Friends Service Committee International Affairs Reports, Vol. VII, No. 7, February 9, 1960. Reprinted by permission; **643** From "The 16 Decisions of Grameen Bank." Reprinted by permission; **646** From "Near the Tipping Point–India must confront dangers of being hyper-populated" by Ashok Ganguly, *The Telegraph* August 12, 2009; **647** From "They're Rounding the First Turn! And the Favorite Is…" by Nicholas D. Kristof. From *The New York Times*, © January 19, 2006, The New York Times. All rights reserved. Used by permission and protected by the Copyright Laws of the United States. The printing, copying, redistribution or retransmission of the Material without express written permission is prohibited; **647** From "Developmental Challenges of the states: Partnership Opportunities" by Shri Prenab Mukherjee, Minister of External Affairs, January 8, 2007; **673** Haiku by Matsuo Basho, reprinted by permission of World Haiku Review; **713** From "Three Gorges Dam- The Great Wall Across the Yangtze" Copyright © International Rivers Network. Reprinted by permission; **716** From "How Japan plans to have more babies," by Takehiko Kambayashi, *The Christian Science Monitor*, September 29, 2009. Reprinted by permission of the author; **716** From "Japan's Population Likely Shrank by Record in 2008, Gov't Says" by Hiroshi Suzuki, www.bloomberg.com, January 1, 2009; **717** From "Japan Must Boost Immigration- Ruling Party Panel" by David Fogarty. All rights reserved. Republication or redistribution of Thomson Reuters content, including framing or similar means, is expressly prohibited without the prior written content of Thomson Reuters. Thomson Reuters and its logo are registered trademarks of the Thomson Reuters group of companies around the world. © Thomson Reuters 2009. Thomson Reuters journalists are subject to an Editorial Handbook which requires fair presentation and disclosure of relevant interests; **784** From "Agent Orange: Birth Defects Plague Vietnam; U.S. Slow to Help" by Jason Grotto, *The Chicago Tribune*, December 8, 2008; **784** From "U.S., Vietnam Face Agent Orange Legacy" by Ian Timberlake, Agence France-Presse, September 23, 2009; **785** From "Report Sees Agent Orange Link to More Illnesses," by Janie Lorber, from *The New York Times*, © July 25, 2009, The New York Times. All rights reserved. Used by permission and protected by the Copyright Laws of the United States. The printing, copying, redistribution or retransmission of the Material without express written permission is prohibited; **831** From "Australian court rules that city of Perth belongs to aborigines" Agence France-Presse, September 21, 2006; **849** From "New Zealand's Nuclear Free Status" by Peter R. Wills, University of Auckland. Peter Wills is an associate of the National Centre for Peace and Conflict Studies at the University of Otago, Aotearoa, New Zealand. Reprinted by permission of the author; **852** From *Tui Tui Tuituia, Race Relations in 2008*, by the Human Rights Commission of New Zealand, March 2009; **853** From "Statement of the Special Rapporteur on the situation of human rights and fundamental freedoms of indigenous people" by James Anaya, August 27, 2009; **R4** From "Selling the Wind" by Michelle Nijhuis. *Audubon*, September/October 2006. Reprinted by permission of the author; **R5** From Storm Swarm: Are Hurricanes Our Fault?" by Paul Rauber. Reprinted with permission from the January/February 2005 issue of *Sierra*, the magazine of the Sierra Club; **R10** From "Puerto Rico, an Island in Distress" from *The New York Times*, © October 23, 2006, The New York Times. All rights reserved. Used by permission and protected by the Copyright Laws of the United States. The printing, copying, redistribution or retransmission of the Material without express written permission is prohibited; **R12** From "Making Sure a Rose Still Smells as Sweet" by Roger Di Silvestro. *National Wildlife*, February/March 2005. Copyright © 2005 National Wildlife Federation. Reprinted by permission of the National Wildlife Federation.

Photo Credits **Cover** (cw from top)Galen Rowell/CORBIS, (1)Tom Stoddart Archive/Getty Images, (2, 5)Penny Tweedie/CORBIS, (3)Paul Chesley/Getty Images, (4)Bethune Carmichael/Lonely Planet Images, (6, 13)Peter Adams/Getty Images, (7)Frans Lemmens/Getty Images, (8)Sally Mayman/Getty Images, (9)Nevada Wier/Getty Images, (10)Andrea Booher/Getty Images, (11)John Beatty/Getty Images, (12)Howard Kingsnorth/Getty Images, (c)Visual Analysis Lab/Goddard Space Flight Center, NASA; **iv** (t)Mark H. Milstein/NorthFoto/ZUMA Press, (b)SuperStock, Inc./SuperStock; **v** (t)Tim Davis/Getty Images, (b)Martin Harvey/CORBIS; **vi** (t)Robert Estall/CORBIS, (b)Philip Gould/CORBIS; **vii** (t)Bob Daemmrich/The Image Works, (bl)Galen Rowell/CORBIS, (br)Robert van der Hilst/CORBIS; **viii** (t)Johner/Getty Images, (b)National Trust Photographic Library/Ian Shaw/The Image Works; **ix** (t)José Fuste Raga/zefa/CORBIS, (b)Richard Nowitz/National Geographic Image Collection; **x** (t, tr)Steve Vidler/SuperStock, (b)Jon Hicks/CORBIS; **xi** (t)Richard Lord/The Image Works, (b)Pixtal/SuperStock; **xii** (t)Keren Su/CORBIS, (b)Arshad Arbab/epa/CORBIS; **xiii** (t)Robert Fried Photography, (b)Yann Layma/Getty Images; **xiv** (t)Sukree Sukplang/Reuters/CORBIS, (b)Macduff Everton/CORBIS; **xv** (t)David Kirkland/Photolibrary, (b)Mark Dadswell/Getty Images; **xvii** Larry Dale Gordon/Getty Images; **RA42** Martin Harvey/Getty Images; **RA43** (t) NOAA/Corbis, (b) The Photolibrary Wales/Alamy; **RA44** (t to b) ThinkStock/SuperStock, Janet Foster/Masterfile, Mark Tomalty/Masterfile, age fotostock/SuperStock, Jurgen Freund/Nature Picture Library; **0-1** Deborah Harse/The Image Works; **2-3** Patsy Davidson/The Image Works; **4** Courtesy Dr. Richard Boehm; **15** (l,c,r)David W. Boles; **16** Mark Christmas/National Geographic Image Collection; **18** Paul Richards/Bettmann/CORBIS; **19** Michael Coyne/Lonely Planet Images; **20** Paul Nicklen/NGS/Getty Images; **22** Charlotte Thege/PeterArnold, Inc.; **23** Jim Reed/CORBIS; **24** Purestock/SuperStock; **25** (bkgd)Robert Laberge/Getty Images, (tr)Paul Nicklen/NGS/Getty Images; **28-29** Jim Sugar/CORBIS; **30** Roger Ressmeyer/CORBIS; **34** Reed Saxon/AP Images; **41** Sandra Teddy/Getty Images; **45** (bkgd)Robert Laberge/Getty Images, (tr)NASA/CORBIS, (cl)Tom Bean/CORBIS; **48-50** Hans Strand/CORBIS; **54** Tim Davis/Getty Images; **60** Vivien Jones/AP Images; **65** (bkgd)Robert Laberge/Getty Images, (tr)Rob Howard/CORBIS, (br)Keren Su/CORBIS; **68-69** Yvette Cardozo/Imagestock; **70** Cris Bouroncle/AFP/Getty Images; **75** Martin Harvey/CORBIS; **79** Justin Guariglia/CORBIS; **82** (t)Will & Deni McIntyre/CORBIS, (b)David Samuel Robbins/CORBIS; **83** (t)Robert Nickelsberg/Getty Images, (b)ML Sinibaldi/CORBIS; **84** (t)CORBIS, (b)Ace Stock Limited/Alamy Images; **85** (t)Atlantide Phototravel/CORBIS, (b)Denis Sinyakov/AFP/Getty Images; **86** (cl)Bettmann/CORBIS, (br)Blank Archives/Getty Images, (tr)Michel Setboun/CORBIS; **87** (t)STR/AFP/Getty Images, (b)Jack Hollingsworth/Getty Images; **88** (t)Ted Streshinsky/CORBIS, (b)CORBIS; **89** (t)Deshakalyan Chwodhury/AFP/Getty Images, (b)Sebastian D'Souza/AFP/Getty Images; **90** (t)Aaron Horowitz/CORBIS, (c)Kenneth Garrett/Getty Images, (b)CORBIS; **91** (t)Muhannad Fala'ah/Getty Images, (b)Madaree Tohlala/AFP/Getty Images; **92** (t)Hardy/zefa/CORBIS, (bl)CORBIS, (br)Comstock Images/Alamy Images; **93** (t)Quique Kierszenbaum/Getty Images, (b)Ted Spiegel/CORBIS; **94** (br)ArkReligion.com/Alamy Images, (cl)Chip Somodevilla/Getty Images, (tr)Ashley Cooper/CORBIS; **95** (t)Maynard Owen Williams/National Geographic/Getty Images, (b)Don MacKinnon/Getty Images; **96** (t)Beth Wald/Aurora/Getty Images, (bl)Ralph A. Clevenger/CORBIS, (br)Martin Harvey/Alamy Images; **97** (cl)CORBIS, (cr)Paul Chesley/Getty Images, (br)Penny Tweedie/Wildlight, (tl)Nicholas DeVore/Getty Images; **98** (tl)Pete Saloutos/CORBIS, (cl)David Ball/CORBIS, (bl)age fotostock/SuperStock; **100** Andreas Gebert/dpa/Landov; **101** Remy De La Mauviniere/AP Images; **102** Shawn Baldwin/CORBIS; **103** Gordon Wiltsie/National Geographic Image Collection; **104** Peter Turnley/CORBIS; **105** Greg Smith/CORBIS; **110** Mug Shots/zefa/CORBIS; **111** (bkgd)Robert Laberge/Getty Images, (br)Evaristo Sa/AFP/Getty Images, (cl)Handout/epa/CORBIS; **114-115** Jeff Titcomb/Getty Images; **116** (l) Sarah Leen/National Geographic Image Collection, (b) Royalty-Free/CORBIS; **116-117** ML Sinibaldi/CORBIS; **117** age fotostock/SuperStock; **118** AP Images; **118-119** Rudy Sulgan/CORBIS; **119** (tl) Philip Gould/CORBIS, (cl) SuperStock, Inc./SuperStock, (bl) CORBIS, **128-129** Richard Sisk/Panoramic Images/National Geographic Image Collection; **130** Philip & Karen Smith/Getty Images; **131** (l)Paul Chesley/National Geographic Image Collection, (r)James Randklev/Getty Images; **134** William Campbell/Peter Arnold, Inc; **135** Will & Deni McIntyre/CORBIS; **136** (bkgd)NOAA/CORBIS, (r)Dan Anderson/epa/CORBIS; **137** (l)EPA/Gerardo Mora/Landov, (r)Mari Darr-Welch/AP Images; **138** Joel W. Rogers/CORBIS; **139** (l)Kraig Lieb/Lonely Planet Images, (r)Stan Osolinski/Oxford Scientific/JupiterImages; **143** (tl)Lloyd Sutton/Masterfile, (bl)Roy Ooms/Masterfile, (br)Thinkstock Images/JupiterImages; **146-147** David La Spina/AFP/Getty Images; **148** Michael J.Doolittle/The Image Works; **150** NASA/Goddard Space Flight CenterScientific Visualization Studio; **153** Richard A. Cooke/CORBIS; **155** Paul A. Souders/CORBIS; **159** Robert Estall/CORBIS; **161** (bkgd, r)Robert Laberge/Getty Images, (l)Lucas Jackson/Reuters/CORBIS; **164-165** Nick Didlick/Bloomberg News/Landov; **166** Kevin Fleming/CORBIS; **171** Jeff Vinnick/Getty Images; **172** Paul Glendell/Peter Arnold, Inc.; **173** (t)Melissa Farlow/Getty Images, (c)Raymond Gehman/National Geographic Image Collection, (b)Kevin R. Morris/CORBIS; **174** Rob Stuehrk/AP Images; **175** WireImageStock/Masterfile; **176** Rich Reid/Earth Scenes; **177** (tl)CORBIS, (tr)Tim Boyle/Getty Images, (bl)Visuals Unlimited/CORBIS; **180** Peter Yates/CORBIS; **181** (t)AP Images, (b)H. David Seawell/CORBIS;

184-185 age fotostock/SuperStock; **186** (br)Kenneth Garrett/NGS/Getty Images, (cl)Layne Kennedy/CORBIS; **186-187** Francesco Muntada/CORBIS; **187** Terry Vine/Getty Images; **188** SuperStock; **188-189** Randy Faris/CORBIS; **189** (t)Wolfgang Kaehler/CORBIS, (c)Hugh Sitton/zefa/CORBIS, (b)The Image Works Archives; **200-201** Mark Cosslett/NGS/Getty Images; **202** (l)Galen Rowell/CORBIS, (r)Robert van der Hilst/CORBIS; **204** Woods Wheatcroft/Lonely Planet Images; **205** Wolfgang Kaehler/CORBIS; **207** Luiz C. Marigo/Peter Arnold, Inc./Alamy Images; **210** Joel Sartore/National Geographic Image Collection; **211** (t)George Grall/NGS/Getty Images (r)MedioImages/Getty Images, (b)Francesco Muntada/CORBIS; **214-215** Aldo Torelli/Getty Images; **216** Steve Elmore/Getty Images; **218** Schalkwijk/Art Resource, NY; **219** Cladio Cruz Valderrama/WORLDPICTURENEWS; **221** Peter Turnley/CORBIS; **222** Paul A. Souders/CORBIS; **224** Danny Lehman/CORBIS; **225** Richard Bickel/CORBIS; **227** David Mercado/Reuters/CORBIS; **229** Gisela Damm/eStock Photo; **231** John Pennock/Lonely Planet Images; **233** (bkgd)Robert Laberge/Getty Images, (b)Martin Alipaz/EFE/CORBIS; **234** (b)Paulo Whitaker/Reuters/CORBIS; **234-235** Jan Butchofsky-Houser/CORBIS; **235** (b)Royalty-Free/CORBIS; **238** Larry Dale Gordon/Getty Images; **239** (tr)John A Rizzo/Getty Images, (c)Brad Barket/Getty Images, (bc)Stephen Chernin/Getty Images; **240-241** The Cover Story/CORBIS; **242** Guy Moberly/Lonely Planet Images; **247** Yuri Gripas/Reuters; **248** (bl)Russell Mittermeier, Conservation International/AP Images, (br)Claus Meyer/Minden Pictures; **248-249** Michael Ende; **250** Brent Winebrenner/Lonely Planet Images; **250** (br)Andreas Salomon-Prym/VISUM/The Image Works; **251** (tr)PNC/zefa/CORBIS, (cr)Gustavo Gilabert/CORBIS, (br)Yann Arthus-Bertrand/CORBIS; **254** Bernard Bisson/CORBIS; **255** (t)Bob Daemmrich/The Image Works, (b)Craig Lovell/CORBIS; **258** Henry Romero/Reuters; **259** (t)Bob Daemmrich/The Image Works, (b)Megapress/Alamy Images; **262-263** Jon Arnold Images/SuperStock; **264** (cl)Atlantide Phototravel/CORBIS, (br)Adam Woolfitt/CORBIS; **264-265** Richard Klune/CORBIS; **265** Charles Bowman/Robert Harding World Imagery/CORBIS; **266** Scala/Art Resource, NY; **266-267** age fotostock/SuperStock; **267** (t)Barry Lewis/CORBIS, (c)Peter Andrews/Reuters/CORBIS, (bl)Keystone/Getty Images, (br)Peter Turnley/CORBIS; **271** Bettmann/CORBIS; **273** age fotostock/SuperStock; **278-279** Altrendo Travel/Getty Images; **280** (bkgd)Jean-Bernard Carillet/Lonely Planet Images, (br)Chris Lisle/CORBIS; **281** SIME s.a.s /eStock Photop; **282** José Fuste Raga/zefa/CORBIS; **283** Skyscan/CORBIS; **285** Prisma/SuperStock; **286** Larry Dale Gordon/Getty Images; **289** (tr)SIME s.a.s/eStock Photo, (bl)Walter Bibikow/Getty Images, (br)Ralf Niemzig/VISUM/The Image Works; **292-293** Jon Davison/Lonely Planet Images; **294** National Trust Photographic Library/Ian Shaw/The Image Works; **296** Thomas Haertrich/Peter Arnold, Inc.; **299** Berndt Fischer/age fotostock; **302** The Art Archive/CORBIS; **304** (l)Picture Contact/Alamy Images, (r)Patrick Gardin/AP Images; **304-305** Jean-pierre VERGEZ; **306** Atlantide Phototravel/CORBIS; **310** Burt Glinn/Magnum Photos; **310-311** Régis Bossu/Sygma/CORBIS; **312** (l)Angelo Cavalli/age fotostock, (r)Royalty-Free/CORBIS; **315** James Doberman/Getty Images **317** (bkgd)Robert Laberge/Getty Images, (br)Owen Franken/CORBIS; **320** (bkgd)Monika Graff/UPI/Landov, (l)Jason Wasmiller Photo, (r)Dave Lowrey's Photo Gallery; **321** Ingram Publishing/SuperStock; **322-323** William Manning/CORBIS; **324** Mark H. Milstein/NorthFoto/ZUMA Press; **328** Chung Sung-Jun/Getty Images; **330** Caroline Penn/CORBIS; **331** Sasa Djordjevic/AFP/Getty Images; **333** Ashley Cooper/CORBIS; **335** (t)Johner/Getty Images, (b)Pete Cairns/Nature Picture Library; **338** (t)IFA Bilderteam/eStock Photo, (b)IT Stock Free/eStock Photo; **339** Murat Ayranci/SuperStock; **340** Michal Cizek/AFP/Getty Images; **342-343** Richard Klune/CORBIS; **344** (cl) John Cancalosi/Peter Arnold, Inc., (br) Scott Warren/Aurora/Getty Images; **344-345** Sarah Leen/National Geographic Image Collection; **345** Peter Blakely/CORBIS; **346** Bettmann/CORBIS; **346-347** Jonathan Smith/Lonely Planet Images; **347** (t) Liba Taylor/CORBIS, (c) Chernysheva Marina/ITAR-TASS/CORBIS, (b) Hulton-Deutsch Collection/CORBIS; **353** Alexander Nemenov/AFP/Getty Images; **354-355** Bruno Morandi/Getty Images; **356 357** Bryan & Cherry Alexander Photography; **357** (tr)Belinsky Yuri/ITAR-TASS/CORBIS; **358** Steve Raymer/National Geographic Image Collection; **360** Bryan & Cherry Alexander Photography; **361** Getty Images; **365** (tr)Topham/The Image Works, (cl)Michel Setboun/CORBIS, (br)Brand X/SuperStock; **368-369** Denis Sinyakov/AFP/Getty Images; **370** Getty Images; **374** Rose Hartman/CORBIS; **376** Dominique Faget/AFP/Getty Images; **376-377** Yuri Kadobnov/epa/CORBIS; **377** Ruslan Alkhanov/AFP/Getty Images; **378** Richard Nowitz/National Geographic Image Collection; **380** Mary Evans Picture Library; **381** (l)Topham/The Image Works, (r)Jacques Langevin/CORBIS; **383** (bkgd)Robert Laberge/Getty Images, (l)Gleb Garanich/Reuters/CORBIS; **386** NASA/AP Images, (bkgd)age fotostock/SuperStock, (l)Belinsky Yuri/ITAR-TASS/Landov; **387** (tl)Fabrizio Bensch/Reuters/CORBIS, (c)Steve Liss/Time Life Pictures/Getty Images; **388-389** José Fuste Raga/zefa/CORBIS; **390** Richard Nowitz/National Geographic Image Collection; **393** Wolfgang Kaehler/CORBIS; **395** Dean Conger/CORBIS; **397** Klaus Nigge/NGS/Getty Images; **399** (l)Vasily Melnichenko/ITAR-TASS/Landov, (r)Zinin Vladimir/ITAR-TASS/Landov; **403** Savin Oleg/ITAR-TASS/CORBIS; **405** TATYANA MAKEYEVA/AFP/Getty Images; **406-407, 408** Steve Vidler/SuperStock; **408** (cl)age fotostock/SuperStock; **408-409** George Steinmetz/CORBIS; **409** Langevin Jacques/CORBIS; **410** Erich Lessing/Art Resource, NY; **410-411** Royalty-Free/CORBIS; **411** (t)Elizabeth Dalziel/AP Images, (c)Jon Hicks/CORBIS, (b)Alfred Hennig/dpa/Landov; **412** AP Images; **413** (l)Royalty-Free/CORBIS, (r)George Steinmetz/CORBIS; **415** Robert Mackinlay/Peter Arnold, Inc.; **422-423** Sandro Vannini/CORBIS; **424** Gary Cook/Alamy Images; **427** Photri/Topham/The Image Works; **429** K.M. Westermann/CORBIS; **430** Keren Su/CORBIS; **432** Doug Scott/age fotostock; **433** (tl)AP Images, (tr)Antoine Gyori/AGP/CORBIS, (br)Charles & Josette Lenars/CORBIS; **436-437** Gary Cook/Alamy Images; **438** (bkgd)Celia Mannings/Alamy Images, (br)John and Lisa Merrill/CORBIS; **440** Lutz Jaekel/Bilderberg/Peter Arnold; **443** Mahfouz Abu Turk (ISRAEL)/Reuters; **448** Menahem Kahanna/AFP/Getty Images; **448-449** AP Images; **449** John Angelillo-Pool/Getty Images; **450** Joe Raedle/Getty Images; **452** Steve Vidler/SuperStock; **454** Phil Weymouth/Lonely Planet Images; **455** (bkgd)Chris Mellor/Lonely Planet Images, (br)Anthony Ham/Lonely Planet Images; **458** age fotostock/SuperStock; **459** Nabeel Turner/Getty Images; **460** Natalie Behring-Chislom/Getty Images; **460-461** AP Images; **461** Massoud Hossaini/Shah Marai/AFP/Getty Images; **462** Efrem Lukatsky/AP Images; **465** Seth Kushner/Getty Images; **467** (t)Jose Fuste Raga/CORBIS, (bkgd)Robert Laberge/Getty Images, (br)Mohsen Shandiz/CORBIS; **470-471** Rebecca Cook/Reuters/CORBIS; **471** (tr)Douglas Johns/Photo Library, (c)Stephen Shugerman/Getty Images, (b)Kayte M. Deioma/PhotoEdit; **472-473** Jose Fuste Raga/CORBIS; **474** (bkgd)Trip/Alamy Images, (b)The Photolibrary Wales/Alamy Images; **480** Peter Turnley/CORBIS; **482** Akwa Betote/H20/CORBIS; **483** NASA; **485** (l)Paul Stepan/Photo Researchers, Inc., (r)Chris Lisle/CORBIS; **488** Wathiq Khuzaie/Getty Images; **489** (t)Kazuyoshi Nomachi/CORBIS, (b)Gerald Holubowicz/CORBIS; **491** (t)2006 Star Tribune. Reprinted with permission of Star Tribune, Minneapolis, MN; **492-493** Liba Taylor/CORBIS; **494** (c)Jon Hicks/CORBIS, (br)Royalty-Free/CORBIS; **494-495** Remi Benali/CORBIS; **495** Guillaume Bonn/CORBIS; **496** George Steinmetz/CORBIS; **496-497** Reuters/CORBIS; **497** (t)Earl & Nazima Kowall/CORBIS, (c)Penny Tweedie/CORBIS, (b)David Turnley/CORBIS; **508-509** Chris Johns/National Geographic Image Collection; **510** (bkgd)Yann Arthus-Bertrand/CORBIS, (br)Remi Benali/CORBIS; **512** Herbert Spichtinger/zefa/CORBIS; **515** Pixtal/SuperStock; **519** (tl)Suzanne Porter/Impact Photos, (bl)Carsten Peter/National Geographic Image Collection, (bc)Michael Nichols/NGS/Getty Images; **522-523** Albert Normandin/Masterfile; **524** Sarah Leen/National Geographic Image Collection; **525** Jacques Langevin/CORBIS; **527** Peter Adams/The Image Bank/Getty Images; **529** Dennis Johnson/Lonely Planet Images; **531** Michele Burgess/CORBIS; **532** Peter Turnley/CORBIS; **534** Ton Koene/Peter Arnold, Inc.; **536** Mark Kauffman/Time Life Pictures/Getty Images; **537** Liba Taylor/CORBIS; **538** Lawrence Manning/CORBIS; **539** Martin Harvey/CORBIS; **540** Stefano Amantini/Atlantide Phototravel/CORBIS; **543** Jean Pierre Kepseu/Panapress/Getty Images; **544** Lee Frost/NGS/Getty Images; **547** Jonathan Shapiro; **549** (bkgd)Robert Laberge/Getty Images, (l)Royalty-Free/CORBIS; **552** (l)Frank Micelotta/Getty Images, (r)Bridgeman Art Library/SuperStock; **552-553** Peter Turnley/CORBIS; **554-555** Gallo Images/CORBIS; **556** David Alan Harvey/Magnum Photos; **559** Richard Lucas/The Image Works; **560** Denis Farrell/AP Images; **562** (l)David Reed/CORBIS, (r)Reuters/CORBIS; **562-563** Heinrich van den Berg/Getty Images; **564** Richard Lord/The Image Works; **567** Frans Lanting/Minden Pictures; **569** Sean Sprague/Peter Arnold, Inc.; **572** Eric L Wheater/Lonely Planet Images; **573** (t)Michael S. Lewis/CORBIS, (b)James Marshall/CORBIS; **575** Amos Gumulira/AFP/Getty Images; **576-577** Patrick Horton/Lonely Planet Images; **578** (cl)Anuruddha Lokuhapuarachchi/Reuters/CORBIS, (br)Keren Su/CORBIS; **578-579** David Sutherland/CORBIS; **579** Richard I'Anson/Lonely Planet Images; **580** Robert Harding/Getty Images; **580-581** Sara-Jane Cleland/Lonely Planet Images; **581** (t)Keren Su/CORBIS, (c)Arshad Arbab/epa/CORBIS, (bl)Hulton-Deutsch Collection/CORBIS, (br)epa/CORBIS; **583** Galen Rowell/CORBIS; **588-589** age fotostock/SuperStock; **590** Gavin Hellier/Robert Harding World Imagery/Getty Images; **592** Bojan Brecelj/CORBIS; **594** (bkgd)DigitalGlobe/Getty Images, (b)Jeremy Horner/CORBIS; **595** (t) Andy Rain/epa/CORBIS, (b) Nayan Sthankiya/CORBIS; **596** Grant Dixon/Lonely Planet Images; **599** Steve McCurry/National Geographic Image Collection; **600** Shehzad Noorani/Peter Arnold, Inc.; **601** (tr)DPA/AA/The Image Works, (bl)Sanjib Mukherjee/Reuters/CORBIS, (br)David Samuel Robbins/CORBIS; **604-605** Rajesh Kumar Singh/AP Images; **606** Noshir Desai/CORBIS; **609** Richard Lord/The Image Works; **610** Nikreates/Alamy Images; **611** Shabbir Hussain Imam/AP Images; **614** Amit Gupta/Reuters/CORBIS; **615** Rafiqur Rahman/Reuters/CORBIS; **616** Jayanta Shaw/Reuters; **621** (bkgd)Robert Laberge/Getty Images, (l)John Henry Claude Wilson/Getty Images; **624** Francois Mor/AP Images; **624-625** Nick Ut/AP Images; **625** J.Garcia/photocuisine/CORBIS; **626-627** Peer Grimm/dpa/Landov; **628** Alison Wright/CORBIS; **631** Lindsay Hebberd/CORBIS; **634-635** Dinodia Photo Library; **635** Gautam Singh/AP Images; **636** Amit Bhargava/CORBIS; **637** Shehzad Noorani/Peter Arnold, Inc.; **641** (l)Anuruddha Lokuhapuarachchi/Reuters, (r)Pallava Bagla/CORBIS; **644 645** David H. Wells/CORBIS; **645** (t)Greg Elms/Getty Images; **648-649** Martin Vincent Robinson/Lonely Planet Images; **650** (cl)Dean Conger/CORBIS, (br)Gavin Hellier/Robert Harding World Imagery/CORBIS; **650-651** Dallas and John Heaton/Free Agents Limited/CORBIS; **651** Jason Lee/Reuters/CORBIS; **652** (b) Chinese School, (17th century)/Bridgeman Art Library; **652-653** Jose Fuste Raga/CORBIS; **653** (tl)Chien-Min Chung/Reuters, (c)Hamid Sardar/CORBIS, (bl)AP Images, (br)Yao Dawei/AP Images; **659** Worldscapes/age fotostock; **660-661** Yann Layma/Getty Images; **662** Karen Kasmauski/National Geographic Image Collection; **665** Xinhua/CORBIS; **667** Wolfgang Kaehler/CORBIS; **671** (l)Pixtal/SuperStock, (r)Roger Ressmeyer/CORBIS, (bc)George Steinmetz/CORBIS; **674-675** Christian Kober/Robert Harding Picture Library Ltd/PhotoLibrary; **676** Randy Faris/CORBIS; **679** Todd Gipstein/National Geographic Image Collection; **681** (l)Greg Elms/Lonely Planet Images, (r)Robert Essel NYC/CORBIS; **683** Bettmann/CORBIS; **686** Setboun/CORBIS; **689** Ted Kawalerski Photography, Inc./Getty Images; **691** (bkgd)Robert Laberge/Getty Images, (c)Jason Lee-Pool/Getty Images; **694** Najlah Feanny/CORBIS; **694-695** Jerry Amster/SuperStock; **695** (l)Yoshikazu Tsuno/AFP/Getty Images, (r)Science Museum/SSPL/The Image Works; **696-697** Robert Fried Photography; **698** (bkgd)Ken Straiton/CORBIS, (br)Royalty-Free/Getty Images; **704** (bkgd)Allen Birnbach/Masterfile, (r)eStock Photo/DigitalVision; **705** Liu Liqun/CORBIS; **706** Liu Jin/AFP/Getty Images; **711** (l)Andrew Holbrooke/CORBIS, (r)Colin Garratt; Milepost 92 1/2/CORBIS; **714** Frank Carter/Lonely Planet Images; **715** (t)B.S.P.I./CORBIS, (b)Eri Morita/Getty Images; **718-719** Kevin R. Morris/CORBIS; **720** (c)Nevada Wier/CORBIS, (b)WEDA/epa/CORBIS; **720-721** Michael S. Yamashita/CORBIS; **721** Karen Kasmauski/CORBIS; **722** John Banagan/Lonely Planet Images; **722-723** Jose Fuste Raga/CORBIS; **723** (t)Keren Su/CORBIS, (c)Tibor Bognar/CORBIS, (bl)Bettmann/CORBIS, (br)Jewel Samad/AFP/Getty Images; **730** Steve Raymer/CORBIS; **731** Andre Maslennikov/Peter Arnold Inc.; **732-733** Bernard Napthine/Lonely Planet Images; **734** Steve Raymer/CORBIS; **736** Mark Lewis/Getty Images; **737** Michael S. Yamashita/CORBIS; **738** Barbara Walton/epa/CORBIS; **739** Royalty-Free/CORBIS; **742** Kham/Reuters/CORBIS; **743** (t)John William Banagan/Getty Images, (b)Guus Geurts/Peter Arnold, Inc., (r)Ludovic Maisant/CORBIS; **746-747** age fotostock/SuperStock; **748** (b)Steve Raymer/CORBIS; **752** Lee Snider/The Image Works; **753** Sukree Sukplang/Reuters/CORBIS; **754** Macduff Everton/CORBIS; **757** Steve Raymer/National Geographic Image Collection; **759** (bkgd)Robert Laberge/Getty Images, (br)Brunei Information/epa/CORBIS; **762-763** Kathleen Voege/Getty Images; **763** (cr)Jackson Vereen/Jupiterimages, (bl)Jeff Greenberg/Alamy Images, (br)Patrick Riviere/Getty Images; **764-765** Andre Maslennikov/Peter Arnold Inc.; **766** Sean Sprague/Peter Arnold Inc.; **768** Angelo Cavalli/Getty Images; **772** Les Stone/Sygma/CORBIS; **772-773** T. J. Casadevall/USGS; **773** Roger Ressmeyer/CORBIS; **774** Paul A. Souders/CORBIS; **775** Jean-Leo Dugast/Peter Arnold, Inc.; **777** Rungroj Yongrit/epa/CORBIS; **779** (tr)Andre Maslennikov/Peter Arnold Inc., (l)David Longstreath/AP Images, (br)Mast Irham/epa/CORBIS; **782** Wally McNamee/CORBIS; **783** (t)Dick Swanson/Time Life Pictures/Getty Images, (b)Tim Page/CORBIS; **784** Les Stone/Sygma/CORBIS; **786-787** Altrendo Images/Getty Images; **788** (cl)Todd A. Gipstein/CORBIS, (br)George Steinmetz/CORBIS; **788-789** Barry Lewis/CORBIS; **789** (b)Paul A. Souders/CORBIS; **790** (bl)Barry Lewis/CORBIS, (br)Hulton Archive/Getty Images; **790-791** Jose Fuste Raga/CORBIS; **791** (t)Reuters/CORBIS, (c)Mark Dadswell/Getty Images, (b)Bettmann/CORBIS; **800-801** H. Takeuchi/PanStock/Panoramic Images/National Geographic Image Collection; **802** Paul Nicklen/National Geographic Image Collection; **804** (tr)B.S.P.I./CORBIS, (bl)Sylvain Grandadam/Robert Harding World Imagery/Getty Images, (br)Gerry Ellis/Minden Pictures; **806** Royalty-Free/CORBIS; **807** (cl)Paul A. Souders/CORBIS, (cr)DLILLC/CORBIS, (bl)Frank Todd/Arctic Photo, (br)Keren Su/CORBIS; **808** Galen Rowell/CORBIS; **809** Peter Johnson/CORBIS; **810** Greg Wood/AFP/Getty Images; **813** (t)Jurgen Freund/Aurora Photos, (b)Medford Taylor/National Geographic Image Collection; **816-817** Grant Faint/Getty Images; **818** Theo Allofs/CORBIS; **820** Krzysztof Dydynski/Lonely Planet Images; **822** Tomas del Amo/Photolibrary; **824** David Kirkland/Photolibrary; **825** Macduff Everton/CORBIS; **827** Owen Franken/CORBIS; **829** (bkgd)Robert Laberge/Getty Images, (r)Simon Baker/AFP/Getty Images; **832** Nick Doan/Icon SMI/CORBIS; **832-833** pbpgalleries/Alamy Images; **833** (t)Royalty-Free/CORBIS, (c)Lester Cohen/WireImage, (b)Clive Brunskill/Allsport/Getty Images; **834-835** Richard Cummins/CORBIS; **836** Stefan Schuetz/zefa/CORBIS; **837** Nigel Dickinson/Peter Arnold, Inc.; **840** Leonard Douglas Zell/Lonely Planet Images; **840-841** age fotostock/SuperStock; **841** Torsten Blackwood/AFP/Getty Images; **842** Oliver Strewe/Lonely Planet Images; **843** Theo Allofs/CORBIS; **846** Jeff Hunter/Getty Images; **847** (l)Colin Monteath/Hedgehog House/Minden Pictures, (r)Matthew McKee; Eye Ubiquitous/CORBIS; **850** Peter Drury/AFP/Getty Images; **851** Medford Taylor/NGS/Getty Images; **852** Lorraine Harris/Index Stock Imagery, Inc.; **R0-R1** Visual Analysis Lab/ Goddard Space Flight Center/NASA; **R4** Lester Lefkowitz/CORBIS; **R6** Volkmar Brockhaus/zefa/CORBS; **R10** Robert Frerck/Odyssey Productions.